THE EXPLORATION FOR PETROLEUM IN EUROPE AND NORTH AFRICA

Proceedings of a Joint Meeting between the Institute of Petroleum and the American Association of Petroleum Geologists (in conjunction with the Petroleum Exploration Society of Great Britain) held at Brighton, England, 29 June–2 July 1969

EDITED BY

PETER HEPPLE

INSTITUTE OF PETROLEUM
61 NEW CAVENDISH STREET
LONDON, W1M 8AR

1969

Sole Agents for Overseas Sales
The Elsevier Publishing Co. Ltd, Amsterdam, London, New York

Made and printed in Great Britain by
ADLARD AND SON LTD, DORKING, SURREY

COMMITTEES

Organizing Committee

D. C. Ion, British General Chairman
Michel T. Halbouty, American General Chairman

D. R. Brown	J. McGinty
D. A. Busch (U.S.A.)	C. J. O. Moorhouse
E. V. Corps	W. G. Ogilvie
J. Hazzard (U.S.A.)	A. N. Thomas
P. W. Hepple	C. H. Maynard (Secretary)
D. A. Hough	

Papers Sub-Committee

A. N. Thomas, Chairman
H. D. Hedberg (U.S.A.)
G. D. Hobson
P. E. Kent
R. E. King (U.S.A.)
J. D. Moody (U.S.A.) Chairman
W. G. Ogilvie
P. W. Hepple (Editor, Institute of Petroleum)
J. McGinty (Secretary)

Excursions Sub-Committee

N. L. Falcon, Chairman
R. Dyk
T. F. Gaskell
C. H. Maynard (Secretary)

Tour Leaders

R. G. W. Brunstrom, Excursion No. 1
D. B. Smith, Excursion No. 2
A. P. Terris, Excursion No. 3
C. A. Fothergill, Air Excursion 'A'
P. J. Walmsley }
B. J. Taylor } Air Excursion 'B'

Contents

Foreword

EXPLORATION is the start of all petroleum industry activity. This meeting illustrated the intense current international interest in Europe and North Africa and also the growing appreciation, both inside and outside the petroleum industry, of the importance of the role of the explorer.

This joint meeting of the Institute of Petroleum and the American Association of Petroleum Geologists in association with the Petroleum Exploration Society of Great Britain was initiated in conversation with Michel T. Halbouty, then President of AAPG, during his visit to London in May 1967.

This book of the proceedings of the meeting held two years later is the successful result of a great deal of effort. First there was the effort of the American Organizing Committee under the enthusiastic leadership of Michel T. Halbouty, with its Papers Committee under John D. Moody, helped by the staff and the Executive Committee of AAPG under President Frank B. Conselman. All these men worked in the closest and most friendly collaboration with the British Organizing Committee, with its Papers Committee under A. N. Thomas and the Excursions Sub-Committee under N. L. Falcon, members of PESGB, including Donald R. Brown and Robert Dyk, and the staff of the IP, led by Colin Maynard. Then there was the effort of the invited authors and the session chairmen and the most admirable help at the last minute of Professor Th. Sorgenfrei and Dr J. Dozy, who stood in for authors unable to attend. Lastly, one must thank the 90 delegates who took an active part in the technical discussions.

The value of a scientific meeting rests not only in the quality of the papers presented but also in the quality and freedom of public as well as private discussion. These proceedings stand for inspection as to the quality of the papers and public discussion. New material and ideas were presented on many areas and this book should be a valuable reference for many years.

The 540 delegates came from 33 countries and there was ample opportunity for private discussions, discussions in which geologists, possibly more than any other scientists, are willing and able to engage, and from which they can derive maximum benefit.

This was indeed a joint meeting. There were 263 delegates and 152 registered guests from the U.S.A., and 122 delegates and 32 registered guests from the U.K., whilst the 155 delegates and 45 guests from 31 other countries created an international gathering of almost 800 people.

Outside the technical sessions, there were two field trips by air during the meeting and three by land after the meeting. Their success well recompensed the organizers and guides for their careful planning and execution. Furthermore, the social and ladies programmes were well attended and included a concert presented by the Mayor and Corporation of Brighton, which over 500 people considered most enjoyable.

In summary, therefore, this book is itself only part of the benefit derived by those who attended a truly joint and international meeting. Indeed, as Kenneth H. Crandall, who started his term as President of AAPG during the meeting, said in his Closing Address, he hoped that this would only be the first such meeting, and that there may be possibly a second in Australia in 1972, when the AAPG, the IP (through its branches in Sydney and Melbourne), and other Australian bodies may look at petroleum exploration in that part of the world.

D. C. ION

Honorary Editor of the IP
and Chairman of the British Organizing Committee

Opening Address

By Dr A. W. PEARCE

(*President of the Institute of Petroleum*)

It is a privilege and a pleasure to welcome you to the joint meeting of the AAPG, the PESGB, and the IP.

The Institute of Petroleum is pleased to have arranged the conference here in the U.K., especially as it is the first time the AAPG has met for a scientific conference outside North America.

The Institute has, however, been holding its Summer Meeting for some years; some have just been IP meetings, some joint meetings with other institutions, and some meetings with non-petroleum industry participation, but in all cases we have welcomed visitors from countries other than the U.K., but never before so many from the U.S.A. Mr Wilson will, I am sure, be pleased to know you are here and the Chancellor of the Exchequer will be very grateful for any contributions you may care to leave behind in the form of greenbacks!

Some of you might wonder what part the IP plays in the petroleum industry here because it is such a very different organization from the API. It was formed in 1913 with the objective of encouraging the scientific study of petroleum and distributing technical information which would help to improve the education levels of those working in the petroleum industry. It is a scientific body which has no commercial or political function, yet it works closely with government and companies in the production of codes of practice covering such diverse subjects as electrical installations, beach cleaning, and safety regulations on offshore rigs. It operates not only in the U.K. but also has branches in Australia, Bahrain, Kuwait, and Trinidad, and its interests range from exploration to marketing.

In 1913 most of the membership were connected with exploration and production. After 1945 marketing and refining came to the fore, but today, with the North Sea programme, exploration and production are again coming back into prominence.

It is most appropriate, therefore, that the Summer Meeting this year should be a joint meeting, the theme being "The Exploration for Petroleum in Europe and North Africa".

Since 1945 the use of petroleum in Europe has increased remarkably, so that by 1968 over 50 per cent of Europe's total energy needs were met by petroleum, and this amounted to a supply requirement of approximately ten million bd. With such a major dependence upon petroleum, it is only natural that there is a great political and economic desire and need to find and exploit indigenous hydrocarbon supplies. The cost of petroleum to most European countries is the largest commodity bill they have to pay, and consequently the operations of the oil industry have an important impact upon the balance of payments of the region which depends so heavily upon petroleum for its energy needs. The refining industry in Europe has also grown remarkably since 1945, so much so that Europe is the largest refining centre in the world. There is now some 12 million bd capacity installed here, which is the equivalent of 27·6 per cent of the total world refining capacity, compared with the U.S. figure of 26·5 per cent. An increasing number of companies, both international and national are participating in the market. In the U.K. alone there are now over 50 companies competing for the market, whereas about three years ago the figure was nearer ten, and product prices and competitive pressures are forcing companies towards lower cost crude supplies. Inevitably, therefore, Europe is looking towards indigenous supplies and also the short-haul North African crudes. I do not think it is also necessary for me to stress the effects of two Middle East crises upon supply strategies and the need to diversify supply sources for the European market.

Your subject, gentlemen, is therefore one which has great interest for Europe, since the results of the exploration and production work now going on will probably determine much of the economic growth and direction of this area for many years to come.

It had been my hope that I could announce this morning that our Minister of Power would have been here on Wednesday to speak at the closing session. Pressure of parliamentary business prevents this, however, but he sends to all his best wishes.

May I, therefore, with the Minister, wish you a highly successful conference with active discussion and also a pleasant social activity outside this room. The Institute staff are here to help wherever they can to make your stay more beneficial and enjoyable. It is my privilege and pleasure, therefore, to open this meeting and now pass the responsibility for the technical session to your Chairman for this morning, Dan Ion.

The Geological Framework of Petroleum Exploration in Europe and North Africa and the Implications of Continental Drift Hypotheses

By P. E. KENT

(British Petroleum Co. Ltd)

SUMMARY

Structurally Europe is tripartite, consisting of the northern crystalline Baltic Shield, a broad median epeirogenic shelf region extending from the Urals westwards to the Atlantic, and the southerly Tethyan belt, coincident with the Mediterranean region, of Mesozoic-Tertiary geosynclines which provided the locus for the Alpine orogeny. North Africa continues the sequence southwards in mirror image, with the Atlas orogenic belt of the Mediterranean Alpine system flanked by the shelf of the Sahara and N Libya, and the emergent crystalline shield beyond.

Oil and gas fields have been worked for a century in the shelf region of N Europe; they were small by world standards up to the time of the major Netherlands and North Sea gas discoveries. The greater part of European oil has been produced from the Tertiary (Neogene) molasse basins flanking the Alpine chains, especially in Austria, Hungary, and Rumania, and active exploration continues, for example, in and around the Adriatic.

In North Africa hydrocarbon production is predominantly from the shelf sediments lapping on to the African shield. The absence of post-Cambrian diastrophism has permitted survival of major oil and gas accumulations in rocks as old as Lower Palaeozoic. Faulted, gently folded structures and reefal stratigraphic traps provide large oilfields in Mesozoic and Lower Tertiary rocks, particularly in Libya.

Exploration is now taking place on the continental margins, on the Atlantic coast and around the Mediterranean, so that the validity of continental displacement has become economically important. Some of the data relevant to this problem are discussed below. Late Mesozoic development of the North Atlantic appears to be supported by the stratigraphic and structural history of NW Europe, but current theories of development of the Mediterranean involving rotation of Africa are not consistent with the geological evidence.

INTRODUCTION

As AN introduction to the exploration problems discussed in the conference, I propose to trace briefly the evolution of the European and North African basins and to refer to the frontiers of exploration where operations are moving into the marine areas, especially on the NW shelves. In particular, I propose to look at some of the problems arising from the Continental Drift hypothesis—a two-way exchange in which Continental Drift bears on our concepts of basin development, and in which the results of petroleum exploration tend to qualify and limit some of the proposals of the drifters.

The review of hydrocarbon bearing formations which follows is made against the background of the broad regional structure—the Old Baltic Shield in the N, the emergent African Shield in the S, each flanked by broad epeirogenic shelves and with the Tethyan geosyncline, now the Alpine orogenic zone, geologically dividing Africa from Eurasia. (Fig 1).

LOWER PALAEOZOIC

Occasional tests have been drilled into the Lower Palaeozoic in NW Europe, *e.g.* on anticlines in the non-metamorphic Silurian of the Welsh border, but traces of hydrocarbons are rare, in most cases are connected with unconformable later cover (past or existing), and the prospects must be regarded as minimal. Elsewhere the Pre-Devonian rocks are too highly disturbed, indurated, or metamorphosed to be considered prospective.

This situation contrasts with the Saharan platform, where vast areas of non-metamorphic Palaeozoic rocks exist with thick reservoir beds (the Acacus and Cambrian sandstones) in conjunction with source rocks such as the Gothlandian, and the development of such oil and gas fields as Hassi Messaoud, Rhourde Baguel, el Gassi, el Agreb (Cambro/Ordovician), and Dôme à Collenias (Ordovician) (Perrodon 1966). (Fig 2). The gas in the Hassi R'mel field may also be of Lower Palaeozoic origin.

Over Europe generally the most obvious influence of the Lower Palaeozoic is seen in the survival of rigid remnants of the Caledonian orogenic belts; in the North, Fennoscandia, Scotland, and Ireland as a rigid mass—interrupted by the major downwarp of the North Sea—which has largely limited the later basins of NW Europe; in mid-Europe by the early Armorican belt and in Spain by the Iberian folding; the latter two exposed as only small remnants of the former chains.

UPPER PALAEOZOIC—DEVONIAN AND CARBONIFEROUS

With the Upper Palaeozoic we enter the range of important hydrocarbon generating rocks within

Europe, and of structures which are themselves exploration objectives.

The Devonian deserves brief mention; its more southerly marine facies may provide source rocks or reservoirs, and both are in fact developed in the Sahara, *e.g.* at Edjele. In its northerly, continental Old Red Sandstone facies it is sometimes strongly bituminous, and it is a matter for the future to discover whether this has contributed to the prospectiveness of the N North Sea by functioning as a source rock. The Devonian in marine facies is a major source of oil in E Russia.

The Lower Carboniferous (Mississippian) shows two clearly distinct facies—a geosynclinal flysch facies known in Britain as "Culm", which extends E–W from Ireland into Central Europe, and a shelf facies, lying to N and S, in which clear water limestones predominate. A marginal delta brought clastic sediments into the shelf sea in N Britain. The shelf facies has been the locus of small gas and oilfields (Hardstoft and Dukes Wood in England; Midlothian and Cousland in Scotland) and it is apparently the main objective of current exploration in Ireland. Promise of large-scale production is, so far, lacking, but the extent of this formation beneath the gently folded region of the N European plain holds some hope of eventual discoveries on the mainland.

The Upper Carboniferous (Pennsylvanian of North America) is characterized pre-eminently by the Coal Measures, which reached maximum development across Britain and the N European plain with thicknesses aggregating 6000 m in the foredeep of the rising Variscan orogenic belt, 2000 m in N France and Britain; 3000 m in Belgium.

It initially achieved a modest interest for petroleum from occasional shows in deep boreholes in the Low Countries, and from the development of ten small oilfields in England in the Namurian (Millstone Grit) and Westphalian. The size of these discoveries did not encourage exploration for Upper Carboniferous oil in the deeper objectives of the N European plain or the North Sea, but the formation now has a new and massive economic importance as the most probable source-rock for the gas reserves of the Netherlands and North Sea region generally. Progressive deep burial of the coal-bearing rocks during Permo-Triassic and later times has released methane in enormous quantities, and this has been trapped by the fortunate conjunction of a first-class Lower Permian reservoir and an evaporite cap rock. Probably still larger quantities of gas were generated in the thicker Coal Measure development across N France, but the strong deformation of the late-Palaeozoic Hercynian folding and the absence of trapping mechanism caused their total loss.

Although the Hercynian orogeny migrated northwards during the various Carboniferous and early Permian phases, it was not expressed by folding N of the Wales–London–Ardennes platform. In more northerly areas the deformation is essentially of epeirogenic type, and a large proportion of the vertical movements are associated with the development of fault-bounded basins (Kent 1966). This applies in the North Sea as well as in Britain. Structures are not lacking, but they are related to movement of basement blocks with a predominant NNW strike in England and—as is now known—a NNW or N–S strike beneath the North Sea in continuation of the Rhenish trend. It is this absence of severe tectonic stresses which has permitted the survival of oil and gas in the Carboniferous of the English Midlands and Scotland, and of the methane resources of the North Sea of N Germany and perhaps also of Poland.

PERMO-TRIAS

Faunally and traditionally the Permian is included in the Upper Palaeozoic, but in terms of tectonics, basin development, and hydrocarbon prospects it belongs to the Mesozoic and Tertiary part of the history of Europe and North Africa.

In NW Europe and the North Sea region the Permian was deposited across planed-off faulted and gently folded Westphalian. It occupies a series of basins almost completely independent of the previous sedimentational history. Among the earliest sediments were the coarse Rotliegendes sands—the prime North Sea basin reservoir; above are the Zechstein dolomites, a further reservoir facies, and the Zechstein evaporites with their importance as cap rock and for their generation of the widespread halokinetic structures which provide structural relief in the later rocks. Collette (1968) has postulated deep crustal adjustments to account for the inception of the North Sea subsidence at this time, but the North Sea area (as it now exists) coincides with only a part of the Permo-Triassic basin complex, and it would seem necessary to apply the same mechanism from Poland at least to the Irish Sea. The last ten years' exploration in NW Europe has again demonstrated that the application of petroleum geology can change the energy balance of a whole region with enormous social and economic implications.

Even in N Europe conditions were not uniformly so good, for during the Lower Permian there occurred a major phase of volcanic activity, so that in extensive areas the reservoir beds are replaced by lavas and their detritus. On the E side of the Russian platform the Lower Permian measures 5000 m, partly marine in the lower part, with an upward transition to a continental facies. The marine facies with its Fusulinid fauna extends across the Mediterranean region, with developments of reefal limestones which might well be prospective in an appropriate tectonic setting.

There is an interest in locating conditions analogous to NW Europe in the more southerly basins, but so far without success. The basal Permian is often in a coarse clastic facies (Verrucano), in part covered by evaporites, but has usually been sufficiently involved in Alpine or other movements to have been compacted and indurated—even metamorphosed—with complete loss of porosity. An effective poroperm

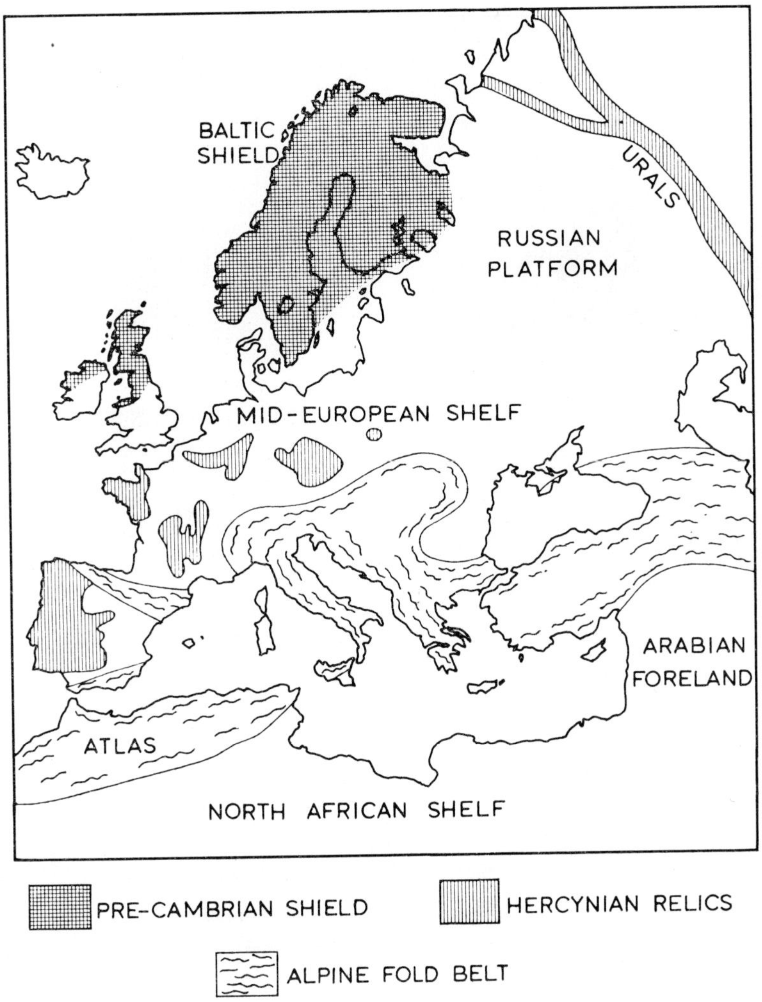

FIG 1. Structural units of Europe and North Africa.

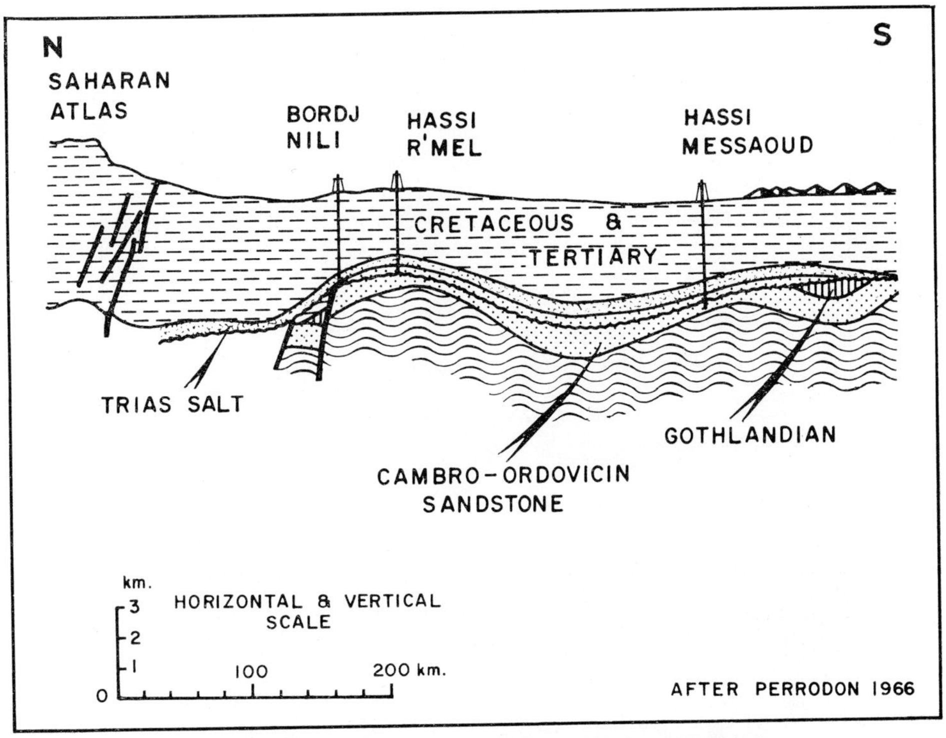

FIG 2. Diagrammatic section through the Saharan oilfield belt.

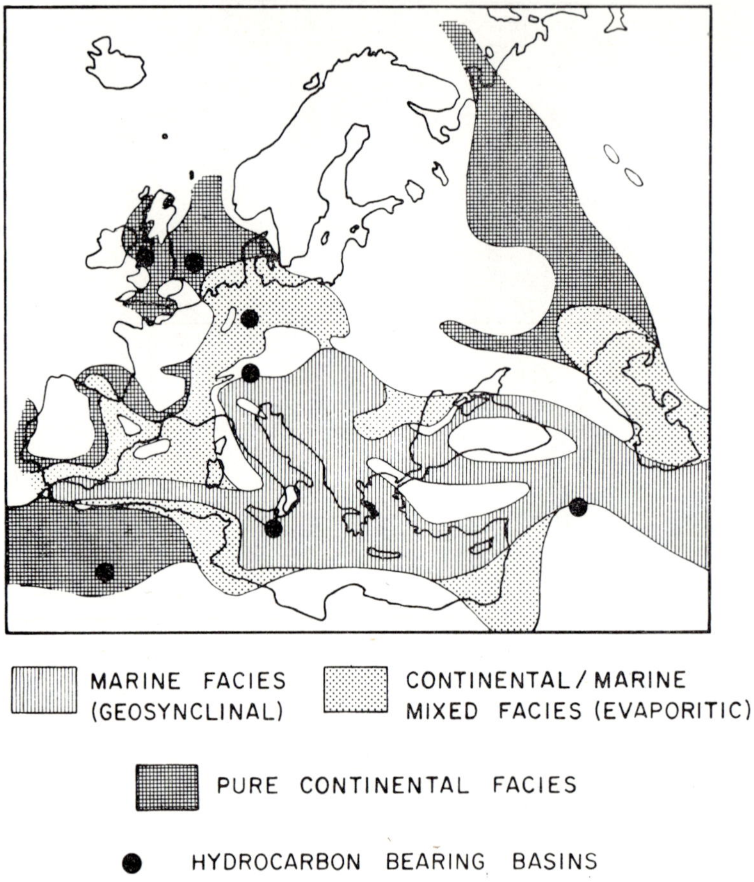

FIG 3. Distribution of Triassic basins.

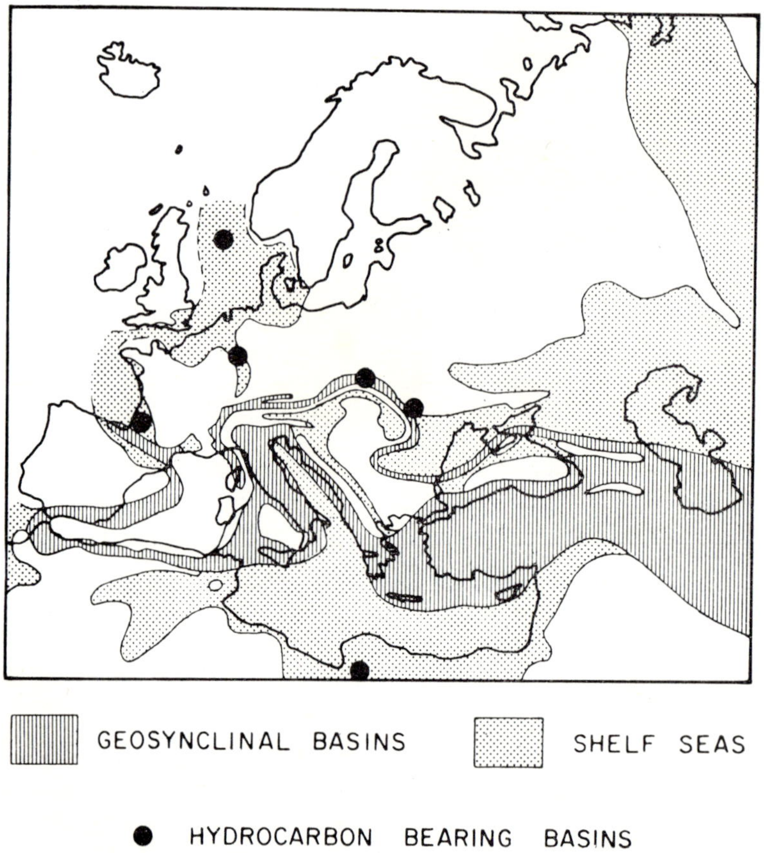

FIG 4. Lower Tertiary (Palaeocene to Oligocene) basins.

development may yet be found, but there is no expectation of a combination of the favourable factors of the North Sea basin.

The European Trias shows two fundamentally different developments—the classic continental facies of the N, and the Tethyan facies with massive dolomites in the Alpine geosyncline and the Mediterranean. (Fig 3).

The N Trias has its importance in the development of cap rocks and reservoir beds; the Bunter sandstones provide large gas fields in Germany and the British North Sea; the Keuper is oil-bearing on a minor scale in England (Formby oilfield) but is mainly a seal, and Rhaetic sandstones are one of the important reservoirs of the oilfields of the Hanover region of NW Germany.

In contrast, the massive Triassic dolomites of the Tethyan region provide an eminently attractive reservoir in a fully marine facies, and have proved to be oil-bearing in the Vienna Basin (Schönkirchen) and in Sicily.

JURASSIC AND CRETACEOUS

With the Jurassic and Cretaceous we come to the oldest important oil-bearing formations in Europe. The broad pattern of basins established across the subsided Hercynian and older crust during the Trias persisted through the remainder of Mesozoic times. In Europe, as in so many other parts of the world, the predominantly continental Triassic basins were flooded by marine waters, and a varied series of source and reservoir rocks deposited.

Many of these productive basins are described in detail by other authors in this symposium, and it suffices to list the areas concerned—the Paris Basin Middle Jurassic (with minor production also on its NW continuation in England); the Netherlands with the small productive Cretaceous basin near The Hague, and from Jurassic and Cretaceous in the Schoonebeck region; the SW German basin, with its numerous productive reservoirs in all the formations from the Rhaetic through the Jurassic to Lower Cretaceous and the Chalk, S France with oil pools in Jurassic and Cretaceous of the Aquitaine Basin and gas at great depth in earlier Jurassic of the Pyrenean foothills, and the basins of N Spain with production on the Jurassic-Cretaceous boundary. There are no developed fields so far in the Jurassic dolomites of the true Tethyan basin. Of the list above, only those of Germany were listed in "The World's Oilfields" by Illing in 1953; the others are relatively recent discoveries.

Much of this oil has no doubt migrated, vertically and laterally, but it is interesting to notice how frequently in Europe it is associated with the highest Jurassic/Lower Cretaceous levels, perhaps because these provide the first good reservoirs above the marine clay source rocks of the Lower and Upper Jurassic.

Reference must also be made to the Mesozoic oilfields on the edge of the African Shield in Libya, where the group formerly known as "Nubian Sandstone" or "Continental Intercalaire" has provided a series of major discoveries in S Cyrenaica. The formations are basal sands, overlapping on to crystalline Basement rocks; in this area they are overlain by marine beds of Albian–Cenomanian date and are now ascribed to the Lower Cretaceous. Here the oil is thought to have migrated from later Cretaceous rocks, either downwards or laterally from adjoining structurally low areas; the environment is widely different from that of the Lower Cretaceous reservoirs of Europe.

TERTIARY

In Europe and North Africa, as in the world generally, the Tertiary provides the prime producing formations, and in consequence the distribution and features of the oil and gas fields have been well documented—the older data are, for example, summarized in "The World's Oilfields" already cited.

The dominant features of the Tertiary palaeogeographical maps are the geosynclinal furrows which originally flanked the rising Alpine chains, the molasse basins which developed as erosion of the high mountains took place, and broader epicontinental seas extending over the shelves to N and S (Figs 4 and 5). The geosynclinal deposits themselves, mainly in flysch facies, have so far yielded only meagre quantities of oil and gas, although their exploration still continues. The important fields are in the epicontinental regions with their more varied rocks and high grade reservoirs, particularly in the Neogene of the molasse basins and on the shelves beyond.

The molasse basins form a double chain across Europe. One trend lies N of the Alpine folds from Bavaria, through the Vienna Basin, the Polish Galician Basin to Moldavia and the Rumanian oilfields, with the Pontic and Caspian basins beyond.

Between the orogenic belts the Po–Adriatic Basin lies between the Appennines and the Dinarides; the Pannonian, Hungarian, and Transylvanian Basins between the Dinarides and Carpathians and the Levant Basin between the E Balkans and the Helvetides. To a greater or lesser extent these basins were involved in complexities arising from the final phases of the orogeny, but they have always provided the bulk of European oil and gas production.

There is now a strong interest in finding true Tertiary oil in the thicker development of the North Sea Basin, where clastic formations total more than 3000 m in thickness and where one field on the borderline of commerciality has been found in W Norwegian waters.

The Tertiary shelf area whose potential now overshadows any discoveries in Europe is that of Libya, S of the orogenic region. There production is mainly from the Palaeocene of the Sirte Basin, from shale-capped carbonate reservoirs which are quite large by world standards. In part—as in the Zelten field—these are normal shelf carbonate reservoirs, in gently folded structures which may be related to Basement faulting. Additionally, the high productivity Idris fields have been found in Palaeocene reefal limestones near the edge of the contemporary basin; these are described in detail later in this volume. Other Tertiary fields have been found in Egypt, and there is still scope for continuing exploration.

THE ATLANTIC MARGIN OF EUROPE

Serious exploration of the marine shelves of W Europe began in the Bay of Biscay, with offshore drilling in search of analogues of the big Parentis oilfield. Elsewhere interest was minimal (although in 1963 one well was drilled in S British waters) until knowledge of the Netherlands NAM discoveries led to realization of the potential of the North Sea, and in due course to the large-scale operations there. This in turn has led to concern about the prospectiveness of other Permo-Triassic basins around the British Isles, as possibly providing a similar combination of favourable factors, and drilling is due to start in the Irish Sea this year (Figs 6 and 7).

Interest has since spread into the largely different regime of the Celtic Sea which extends from NW France to S Ireland, due to be covered by major seismic survey extending out to the edge of the shelf, and to the Portuguese offshore, where a similar exercise is planned.

With this developing interest, the history of the Atlantic margin becomes of major importance in understanding stratigraphical and structural trends, and it is necessary to look at the implications of continental drifting. (Fig 8).

In the Lower Palaeozoic Dewey has recently demonstrated how closely the detailed features of the Caledonian orogen on opposite sides of the North Atlantic can be correlated, providing strong support for the concept of their original contiguity, without prejudice as to acceptance of his further postulate that an Atlantic ocean closed and opened again before the Devonian. (Fig 9).

Whether or not the Caledonian chains will be found to be continuous beneath the North Sea, or whether this is a miniature Atlantic (a sphenochasm) produced by lateral parting of the opposite sides, is an associated interesting problem on which further exploration should supply important clues. So far the evidence (mainly geophysical) seems to be in favour of continuity, with the Caledonian metamorphic rocks of E Scotland apparently extending over a broad area 100 miles eastwards of Aberdeen, and a probable metamorphic core to the Ringköping-Fyn high extending a similar distance west from Central Denmark.

Subsequently, the limited flooding of the Permo-Triassic basins by marine waters, the enclosed gulf-

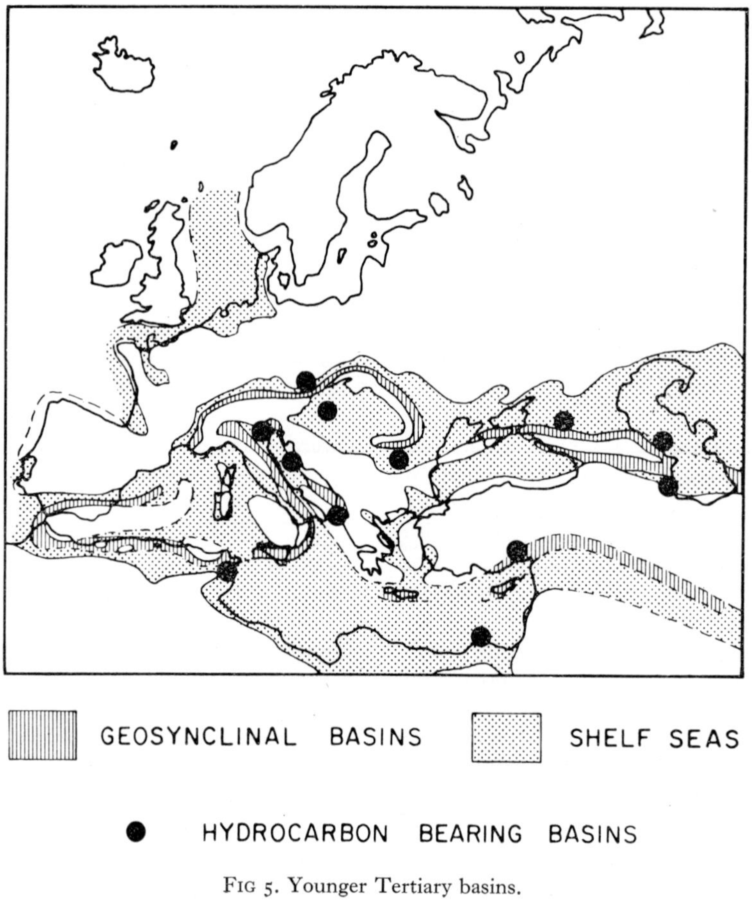

FIG 5. Younger Tertiary basins.

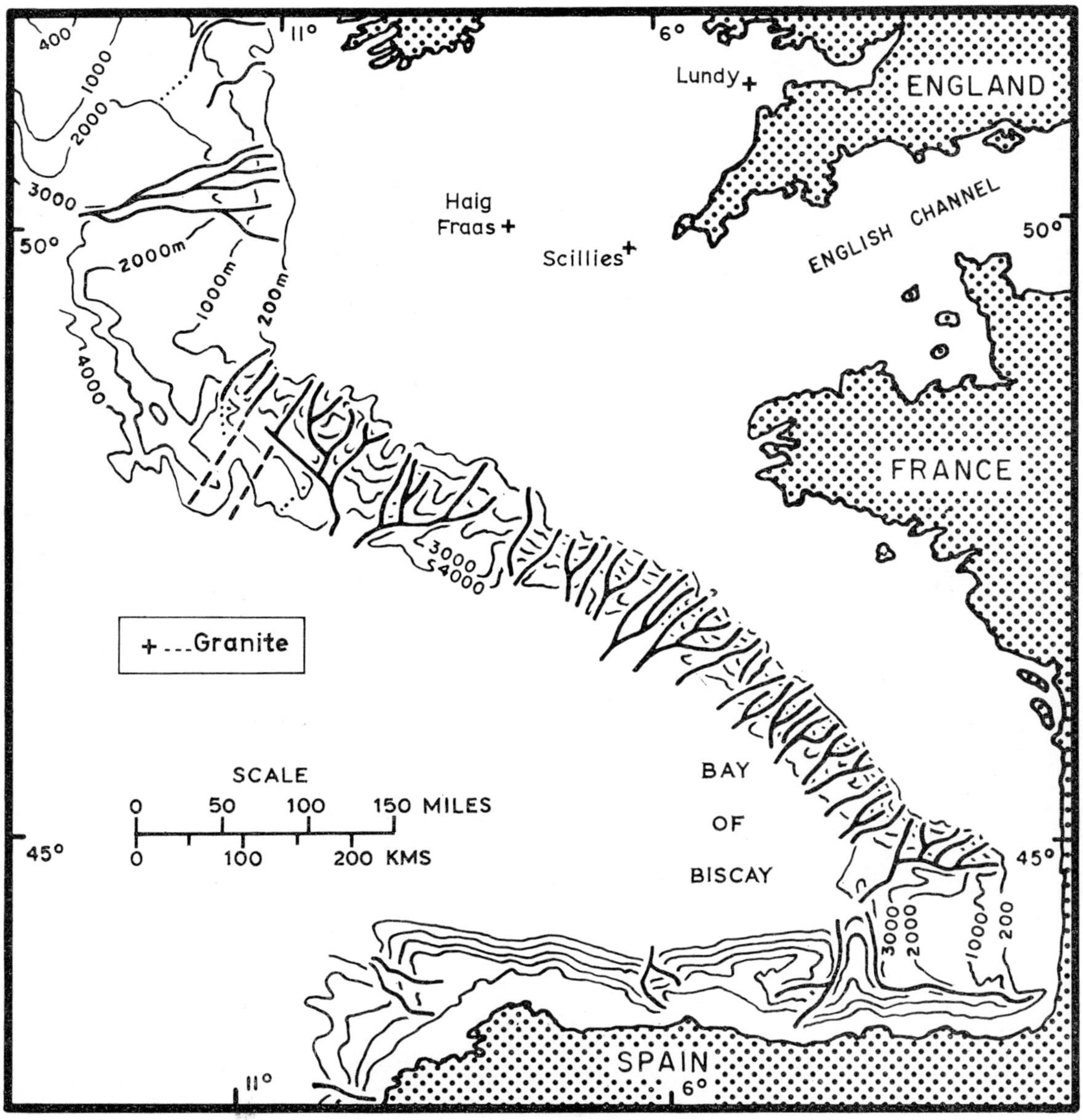

Fig 6. Continental Shelf SW of Britain (after Cholet *et al.*).

like structure implied by the repeated massive evaporite phases in Poland, Germany, the North Sea, NW England, the Irish Sea, and even Northern Ireland are virtually impossible to reconcile with a continental margin anything like the present outline of Europe, and provide obvious support for the Drift hypothesis. In Britain and (as is now known) in the North Sea, the deepest Permo-Triassic basins tend to have faulted margins, or relatively steep edges probably related to deep basement faulting. This faulting could well be linked with the structural failure which subsequently led to formation of the North Atlantic Ocean. The displacements are, however, essentially vertical with an absence of evidence of lateral extension on a regional or continental scale, so that it appears that the expression in the crust of the continental opening was normal epeirogenic subsidence.

The later history of the North Sea area is directly relevant to an understanding of Mesozoic and Tertiary oceanic connections. The first marine transgression in Britain—that of the Rhaetic—came from the W and SW, for the sequence in SW England is much more complete than that of either E England or the North Sea. In the Lower Jurassic a full marine connection was established across England and N Europe; this persisted throughout the Lower Jurassic. In the Middle and early Upper Jurassic, however, the seas retreated southwards, so that the marine element becomes subordinate in N England and in the North Sea alike, while fully marine conditions persisted in S England, the Paris Basin

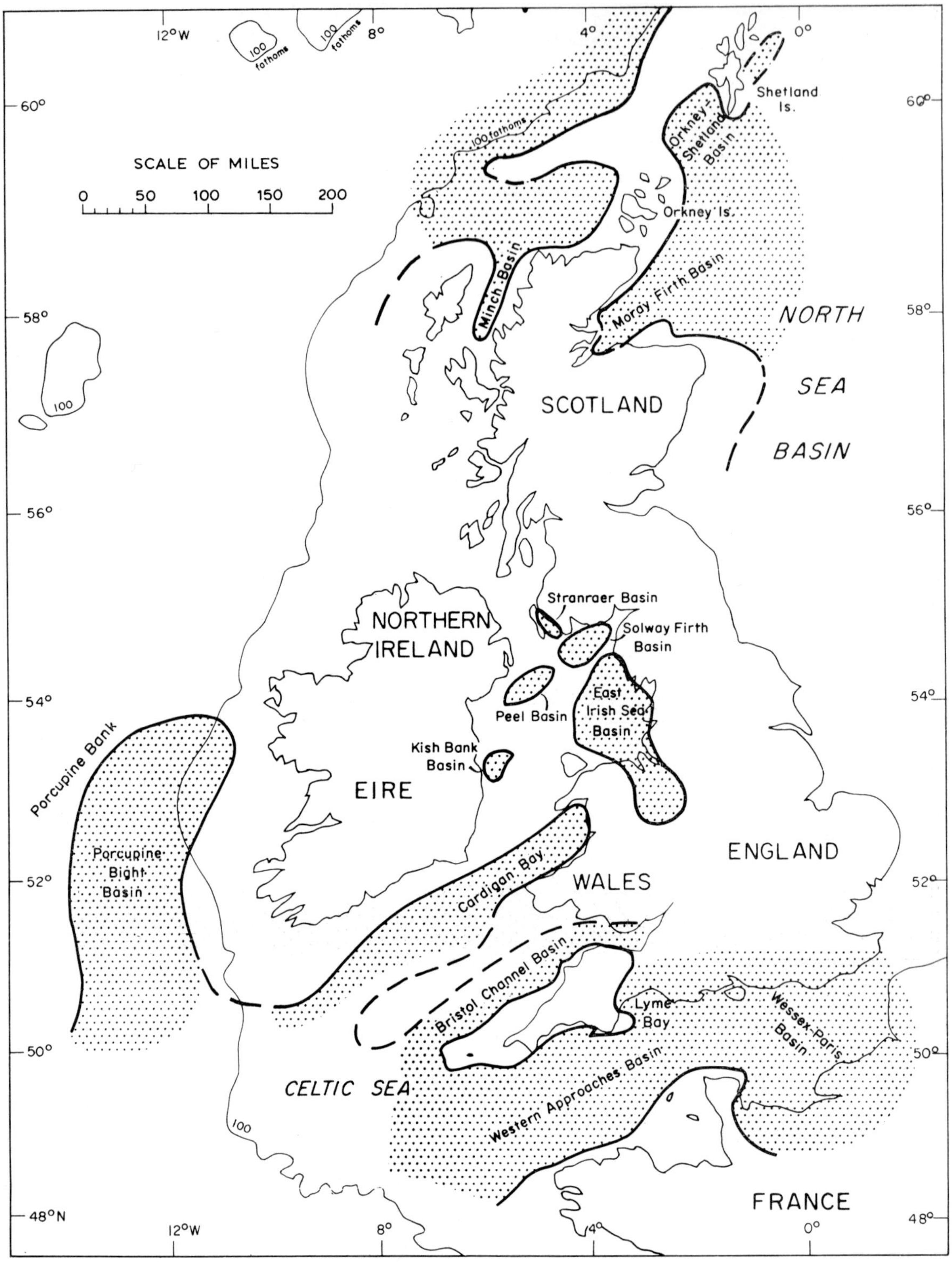

FIG 7. Sedimentary basins around the British Isles.

and English Channel region, and southwards along the Atlantic coast. During the earlier Upper Jurassic (the Oxfordian) marine conditions were re-established in the N, but with waters which were relatively shallow, whereas towards the end of the Jurassic the southern connection appears to have become less dominant; the faunas of E England have a boreal aspect, and we find that in the North Sea there was a widespread transgression during the Lower Cretaceous which seems unconnected with the rocks of S Britain. One may suggest that this was related to the first opening of the N North Sea basin with its oceanic connection and hence of the northernmost North Atlantic, for the Upper Cretaceous chalk sea extended over almost the entire area, and was directly succeeded by the Tertiary with the major northerly basin open (as far as we know) to the Atlantic between Norway and Scotland. A precisely similar conclusion has been independently reached by Allen (1969) from sedimentological evidence.

In the absence of deep borings the evidence of the history of the outer shelves is derived from bottom sampling and the inferential interpretation of seismic velocities in different layers. On the English Channel floor W. B. R. King mapped outcrops of marine Jurassic, Cretaceous, and Tertiary rocks (1954) and these have been shown to extend to the outer edge of the continental shelf (Smith *et al.*, 1965). Gravity and seismic surveys have defined a deep basin of post-Palaeozoic sediments W of Ireland, with a broad granite bastion of Porcupine Bank beyond (Stride 1969). In the Bay of Biscay the continental shelf is known to be built up of gently seaward dipping sediments, in lateral and deeper water continuation of the Mesozoic and Tertiary of SW France, which have in part suffered slumping down slope towards the ocean (Cholet *et al.*, 1968). Deep water Maestrichtian sediments are known on the deep sea floor beyond (Jones and Funnell, 1969). W of the Iberian peninsula investigation of the Vigo seamount shows that fairly shallow water Jurassic and Lower Cretaceous sediments similar to those of the coastal region have subsided into oceanic depths (Black *et al.*, 1964), demonstrating that there also was a build-up of late Mesozoic rocks, but that the modern shelf has been modified by down-faulting.

All the evidence so far thus agrees with a post-Triassic inception of the Atlantic, with a seaway which may have been narrow and close to the European coast in Jurassic times but became progressively more important in late Cretaceous and Tertiary. There is evidence of vertical movement in the development of the Permo-Triassic troughs, and later in the subsidence (downfaulted?) features on the shelf edge;

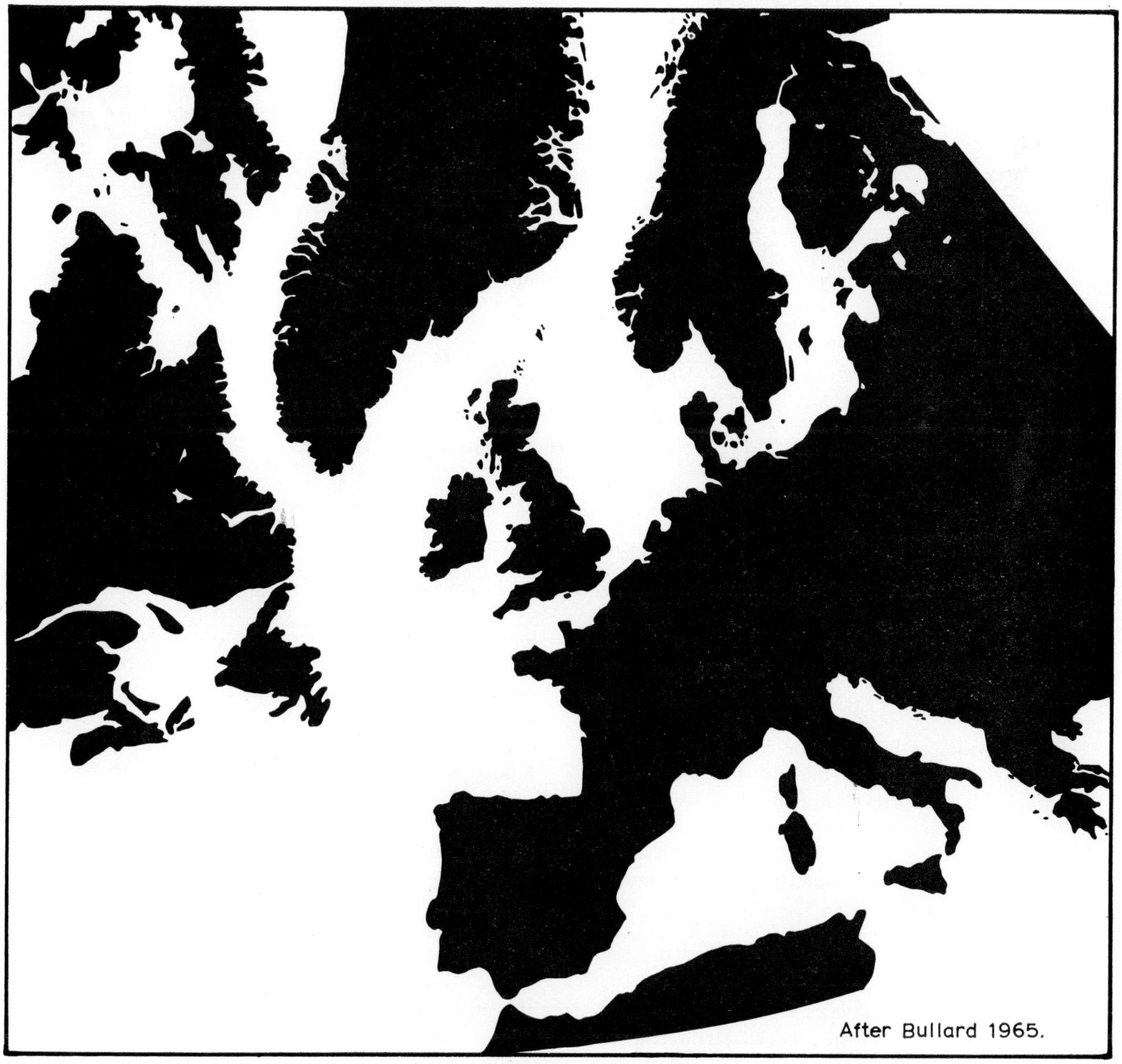

FIG 8. North Atlantic Continents before Continental Drift (traditional arrangement).

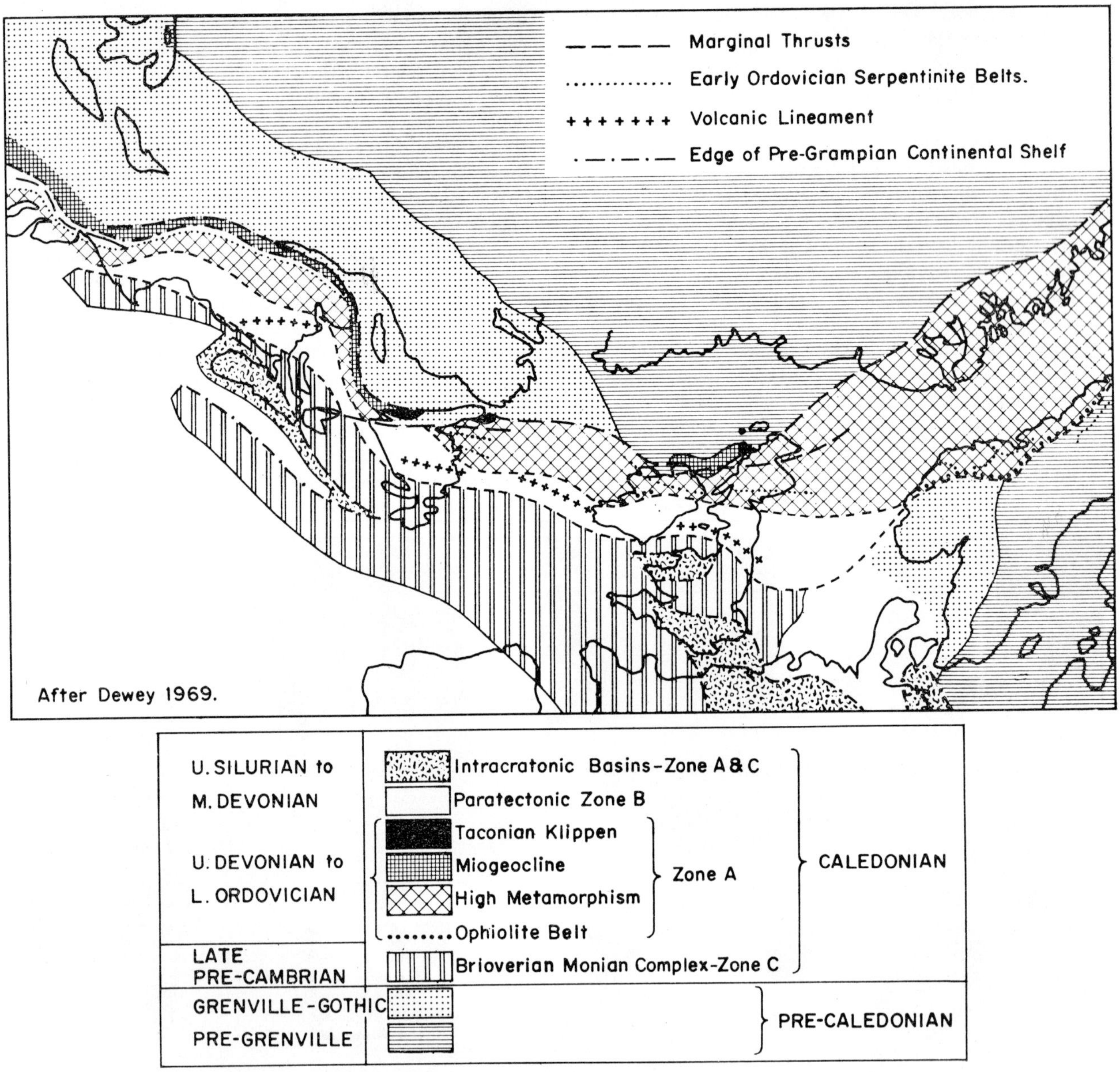

FIG 9. Palaeozoic trends in the North Atlantic Region before Continental Drift (after Dewey).

these could in fact be related either to tensional adjustment related to the lateral stresses of continental drift, or alternatively to direct subsidence of the ocean floor.

This provides at least a hope that shallow water beds will be found through much of the sedimentary column, with satisfactory alternation of sealing and reservoir beds, and that there may well be lateral transitions down dip from a littoral facies with clastic sediments to deeper water source rocks—a potentially favourable combination.

THE RELATIONS OF EUROPE AND AFRICA

I would like to conclude by referring to problems involved in the origin of the modern Mediterranean. This region is, of course, clearly recognized as a long-standing geosynclinal belt, closely associated with the Alpine ranges to N and S, an area which—on the evidence of compressional folding and thrusting—appears to have been greatly narrowed from its original extent. (Fig 10). It now provides a water-hidden belt which is currently supposed to conceal major trans-current faults (affecting, for example, the former lateral relation of Spain to North Africa) and the remains of an enormous gulf supposed to have crossed the E Mediterranean and the Middle East, which has been closed by the pivoting of Africa relative to SW Europe.

These hypotheses were initially interesting academic exercises; they now impinge on real life in that the offshore waters of Algeria, Spain, Italy, Sicily, Libya, and Egypt, and perhaps the deeper waters beyond the shelves, are in varying degree prospective for hydrocarbons. The structures known on land (*e.g.* in Sicily) are believed to continue on to the sea floor; Malta, for example, stands on a shelf linking Italy and North Africa, both distinct hydrocarbon provinces, and exploration of the diapirs of the deep Ligurian–Balearic sea (believed to be salt domes) is envisaged for the future. Dr Burollet deals with the westerly area in more detail later in this volume. My much briefer review was written without knowledge of his views and conflicts with them in various respects, but is submitted as an alternative approach to the problems involved.

At this stage of grossly imperfect knowledge it would be as foolish to disregard the less likely drift hypotheses as to accept them uncritically. Fig 11 shows the extreme case, with major lateral displacement in the W Mediterranean and a broad notional ocean—now lost—in the E. I would, however, here list several points which may give us pause.

First, it is most difficult to accept a late lateral displacement between the opposite sides of the Straits of Gibraltar (Fig 12), as Bullard and others have suggested to optimize the Atlantic fit. There is a high degree of continuity of structural trends, certainly Alpine, apparently also Hercynian, which tend to exclude this as a possibility. Similarly, the shelf connecting Sicily to Tunisia is not only a topographical feature but joins two areas of similar and highly characteristic Jurassic to Tertiary structural features; any hypothesis denying their original continuity must be questionable.

Secondly, problems are created by the neat explanation of the Bay of Biscay by the anti-clockwise rotation of Spain. Between the Balearic Islands and Corsica is the striking N–S Ligurian graben filled with diapirs (Fig 13) (Glangeaud *et al.*, 1966). This feature is precisely what one might imagine a "sphenochasm" to be. It has recently been correlated with the central compressional zone of the Alpine orogeny, but nothing like it is found in the Alps.

If the diapirs are Triassic salt, as Glangeaud suggested, and as is indicated by the circumstantial evidence of Trias salt in the surrounding land areas, not only in N Spain and Italy but also in N Sardinia, it must far pre-date the Alpine orogen. The theory that it reflects a gap left by the post-Alpine rotation of Sardinia and Italy to the NE seems now to be entirely ruled out by unpublished recent surveys which show the Cretaceous to be only gently folded—not crushed or even overthrust—in the Adriatic. A possible alternative is that the graben is due to an early (Pre-Jurassic) clockwise relative rotation of the Iberian isthmus; this suggestion is admittedly far-fetched, but may receive some support from the evidence that palaeomagnetic data in Spanish rocks from the late Palaeozoic to the Tertiary give decidedly conflicting results, and are not compatible with a simple history of anti-clockwise rotation (Van der Voo, 1969).

Thirdly, the postulated major rotation of Africa is now taken by some to imply the former existence of a 2000-mile wide ocean (again in a conveniently water-obscured belt) on the site of the present Persian Gulf (Girdler, 1965). As it is stated, this arrangement is quite unacceptable to all geologists familiar with the detailed structure of the Gulf and with the scores of deep offshore well sections in the marine area, which prove beyond doubt that Arabia, the

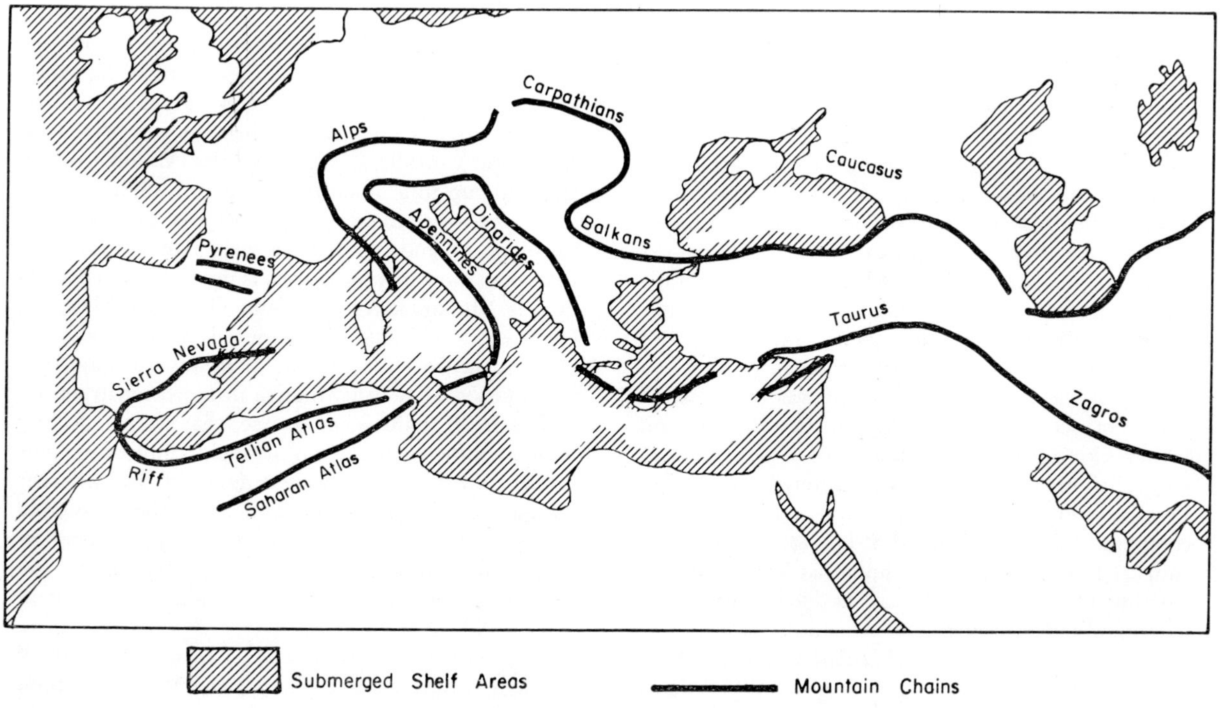

FIG 10. Trend lines and marine shelves in the Mediterranean.

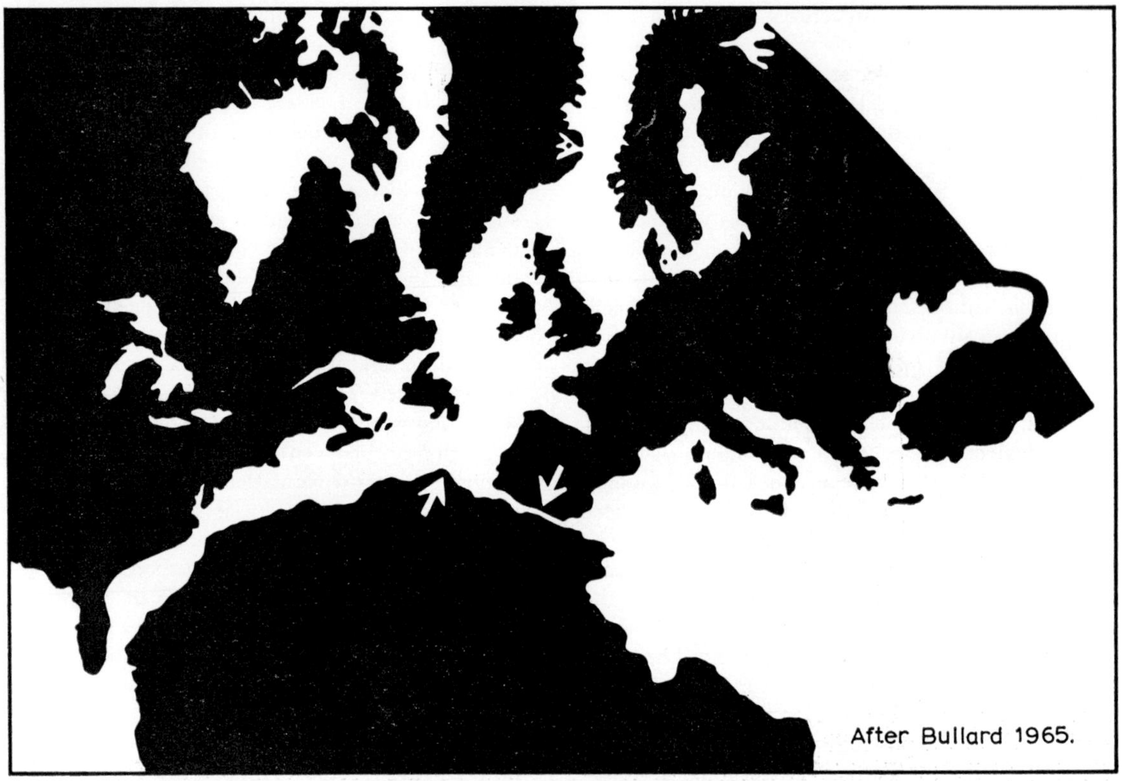

FIG 11. Atlantic Continental Drift assuming major rotation of Africa. (The arrows indicate the position of the Straits of Gibraltar.)

Persian Gulf, and SW Iran are a single tectonic entity, without possibility of discontinuity of the kind required (Fig 14). If we adjust the theory to the minimum degree required by the evidence, at best the presumed former ocean becomes identified with the ophiolite zone which extends through Oman, inland along the Zagros thrust belt, into Turkey and Cyprus. Here there is clear evidence of abyssal waters and oceanic-type deposition at the end of the Mesozoic (Campanian radiolarites), followed by particularly severe compression as the presumed oceanic deep closed up. Again, however, we have severe difficulties: a number of formations on opposite sides of this hypothetical ocean—varying from the Cambrian to the Lower Miocene—are so stratigraphically and faunally similar that former separation by thousands of miles seems unlikely, if not impossible, and there is clear evidence (at least in Iran) that the tectonized belt itself was one of shelf seas—not oceanic depths—through the Jurassic and Lower Cretaceous. Even Oman was a positive area with shallow water carbonates through the Upper Permian and Trias (Reinhardt, 1969). Finally, if we compare times of movement at opposite ends of the supposed African lever we find that in the W the opening of the Bay of Biscay has to be pre-Maestrichtian and is thought to be end Jurassic/Lower Cretaceous (Jones and Funnell, 1969), whereas in the E the closing of the Oman gap post-dates the Campanian deep water sediments. Additional data on accurate timing of the movements will clearly be critical.

We could conclude that if it is considered valid to make a palinspastic reconstruction by de-telescoping the folded and thrust compressional zones it is logical to stretch out not only the Oman–Zagros–Cyprus belt but also the whole Alpine trend, including the Tyrrhennian Sea, from which the exotic blocks of the Argile Scagliosa were spewed across the Appennines, the Alps themselves, the Pyrenees, the Baltic Cordillera, and the Atlas Mountains. All this adds up to a much more nearly N–S translation of Africa—the traditional movement of the older tectonicists—and it can be argued that such a large rotational movement of Africa is not only unnecessary but is an unlikely postulate. The pre-Mesozoic continental fit could be achieved with Africa still oriented nearly N–S, by the alternative of adjusting the position of South America in the reconstruction.

The main point is, of course, that we are dealing with hypotheses based on strongly conflicting lines of evidence; that the weighting given by different authors differs strongly; and that the economic geologist must still keep an open mind, being as ready to consider exploration problems in the Mediterranean on a basis of no significant drift as in the light of the other extreme.

As petroleum geologists we, with our geophysical colleagues, have made the most accurate contribution of all to deep knowledge of the sedimentary basins,

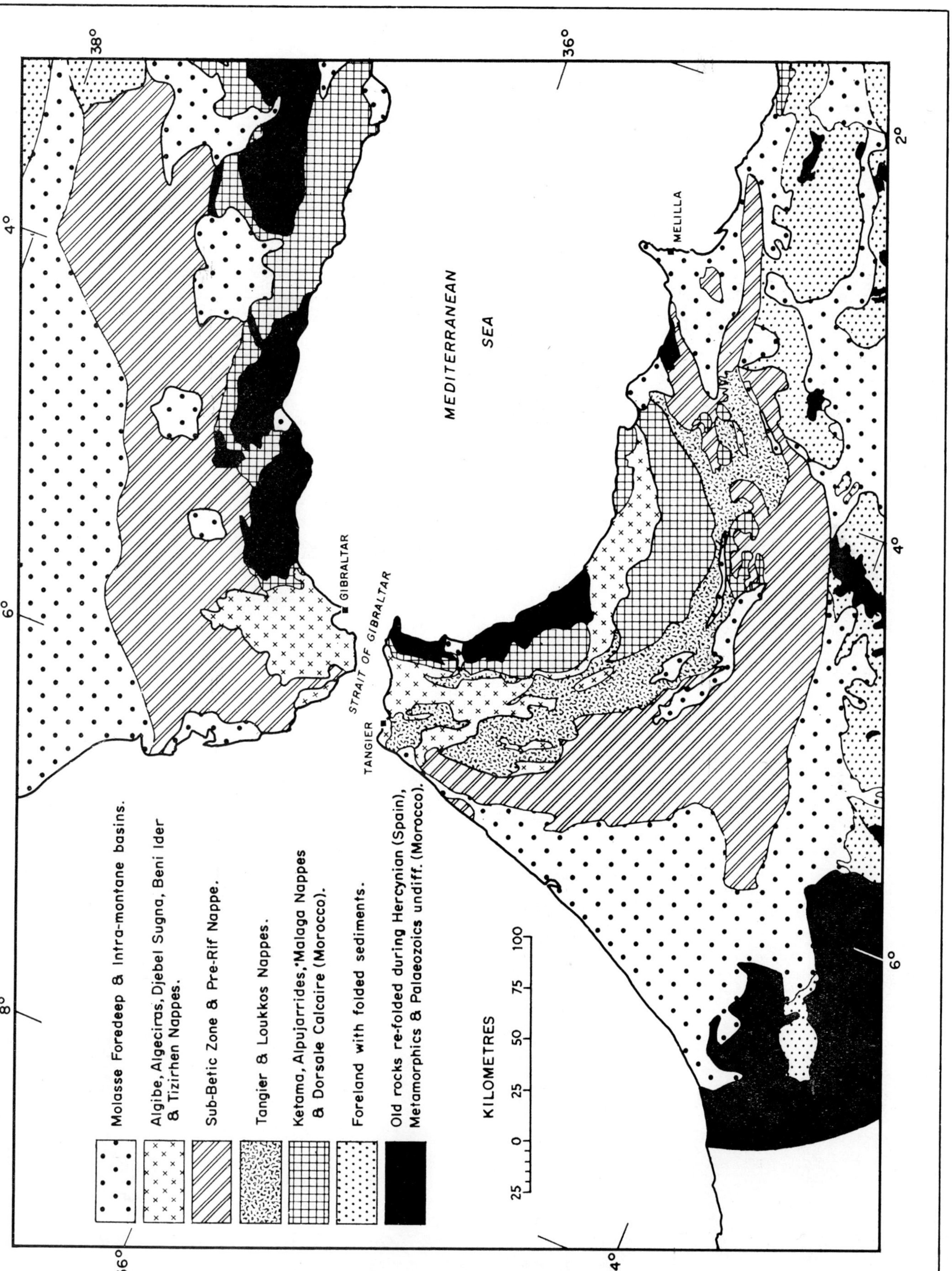

FIG 12. Straits of Gibraltar, showing Alpine and older structural trends. (Based on the Tectonic Map of Europe.)

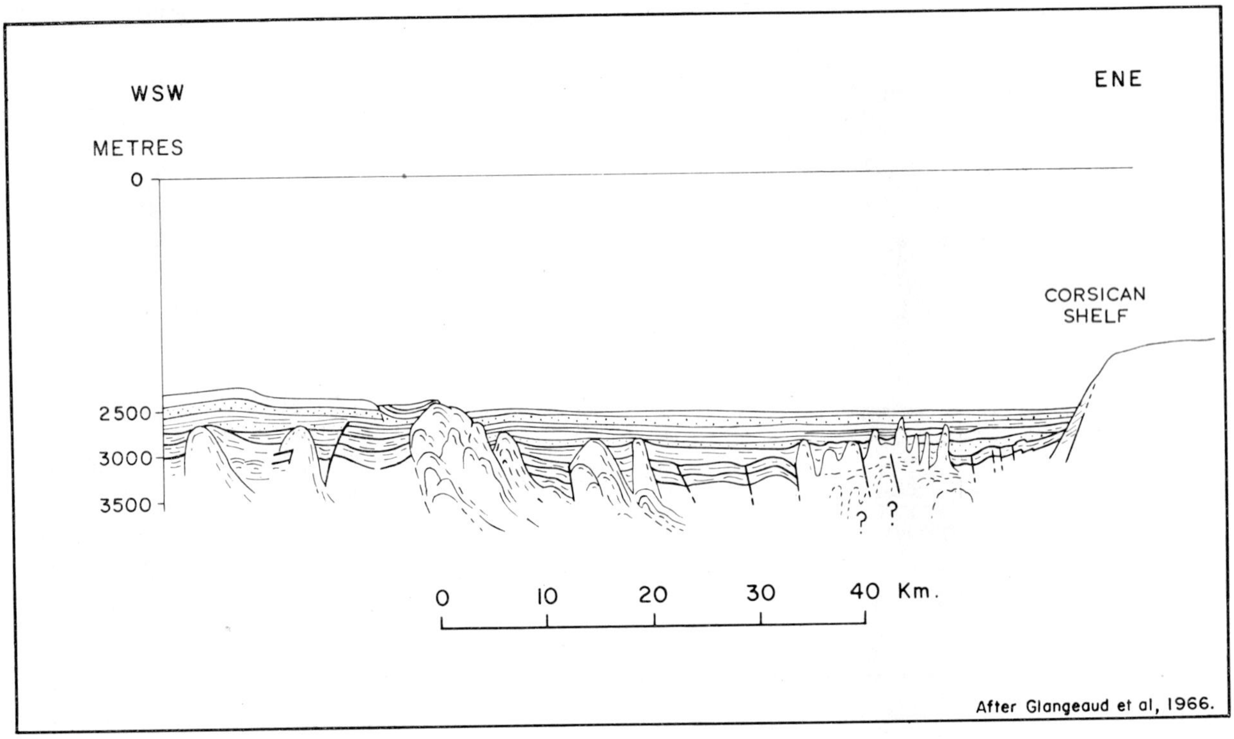

FIG 13. Diapiric structures in the deep basin of the Ligurian Sea, W of Corsica.

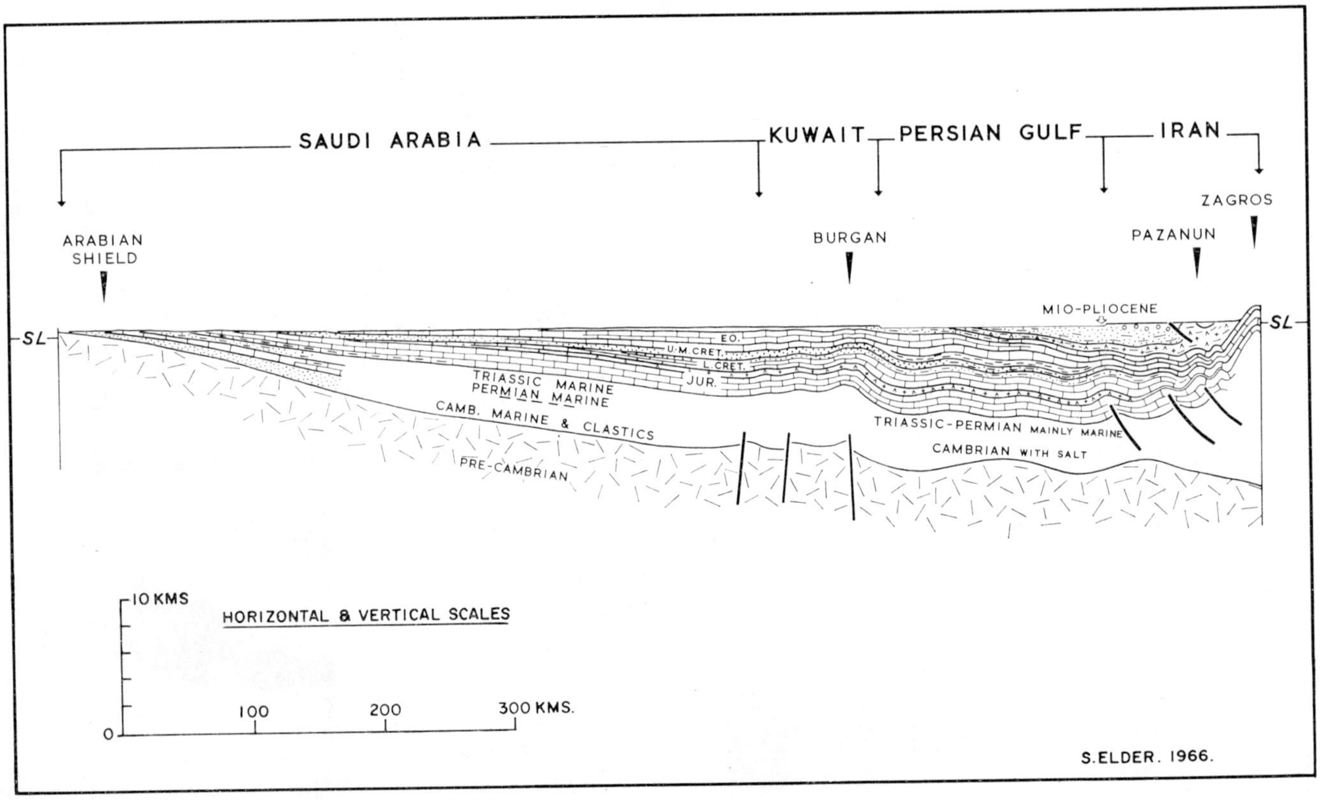

FIG 14. Cross-section of the Persian Gulf region, illustrating the stratigraphical and structural continuity between the Arabian Shield and the Zagros mountains.

the structure of continental shelves and the depths of mountain ranges. The solution of these other problems, no longer only academic, will be in a major degree influenced by our activities in future exploration.

ACKNOWLEDGMENTS

The writer is indebted to R. G. W. Brunstrom and other colleagues for critical reading and constructive criticism of the draft, and to The British Petroleum Company for permission to publish this outline survey.

POSTSCRIPT NOTE

While this paper was passing through the Editor's hands, a major review of the origin of the W Mediterranean was published in The Netherlands ("Symposium on the Problem of Oceanization in the Western Mediterranean", *Verh. ned. geol.-mijgb. Genoot*, 1969). This work includes detailed discussions by a number of authorities of the geological and geophysical features, genesis of the sea, and the disappearance of crust. It again illustrates the range of possibilities and the inconclusive nature of present evidence.

BIBLIOGRAPHY

Allen, P. *Nature, Lond.*, 1969, **222** (5194), 657–8.

Black, M. *et al. Q. Jl. Geol. Soc. Lond.*, 1964, **120** (4), 477–577.

Brinkmann, R. "Geologic Evolution of Europe" Stuttgart, Ferdinand Enke Verlag, 161 p, 1960.

Bullard, E. *et al. Phil. Trans. R. Soc. ser. A*, 1965, **258** (1088), 41–51.

Carte Tectonique Internationale de l'Europe 1 : 2,500,000, 1964, Moscou, Congres Geologique International, Sous-Commission de la Carte Tectonique du Monde.

Cholet, J. *et al. Rev. Inst. fran. Pétrole*, 1968, **23**, 1029–45.

Collette, B. J. "On the Subsidence of the North Sea Area", *in* D. T. Donovan, ed., "Geology of Shelf Seas". Edinburgh and London, Oliver and Boyd, 160 p, 1968.

Dewey, J. F. *Nature, Lond.*, 1969, **222** (5189), 124–9.

Girdler, R. W. *Nature, Lond.*, 1965, **207**, (4995), 396–8.

Glangeaud, L. *et al. Bull. Soc. Geol. Fr.*, 1966, ser. 7, **8** (7), 921–37.

Illing, V. C., ed. "The World's Oilfields—The Eastern Hemisphere", *in* "The Science of Petroleum", vol. 6, pt 1. London, Oxford University Press, 174 p, 1953.

Jones, E. J. W., and Funnell, B. M. *Deep-Sea Res.*, 1968, **15**, 701–9.

Kapounek, J., and Horvath, S. *Erdol-Z. Bohr-u. Fordertech.*, 1968, **84** (11), 397–407.

Kent, P. E. *Proc. Yorks. Geol. Soc.*, 1966, **35**, 323–52.

King, W. B. R. *Q. Jl. Geol. Soc. Lond.*, 1954, **110** (1), 77–101.

Perrodon, A. "Géologie due Pétrole". Paris, Presses Universitaires de France, 437 p, 1966.

Reinhardt, B. M. *Schweiz. miner. petrogr. Mitt.*, 1969, **49** (1), 1–30.

Smith, A. J. *et al.* "The Geology of the Western Approaches of the English Channel." IV. A Recently Discovered Variscan Granite WNW of the Scilly Isles," *in* W. F. Whittard and R. Bradshaw, eds, "Submarine Geology and Geophysics." Butterworths, 464 p, 1965.

Stride, A. H. *Phil. Trans. R. Soc., ser. A*, 1969, **264** (1148), 31–75.

Van der Voo, R. *Tectonophysics*, 1969, **7** (1), 5–56.

Wills, L. J. "Palaeogeographical Atlas of the British Isles and Adjacent Parts of Europe." London, Blackie, 64 p, 1951.

MS received 23 July 1969

Petroleum Geology of the Western Mediterranean Basin

By P. F. BUROLLET

(*Compagnie Française des Pétroles, Paris*)

INTRODUCTION

THE Western Mediterranean has a complex structure compared with that of the Eastern Mediterranean. The Alboran Sea and the southern Balearic Sea represent the "tectorogenic axis", from which the alpine folds diverged over the Balearic area, Andalusia, Morocco, Algeria, Tunisia, and Sicily. This axis has collapsed 2000 m deep below the sea and is an area of oceanic type sea bottom sedimentation.

The Tyrrhenian Sea is part of this "tectorogenic axis" that has been bent sharply round to the NW with expulsion of the sedimentary series to the NE (Italy) and to some small extent to the west (Corsica). This axis links up with the western Alps. Corsica and Sardinia could be "external zone" crystalline massifs. In front of them the Balearic and Ligurian Seas, although part of the Alpine area, did not have the full effect of the Alpine movements. These seas are almost an "oceanic" trough with a fairly thin sedimentary cover cut by numerous salt domes. The NW corner of the Balearic Sea and the Gulf of Lion are the sunken remains of the Hispano-Provençal craton.

If we leave aside the petroleum prospects of the deep water sediments, associated with the salt domes, the other prospects of the Western Mediterranean are limited to several fairly narrow continental platforms and to coastal basins.

Past exploration in Tunisia has not had much success, although another phase is starting in the more stable area of the Pelagian Sea to the east of the country.

Oil shows and minor oil discoveries have long been known in the Tellian Alps of Algeria (Si Aissa in the east and Tiliouanet in the Chelif Basin) without there having been much economic production. The Miocene foredeep in Morocco has a small oil production near Petit-Jean.

In Spain exploration has been disappointing in the Andalusian Neogene foredeep, as well as in the Cantabric mountain chain and the Ebro lagoonal trough. The southern border of the Pyrenees and the area of the Ebro delta could still be of interest.

In the French areas the Languedoc is over-faulted and the Camargue Tertiary lacks reservoirs. Exploration results have been deceptive despite a major effort over a long period and the excellent shows at Gabian and Gallician. Exploration of the continental shelf of the Gulf of Lion is taking place and the recent well Mistral No. 1 proved a thick marine Neogene section.

The violent tectonics and narrowness of the continental platform of the Tyrrhenian Sea offshore give but the slightest of hopes for economic petroleum production, despite which there is sure to be some exploration activity now the zone is open to bids.

Sicily is the only area of the Western Mediterranean where exploration has been crowned with some success, there being gas in the autochthonous Neogene (under glided nappes) and oil in the Trias of Ragusa and Gela.

PHYSIOGRAPHIC FEATURES

The Mediterranean is almost a closed sea with an area of about 1,150,000 square miles. It is connected with the Atlantic Ocean by the Strait of Gibraltar; otherwise the only other outlets are the Bosphorus, to the closed Black Sea, and the man-made Suez Canal, to the Red Sea.

It is divided into two basins, differing as much geologically as geomorphologically, which are separated by the Strait of Sicily (between Tunisia and Sicily) and the Strait of Messina (between Sicily and Calabria).

The Eastern Mediterranean has a simple and dissymetric structure with the northern border of the main basin (Ionian Sea and Cyrenaïcan Sea) formed by the alpine thrust front of the Hellenide and Tauride belts. To the south of these is the Hellenic trough, marked by a strong negative gravity anomaly. Parallel and south of the trough is the "Mediterranean Ridge", lying at a depth of 2000 m. Further south again there is a deep zone which shallows progressively towards the African platform. Regions of great depth are known in the Eastern Mediterranean, but the abyssal zones are very small in area (the greatest depth in the Mediterranean is 5092 m in the Hellenic trough to the west of the Peloponnese and there is another deep of 4070 m in the Ionian Sea).

To the north of the Eastern Mediterranean two extensions are developed. One, an infra-alpine furrow, is the Adriatic Sea, whilst the other is the Aegean Sea, which extends by the Sea of Marmara to the Black Sea. Both are drowned areas that sank across the lines of the alpine chains.

To the west, between Sicily, Tunisia, and Tripolitania, a vast continental platform was developed and is cut only by the deep Pantellaria Channel. This area is called the Pelagian Sea and is part of the African platform. Its tectonics will be discussed later.

The Western Mediterranean, on the other hand, is quite different. Although not very deep (maximum 2866 m between the Balearic Islands and Sardinia) the abyssal plains are more extensive and the continental slopes are steep and cut by canyons. The major part of the Western Mediterranean represents a zone, inside the alpine folding, with a "tectorogenic" axis which gave rise to areas of folding to the north (Balearic and Andalusian folds) and the south (Tellian Atlas and Sicilian folds). A tectorogenic axis is understood to be an axial zone of alpine folding which has acted either as a eugeosyncline or as the central part of an orogenic area with divergent folding.

Two large mountainous islands, Corsica and Sardinia, separate the Western Mediterranean into two parts. Between these islands and Italy is the Tyrrhenian Sea, with numerous seamounts piercing the central plain with others forming islands along the continental slope. This area is also an axial zone inside the alpine folding. To the west of Corsica and Sardinia is the Balearic Sea with a broad flat abyssal plain, of 30,000 square miles, at depth between 2700 and 2900 m. Whilst its northern edge is marked by the large Rhone Delta fan, this abyssal plain extends NE into the Ligurian Sea between Corsica and the French and Italian Rivieras. To the SW between Andalusia and Morocco the area is called the Alboran Sea. Both the Central Balearic and Ligurian Seas are pierced by numerous anomalies thought to be deeply-buried salt domes.

GEOLOGICAL STRUCTURE OF THE WESTERN MEDITERRANEAN

As has been indicated above, the structure of the Mediterranean is linked with the Alpine orogeny. Fig 1 shows the different unities that are a consequence of geosynclinal evolution, and form the major elements of the Alpine overthrust zones: the figures does not distinguish the areas of stable platforms, forelands, epicontinental seas, intracratonic basins (not even those where the sediments were subsequently folded as a consequence of different alpine paroxysms, *e.g.* French pre-Alps, Provence and Languedoc folds, the chains of the Iberic mountains, Saharan Atlas, eastern Tunisia, Jebel Akhdar in Cyrenaica, etc.).

There is a fundamental structural difference between the Western Mediterranean, which corresponds to a tectorogenic axial zone, and the Eastern Mediterranean, which is an external trough to the alpine folds and has a relatively simple structure. The difference of position relative to the axis of the alpine chain of these two parts of the Mediterranean is due to the large dislocation of the African block as marked by the Jeffra–Malta axis which separates the Pelagian from the Ionian Sea.

The structure of the Eastern Mediterranean is relatively simple, lying as it does between the African platform and the southern front of the Alpine Chain. The structure of the Alboran Sea and the southern Balearic Sea is also fairly simple, being the result of a final collapse of the "tectorogenic" axis of alpine

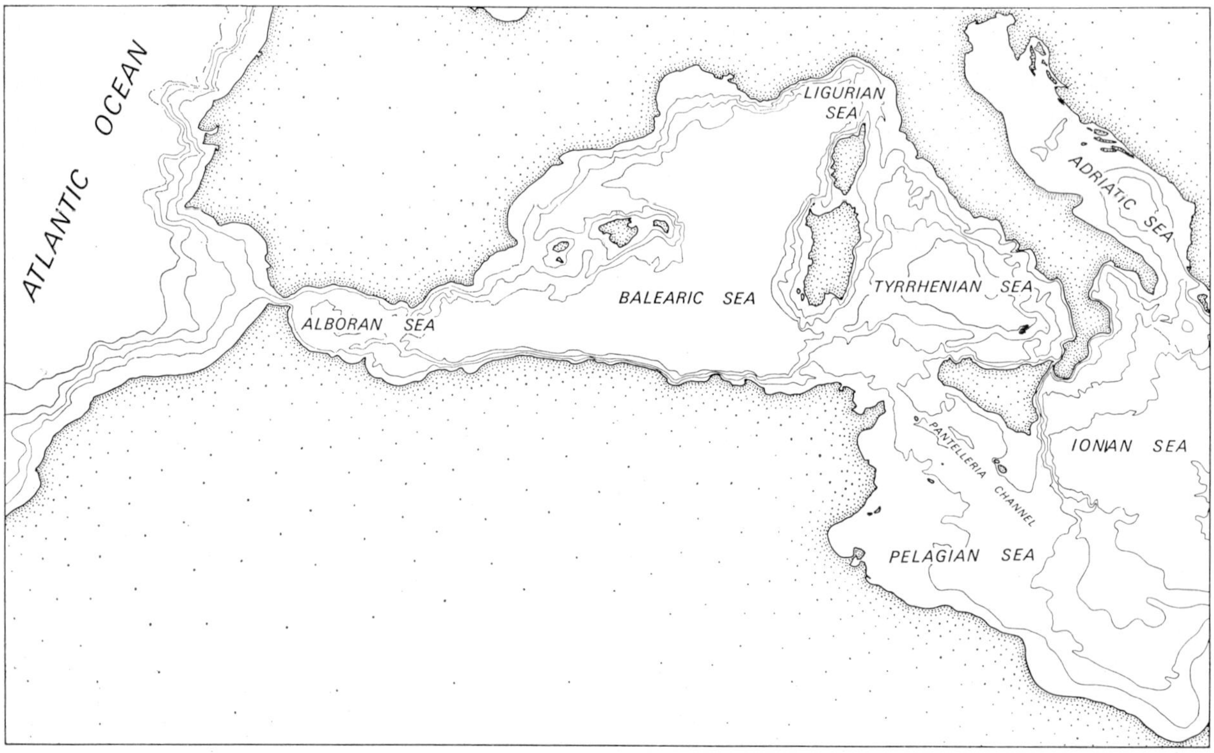

FIG 1.

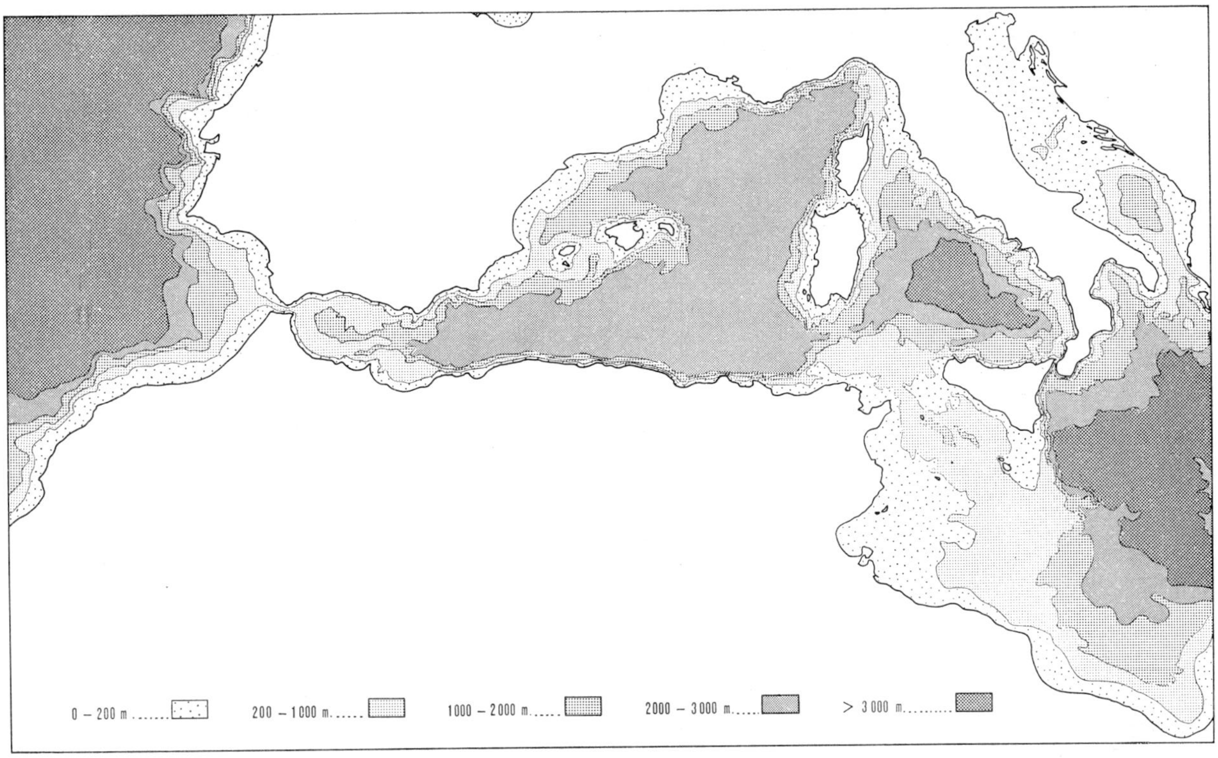

FIG 2.

folding. In Central Europe, despite the complications where the northern flank of the Central Alps poured its sediments into the mollassic troughs of the fore-deeps, there is a clear overall logic to the structure. One could almost say the same thing for the western flank of the French Alps, despite there being a tighter geometry and more limited thrust movements. This tectonic framework could correspond to the old idea of squeezing the alpine geosyncline between the European and African platforms.

But the centre of the area that we are discussing, formed by the Adriatic, Italy, Tyrrhenian Sea, Corsica, Sardinia, and the northern part of the Balearic Sea, shows the alpine folds turning back on themselves and becoming convergent around the Po plain and the negative axis of the Western Adriatic. It is this bend of the alpine trends, northward around the Adriatic, which has created the complexity of the geology of the Tyrrhenian and North Balearic seas. A certain number of explanations can be given, but none of them are entirely satisfactory. None the less, the research work done on the deep structure (Fahlquist, 1963; Berry and Knopoff, 1967) and the geological nature of the sea bottom of the Ligurian Sea (Glangeaud *et al.*) gives us arguments upon which to build some regional ideas. There were probably two reasons for the structural complexity of this area. First, a NW differential rotation of the African block, with the Maghreb, Pelagian, and Ragusa platforms going much further N than the Egyptian and Cyrenaican platform, which was more stable. This phase started in Cretaceous times and reached its climax and finished during the Oligocene. The limit between the two areas is traced by a major fault zone running from the Tibesti to the east of Sicily, marking the boundary between the Pelagian and Ionian Seas, along which, from Tibesti to Etna, there is a considerable amount of volcanism.

The second and main cause may be be found in an E to W trending compression across the Tauride and Hellinide belts. This could be compared to a wave of long amplitude which caused the progressive orogeny of the axial and internal zones during Miocene times. According to Glangeaud, the major part of W Mediterranean was uplifted and emerged during the general Pontian regression.

The tectonic outlines of Fig 3 show in a simplified form the structural zones around Italy. Part of the Alpine thrust front is shown by a broken line, as even now such an important limit cannot yet be drawn everywhere with certainty.

A synclinal zone, marked by negative gravity anomalies, represents the foredeep, as well as the folds of the Appenines, the Bergamasc Alps, the Venetian Alps, and the Dinarides. This is the trough of the Po plain, of the Molise, and in general of the Adriatic. There has been a progressive migration of this trough from west to east from the Miocene to Quaternary times. Bending around this trough is the divergent or "tectorogenic" axis, which runs up the middle of the Tyrrenian Sea (passing between Corsica and the island of Elba). It turns around the Po plain, where "Flysch à Helminthoïdes" and "schistes lustrés" are seen as remnants of the eugeo-syncline. It then follows a complicated path through

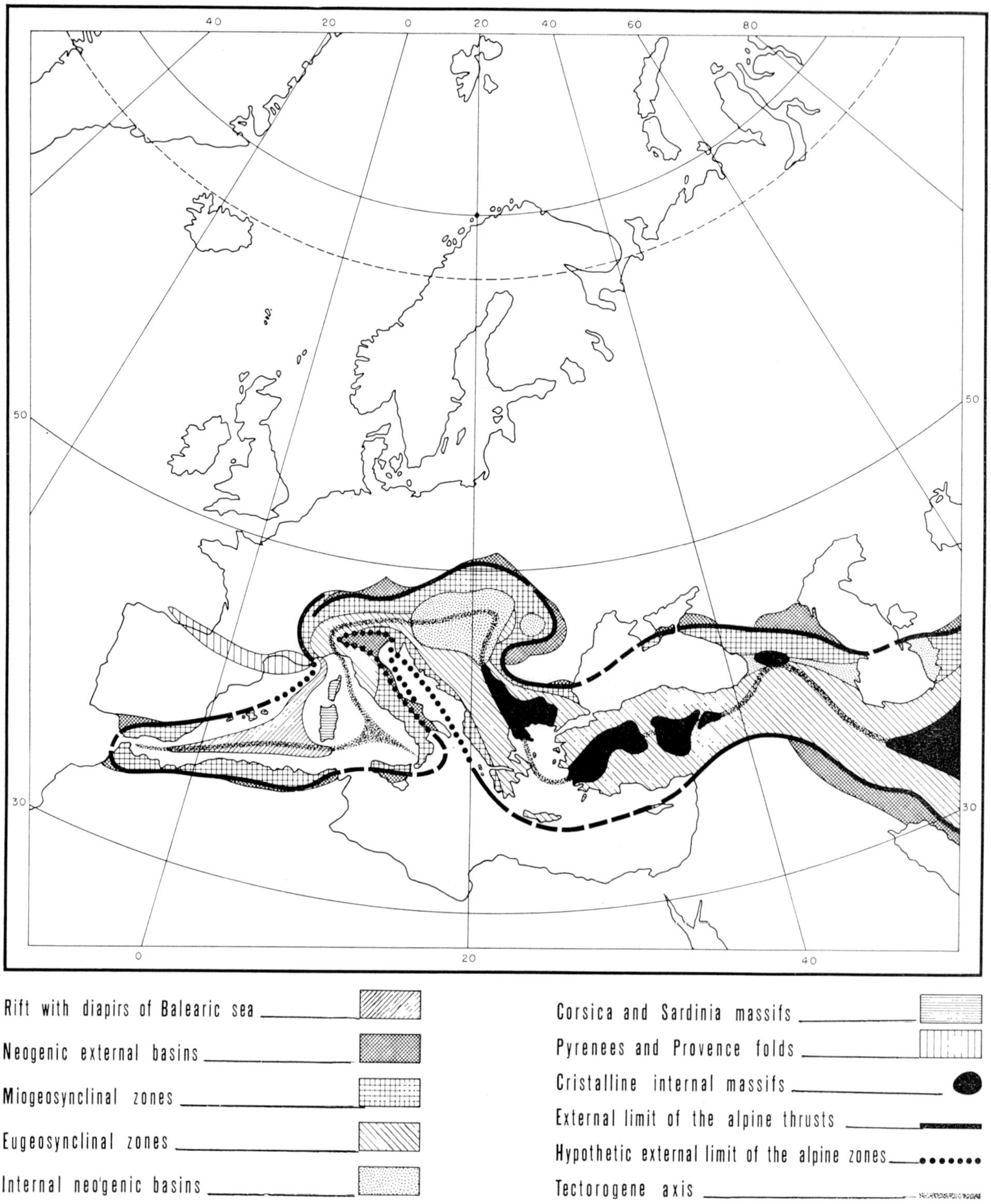

FIG 3. Tectonic sketch of the Alpine ranges.

the axial zone of the Carpathian chain before joining the Dinaride axis.

One is often tempted to take Corsica and Sardinia as forming part of a Hercynian framework which includes the Maures, the Esterel, the Catalan chain, and even the axial zone of the Pyrenees. It is certain that the northern part of the Balearic Sea has not really played a part in the Alpine movements and that the resistant massifs of the Esterel and Maures were partially the cause of the eastward swing of the French Alps in the Maritime Alps region. On the other hand, the alpine thrust of Corsica has only "internal zone" elements equivalent to the "Flysch à Helminthoïdes" and the "schistes lustrés", whilst the north of the Balearic Sea and the Ligurian Sea have an "intermediate zone" crystal structure similar to

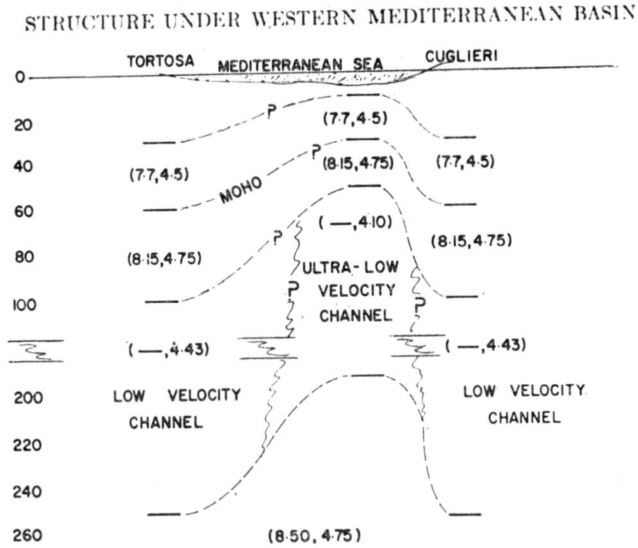

Schematic profile of upper mantle structure along the line Tortosa-Cuglieri. A discontinuity is found in the depth scale at 100 km.

FIG 4. Section under the Western Mediterranean: Spain—Sardinia (after M. J. Berry and L. Knopoff, *J. Geophys. Res.*, 72).

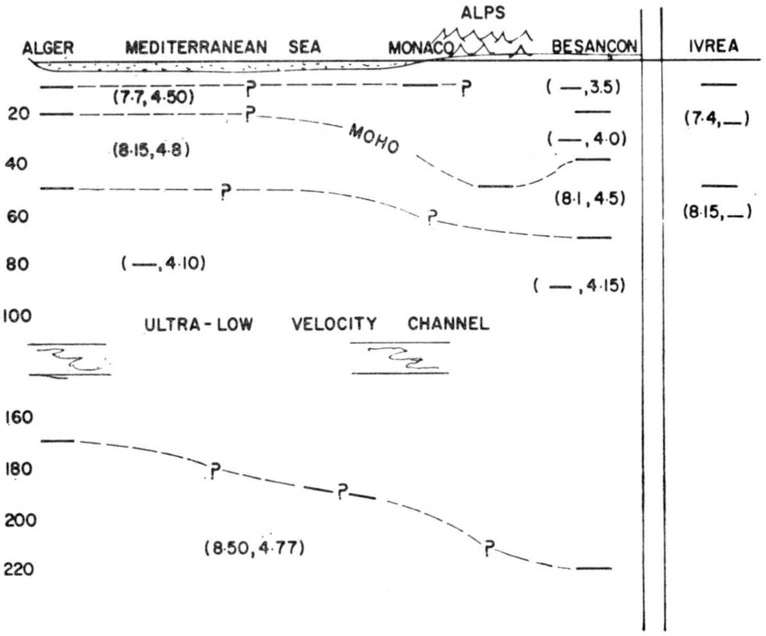

Schematic profile of upper mantle structure along the line Alger-Monaco-Besançon. A discontinuity is found in the depth scale at 100 km. The structural section of the Ivrea zone is found to the west of the Alps in the main section.

FIG 5. Section under the Western Mediterranean: Algiers—Monaco—Besançon (after M. J. Berry and L. Knopoff, *J. Geophys. Res.*, 72).

that of the Briançonnais area. Those oceanic areas limited to the north by the Hispano-Provençal craton have thin sedimentary cover, and contain numerous deep salt domes of unknown age typical of an area with unstable deep-seated tectonics. For reasons unknown, these areas have not been folded by the Alpine movements and it seems that Corsica and Sardinia acted as crystalline massives of the "external zone" in this area, where the alpine movements were so imperfectly developed as to be almost non-existent.

The sharp bend to the SE of the divergent "tectorogenic" axis (W of Calabria) and the mobility of the Appenine geosynclines, which sucked in the

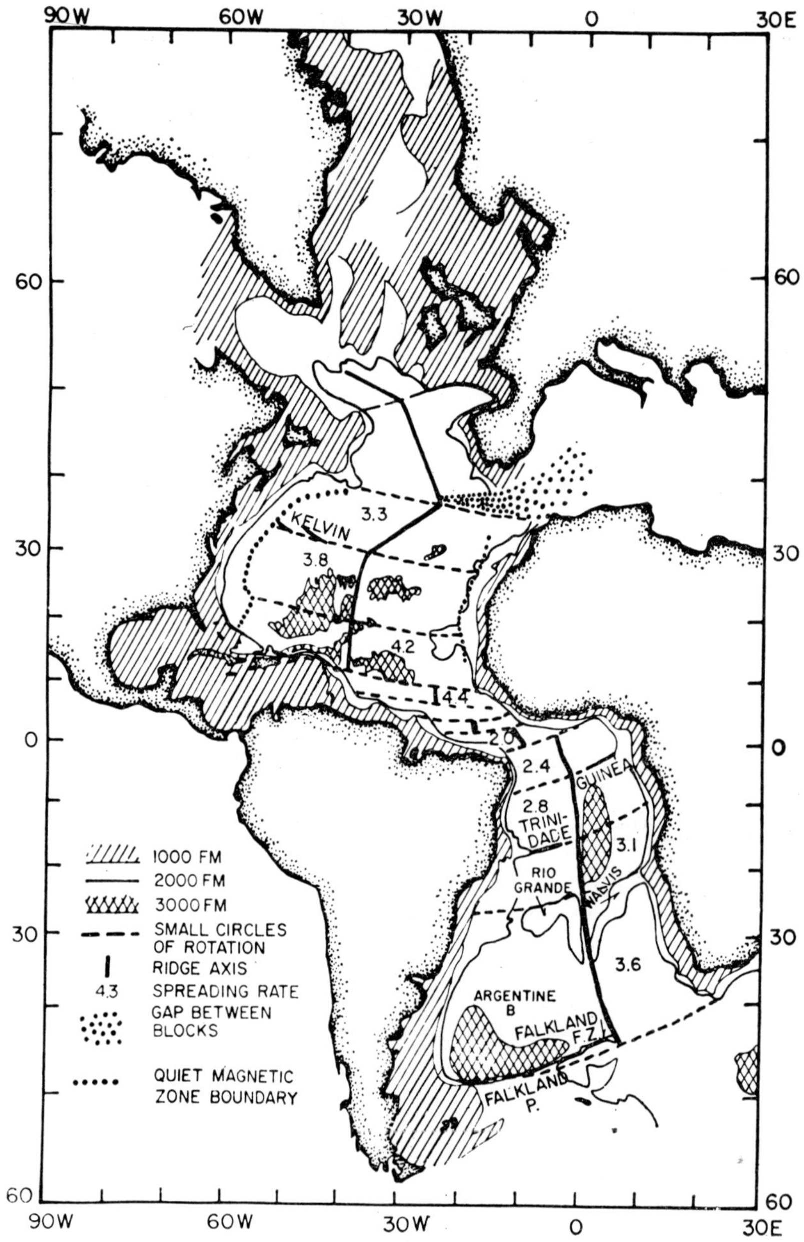

FIG 6. The Atlantic at the time of anomaly 31 (after X. Le Pichon, *J. Geophys. Res.*, **73**).

major part of the material expelled by the Tyrrhenian orogeny, explains the absence of alpine folding in Sardinia and the reduced effect of these phenomena in Corsica, where the tectonic activity is limited to sheet thrusting of units of the "internal zone" and a slight overthrusting of the Corsides over the Ligurian trough (A. Mauffret, 1966; L. Glangeaud, J. Alinat, J. Polveche, A. Guillaume, and O. Leenhart, 1966).

Figs 4 and 5 (taken from Berry and Knopoff) show the deep structure of the Balearic Sea compared with that of the French Alps. Fig 6 of X. Le Pichon shows the probable relative positions, in Palaeocene times, of the African and European blocks on either side of the Western Mediterranean basin. It also illustrates the importance of the rotation of the NW part of the African craton relative to Europe during the Tertiary era.

It must be noted that the NW corner of the Western Mediterranean is outside the Alpine area. To the north of a line from the Balearic Islands through the Maures and Esterel is a piece of the Catalan and Pyreneo-Provençal framework which collapsed during Tertiary times. The major part of the movement took place during the Pliocene and the greater part of the area is covered by the thin, fine-grained sediments of the Rhone deep alluvial fan.

REGIONAL AND PETROLEUM GEOLOGY

Tunisia

The Alpine areas of NW Tunisia have not been touched by oil exploration since the SNAP concessions (Gulf-SEREPT) were surrendered in 1954, and SEREPT stopped work in the Mabtouha area.

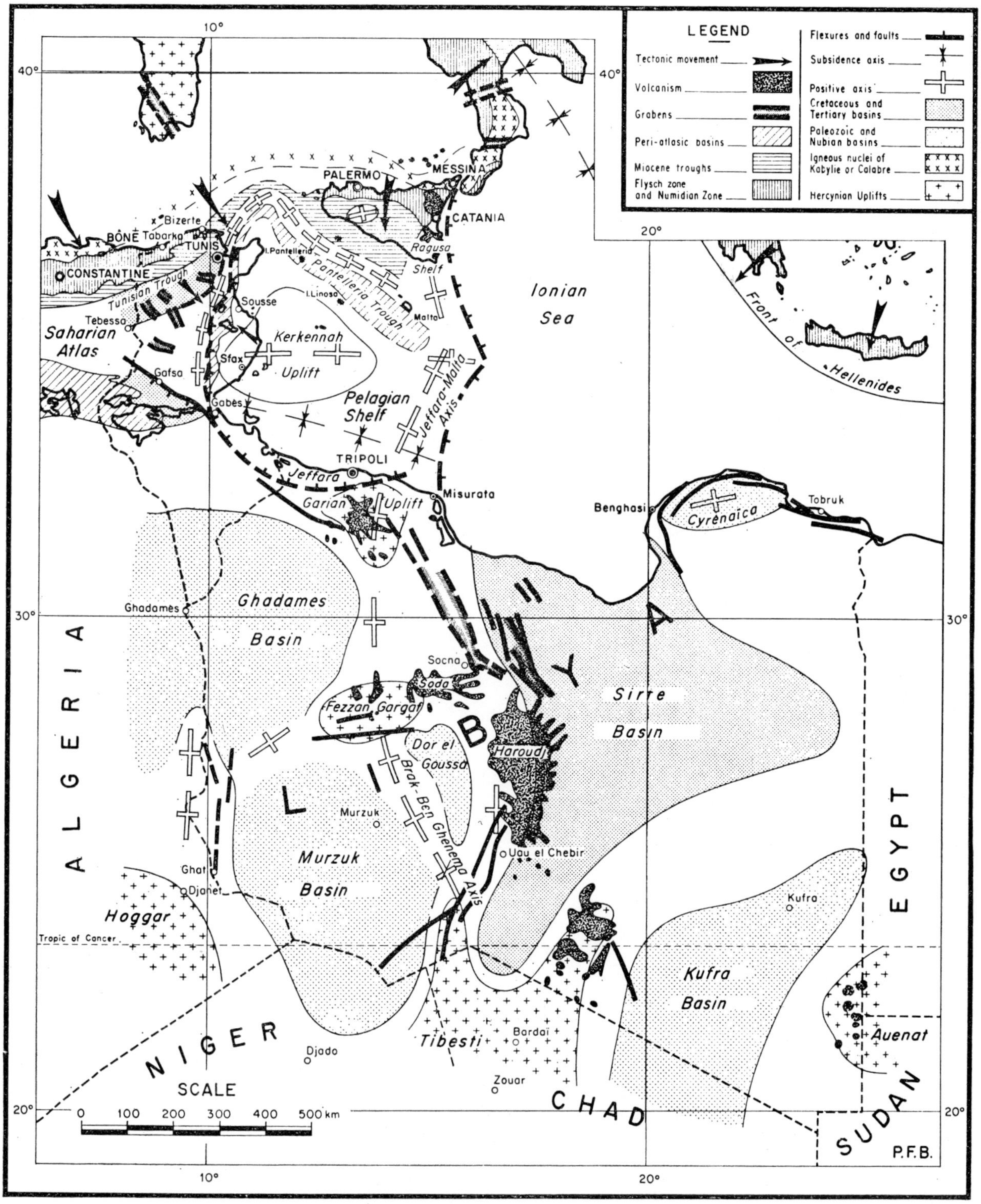

FIG 7. Tectonic sketch of Tunisia and Libya.

The tangental movement area, and in particular the Numidian flysch zone, would appear to offer poor exploration prospects, although the oil seeps of the Miocene foredeep areas were occasionally the object of exploration work before and after the Second World War (Ain Rhelal, El Haroun No. 1 and 2, Sloughia). These oil shows are restricted to the areas along the regional faults. Some seeps come from Miocene or Eocene beds, whilst others are certainly from the Lower or Middle Cretaceous (Kef bou Debbous near Teboursouk, Zag and Tyr near Le Kef, etc.).

The Tunisian Atlas has been regularly prospected for objectives in the Lower Cretaceous. The Serdj formation (Albo-Aptien) at Douleb, Semmama, and Tamesmida has since 1958 a small oil production of 5000 bd, with evacuation of the oil by 6-inch pipeline to La Skhira.

At Cap Bon there is a small gas production (890 Mcf/day) from the Lower Cretaceous sands, which supplies the town of Tunis.

Eastern Tunisia (Sahel and Basses Steppes) is part of the Pelagian platform. Exploration is in an active phase. Since Amerada and Rimrock abandoned their concessions, the areas have been taken up by SITEP (AGIP) in Jeffera, SNPA, ELF, and CFP in the Sahel and offshore areas.

The Sahara areas of Tunisia are still held by SITEP, with El Borma as the only field in the area. This field, with 50,000 bd, is by far the most important Tunisian producer. But here we are already beyond the strict limits of the Mediterranean area.

Algeria

In the Tellian nappes zone there is the small Sidi Aïssa field, which was for many years the only producer in Northern Algeria. The Chelif Mio-Pliocene basin in the Oran Department has surface oil shows at Tiliouanet and elsewhere. There has been no activity since several years' exploration was done at the end of the Second World War.

The Saharan Atlas, because of its Lower Cretaceous and Jurassic objectives, is a centre of exploration. Apart from the old small Jebel Onk fields, there have been no discoveries of consequence in the area. There is little hope of exploration in the Mediterranean sector of Algeria, where there is almost no continental shelf, whilst the major production of the country continues to come from Algerian Sahara.

Morocco

The Petit-Jean area in the Miocene foredeep has been a source of oil for a long time. The Mediterranean continental shelf is narrow, lies in the middle of a highly tectonized zone, and would not seem to be an attractive area for oil exploration.

Southern Morocco has frequently been prospected at different times. An attractive area seems to be the Atlantic offshore, especially that held by Esso.

Spain

Exploration in Spain is concentrated in the Andalusian foreland area, in front of the Alpine overthrusts. It has not been a notable success up to now. In the western part of the Cantabric range and the southern Pyrenees areas, the developing Ayoluengo field, and the wells of Polientes, Tozo, and Huildobro, with good oil shows, underline the petroleum prospects of the eastern edge of Asturia.

Despite the many surface seeps, oil has not been found in commercial quantities in the area of the southern Pyrenees. It is the same for the Ebro delta, the Ebro lagoonal Oligo-Miocene foredeep, and several plateau zones of central Spain.

France

Exploration activity, though slow, continues in the Mediterranean Languedoc. A complicated broken tectonic pattern, together with very thick compact Jurassic beds, explains the many disappointing results. However, there are encouraging oil shows that have been known for a long time, such as Gabian (Herault), where a large fault cuts a Triassic section, and Galician, in the lagoonal Tertiary series of the Camargue. The prospects lie seawards on the continental platform, where the recent well Mistral No. 1 traversed a thick Neogene marine series.

Italy

The Tyrrhenian slope along the Italian peninsula is not a particularly encouraging area. Placed as it is in an orogenic axial zone, one can only expect the post-orogenic series, such as the Pliocene, to be of any interest. The oil exploration of this narrow continental shelf is beginning now as the new oil laws open this offshore area for concession applications.

Of all the areas around the Western Mediterranean, only in Sicily have major accumulations of hydrocarbons been found. Oil and gas accumulations exist in the overthrust zones and intraformational gliding zones, as well as in the SE stable platform area (Ragusa). In the north central area there are Miocene gas-bearing sands and shales of the Collesano formation, etc., underlying complicated allochthonous beds. To the south it is the dolomitic Triassic which forms the reservoir at Ragusa, Gela, and Ponte Dirillo. This latter field also produces heavy oil (14° API) from limestones intercalated in the black shales which are the Triassic cap rock.

CONCLUSIONS

Could the marine zones be the area for future petroleum exploration? The major continental shelves are being explored at present—Gulf of Lion, Ebro delta, Spanish Levant shelf, and the marine areas of the Balearic Islands. Elsewhere the continental shelves are narrow and the continental slopes cut by canyons. These narrow shelf areas are axial orogenic zones where only the post-orogenic beds can be objectives.

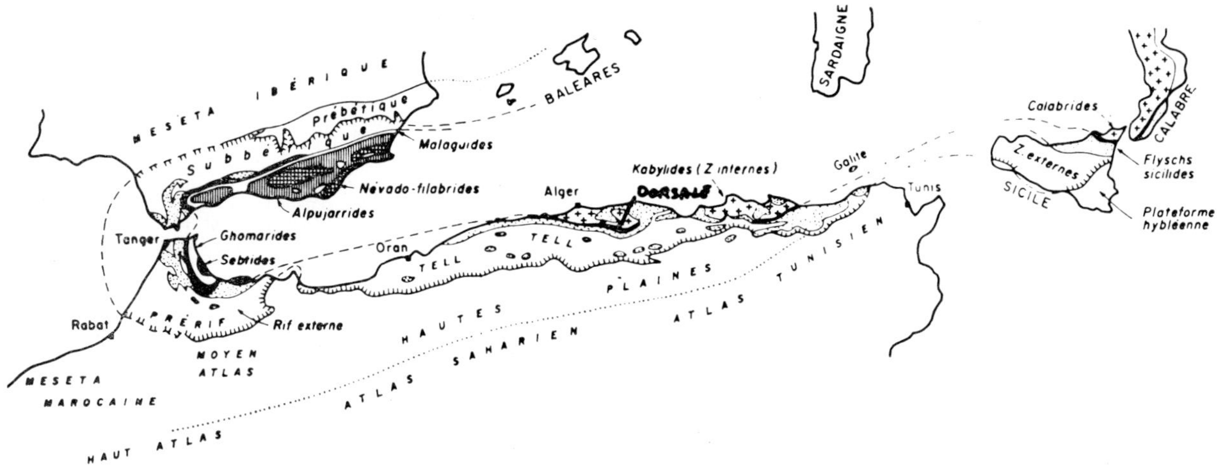

Peri-Mediterranean Alpine Orogenic Framework.

Internal Zones. *Spain: Malaguide, Nevado-Filabride, and Alpujarride zones; Morocco (Rif): Ghomaride and Sebtide zones; Algeria: Kabylie and Dorsale zones; Sicily-Calabria: Calabride zone (crosses).*

Cretaceous and Paleogene Allochthonous Flysch *(stippled).* External Zones. *Thrust front (hachured); contact of folded Prebetic and Meseta (dashed).*

North Africa: *Contact of rigid High Plains (Plateaux) and folded Saharan-Tunisian Atlas (dotted).*

FIG 8. Peri-Mediterranean Alpine orogenic framework (after M. Durand-Delga, "Guide Book of Tunisia", P.E.S.L., 1967).

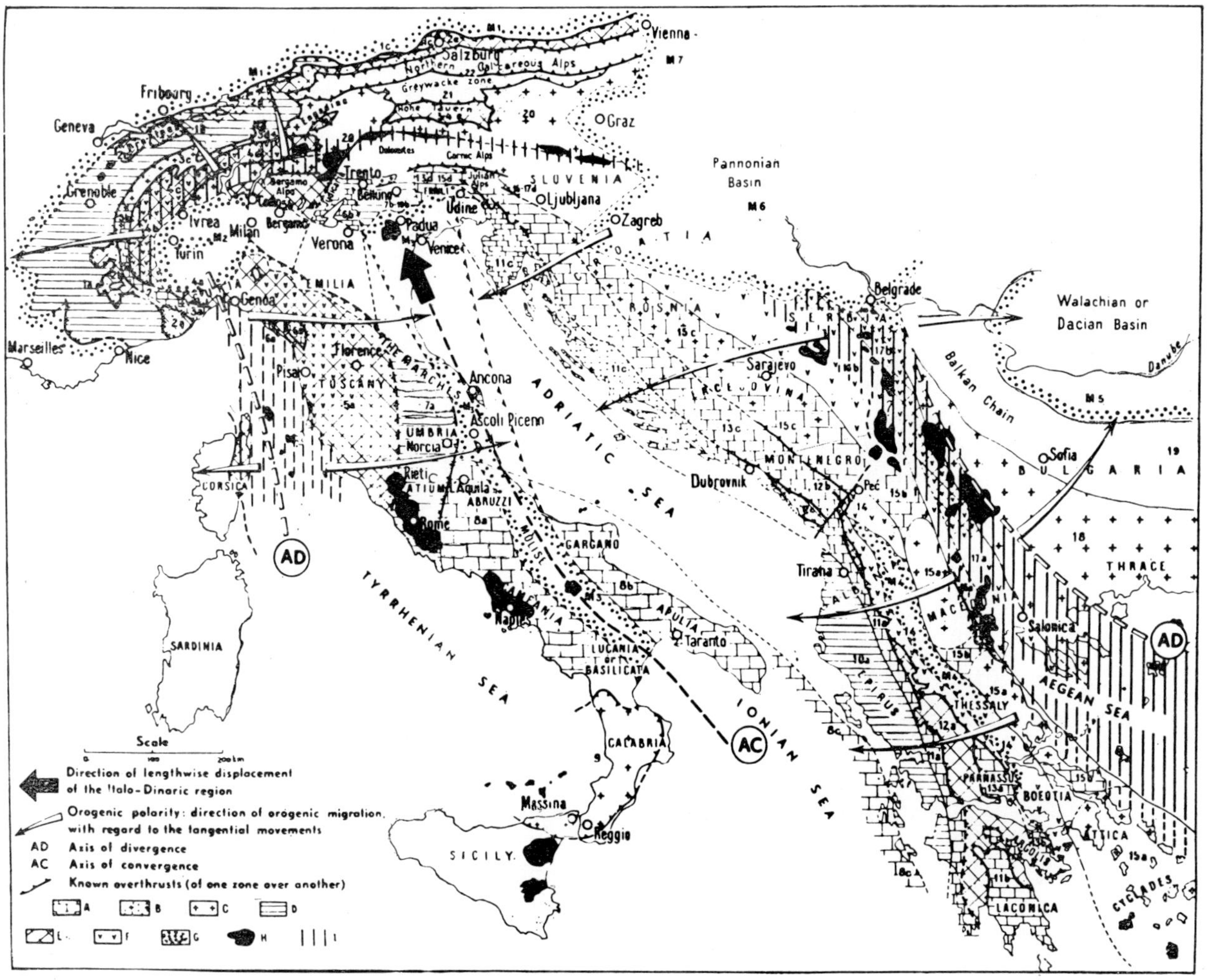

Outline of the paleogeographic and structural relationships of the Italo-Dinaric system and the Alpine arc.

FIG 9. Outline of the palaeogeographic and structural relationships of the Italo-Dinaric system and the Alpine arc (after D. W. Baird, "Guide Book of Tunisia", PESL, 1967).

We have already spoken of the Pelagian Sea to the east of Tunisia, which geographically is not part of the Western Mediterranean. But being west of the major fault line delimiting the Ionian Sea, it is geologically a transition zone in which the petroleum prospects of the Miocene, Oligocene, Eocene, Upper and Lower Cretaceous, Jurassic, and Trias are worth consideration. Politically these areas belong to Tunisia, Libya, Italy, and Malta. Some of the shoals are separated from the various mainlands and islands by water over 200 m deep and in consequence their international attributions will be difficult to settle.

Along the southern edge of Sicily, to the north of the Pantelleria deep, some shoals could have Sicilian-type oil prospects, although recent volcanism could have played an important role in the area.

The deep water zones cannot yet be commercially exploited. Those areas overlying the "tectorogenic axis" would not appear to be worth considering. But the salt dome areas of the Ligurian and Balearic seas in the north and central areas of the Western Mediterranean could be considered as possible exploration areas. The scattering of the fine-grained sediments of the Rhone deep fan does not leave much hope of good reservoir development. However, nearer to the old massifs (Catalan, Pyrenean, Corsican, Sardinian) one could expect sandstone intercalations in the Pliocene shales.

In conclusion, the Western Mediterranean is a recent collapse basin (basically of Pliocene and Quaternary age) sited for the main part over the "tectorogenic axis" of the western end of the Alpine chain. The major part of the Mesozoic and Nummulitic sediments, which were laid down in geosynclinal conditions, were ejected and spread over the forelands of the surrounding areas. One must therefore look to the Tertiary sediments for the oil prospects, even though for the most part they are disappointing.

The narrowness of most of the continental platforms limit their possibilities. Only the NW corner of the Mediterranean, near to the relatively stable Spanish and French Languedoc platform, has a continental shelf wide enough to be of interest, and this is at present being explored. The deep water areas could be of interest in future years if there is a good thickness of sediments over some of the salt domes.

BIBLIOGRAPHY

Beneo, E. *Bull. Soc. Geol. Ital.*, 1949, **68**, 66–80.

Berry, M. J., and Knopoff, L. *J. Geophys. Res.*, 1967, **72** (14), 3613–26.

Boni, A. *Bull. Soc. Geol. Fr.*, 1962, **4** (7), 644–56.

Broquet, P., Caire, A., and Mascle, G. *Bull. Soc. Geol. Fr.*, 1966, **8** (7), 994–1013.

Brown, J. H. *Rev. biog. Phy. ct biol. Dyn.*, 1967, **9** (2), 17–34.

de Bruyn, J. W. *Geophys. Pros.*, 1955, **3**, 1–14.

Burollet, P. F. "Contribution à l'Etude Stratigraphique de la Tunisie Centrale". Ann. Mines. Geologie No. 18, Tunis, 1956.

Burollet, P. F. "Signification geologique du Détroit de Sicile". 87e Congrès Sociétés Savantes, Poitiers, pp. 849–53, 1962.

Burollet, P. F. "General Geology of Tunisia", in Petroleum Exploration Society of Libya, "Guidebook to the Geology and History of Tunisia", pp. 51–58, Tripoli, 1967.

Caire, A. *Bull. Soc. Geol. Fr.*, 1962, **4** (7), 774–84.

Castany, G. "Essai de synthèse du territoire Tunisie-Sicile". Ann. Mines et Geol. 16, Tunis, 1956.

Commission for the Geological Map of the World. International Geological Congress—1964—Explanatory note to the International Tectonic Map of Europe, Scale 1 : 2, 500,000. Ed. A. A. Bogdanoff, M. V. Mouratov, and N. S. Schatsky, Moscow.

Durand-Delga, M. "Structure and Geology of the North-East Atlas Mountains", *in* "Guidebook to the Geology and History of Tunisia". Petroleum Exploration Society of Libya, Tripoli, pp. 59–83, 1967.

Durand-Delga, M., and Foucault, A. *Bull. Soc. Géol. Fr.*, 1968, **10** (7), 65–74.

Elter, P. *Bull. Soc. Geol. Fr.*, 1962, **4** (7), 668–74.

Fahlquist, D. A. "Seismic Refraction Measurements in the Western Mediterranean Sea", Ph.D. thesis, Massachusetts Institute of Technology, Cambridge, 1963.

Fallot, P. *Bull. Soc. Geol. Fr.*, 1932, **V** (2), 533–52.

Fuchs, K., Mueller, St., Peterschmitt, E., Rothe, J. P., Stein, A., and Strobach, K. "Krustenstruktur der Westalpen nach refra kionsseismischen Messungen". Gerlands Beitr. Geophysik, **72**, 149–69, 1963.

Giannini, E., and Tongiorgi, M. *Bull. Soc. Geol. Fr.*, 1962, **4** (7), 682–90.

Glangeaud, L. "Paléogéographie dynamique de la Méditerranée et de ses bordures—Le rôle des phases Ponto-Plio-Quaternaires". Océanographie Géologique et Géophysique de la Méditerranée Occidentale, pp. 125–65, CNRS, Paris, 1962.

Glangeaud, L. *Bull. Soc. Geol. Fr.*, 1967, **9** (7), 426–49.

Glangeaud, L., Alinat, J., Polveche, J., Guillaume, A., and Leenhardt, O. *Bull. Soc. Geol. Fr.*, 1966, **8** (7), 921–37.

Glangeaud, L., Boher, C., and Bellaiche, C. *C.R. Acad. Sci., Paris*, 1967, **265**, 1672–5.

Glangeaud, L., and Rehault, J. P. *C.R. Acad. Sci., Paris*, 1968, **266**, 60–63.

Goguel, J. "Les Alpes Françaises", *in* "Commission for the Geological Map of the World". *Loc. cit.*, pp. 317–21, 1964.

Grandjacquet, Cl., and Glangeaud, L. *Bull. Soc. Geol. Fr.*, 1962, **4** (7), 760–73.

Klemme, H. D. *Bull. Amer. Ass. Petrol. Geol.*, 1958, **42** (3-I), 477–512.

Knopoff, L., Mueller, St., and Pilant, W. L. *Bull. Seismol. Soc. Amer.*, 1966, **56**, 1009–45.

Lanteaume, M. *Bull. Soc. Geol. Fr.*, 1962, **4** (7), 627–43.

Le Pichon, X. *J. Geophys. Res.*, 1968, **73** (12), 3661–97.

"Livre a la Memoire du Professeur P. Fallot". Ed. M. Durand-Delga, Soc. Geol. Fr. Paris, t. I and II, 1960–62.

Manfredini, M. "Italie", *in* "Commission for the Geological Map of the World". *Loc. cit.*, pp. 321–32, 1964.

Marchetti, M. P. Summary introduction to "Geology of Sicily—Excursion in Sicily". Petroleum Exploration Society of Libya, pp. 17–60, Tripoli, 1960.

Mauffret, A. "Etude des profils sismiques obtenus au cours de la campagne Géomède 1 au large des Baléares et de la Mer Ligure". Thèse 3e Cycle, Paris, 1967.

Routhier, P. *Bull. Soc. Geol. Fr.*, 1968, **10** (7), 13–35.

Trevisan, L. *Bull. Soc. Geol. Fr.*, 1962, **4** (7), 675–87.

MS received 12 February 1969

Discussion

Chairman: D. C. ION

M. D. Beltrandi (Gulf Italia): M. Burollet mentioned in his paper the existence of "salt domes" in the Ligurian and Balearic seas. What geological or geophysical evidence does he have for calling those features salt structures?

If they are deformed by salt, could this salt be part of the Triassic evaporitic sequence?

And in this case could the salt dome tectonics also be a pre-Alpine structural event and presently buried under the overthrusted Alpine belt? The presence of the salt in the Alpinic tectonized areas could be a good lubricant for the horizontal displacements.

P. F. Burollet: The first is the existence of the salt domes. There have been several seismic surveys of the sea; you have seen one section and Dr Kent has presented another. I think that we have reasonable evidence of the possibility of diapiric structures, but at present no proof has been found from drilling or from sampling. I have heard that very soon some exploration will be made by drilling in the zones of the domes in order to prove the existence of salt. Personally, I think that seismic sections are likely proofs of the existence of salt domes.

The second is the age of the evaporites. Evidently Triassic or Lower Liassic salt is known in numerous places around the Mediterranean Sea. The first idea was to date the possible salt domes as Triassic or Lower Liassic, but in fact it seems that the Ligurian and Balearic sediments correspond to a post-orogenic deposition.

There was a seismic survey made by the crew of Professor Glangeaud which shows two horizons, one on top of the Pliocene, and another which seems to be at the base of Pliocene or at the top of Pontian beds and may correspond to an emersion. There are erosion features and that is why Professor Glangeaud thinks that the Ligurian Sea and the Alboran Sea have been definitely uplifted at the end on Miocene and probably were completely out of water, and it was the final collapse during the Pliocene which caused the new invasion by the sea. So in that case the best thing is to suppose that the salt is "post-invasion" and probably Pliocene. In some place we may imagine that the central part of the sea was not pushed as high as the borders of the Ligurian and Alboran Seas, and that a furrow remained with the salt deposited before the Pliocene, so in that case the salt may be Miocene. In conclusion, Miocene or Pliocene seems to be more probable than Triassic. Evidently this is just hypothetical at the present status of our knowledge.

In the case in which there was Triassic salt evidence, we might imagine the existence of salt domes below some Alpine folds, but we know, for example, that in Algeria or Tunisia at last the salt or mixed salt clay and gypsum diapirs of the Triassic are parts of the Alpine folds in the nappe zone but also in the platform style zones such as the Saharan Atlas. So it seems that the evaporitic series existing in North Africa are involved in the Alpine folding, acting as diapiric structures, and it is difficult to imagine that there may be salt basins remaining below the main part of the Alpine folds.

In the latest issue of *Bulletin de la Société Géologique de France*, 1969, **7** (4) 497–509, O. Leenhardt gives new information about the domes. After him, as after D. A. Fahlquist and J. B. Hersey, the layers with a 4.5 km/s velocity, which may be the main bed of evaporites, are nearly 1500 m below the seismic horizon—(probably corresponding to the base of Miocene).

Contrary to the reply I gave Mr Beltrandi, I guess that these new data support an ante-Miocene age for the salt. An Oligocene age cannot be excluded, but a Triassic or Liassic age seems to be the most plausible. In the Gulf of Mexico a deep salt dome was drilled by the Soides at Sigsbee Knoll and the salt is reported to have approximately a Lower Jurassic age.

John D. Moody (Mobil): Would M. Burollet enlarge on the reasons why Corsica and Sardinia are considered as Alpine massifs rather than foreland Hercynian massifs?

P. F. Burollet: It is just a geological speculation, but there are two reasons. One is that Corsica is over-thrusted by some part of Alpine material, and that kind of material forming the NE of Corsica is just made of internal zones, so we do not have in Corsica external features such as in the Alps or the Atlas Mountains. The problem is to know where are the external features or why they are lacking, and my idea is to suppose that this material was W of Corsica and not E. On the other hand, some seismic profiles made by the crew of Professor Glangeaud gave some information NW of Corsica with the possibility of reverse faults and a sort of thrusting of Corsica NW above the Ligurian Sea material. This interpretation given by Glangeaud was published two years ago in the *Bulletin de la Société Géologique de France.*

Professor A. J. Whiteman (University of Ibadan): Would M. Burollet care to comment upon the fault pattern running from Tibesti–Maroudj–Misurata? This, with its associated volcanics, looks like a rift structure in parts. As is well known, Furan, on very flimsy evidence, has inserted a major fault running from Cameroun to Tibesti (Curte Pinisoire structurale d'Afrique 1 : 10,000,000).

Is there any definite evidence to support this idea in Tibesti?

P. F. Burollet: It is a very complicated question which would need, I suppose, hours of discussion to be answered, and personally I am more or less in favour of the very large fractures of the African continent, but I know the area N of Tibesti and avoided extending that kind of feature S of Tibesti. However, I suppose it is possible, and I am more or less in favour. In all cases large fault zones separate the NW part of Tibesti shelf from Murzuk Basin.

You have seen that probably the tension of the African block has caused a kind of break in numerous places like the Great Rift Valley in East Africa, and some of the "graben" known in Libya, for example. Small "grabens" known in Tunisia have the same orientation and are probably results of the same kind of tensions of the African block, but one of the most important is the large fracture between the Pelagian Sea on the W side and the Ionian Sea on the E side, with the present difference of perhaps about 2000 or 3000 m and that kind of fracture seems to have appeared probably during the Pliocene. It is like the Straits which appeared probably during the Pliocene too. It seems that between Tunisia and Sicily, for example, at the end of the Miocene the Alpine Range was continuously folded and broken by Pliocene large faults forming the Strait of Sicily and probably the graben of Pantellaria in the N part of the Pelagian platform. This kind of fracture, with an orientation from NW to the SE, belongs probably to the main African directions.

Dr Robert E. King: Dr Kent effectively showed how difficult it is to reconstruct past positions of continental masses. If one shows the Iberian Peninsula rotated toward Brittany in the late Mesozoic then it is a problem to account for Africa with respect to Eurasia. It seems illogical to separate Eurasia and Africa along the Strait of Gibraltar, as shown in M. Burollet's Fig 6 after Le Pichon, because this splits the Balearic–Betic–Rif arc, which is a coherent unit. Perhaps the split occurred along the S Atlas rift, which reaches the Atlantic coast S of Essaouira, Morocco. However, if we trace the S Atlas Fault E to Tunisia it seems unlikely that this zone represents a major displacement.

Many of us are willing to accept the general principles of the "New Global Tectonics", but reconstructions of the fit of continental masses that have so far been made are unsatisfactory. Perhaps this is because they fail to take account of major strike-slip fault displacements, crustal shortening in orogenic zones, and downward absorption into the mantle of internal zones of eugeosynclines such as has been postulated in the Alps.

P. F. Burollet: I agree absolutely with Dr King. We are far from the true proof of all the displacements.

Regarding the Straits of Gibraltar, we have to separate the different successive phases. (1) Compression of the alpine geosyncline between African and Spanish craton, from Cretaceous to Oligocene. The width of the Western Mediterranean Sea, as described on the sketch of Le Pichon (Palaeocene time) fits in correctly with this first kind of movement. (2) Miocene orogeny mainly due to the compression wave along the axis from E to W. This Miocene orogeny gave the continuous tectonic shape of the Rif and Betic Arc which was really a coherent unit at the end of Miocene. (3) Final Pliocene collapse of the Alboran Sea and formation of the Straits by faults.

Abu Amr (Sonatrach): Does any stratigraphic relationship exist between the Nubian sand and the continental intercalaire and is there enough subsurface evidence of Permian in Arabia and which facies?

Dr P. E. Kent: In fact, the group of sandstones at the base of or between the Cretaceous and the Palaeozoic probably includes beds of considerable range of different dates, and I do not think one can be specific about the precise age. I was speaking of course in my talk about the Basil sandstones, the oil reservoirs of Libya where the beds do grade up and pass laterally also into argillaceous beds which are known to be Lower Cretaceous, so that there the Basil sands are of Lower Cretaceous date, but this does not in any sense imply that this is a universal dating.

Dr Rudolf Martin (written contribution): The term "tectorogenic axis" is introduced in this paper in place of more widely used concepts such as "medium mass", "intramontane basin", "stable platform", etc. The word "tectorogenic" suggests movement and would thus appear to be a rather unhappy choice, considering that these masses are commonly thought of as stable during the alpine orogenesis.

The term "intra-alpine furrow", as applied to the Adriatic Sea, is equally unclear. The Adriatic is probably more complex than is commonly believed, and may consist of several discrete tectonic units.

The Jeffara–Malta axis may well be part of a major break in the earth's crust that extends as far S as Cameroun and as far N as the Vienna Basin. From Cameroun southward, it became part of the W coast of Africa at the time of formation of the Atlantic rift. N of Sicily, it separates the Gargano–Murge block from the Apennines, is overridden by the Dinarides, and then reappears in the Vienna Basin area as the eastern boundary of the Bohemian Massif, the Western boundary of the Pannonian Basin, and a line of major displacement in the Alpine–Carpathian fold system.

P. F. Burollet: The term "tectorogenic axis" is introduced in this paper as a direct transcription of the French "tectorogene" of several authors and especially of Professor Glangeaud. It is exact that this word suggests movement as it is the axis of surrection of the geosynclinal zone; it is by definition the most unstable part of the orogenic complex and has nothing in common with a "medium mass" or a "stable platform".

I agree that the Adriatic is very complex and that several tectonic units have to be discovered there. But, as a general feature, during Miocene and Pliocene times, the Adriatic furrow was continuously a subsiding zone with convergence of the thrust units coming from both sides—Apennines from SW, Dinarides from NE.

I agree with the African extension of the Jeffara–Malta axis. But, on the contrary, it is difficult for me to imagine a relation of this line with any feature of the European Craton N of the main alpine geosyncline which was a fundamental transversal unstable break between the two different cratonic blocks.

The Idris "A" Bioherm and Oilfield, Sirte Basin, Libya—its Commercial Development, Regional Palaeocene Geologic Setting and Stratigraphy

By C. E. TERRY and J. J. WILLIAMS

(*Occidental of Libya, Inc.*)

SUMMARY

Occidental of Libya discovered a major oilfield and a new oil province in the Sirte Basin of Libya in May 1967, when its A1–103 was tested for an estimated 43,000 bd. The well penetrated 959 ft of oil pay in a 1200-ft thick bioherm of upper Palaeocene age. The Idris "A" bioherm, one of six known isolated bioliths, is oval-shaped in plan view, and measures about 2½ by 3 miles at its greatest lateral dimensions.

The Idris bioherm province developed on the crest and flanks of a broad, low relief anticline in Concession 103. The bioherms began as foraminiferal banks that built on local mounds on the floor of a shallow sea. Algae helped to bind the foraminifera loosely. Later, sparse coral growth bound a predominantly coral detritus until a more favourable environment allowed a richer coral growth which produced a massive coralline reef. A major transgression killed reef growth with an argillaceous deposition and marked the end of the Palaeocene epoch.

Over 2 billion brl of oil in the Idris "A" bioherm are being produced by 17 wells, while reservoir pressure will be maintained by about 25 water input wells. Through May 1969 the field has produced over 175 million brl of oil.

INTRODUCTION

Geography

The Idris "A" field is 220 miles (350 km) SSE of Benghazi and 525 miles (840 km) SE of Tripoli, Libya. The field is located in Concession 103 in NE-Central Sirte Basin, a Late Cretaceous and Tertiary sedimentary basin in N and E-Central Libya (Fig 1).

Topographically, Concession 103 is a very gently rolling, sandy plain. The surface section is a thin veneer of Quaternary sands and fine gravels overlying Miocene sandy limestones. Surface elevations range from 460 ft (140 m) in the S part of the concession to 270 ft (80 m) above sea level in the N part. In the Idris "A" field, the elevations average about 330 ft (100 m).

Ground water in sufficient quantities for drilling and camp needs is available at depths of 450–500 ft (135–150 m). Sodium chloride in this water averages 900 ppm.

Exploration History

Concession 103 was awarded to Occidental of Libya, Inc. by the Libyan Ministry of Petroleum on 29 March 1966. The concession, with an area of 465,000 acres (1880 sq km) was formed by the consolidation of two tracts of acreage relinquished by previous concessionaires. A dry exploratory well in one of these tracts had been abandoned at a total depth of 12,414 ft (3780 m).

Occidental began seismic operations in Concession 103 early in 1967. By good fortune, the seismic crew shot its second line across the crest of the Idris "A" bioherm. Shortly after, the company drilled the A1-103 discovery well, which topped the pay at 9417 ft (2870 m). A biolith 1193 ft (365 m) thick contained 959 ft (290 m) of oil saturation. Porosity averaged 21 per cent through the upper 583 ft (180 m) and 27 per cent through the lower 376 ft (115 m) of the oil column.

Representatives from Occidental and the Libyan Petroleum Ministry tested the A1 well in May 1967 for an estimated flow rate of 43,000 bd from 690 ft (210 m) of open-hole section below 7-inch (17½-cm) casing.

Development History

Following the completion of the discovery well, Occidental rapidly developed the Idris "A" field. By the end of 1967, eight wells were shut-in awaiting production outlets, and four rigs were drilling. During 1968, an additional nine wells were completed and a total of 17 wells were put on stream. Three rotary rigs were active at year's end.

Development data have shown the Idris "A" field to be a NW to SE oriented, oval-shaped body in plan view with maximum proven dimensions of 2½ by 3 miles (4 by 5 km). The oil column in wells drilled through 1968 has a maximum thickness of 1000 ft (305 m) and a minimum of about 400 ft (120 m) near the outer edge of the bioherm. The Idris "A" bioherm is about 1200 ft (365 m) thick at its crest and 640 ft (195 m) thick in flank wells (Fig 2). The zero isopach line of the complex has not been defined by wells drilled through 1968.

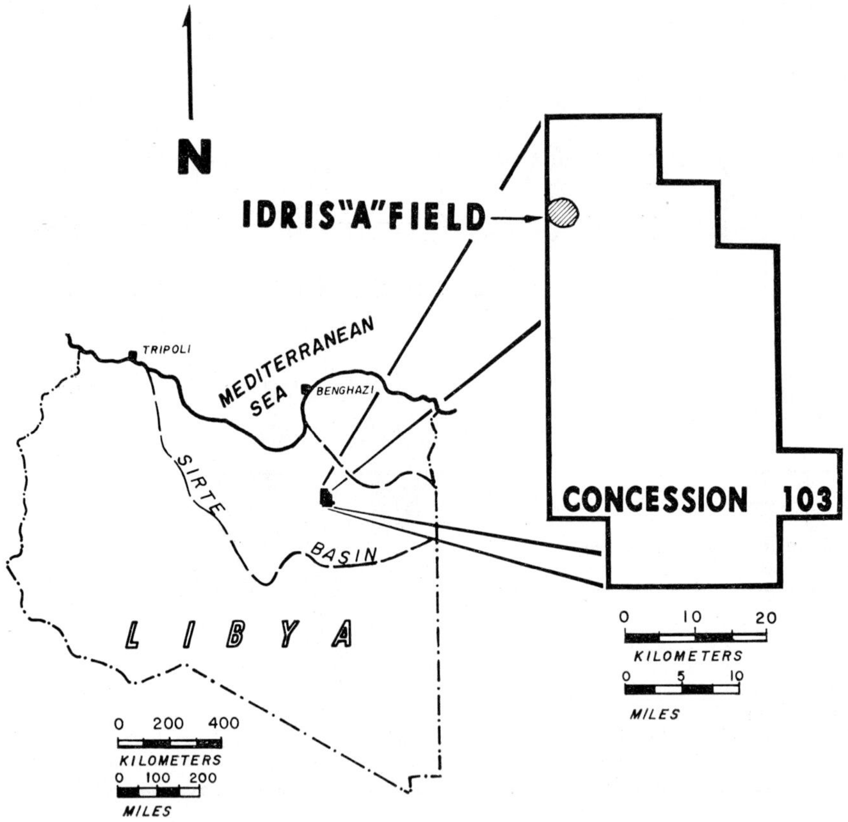

FIG 1. Map of Libya showing the Sirte Basin, Concession 103, and the Idris "A" field.

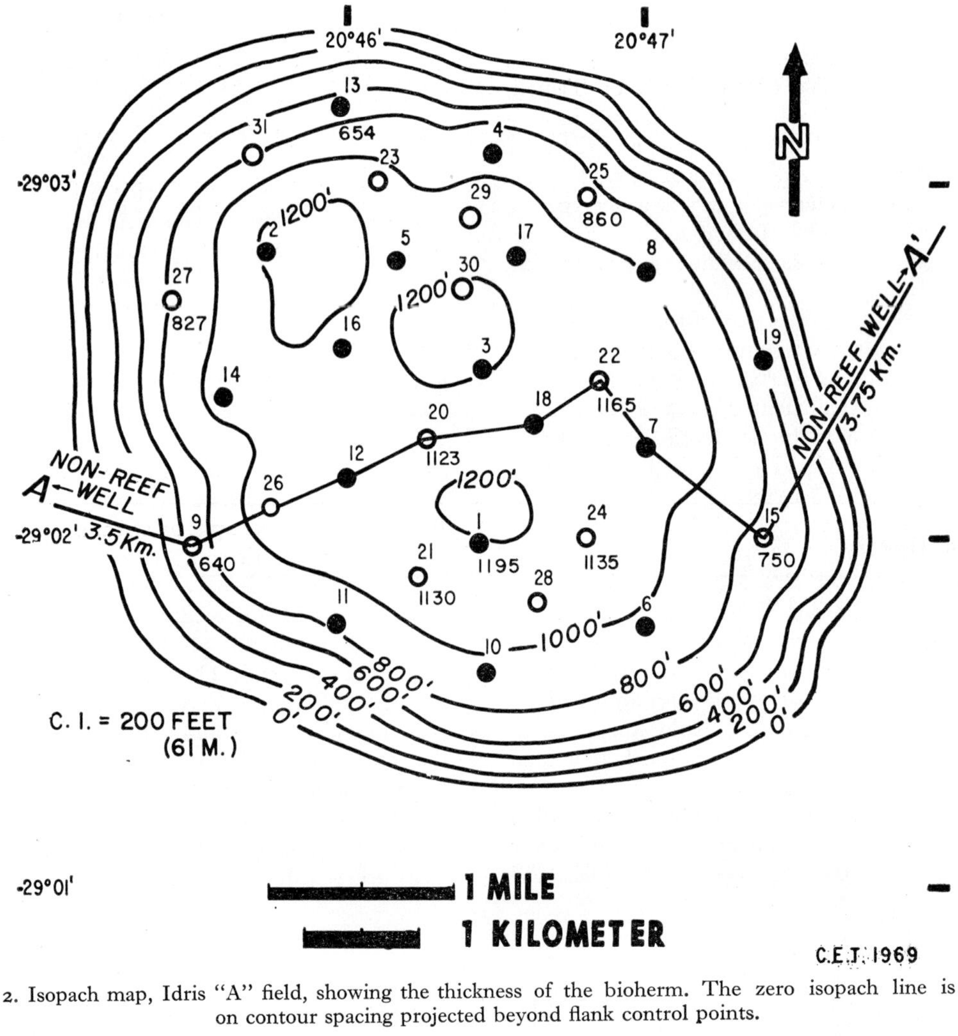

FIG 2. Isopach map, Idris "A" field, showing the thickness of the bioherm. The zero isopach line is based on contour spacing projected beyond flank control points.

GEOLOGY

Introduction

The Idris "A" formation includes three members, only one of which may be classified as true reef by current definition. The entire Idris unit is considered a bioherm and/or biolith. For the purposes of this report the following definitions will apply:

Biolith—"any rock body formed predominantly by the skeletal remains of organisms" (Grabau, 1913).

Bioherm—"a reef, bank, or mound—of strictly *organic* origin, embedded in rocks of different lithology" (Cummings, 1932).

Biostrome(s)—"purely structures such as shell beds, crinoid beds, coral beds, etc., consisting of and built mainly by sedentary organisms, and not swelling into mound-like or lens-like forms" (Cummings).

Bank—a bioherm "formed by organisms that do not have the ecological potential to erect a rigid, wave-resistant structure" (Nelson *et al.*, 1962).

Reef—a bioherm that "is the product of the actively building and sediment-binding biotic constituents, which, because of their potential wave resistance, have the ability to erect rigid, wave-resistant topographic structures" (Lowenstam, 1950).

We use the Folk Carbonate Classification in discussions of rock units identified by petrographic analysis; otherwise the carbonate lithologies are referred to in simplified terms (Folk, 1959).

REGIONAL PALAEOCENE GEOLOGIC SETTING

General

The principles of shallow and clear-water sedimentation upon broad, flat to low-dip shelves suggested by M. C. Irwin (1965) apply remarkably well to the Sirte Basin of early and middle Tertiary.

Irwin states that clear, shallow epicontinental seas covering broad, relatively flat shelves deposit carbonate rock types according to energy levels in the following zones.

1. A very low energy zone below wave and current activity receives deposits of shale, clay, and very fine derivatives from higher energy zones; all units have a high percentage of plankton.
2. A "high" energy zone, where wave base and current energies intersect the sea floor, has energy levels ranging from low to high according to water depth and current and wave energies. Deposition ranges from fine-grained precipitates with planktonic and sparse benthonic organisms in the lower energy zone to skeletal limestones in the moderate to high energy zones.
3. A low energy restricted zone exists on the upper shelf, where there is very little circulation of water, and wave action is limited to occasional, perhaps local storms. In this sedimentary environment, bioclastics and chalks from the high energy zones intermix with dolomites in the down-shelf region and grade shoreward to pure evaporates.

In the central Sirte Basin during Palaeocene time, there were few or no beach deposits. Terrigenous material was possibly introduced from the north and may have been transported for great distances as suspended argillaceous material.

Lithologies of Palaeocene age in the Sirte Basin were controlled by transgressions and regressions of an epeiric sea across a broad shelf region that dipped gently into a deep and persistent shale basin. This basin was located N and NW of Concession 103.

There were two major marine transgressions across the Sirte Basin shelves during Palaeocene time; one occurred midway through the series, the second at or near the close. Shale deposits, generally confined to the very low energy zones in the basin and adjacent re-entrants, blanketed much greater areas during transgressive phases. During the normal regressive periods, carbonates were deposited on the shelves. Varying energy levels determined the type of carbonate deposited, and these levels were controlled by water depths, sea bottom configuration, and the type, intensity, and patterns of water turbulence.

The Palaeocene Series in and adjacent to Concession 103 has been zoned palaeontologically. This zonation is based upon the stratigraphic ranges of various species of planktonic foraminifera, in particular the globorotaliids. The relationship between pre-established Palaeocene Stages and Sirte Basin rock units are presented in Fig 3. Specific zonations are not included in the diagram but are compatible with previous work (Berggren, 1965). Although the chart has been modified from those previously published, it does not violate precedence or the consensus of opinion at the First International Conference on Planktonic Micro-fossils, Geneva, Switzerland, 1967.

First Cycle-Palaeocene

During early and continuing into middle Palaeocene time a major basinal re-entrant existed throughout the central part of the Sirte Basin. A thick section of Heira shale was deposited in this re-entrant and in the shale basin to the north. An abrupt lateral facies change from shale to Heira carbonate eastward marks the margin of the shelf during this time and suggests a delineating hinge-line or down-to-west fault (Fig 4).

The first major marine transgression, occurring in very early Upper Palaeocene time, blanketed most of the Heira carbonates with a thin veneer of shale. Certain areas of the sea floor stood high relative to their surroundings and these areas continued to receive carbonate deposition contemporaneously with adjacent shale deposition during the encroachment and consequent deepening of water. Central Concession 103 was one such area where energies were sufficient to deposit micrite, fine-grained planktonic biomicrite, and biopelsparite even during maximum transgression. This incursion of the sea marked the end of the first Palaeocene cycle.

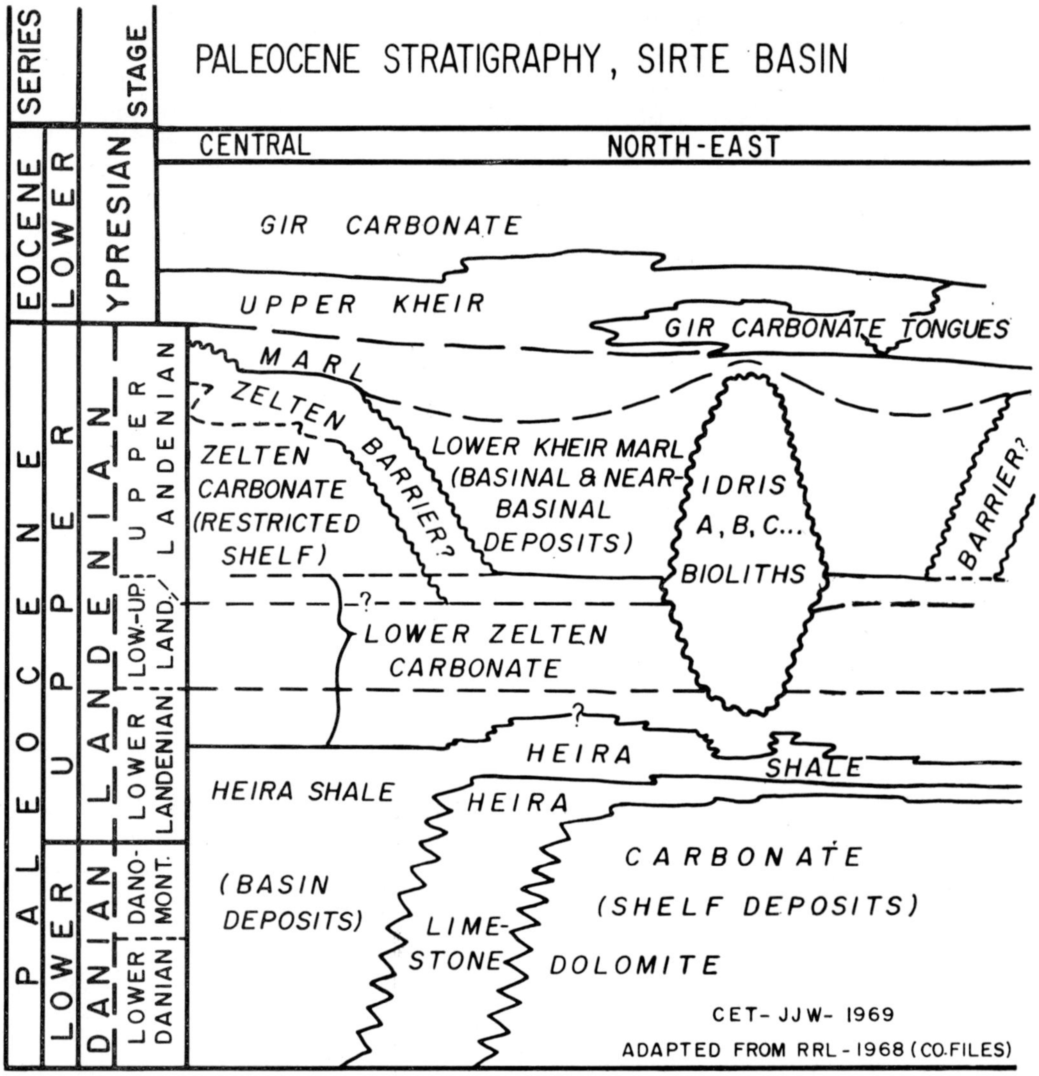

FIG 3. Palaeocene stratigraphic diagram showing intra-Occidental Palaeocene nomenclature for Central and NE Sirte Basin, Libya. The rock units were named mostly through prior usage by various operating oil companies in Libya. Therefore it is not presumed that this nomenclature should become standard or that the authors should be considered originators.

Second Cycle-Palaeocene

During the remaining part of the Upper Palaeocene, the principal events that controlled deposition and lithologies in NE Sirte Basin were:

1. A very slow and oscillating retreat (and consequent shoaling) of the sea that accounted for an increase in biotic constituents in the upper part of the Heira shale.
2. A banking of benthonic fauna in carbonates above Heira shale in select areas.
3. A realignment of the basinal re-entrant, the axis of which was shifted to eastern Concession 103.
4. A coral growth that continued over the pre-existing banks within the re-entrant area.
5. A final major transgression that killed the bioherms and closed the Palaeocene epoch.

From maximum transgression, the sea began a very slow and oscillating retreat from the forelands, which were far removed from the area of interest. The effect of this retreat in NE Sirte Basin was a very gradual but intermittent lowering of sea level, which in turn increased depositional energy levels preferentially according to sea bottom topography. The result over the high areas was a deposition of micrite, fine-grained biomicrite, and some biosparite that graded laterally down-dip to argillaceous micrite and shale. These lithologies form the Lower Zelten carbonate grading down-dip to uppermost Heira shale. As water continued to shoal, similar depositional conditions were expanded in ever broader patterns outward from a high area in central Concession 103. Under the proper local circumstances isolated bioherms were born. These quickly generated their own preferential environment by building into higher energy environments and thus perpetuated themselves while the surrounding deposition was an argillaceous micrite with only

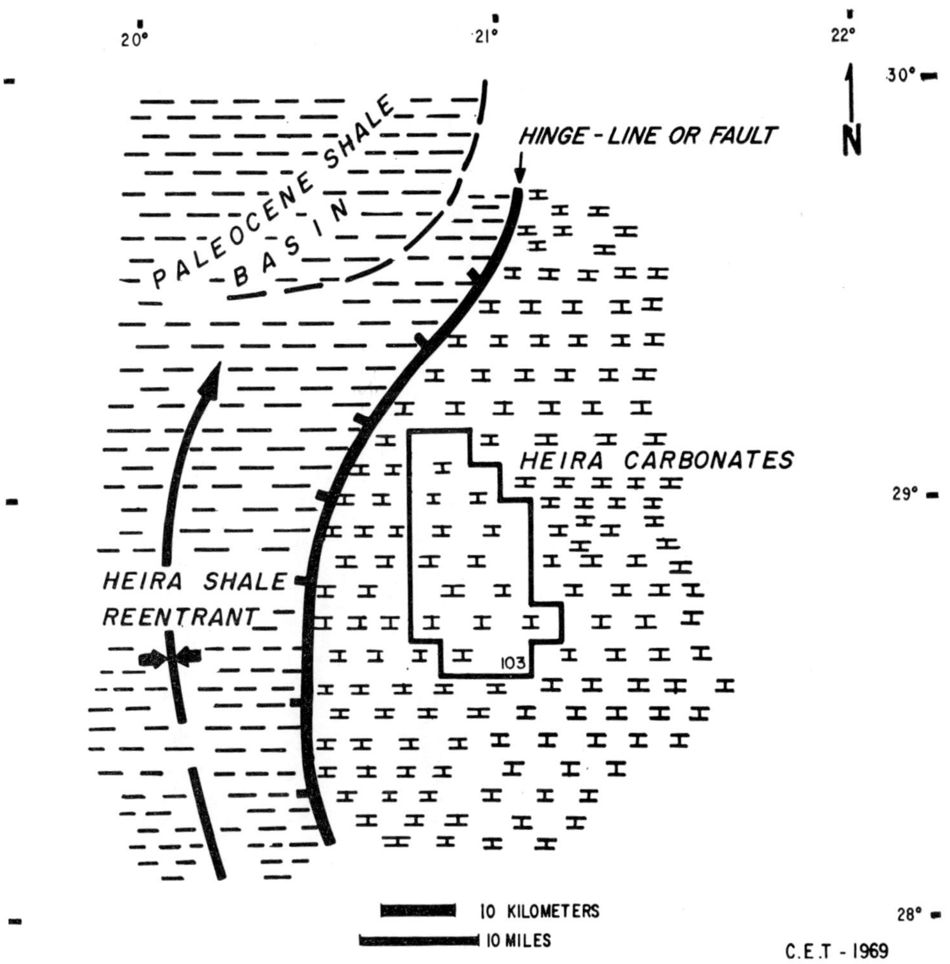

FIG 4. Early Upper Palaeocene lithofacies map of NE Sirte Basin prior to the first major Palaeocene marine transgression.

a poor, dominantly planktonic faunal assemblage.

Further shallowing of the sea allowed the deposition of a bedded biomicrite over a broad area of NE and central Sirte Basin. By this time the isolated banks that form the lower part of the Idris bioherms were well along in their growth; they probably attained their maximum circumference at the end of the Lower Zelten biomicrite deposition. Since the Idris banks fuse with the bedded biomicrite, the boundaries between the two are arbitrarily placed at the flat bedding on the mound peripheries, even though bank detritus may spread over much wider areas within the bedded biomicrite. These banks stood higher on the sea floor than the surrounding deposits of the biomicrite. As a consequence, the banks were thoroughly winnowed of lime muds and very good intergranular porosity resulted. Solution processes subsequently enhanced this original porosity.

While the banks and later reefs are most important economically, they represent only a very small percentage of the total area of the biohermal province. There are six known Idris bioliths, all similar in size and shape to the Idris "A" bioherm, and three contain appreciable volumes of oil.

Fig 5 is a diagram of theoretical depositional conditions in the Concession 103 area from the transgressive phase of the first Palaeocene cycle through the first phase of the second cycle. The pre-existing high in the vicinity of the Idris "E" area is the centre of the Idris bioherm province. It is conjectured that the province expanded outward as similar depositional factors were attained at down-dip levels through a critical time period. However, the known bioherms all developed more or less simultaneously within a bed of Lower Zelten Carbonate that was approximately flat at the time of its deposition. Local, underlying conditions, discussed later, determined the preferential environment for bank and reef development.

Following the very stable conditions attending the deposition of the Lower Zelten biomicrite, readjustment along a N–S trending fault in Eastern Concession 103 produced a shallow re-entrant over Concession 103 and adjacent areas to the W and N. The deeper water that resulted introduced an increase in clay deposition in the re-entrant and halted the Lower Zelten Carbonate deposition. Within the re-entrant, marls, interbedded argillaceous planktonic micrites, and some fine-grained biomicrites were deposited. This section is termed Lower Kheir Marl. It is postulated by the authors that on the flanks of the re-

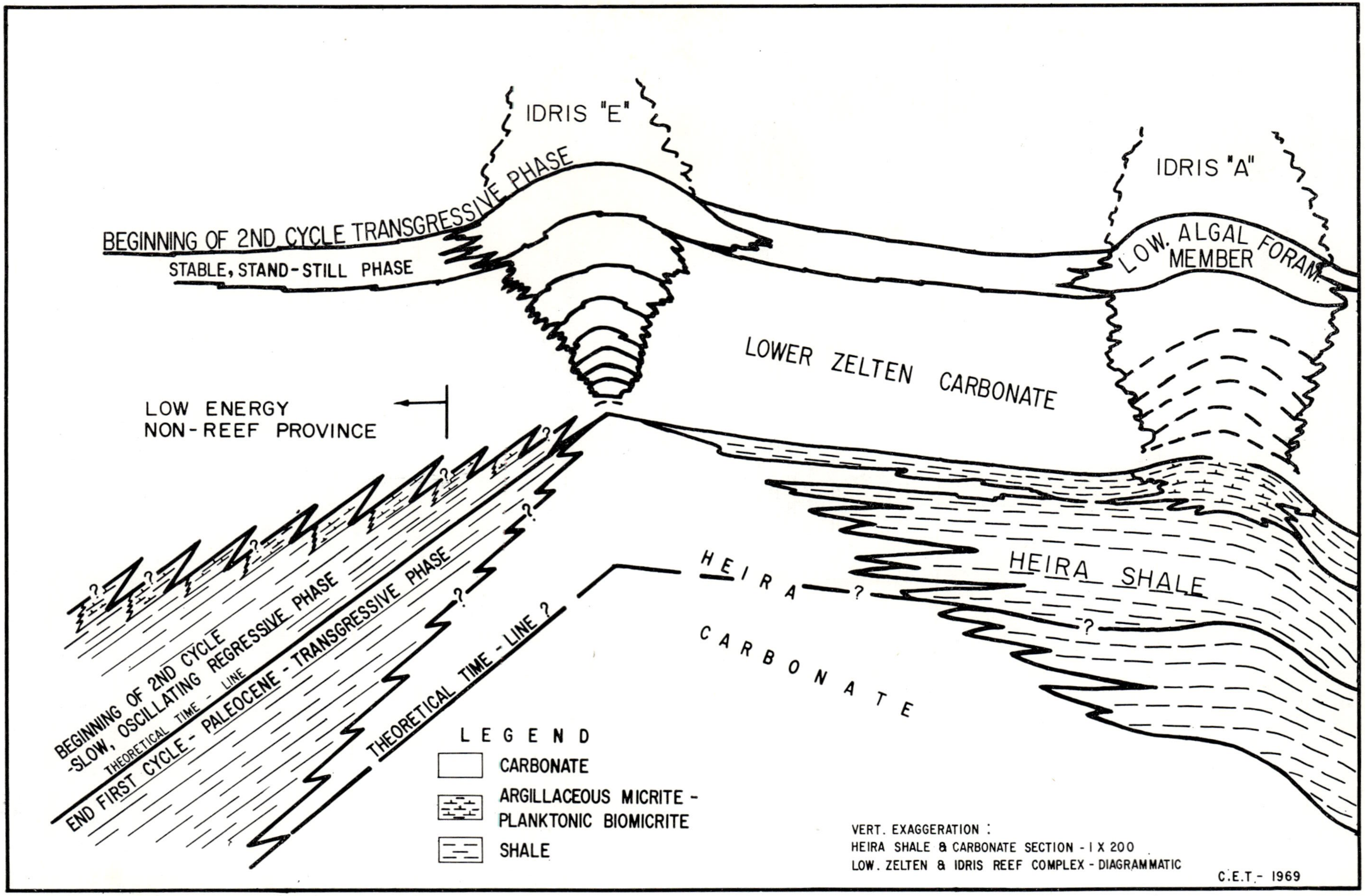

FIG 5. Diagram of theoretical time-lines and facies relationships critical to the definition of the Idris bioherm province and to the evolution of the individual bioherms.

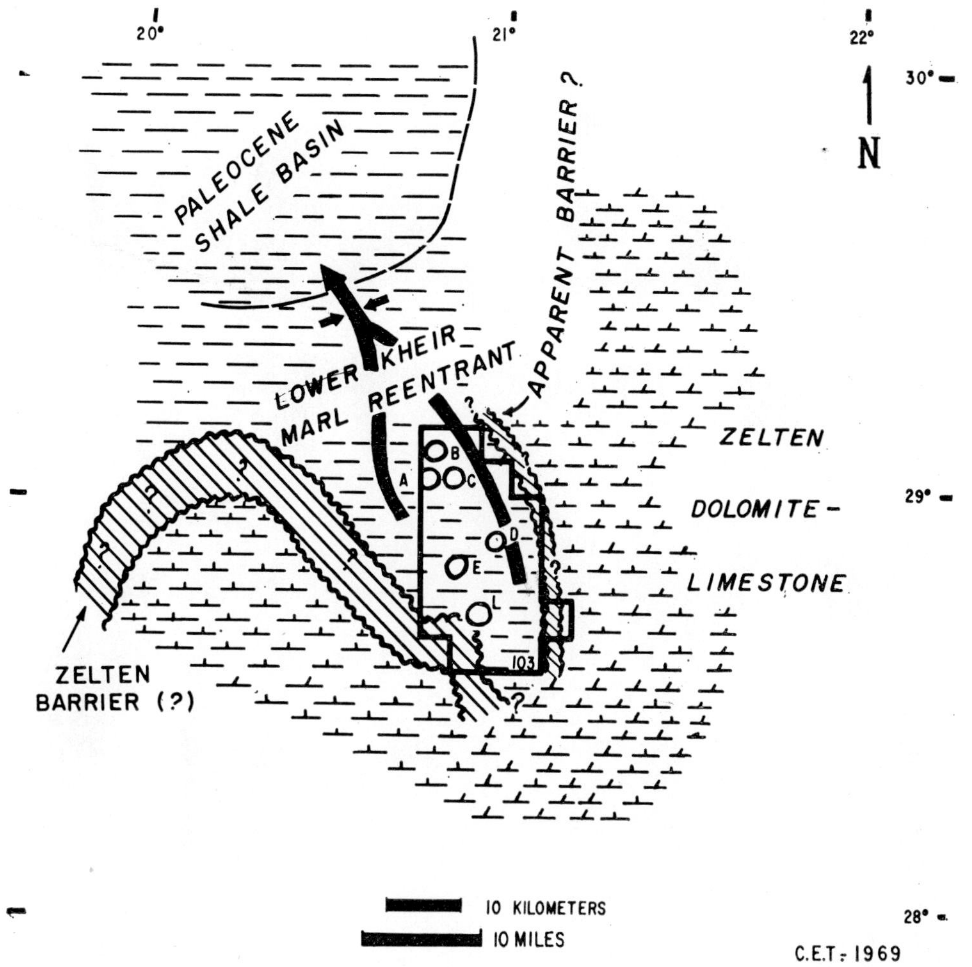

FIG 6. Upper Palaeocene lithofacies map of Central and NE Sirte Basin prior to the second major marine transgression. The transgression deepened seas over the area shown, resulting in the deposition of Upper Kheir marl over the illustrated lithofacies.

entrant a barrier reef or linear bank probably grew upward from the Lower Zelten Carbonate. Behind this postulated barrier, Middle and Upper Zelten Carbonate was deposited on broad shelves to the W, S, and E. This carbonate is predominantly a dolomite.

Significant exceptions to the general marl deposition in the re-entrant occurred over the pre-existing banks. Having an elevated topography relative to the surrounding sea floor, the banks, comprised predominantly of foraminifera and algae, continued to build upward into the higher energy levels. With time, upward growth, and gradual increase in coral development, reefs eventually thrived.

The close of the Palaeocene was marked by a major incursion of the sea. The consequent deepening of the water in the Concession 103 area drowned both barrier and isolated Idris reefs and capped them with Upper Kheir Marl (Fig 6).

PALAEOCENE STRATIGRAPHY—IDRIS "A" BIOHERM

The stratigraphy below, in, and above the bioherm is based on data from 29 Idris "A" wells which have been drilled at the time of writing. Robertson Research personnel examine thin sections of ditch cuttings at 10- to 20-ft (3–6 m) intervals and available cores at one-ft ($\frac{1}{3}$-m) intervals. Ditch cuttings generally are poor. The A3 and A22 wells together represent continuous core through the biolith. To date 13 of the "A" wells have been drilled through the bioherm but only eight have been drilled into Heira shale.

Formations Beneath the Bioherm

Fig 7 shows that Lower Zelten carbonates, Heira shale, and Heira carbonate underlie the bioherm.

Heira carbonates beneath Idris "A" consist of about 2000 ft (610 m) of dolomite overlain by approximately 100 ft (30 m) of biomicrite and biopelsparite which was deposited in a shallow marine, low energy environment prior to deposition of the Heira shale.

The slightly dolomitized biomicrite and biopelsparite grade downward into dolomite. The original nature of the rocks has been destroyed by dolomitization. The rocks have been converted to a finely crystalline dolomite consisting of an anhedral mosaic of dolomite crystals. Vugs are infilled with a medium-grained dolomite mosaic and amorphous anhydrite.

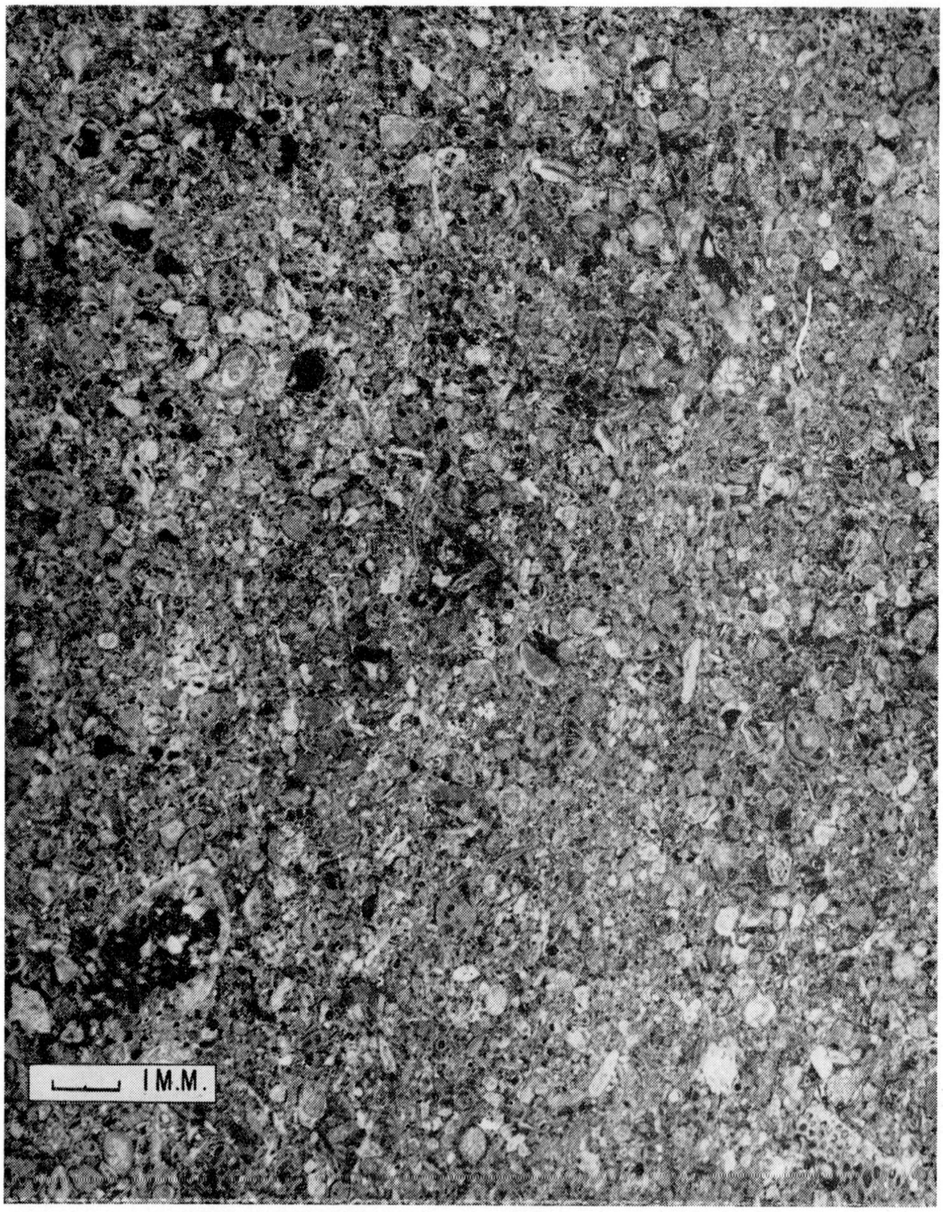

FIG 8. Negative photograph, thin section, A22–103, 10,540 ft, illustrating typical lithofacies of the porosity transition zone. This zone represents the embryonic stage of the Idris "A" bioherm and is characterized by a marked increase in biotic constituents within the Lower Zelten carbonate. The zone is a well sorted, grain-supported, algal foraminiferal biomicrite. Small rotaliids are abundant and algal nodules are common. In addition, there are fragments of large foraminifera, echinoids, crinoids, and coral. A few small buliminids are present. The scattered dark areas represent porosity which averages 20 per cent through the section illustrated. However, it ranges from less than 5 per cent at the base to 25 per cent at the top of the zone, where it grades into the algal-foraminiferal member.

The Heira shale is calcareous with mainly planktonic foraminifera but also with calcareous benthonics and ostracods. The Heira shale in the vicinity of Idris "A" was deposited in a fairly shallow sea under a low energy environment with a dominant open marine influence.

The Heira shale grades upward into the Lower Zelten carbonate, a dense, fine-grained, mud-supported foraminiferal biomicrite. The faunal content is mainly small rotaliids with some planktonic foraminifera. Algae and echinoderm debris are abundant underlying the reef and occasional fragments of bryozoa and corals occur. The rock unit is stylolitic and well cemented and has very low porosity. The Lower Zelten carbonate was deposited in a fairly shallow, low to moderate energy sea with a moderate open ocean influence.

Bioherm

The Idris "A" bioherm is divided into the following mappable units (Fig 7). From bottom to top they are:

1. Porosity transition zone.

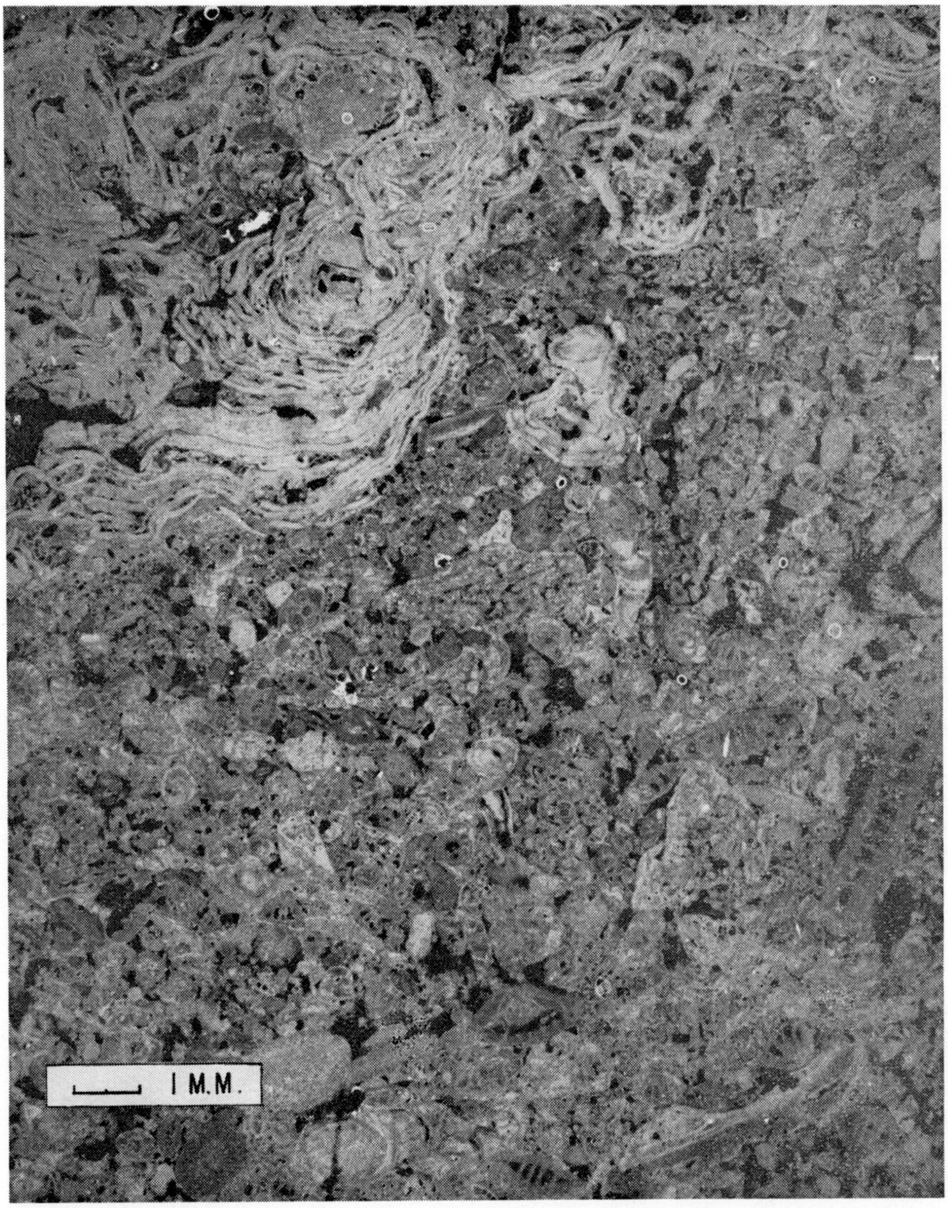

FIG 9. Negative photograph, thin section, A3–103,10,097 ft, illustrating the biofacies of the algal-foraminiferal member of the Idris "A" bioherm. The "ropy" textured material in the upper part of the illustration is calcareous red algae, which encrusts and binds rotaliid and nummulite foraminifera. Porosity in this section is only 22 per cent by core analysis; however, this is not representative of the member as a whole. Although this rock is basically an algal-foraminiferal biomicrite, this specimen contains a degree of sparite which was deposited within pre-existing very high intergranular, intrafossil, and solution porosity.

2. Algal-foraminiferal member.
3. Coral reef member.
   (*a*) Coralline biomicrite unit.
   (*b*) Reef limestone unit.

The above units are bioherm-wide in lateral extent and are discernible by thin section examinat on. They can be correlated on wire-line logs, particularly gamma-ray and porosity logs.

1. **Porosity Transition Zone** (Fig 8)

The Lower Zelten carbonate grades upward into an interval of transition from dense, non-porous biomicrite into high porosity algal-foraminiferal biosparite/micrite. The Porosity transition zone is 50 to 120 ft (15–37 m) thick and is a grain supported, medium-grained, algal-foraminiferal biomicrite composed primarily of algae, operculinids, and rotaliids. Crinoid debris is abundant and coral debris is common. Planktonic foraminifera are nearly absent. The Porosity transition zone has solution vugs usually filled with calcite cement and the rock is generally of low porosity. However, streaks of higher porosity occur where vugs are not cement-filled.

Although the Porosity transition zone is considered the lowermost sub-division of the bioherm, it can be

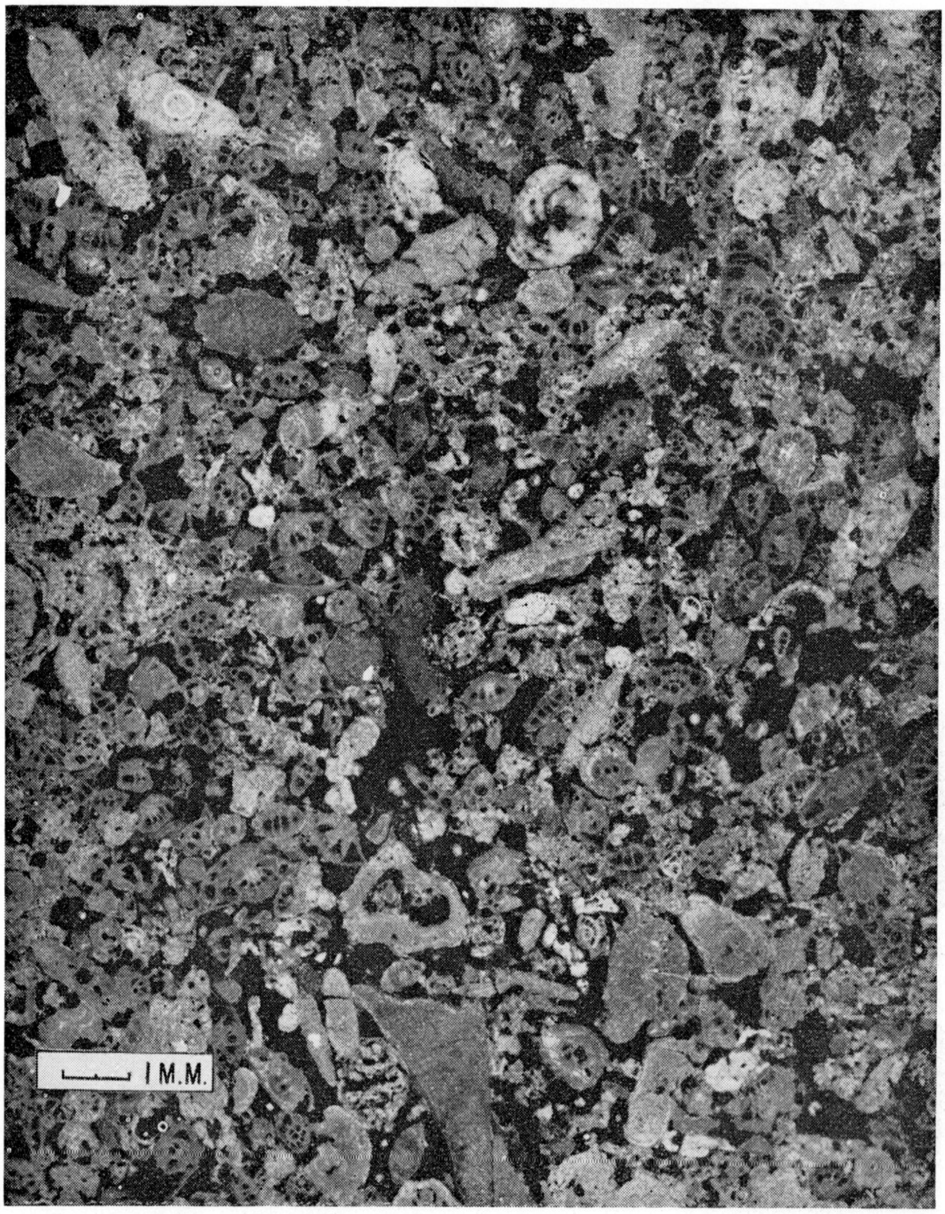

FIG 10. Negative photograph, thin section, A2–103, 10,134 ft, illustrating the typical high porosity of the algal-foraminiferal member of the Idris "A" bioherm. The rock is a well sorted, grain-supported biosparite containing rotaliid, nummulite and discocyclinid foraminifera, as well as fragments of echinoderms, algae, and possibly bones and teeth. The faunal fragments indicate some transportation; however, the lime muds generated in the fragmentation process have been completely winnowed from this specimen. Porosity, which in the illustrated section is 26 per cent, is intergranular and intrafossil, enhanced by solution. There is very litttle cementation.

correlated laterally into off-reef wells. However, lithofacies and biofacies between reef and off-reef sections differ; in off-reef wells the member is composed of fine-grained, argillaceous, planktonic micrite and biomicrite.

2. **Algal-Foraminiferal Member** (Figs 9, 10, and 11)

The vertical gradation from the Porosity transition zone (low porosity algal-foraminiferal biomicrite) to the high porosity Algal-foraminiferal member is well defined on resistivity and porosity logs and can be observed in cores from the A22 well.

The Algal-foraminiferal member was deposited as mound, $2\frac{1}{2}$ by 3 miles (4 by 5 km) wide and 344 to 669 ft (105–200 m) thick. Steep flanks of the build-up plunge rapidly to a zero thickness line. The lower 275 ft (85 m) of the high porosity member is composed predominantly of rotaliids and nummulites which are larger than in the underlying Porosity transition zone. Algae are also very abundant and generally well distributed; crinoid bryozoa and coral fragments are common. Excellent vugular porosity is present, the result of solution. Zones within the lower 275 ft (83 m) contain layers up to 30 ft (10 m) thick with apparent *in situ* coral and abundant red algae.

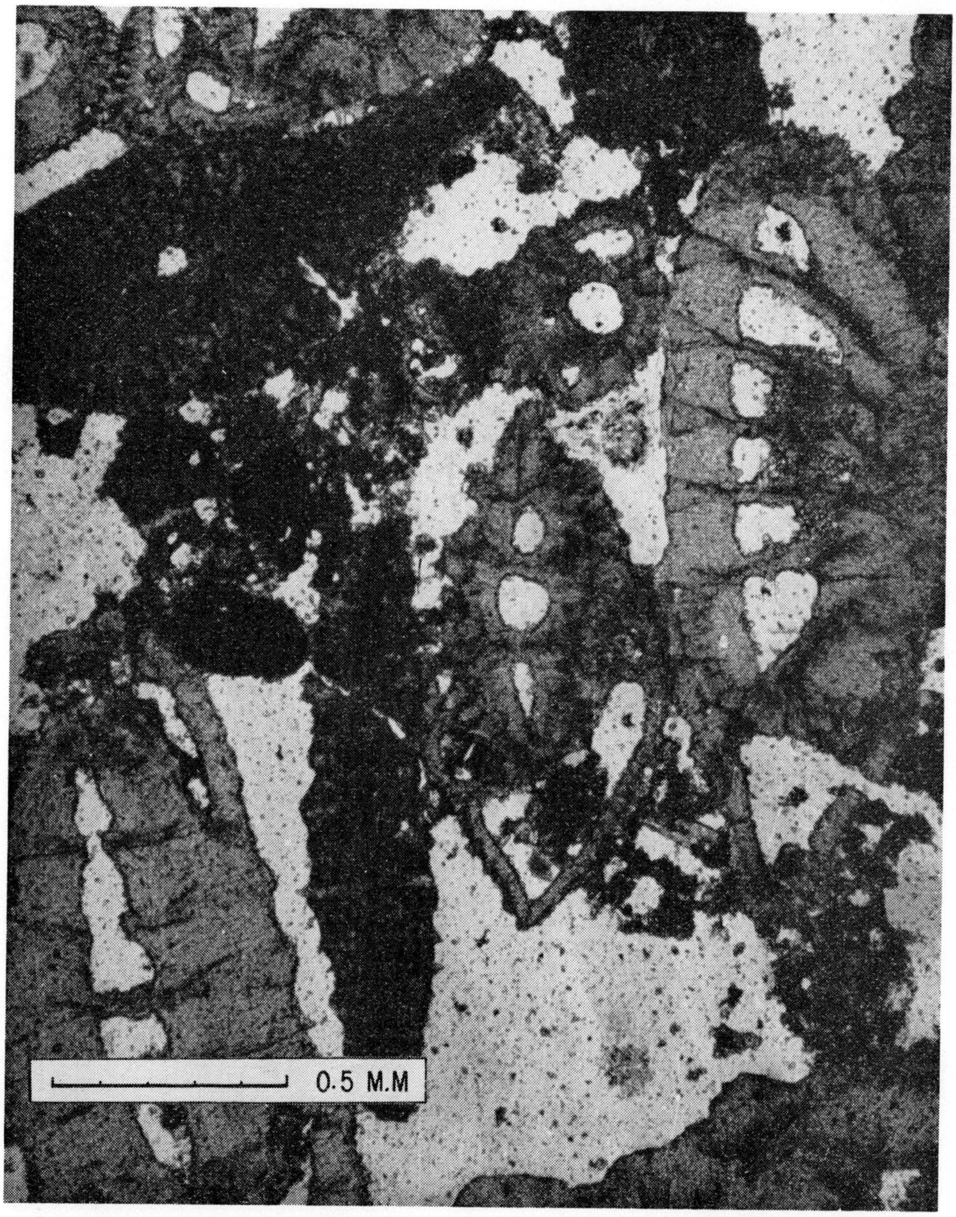

FIG 11. Photomicrograph, thin section, A2–103, 10,122 ft, illustrating the typical high porosity and biofacies of the algal-foraminiferal member of the Idris "A" bioherm. Fauna are large foraminifera, discocyclinids and nummulites. Porosity is inter- and intra-granular, enlarged by solution. The sparse matrix is entirely secondary spar.

Coral is especially abundant on the SE side of the bioherm.

The upper part of the algal-foraminiferal deposit is biomicrite and biosparite with large nummulites and red algae predominating. Rotaliids and discocyclinids, as well as coral and crinoid fragments, are common. Both branching and encrusting algae are present. The excellent porosity, which averages 27 per cent, is primarily inter- and intra-granular which has been enlarged by solution, as illustrated in Figs 10 and 11.

3. **Coral Reef Member** (Figs 12 and 13)

Overlying the Algal-foraminiferal member is the coral reef member, which is divided into a coralline biomicrite unit and an overlying reef limestone unit. The coralline biomicrite unit is 300–350 ft (90–105 m) thick in the central parts of the complex, but very thin at the outside edges, where the unit grades rapidly to an off-reef facies. This unit is a very fine-grained to silt-sized micrite with abundant coral fragments. Some layers have an increased content of large foraminifera and nodular red algae, and other layers have apparently *in situ* coral. Coral fragments were deposited in a micrite matrix which also infilled the coral chambers. Solution processes removed the aragonite coral skeletons and only micrite remained.

Porosity averages about 15 per cent in this section. Permeability is also low as this type of moldic solution porosity is effective only if the dissolved coral frag-

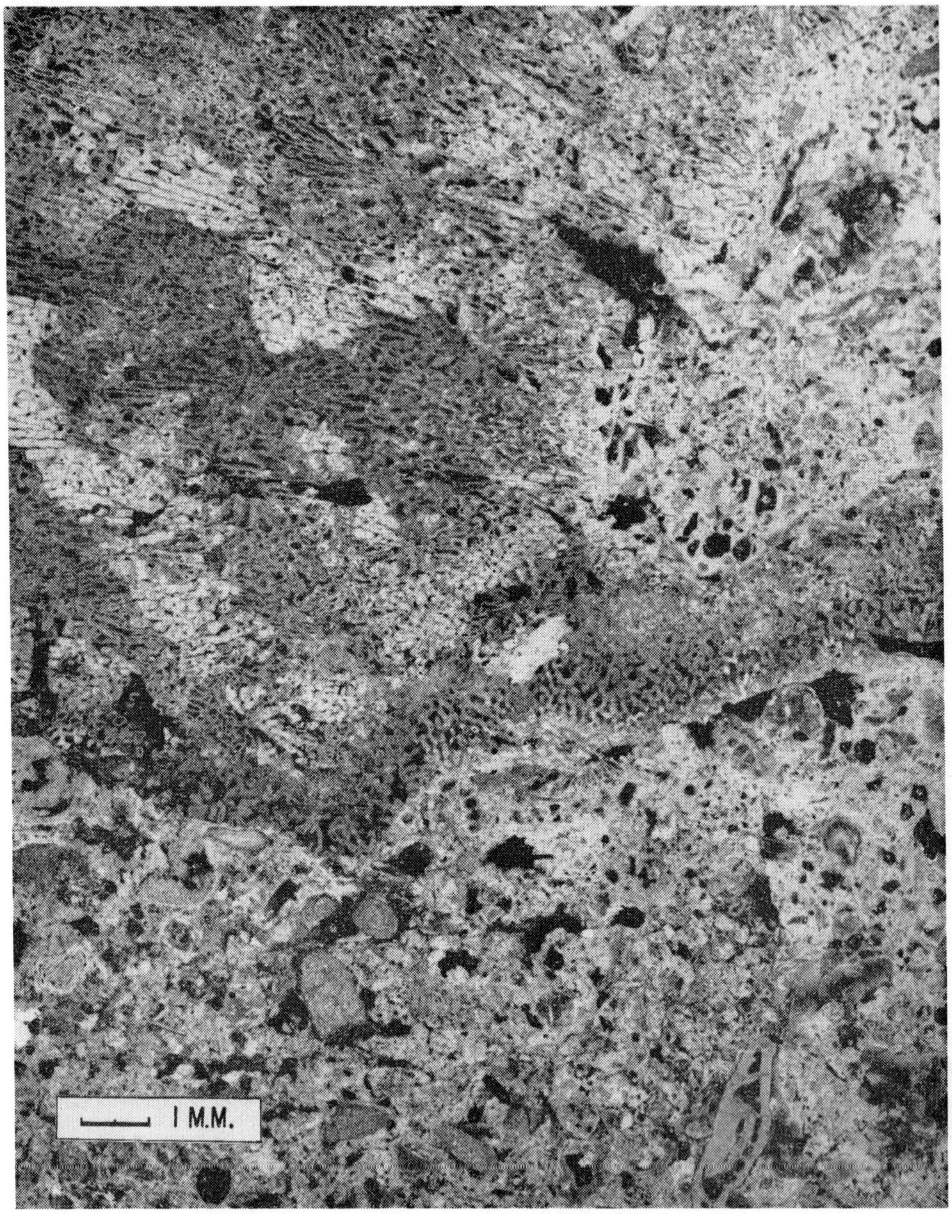

FIG 12. Negative photograph, thin section, A2–103, 9504 ft, illustrating the typical biofacies of the reef limestone unit in the upper part of the Idris "A" bioherm. A part of a massive colonial coral occupies the upper two-thirds of the illustration, while intra-reef detritus comprises the lower one-third. Large foraminifera, echinoderm fragments, and minor algae in a micrite matrix constitute the bulk of the intra-reef detritus. Intra-fossil and solution porosity average about 23 per cent through this part of the reef.

ments were in contact with each other. In most wells excellent gamma-ray markers, reef-wide in latersl extent, exist within this member. This probably indicates deposition on a relatively flat surface which extended throughout the reef area. Edge wells are difficult to correlate due to the increase in cementation or dolomitization.

Bands of dolomite, one to 20 ft ($\frac{1}{3}$–6 m) thick occur throughout the unit, particularly along the western side of the bioherm. The boundary between the coralline biomicrite unit and the overlying reef limestone unit is marked by a dolomite bed in the western side of the biolith. The dolomites are fine-grained, with equigranular crystals of enhedral and sub-hedral dolomite, generally forming an interlocking mosaic with no apparent intercrystalline porosity. However, dolomite streaks often contain vugs up to $\frac{3}{8}$ of an inch (1 cm) in diameter.

The reef limestone unit is 335 ft (100 m) thick in the central part of the bioherm and over 500 ft (150 m) thick in the SE side, where the unit apparently directly overlies the algal-foraminiferal member. The unit thins to a zero line around the edges of the biolith where it grades to a much thinner off-reef facies. The reef unit is dominantly compound coral, red corralline algae, bryozoa, and encrusting foraminifera. Detrital biomicrite both infills coral chambers and, in conjunction with macro- and microfossils and reef debris, fills

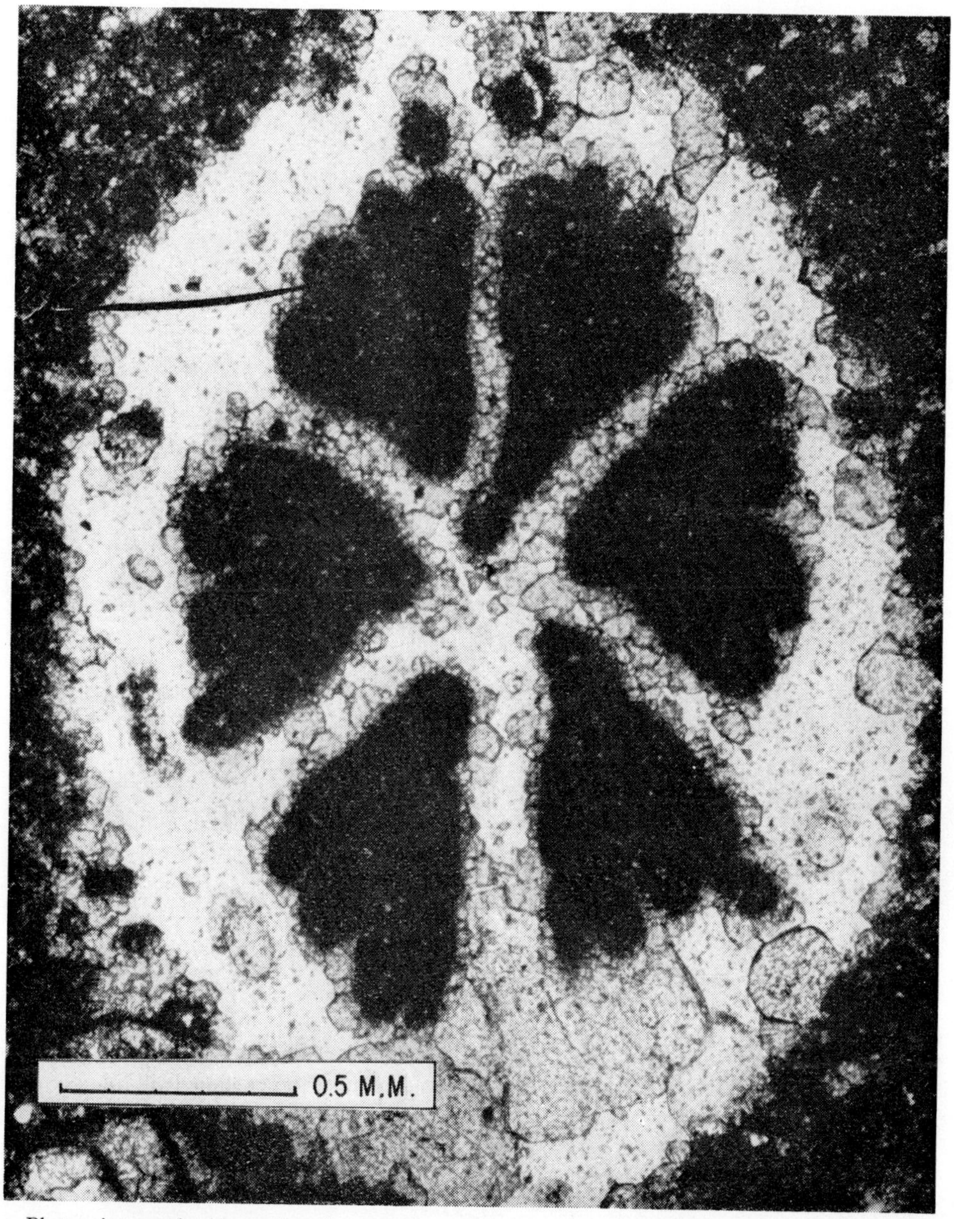

FIG 13. Photomicrograph, thin section, A3–103, 9900 ft, illustrating a fossil mould produced by the dissolution of an aragonitic coral skeleton and the subsequent partial infilling of this mould with a drusy mosaic of calcite. Biomicrite, surrounding and filling the original skeleton, was affected little by the solution processes that produced the skeletal mould. Secondary spar growth to the degree here illustrated is atypical of the Idris reef limestone and of the general biohermal section.

the coral framework (Fig 12). Porosity and permeability are generally high, though streaked, as indicated by core analysis and porosity logs. Intergranular porosity is high, averaging 22 per cent, and primary pores are enlarged by solution. Aragonite coral skeletons were dissolved from both framework builders and coral debris (Fig 13).

The reef limestone unit is much less micritic and has more abundant *in situ* coral material than the underlying coralline biomicrite unit. The marked decrease in micritic content of the upper unit is the main basis for separating it from the coralline biomicrite unit.

Formations Overlying the Idris "A" Bioherm

Overlying the bioherm is 50–100 ft (15–30 m) of dense, non-porous biomicrite with a predominance of benthonic foraminifera, although planktonics are also present. Above the non-porous biomicrite is the Upper Kheir marl, a planktonic micrite and calcareous shale which marks the regional rapid transgression of the sea. Over the west-central part of the bioherm, the typical shale and marl of the Upper Kheir grades laterally to a planktonic micrite facies. This micrite was deposited over the palaeo-high of the bioherm.

DEPOSITIONAL HISTORY OF THE IDRIS "A" BIOHERM

Influence of Underlying Formations on Future Bioherm Growth

Fig 14, isopach and shale-limestone ratio map of the

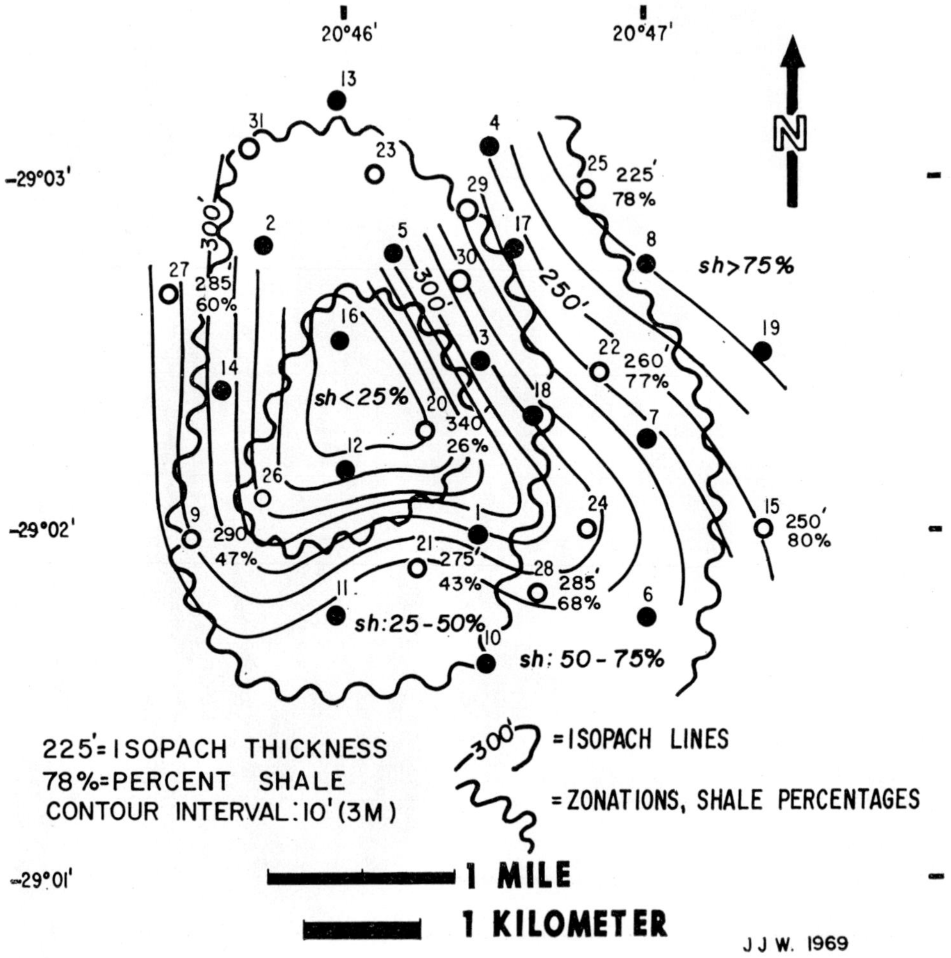

FIG 14. Palaeo-structure map, Idris "A" field; datum: top Lower Zelten carbonate.

interval from top of the Lower Zelten carbonate to base of Heira shale, shows that both interval thickness and ratio of shale to biomicrite vary, but a distinct and consistent pattern exists. There is a thickening of gross interval from the outside edges toward the centre of future reef growth. The ratio of biomicrite to shale also increases in the same direction. Toward the centre, where the gross interval is thickest, the biomicrite is also thickest and the shale thinnest.

The thin shale-thick biomicrite under the central part of the reef is primarily a facies change. This suggests an increase in energy over the central site of future reef growth, probably due to shoaling. The thickening of section takes place in the relatively higher energy biomicrite deposits; the amount of rock deposited as a benthonic biomicrite was greater than the laterally equivalent planktonic shale. Therefore, the embryonic stage of future reef growth was probably contemporaneous with uppermost Heira shale and Lower Zelten carbonate deposition. Throughout the province of reef growth the shale grades upward into a benthonic biomicrite. In the basinal non-reef province the equivalent section is a planktonic and argillaceous micrite. This benthonic to planktonic increase is even more apparent directly beneath the "A" reef than in off-reef wells within the bioherm province.

Palaeo-structure of the top of Lower Zelten carbonate (Fig 15) shows the roughly circular-shaped mound which existed prior to and in the early stages of deposition of the porosity transition zone.

Members Within the Bioherm

The rocks of the porosity transition zone in the "A" area were deposited on the underlying Lower Zelten carbonate and partially represent lateral facies of that unit. The porosity transition zone was deposited in a fairly shallow sea under low to moderate energy and with very little open ocean influence. Thus it represents a continued shallowing of the sea from that in which the older formations were deposited. The size of the rotaliids became progressively larger upward from the base to the top of the zone, implying that conditions for biotic life progressively improved. When ideal conditions were attained, the high-porosity algal-foraminiferal member was deposited.

The algal-foraminiferal member was deposited as a bank under fairly shallow water, shoaling, moderate energy conditions. The high porosity algal-foraminiferal shoal deposit was the culmination of the mound of benthonic tests which had been perpetuating itself since the latter stage of Heira shale deposition.

At times during deposition of the algal-fora-

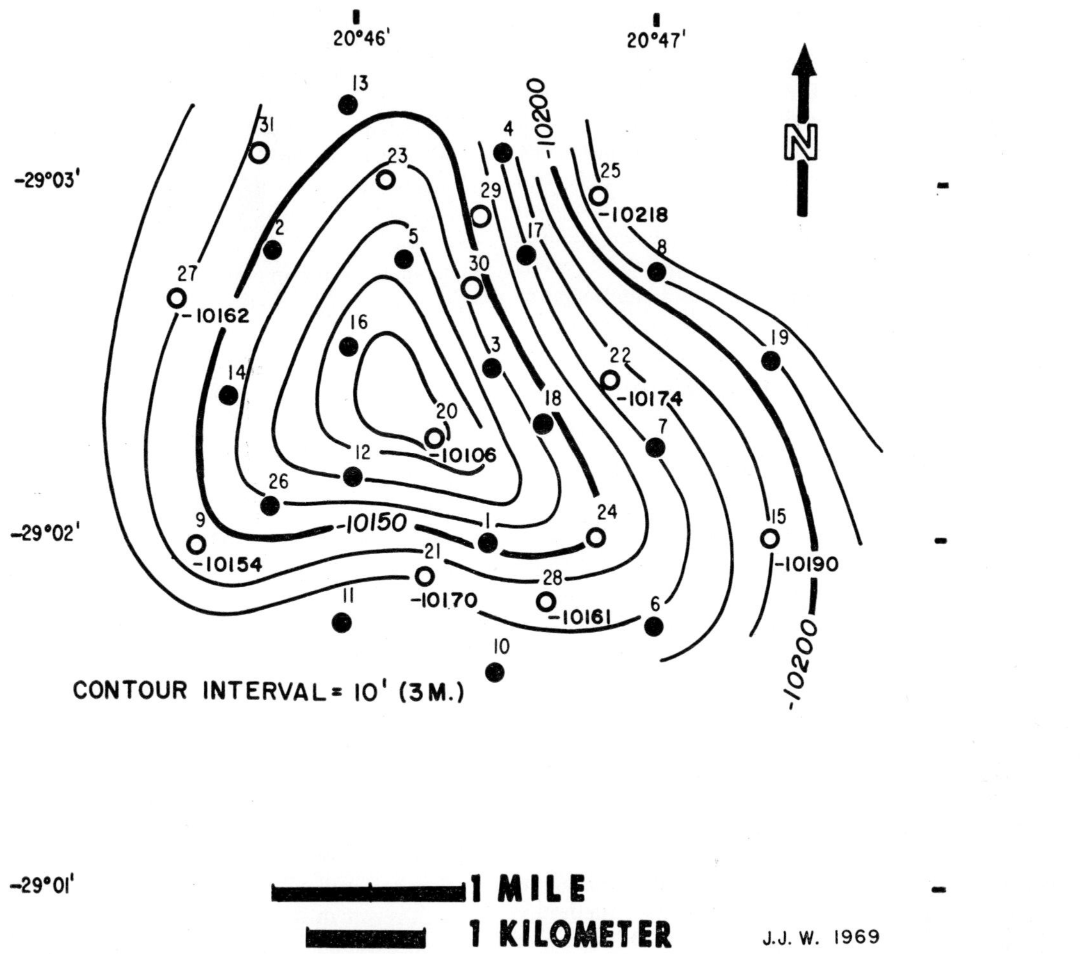

FIG 15. Isopach and shale/limestone map of Idris "A" field; interval top Lower Zelten carbonate to base Heira shale. The isopach patterns partially represent lateral facies change between the dense Lower Zelten carbonate and the basal unit of the bioherm. This facies change is illustrated on Fig 7, cross-section A-A′.

miniferal-rich rocks, conditions were favourable for patch reefs to grow and they were probably widespread across the mound. Later conditions may have been slightly unfavourable for the continued growth, of coral, but bottom dwelling foraminifera and algae flourished and the build-up continued. During deposition of the algal-foraminiferal member, the SE side of the bank was more favourable for coral growth as evidenced by a thin algal-foraminiferal deposit and a thick coral reef in this portion of the bioherm.

The base of the coral reef member marks the time during the history of the biotic build-up when conditions became favourable for coral growth. The influence of algal-foraminiferal constituents decreased as coral reef became the dominant building mechanism. Patches of both compound and solitary corals began growing at many places across the bank, predominantly on the SE side. Subsidence was very slow and corals were continually being broken into very fine detrital material. Coral fragments were deposited with micrite, the dominant rock type of the coralline biomicrite unit. Deposition was under a low energy environment as there was no effective winnowing of mud. As the coralline biomicrite was deposited, the coral reef became more abundant until all biomicrite patches coalesced into one massive coral reef body. This formed the upper unit of the coral reef member, the reef limestone unit. As previously implied, the coralline biomicrite unit is absent in the SE part of the bioherm and the reef limestone unit directly overlies the algal-foraminiferal member.

Depositional Summary

The Idris "A" bioherm was deposited within a province favourable for reef growth in an epicontinental sea, and its site of growth was probably determined by a subtle local upwarp on the sea floor. The reef is surrounded by rocks deposited under definite open ocean conditions. Regionally, Idris "A" is a biotic build-up, the upper and middle parts of which lie within a U-shaped re-entrant. The shape and outside dimensions of the build-up were probably determined by the shape of the initial mound of benthonic foraminifera that formed during the time of uppermost Heira shale deposition. The fact that this original shape continued for the life of the bioherm suggests relatively low current and wave energies and/or variable wind and current directions. Otherwise the shape of the bioherm would have been more elongate.

PRODUCTION

History

Occidental placed the Idris "A" field on production in February 1968, following the completion of a 40-inch (1-m) pipeline from the field to Zueitina, the Mediterranean coast terminal, 130 miles (210 km) NNE. Initial production was 100,000 bd from two wells; however, daily volume was increased to over 548,000 bd in July 1968, after all 17 producing wells were put on-stream. Less than nine months after production began, the field had produced 100 million brl of oil. Cumulative oil production at the end of 1968 totalled 123,180,000 brl. (Through May 1969, 175,636,492 brl.)

Production Techniques

Due to the hazards of lost circulation while drilling the Idris "A" reservoir, it is necessary to set producing casing in the top few feet of the bioherm. Occidental usually drills the producing wells to within 200 ft (60 m) of the oil-water contact and completes them through an open hole interval below $9\frac{5}{8}$-inch (24-cm) casing. The exception is the A1 discovery well, which produces through 7-inch ($17\frac{1}{2}$-cm) casing. By virtue of the excellent reservoir characteristics, primary recovery percentage will be high; however, reservoir pressure is maintained by injecting water into the lowest part of the bioherm. As a result, a very high recovery percentage of the original oil in place is expected. The well programme for pressure maintenance follows one of two procedures. One method is to take water from wells drilled to an Oligocene or Eocene aquifer and inject this water into the section below the oil column. The other procedure is to drill several hundred feet into the Heira carbonate section after a 7-inch ($17\frac{1}{2}$-cm) liner is set through the reservoir section and into the top of the Heira carbonate. The liner is then perforated opposite porous section below the oil reservoir, and water from the dolomite section is allowed to flow into the bioherm aquifer. Substantial water volumes are injected into the reservoir by both methods.

SUMMARY AND CONCLUSIONS

Occidental of Libya discovered a major oilfield when its A1-103 well penetrated the first of several known isolated bioherms. This discovery alerted the oil industry to a hitherto unknown oil province in Libya. This reef province is within an Upper Palaeocene shale re-entrant that opens and plunges NW into a thick shale basin in N Central Sirte Basin. The centre of the province is in the vicinity of the E1-103 well in central Concession 103, where a palaeo-high existed prior to the biohermal development. The shape of this original palaeo-structure was a controlling feature for future growth over which the Heira shale-Lower Zelten carbonate interface was draped in a slightly higher energy environment than the surrounding region. The increased energy over the high allowed benthonic fauna to flourish. Local anomalous areas within the province were greatly enriched by these fauna. Mounds of benthonic deposits, generally small rotaliids, rose above the surrounding open ocean deposits and were self-perpetuating. Algal-foraminiferal banks evolved from these mounds. Algal and sparse coral growth on the early foraminiferal banks provided some sediment-binding but was not profuse enough to form a wave-resistant framework of a true reef.

Since these banks were high on the sea floor and in a relatively high energy zone, they were winnowed of lime muds. A high intergranular porosity resulted that was enhanced by solution processes that both enlarged the pre-existing porosity and dissolved aragonite skeletons.

Coral growth increased on the algal-foraminiferal banks and gradually coalesced into a massive coral reef which thrived until a late Palaeocene transgression introduced shales that blanketed the region and drowned the reefs.

The bioherms were probably never more than 200 ft (60 m) above the surrounding sea floor. Open ocean, low energy shale and micrite deposits were concurrent with the biotic growth on the mounds. These inter-reef lithologies formed lateral seals to the biohermal reservoirs.

The bioherms are $2\frac{1}{2}$ by 3 miles (4 by 5 km) in diameter and up to 1200 ft (365 m) thick. The Idris "A" bioherm represents a major oil reservoir with over 2 billion brl in place which are being produced from 17 wells. The pressure maintenance programme will result in additional recoverable reserves and will be accomplished with about 20 water input wells, 12 of which have been completed at the end of 1968. The net result will be a multi-billion barrel oilfield fully developed with a total of only about 42 wells.

To search out similar pools requires parallel seismic lines no more than 1 to $1\frac{1}{2}$ miles ($1\frac{1}{2}$ to 2 km) apart. Few areas of Libya have been surveyed with this seismic density but this should be done in any suspected reef province. Such a programme is obviously justified economically if the existence of another bioherm similar to the Idris reefs is suspected.

Regional Palaeocene exploration in Libya should begin with a detailed study of depositional environments and lithologic zonations. Historically, much exploration success has resulted from the discovery of faunal anomalies on isolated down-shelf structures that were high enough to reach into increased energy levels associated with a shoaling environment. There are no doubt many such structures that have not yet received the attention they deserve.

In addition, there are vast up-shelf areas where significant energy zones intersected the sea bottom in long linear bands. These areas should be investigated thoroughly.

ACKNOWLEDGMENTS

The authors express their gratitude to Occidental Petroleum Corporation management for their permission to publish this report. Drs W. W. M Brown and E. B. Wolfenden and M. C. Cater, all of Robertson Research Laboratories Ltd, conducted the bulk of the petrographic analysis of the Idris "A" bioherm, worked out the Palaeocene stratigraphic zonation of the area of interest, and provided much of the interpretation attending the biofacies and lithofacies discussion. Drs F. T. Barr and A. S. Campbell and F. G. Engleberts, all of Oasis Oil Company of Libya, and L. F. Ivanhoe, consultant, lent valued editorial assistance. The entire Occidental of Libya staff generously assisted in many areas. Miss Bernadette Mulligan typed the manuscript. To all of these go our sincere appreciation. However, the authors assume full responsibility for the contents of this text.

BIBLIOGRAPHY

Articles

Berggren, W. A. *Bull. Amer. Ass. Petrol. Geol.*, 1965, **49**, 1453–72.

Cummings, E. R. *Bull. Geol. Soc. Amer.*, 1932, **43**, 331–52.

Folk, R. L. *Bull. Amer. Ass. Petrol. Geol.*, 1959, **43**, 1–38.

Irwin, M. L. *Bull. Amer. Ass. Petrol. Geol.*, 1965, **49**, 445–59.

Lowenstam, L. A. *J. Geol.*, 1950, **58**, 439–87.

Nelson, H. F., Brown, C. W., and Brineman, J. H. *Bull. Amer. Ass. Petrol. Geol.*, 1962, 224–52.

Stanton, R. J. *Bull. Amer. Ass. Petrol. Geol.*, 1967, **51**, 2462–67.

Books

Grabau, A. W., "Principles of Geology". New York, N.Y., Seiler, 1913.

Johnson, J. H. "Limestone Building Algae and Algal Limestones" Golden, Colorado, Colorado School of Mines, 1961, p. 297.

MS received 6 February 1969
(*Updated to* 1 June 1969)

Discussion

Victor Church: Was there sufficient geological information on the Sirte Basin prior to Occidental's seismic programme so that you knew what you were looking for on the seismic lines, *i.e.* were you looking for reefs? If not, what were you looking for with this programme.?

C. E. Terry: At the time we commenced our seismic activity in Concession 103 we were looking for structures. We had not considered that there would be the isolated bioherms. We felt that in the S end of the concession we would probably encounter the barrier shoal or reef section which would be the intervening section between our shelf carbonate area to the S and our basinal shales to the N, but to find the bioherms in the fore-reef environment was a surprise. We had not thought of this, and I think few people in Libya had.

Dr Rudolf Martin: In many examples of reefs of the size of Idris "A" there is a peripheral rim, or reef wall, consisting of high porosity "framework" carbonates, surrounding a central depression filled with lower-porosity "lagoonal" sediments. In the diagram, Fig 7, it appears that the algal-foraminiferal member exhibits such a thickening around the edges and a corresponding thinning in the centre but this is not the case with the coral reef member. The diagram gives the impression that the coral reef grew from the centre upwards and outwards, and that much of the material surrounding this core is a detrital mantle, which would explain the absence of a rim effect. Is this impression correct?

C. E. Terry: Yes, to a large extent. I believe the algal-foraminiferal bank, although indicated by a bounding line on the cross-section, may not entirely represent a unit in time in the upper part. The coral reef core in the SE sector of the bioherm may have developed on the crest of the algal-foraminiferal bank simultaneously with the continued deposition of the foraminifera and algae down-flank.

I believe the bioherm area was more or less flat when the coral began to proliferate above the algae-formaniferal bank, but part of this configuration was due to the reef detritus which spread outward from the central coral core. Even in the detritus section we have observed patches of *in situ* coral, which indicates it was trying to be a reef. These isolated patches of coral also probably helped to bind and hold the detritus over the bank.

Said Ahmed (Sonatrach): Which kind of seismic technique was used? What kind of topography was on surface? Did you know before drilling that you would find a reef?

C. E. Terry: As regards the type of seismic work we did across the Idris reefs, it was digital, six-fold CDP, dynamite shooting across this area.

The surface of the ground over this part of the Sirte Basin is flat, sandy, very gently rolling hill country with no vegetation.

We did not know for certain that we would find a reef before drilling, but we felt that we would. The reef parameters were on the seismic records and we felt very good about this.

D. H. Peel (British Petroleum Co. Ltd): Are the other commercial reefs in Concession 103 considered to have been formed contemporaneously with the reef "A" discussed in the paper, and hence would you expect to find that the quality of the crude oil would be the same in each reef?

C. E. Terry: We have two other reefs that are commercially productive. There are three more reefs in our concession that have only small oil accumulations at the crests. Oil quality, gravities, and gas-oil ratios differ from reef to reef.

The reefs grew approximately contemporaneously. The "A" reef and those to the N may have been just a bit younger than those in the highest central part of the mid-basin platform where the first reefs started developing.

Michel T. Halbouty: I notice that in your paper you state that the field will recover "multi-million barrels of oil". Could you be more specific?

C. E. Terry: Our computerized log of the A1–103 showed that we had over 1000 brl of oil per acre-ft in the reef; that amounts to over 1,000,000 brl of oil per acre if all parameters are the same away from the A1–103 well. Of course, the thickness of section diminishes down the flanks. It is estimated that there are over two billion brl of oil in place.

Michel T. Halbouty: How much of this two billion brl is recoverable?

C. E. Terry: From Degolyer and MacNaughton and others we have estimates ranging from 40 to 75 per cent. We believe something between these figures would be fairly accurate.

H. V. Dunnington (Iraq Petroleum Co. Ltd): Could Mr Terry give some indication of the authors' opinion as to the source of the oil and as to the time at which the oil entered the reef? Also, could any information be given on the degree of under-saturation of the oil?

C. E. Terry: Regarding your last question, no, I cannot give you any information about the under-saturation of the oil.

Regarding your first question about when the oil was generated and when it went into the reefs, either the reef section itself was the source, or else the oil came from the flanking argillaceous material which we term the Lower Kheir marl. I suspect the bulk, if not all the oil came from the Lower Kheir marl. We have some evidence that the reefs were subjected to a sub-aerial environment at various times during their growth—the high degree of solution porosity suggests this. If this is the case, it would probably eliminate any possibility that the reef was its own source. We believe the oil migrated into the reefs shortly after burial or in very late Palaeocene and early Eocene times.

J. D. Moody (Mobil): How thick is the shale over the top of the reef?

C. E. Terry: In one area over the reef there is no shale but instead there is a micrite continuously from the reef up to the Lower Eocene carbonate section. This micrite is a very tight lime mud and it grades into an argillaceous material laterally over a part of the crest and down on the flanks.

P. W. J. Wood (Humble Oil): Are you able to high-grade the Heira carbonate high for the postulated ancestral highs on which subsequent reef growths might be found?

C. E. Terry: I believe we have a fairly good definition of where the regional high is and how far down structure our reef province extends. There are places within this province where we question the ability of our $1\frac{1}{2}$-km seismic grid to map all of our smaller reefs.

The reefs are prolific when they come in successfully. If we can produce a 2-billion brl oil pool with 17 producing wells, we must pursue all our reef possibilities actively, for they can be extremely valuable.

G. T. Stephens (Flinders Petroleum N.L.): Have you tried any gravity methods of exploration for these biohermal bodies and, if so, do they have any significant gravity expression?

C. E. Terry: The gravity expressions that we have seen in Concession 103 are probably related to faulting and basement configuration and we have recognized no gravity indications that the reefs are there.

R. B. Holroyd (BP Italiana): Do you have any problems in producing enough water for your water injection programme?

C. E. Terry: Yes, we do. We first drilled "dump flood" wells into the underlying Heira carbonates, which allows water to move upwards in response to the reduced pressure in the reef. We sometimes drilled 1000 to 1500 ft of Heira dolomite section in order to get sufficient production index to put adequate quantities of water into the reef. Now, in addition, we are injecting water derived from surface wells.

J. B. Apfel (Amoseas): Why are some of the Idris type reefs oil-productive and some water-wet?

C. E. Terry: There are several concepts. I feel that it is probably a fracture of the cap-rock that lets the oil get away. I believe Dr Conselman's feeling is that perhaps some of these reefs have permeable "fingers" extending up through the cap-rock which let the oil escape. This is very possible and has been discussed in our shop ever since Dr Conselman was in Tripoli.

Petroleum Exploration in Spain

By RAMON QUEROL

(*Kuwait Spanish Petroleum Company*)

SUMMARY

Petroleum consumption in Spain increased sharply after World War II and reached 17·5 million tonnes in 1967. Petroleum exploration has been moderately active since the mid-1950s and reached a peak in the early 1960s. The only producing field so far is Ayoluengo, with a production of about 69,000 tonnes in the first half of 1968.

The geological features in Spain show a certain symmetry with respect to the stable Spanish Meseta. The Mesozoic and Eocene sediments (all of which constitute petroleum objectives) become more marine and are disturbed by more complicated tectonics away from the Meseta. The structure on these sediments is the result of block faulting, flowage of Triassic salt, and, in the Pyrenees and Cordilleras Beticas, gravity sliding. Marine Miocene sediments found in eastern and southern Spain also constitute petroleum objectives and are tectonically less disturbed.

In spite of the mediocre results obtained so far, exploration in Spain should continue actively, since the cost of foreign crudes represents about 7 per cent of Spain's imports. It appears likely that future activities will be directed towards objectives inaccessible to the older exploratory techniques, such as gas in deep structures, offshore plays, areas of difficult geophysical mapping, and stratigraphic traps.

THE PETROLEUM INDUSTRY IN SPAIN

ENERGY requirements in Spain have increased substantially since the Second World War (Table I). While coal production has been stabilized during the last ten years, the production of hydroelectric energy has increased sharply and the consumption of petroleum has climbed spectacularly (Table II).

Petroleum exploration has been moderately active since the mid-1950s and reached a peak in the early 1960s. Fig 1 shows the exploration permits in Spain, as of 1 July 1968. The only production obtained so far

TABLE I

Energy consumption in Spain

Year	Total consumption of energy in equivalents of millions of tonnes of coal	% coal	% petroleum	% hydroelectric energy
1950	17·3	74·2	9·1	16·7
1955	23·9	57·7	18·4	23·9
1960	34·4	44·1	30·0	25·9
1961	34·4	44·1	29·6	26·3
1962	36·9	42·1	33·2	24·7
1963	41·0	38·4	33·1	28·5
1064	42·8	35·0	38·8	26·2
1965	43·6	34·4	39·0	26·6
1966	46·8	30·8	41·8	27·4
1967	50·0	30·3	47·3	22·4

TABLE II

Consumption of crude oil in Spain

Year	Consumption of crude oil in millions of tonnes	Year	Consumption of crude oil in millions of tonnes
1945	0·7	1957	5·5
1946	1·1	1958	6·4
1947	1·5	1959	6·3
1948	1·5	1960	7·4
1949	1·8	1961	7·3
1950	2·1	1962	8·8
1951	2·2	1963	9·7
1952	2·6	1964	11·8
1953	3·0	1965	12·1
1954	3·7	1966	14·0
1955	4·3	1967	17·5
1956	5·1		

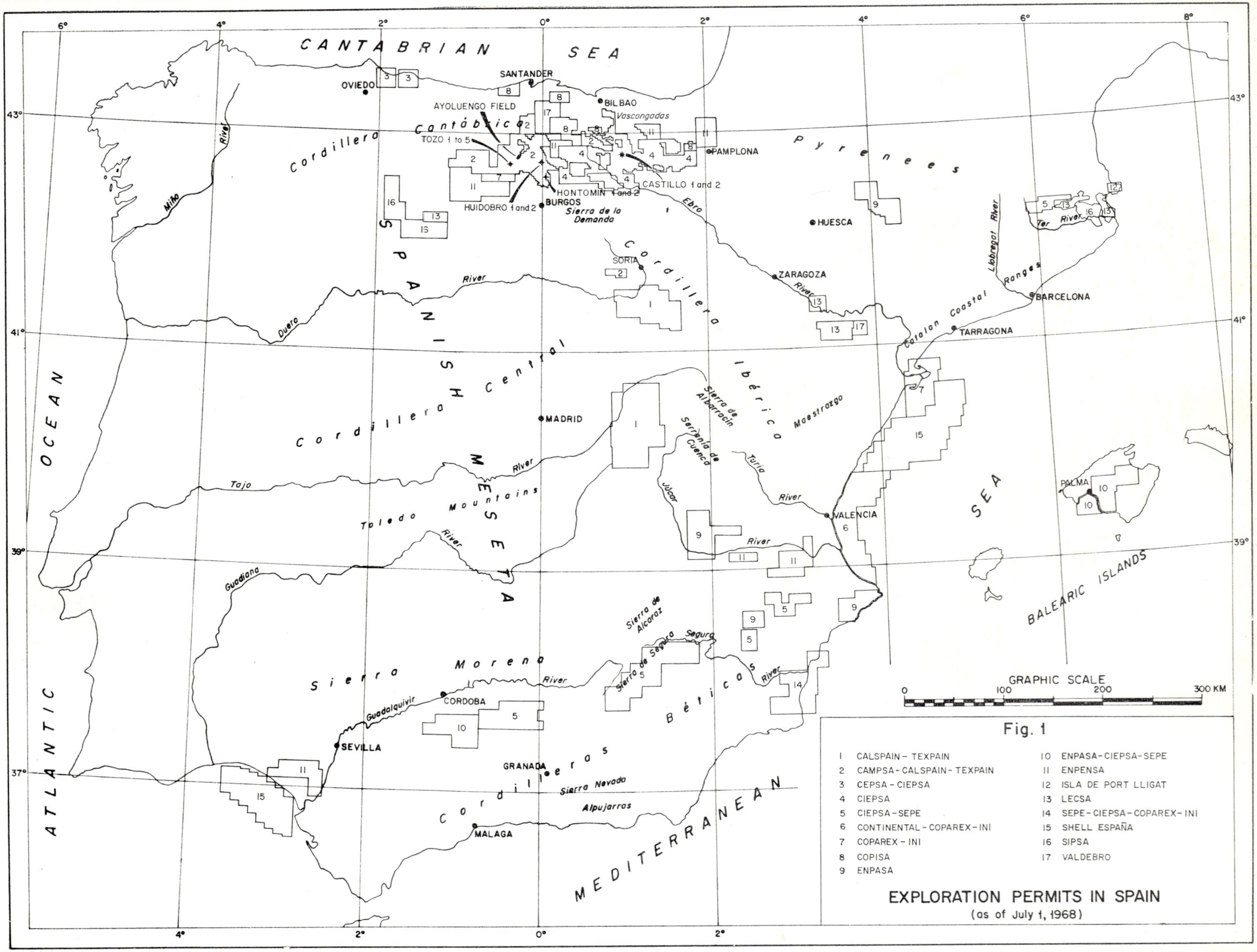

Fig. 1

1 CALSPAIN - TEXPAIN
2 CAMPSA - CALSPAIN - TEXPAIN
3 CEPSA - CIEPSA
4 CIEPSA
5 CIEPSA - SEPE
6 CONTINENTAL - COPAREX - INI
7 COPAREX - INI
8 COPISA
9 ENPASA
10 ENPASA - CIEPSA - SEPE
11 ENPENSA
12 ISLA DE PORT LLIGAT
13 LECSA
14 SEPE - CIEPSA - COPAREX - INI
15 SHELL ESPAÑA
16 SIPSA
17 VALDEBRO

EXPLORATION PERMITS IN SPAIN
(as of July 1, 1968)

is from the small Ayoluengo field. Nearly all the crude requirements are met by imports.

Seven refineries operate in Spain. Total refining capacity (including installations already authorized to be enlarged) is near 40 million tonnes/year. A large proportion of the output is for export.

Nationalization of marketing took place in 1927, with CAMPSA as the agency administering the monopoly.

GEOLOGICAL SUMMARY

(Figs 2 and 3)

The average elevation of Spain (660 m) is the second highest in Europe, after Switzerland. However, most of the Iberian peninsula is not mountainous. The Meseta, or high plateau, comprising three-quarters of the peninsula, is a flat country, bordered by mountains to the north, east, and south, and it slopes towards the Atlantic to the west.

The Meseta contains outcrops of strongly folded Precambrian and Palaeozoic sediments, with various degrees of metamorphism, intruded by common igneous rocks. The Meseta was a very stable area during the Mesozoic and the Tertiary. It was above sea level, was peneplained a number of times, and most of it is still a region of low relief, the only exceptions being the Cordillera Central and the Toledo Mountains. The overlapping Mesozoic rocks vary from continental to shallow marine. The Tertiary is completely continental and towards the west onlaps directly on the Palaeozoics.

The Meseta is rimmed with mountain ranges of different height. They are:

To the north the Cordillera Cantabrica (several peaks above 2500 m).

To the east the Cordillera Iberica, which includes, from north to south, the Sierra de la Demanda, mountains north and west of Soria, Sierra de Albarracin, Serrania de Cuenca, Maestrazgo, and mountains west of Valencia. The southernmost extension of the Cordillera Iberica s.l. could be the Sierras de Alcaraz and Segura. Many peaks reach elevations between 2000 and 2500 m.

To the south the Sierra Morena, with elevations everywhere lower than 2000 m.

The mountain ranges around the Meseta represent folded Palaeozoic and Mesozoic sediments uplifted and compressed against the Meseta. Both Hercynian and Alpine movements can be proven on the Cordillera Cantabrica and Cordillera Iberica, while only Hercynian folding can be observed in Sierra Morena. Isoclinal folding predominates in the Palaeozoics. Block faulting and high-angle reverse faults of Alpine age are common in the Cordillera Cantabrica and Cordillera Iberica s.l. Diapirism of Triassic salt is common away from the Meseta.

The mountain ranges around the Meseta are flanked to the northeast and to the south by large topographic and structural depressions drained by the Ebro and Guadalquivir rivers. They are filled largely with relatively undisturbed Tertiary deposits which are continental in the Ebro Valley and marine in the Guadalquivir Valley.

On the external side of the Tertiary depressions (Ebro and Guadalquivir Valleys) the Alpine mountains of Spain reach elevations well above 3000 m. They are the Pyrenees to the north and the Cordilleras Beticas to the south. Both ranges contain very disturbed Palaeozoic sediments, metamorphosed and intruded by igneous rocks. The Mesozoic and Tertiary are also strongly folded, with development of gravity slides and nappes.

The northern part of the Cordilleras Beticas is known as the sub-Betic trough. The structure is very complicated as a result of gravity sliding and diapirism of Triassic salt. The Jurassic, Cretaceous, and Eocene sediments are marine, often exhibiting deep-water facies. South of the sub-Betic trough is the realm of the Betic nappes. All contain sedimentary rocks of pre-Silurian (?) to Triassic age, with increasing metamorphism in the lower nappes. The Betic nappes appear to have been displaced from south to north during the Alpine Orogeny. It is significant that the Triassic in the Betic nappes exhibits marine facies (Alpine type) as opposed to the Germanic facies found elsewhere in Spain.

The Jurassic, Cretaceous, and Eocene sediments cropping out in the Pyrenees exhibit shallow-water marine facies. The structure is not as complicated as in southern Spain; however, well-developed gravity slides and diapirism of Triassic salt are present.

The Catalan Coastal Ranges, composed of Palaeozoic to Tertiary rocks, separate the Ebro Valley depression from the Mediterranean. The Palaeozoics are strongly folded and have been intruded by igneous rocks. Block faulting with associated folding predominates in the Mesozoics.

It can be seen that the geological features in Spain have a certain symmetry around the stable Meseta. The Mesozoic and Eocene sediments become more marine and are disturbed by more complicated tectonics away from the Meseta.

STRATIGRAPHY

Palaeozoics. Precambrian and Palaeozoic rocks crop out extensively in the Iberian Peninsula. They are strongly folded and faulted, and are commonly intruded by igneous rocks. Only the coal-bearing Carboniferous can be considered a possible objective for hydrocarbons (gas).

The distribution of Palaeozoic rocks in Spain is interesting, as it delineates areas of different geological history which formed a framework for the tectonic developments that continued through the Mesozoic and the Tertiary. The following zones can be distinguished (Fig 4):

1. The Precambrian shield, occupying the central and northwestern part of the Meseta, contains a relatively thin cover of Cambrian, Ordovician,

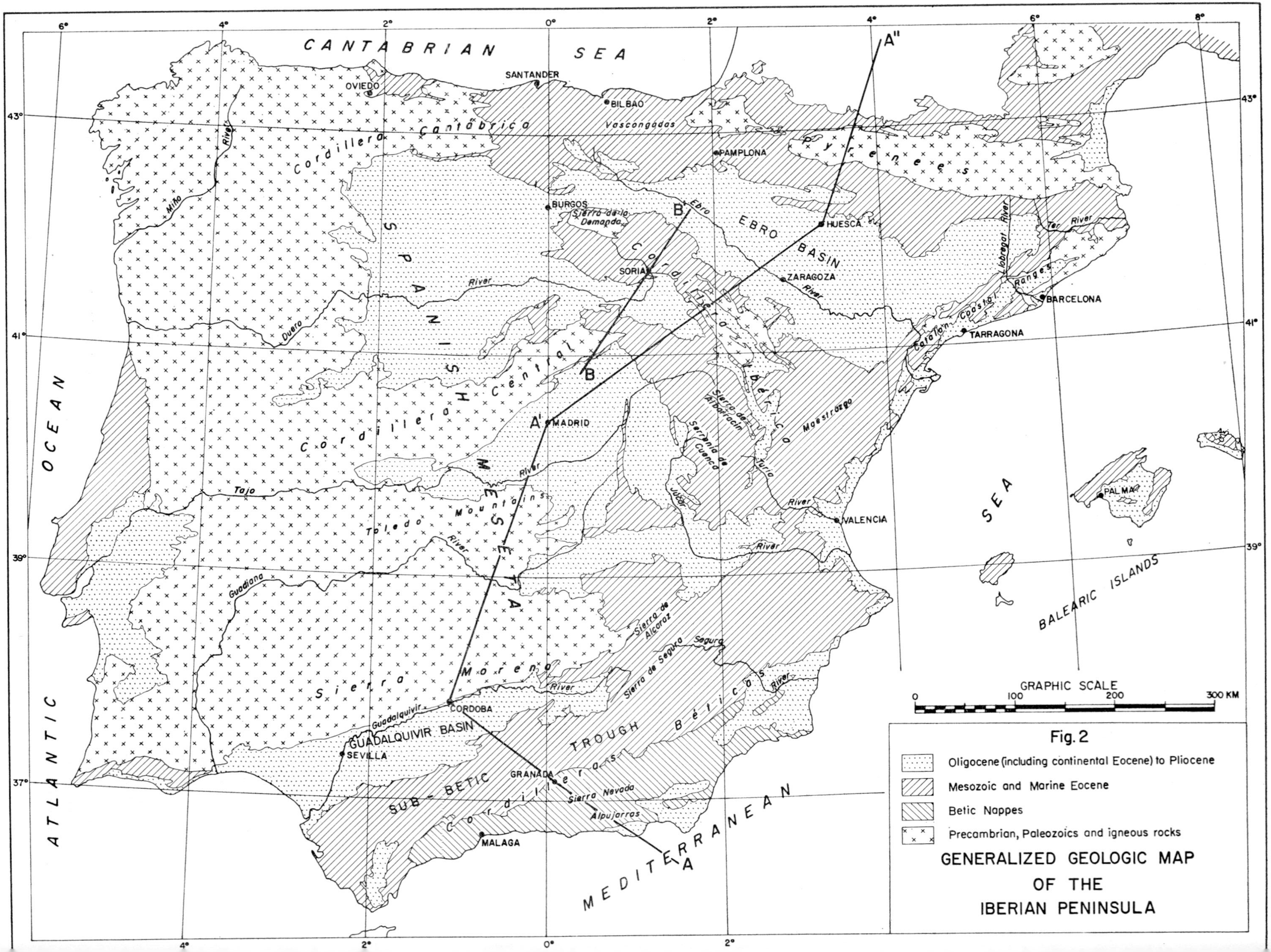

Fig. 2

GENERALIZED GEOLOGIC MAP OF THE IBERIAN PENINSULA

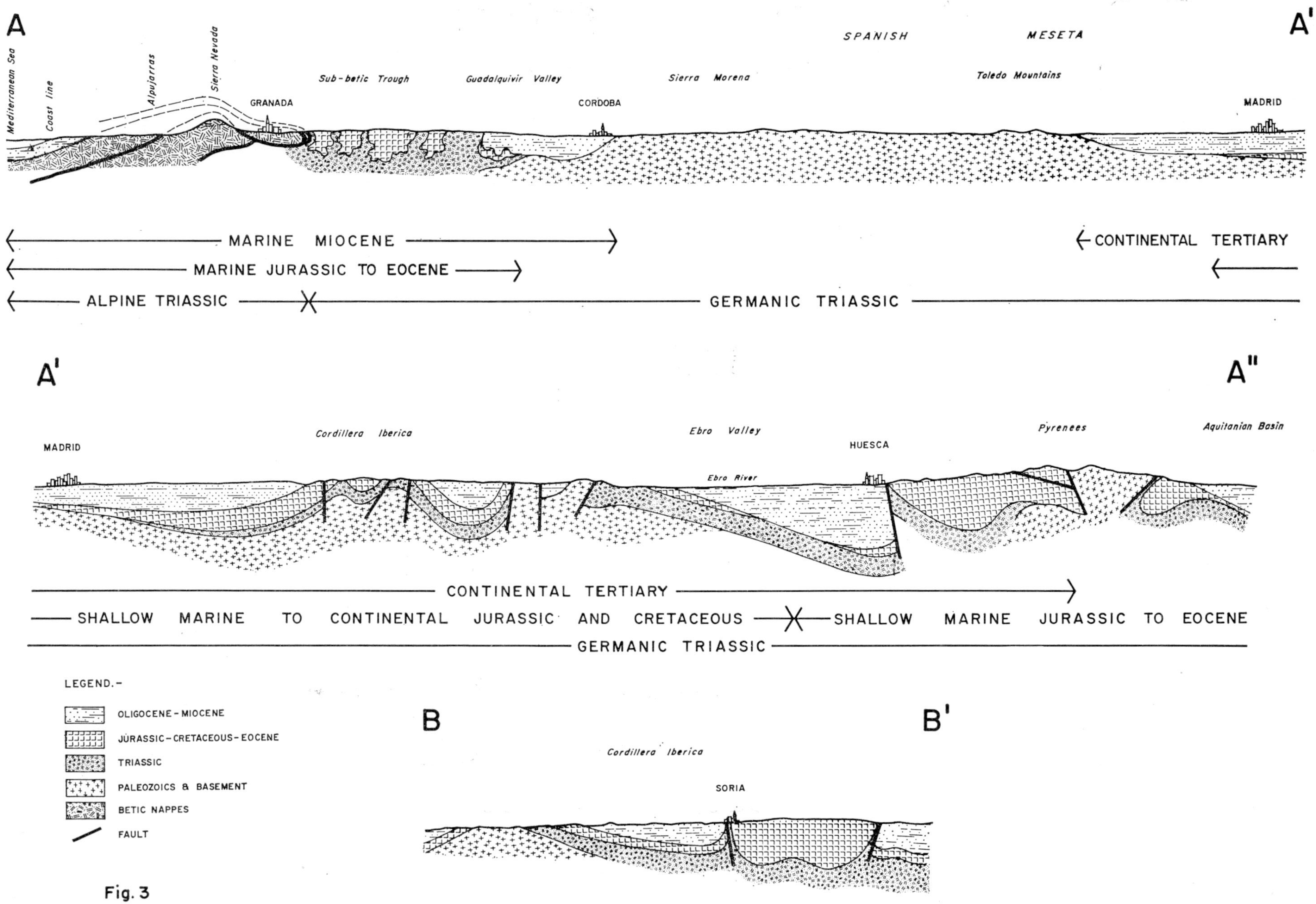

Fig. 3
REGIONAL CROSS SECTIONS
SPAIN
(See locations in Fig. 2)

and Silurian sediments, with slight metamorphism, generally of the phyllitic type.

2. Two Palaeozoic troughs border the Precambrian shield. Both are characterized by a thick fossiliferous Cambrian and by more or less continuous sedimentation in the Ordovician, Silurian, Devonian, and pre-orogenic Carboniferous. Shallow-water marine facies predominate. The southern trough contains abundant acidic igneous intrusions of Hercynian age. The metamorphism of these rocks is slight (development of sericite) or non-existent. In particular, the Palaeozoics in the Cordillera Cantabrica are only locally metamorphic.
3. The Pyrenees also contain thick Palaeozoic rocks, which in the Devonian appear to have been deposited in deeper water than the troughs bordering the Precambrian shield. Granitic intrusions are common, and have noticeably metamorphosed the sediments.
4. The Palaeozoics east of Oviedo show evidence of an important unconformity, with Lower Carboniferous sediments overlying the Ordovician. Similar conditions exist west of Tarragona, where the Lower Carboniferous rests on the Silurian or on the Lower Devonian. The pre-orogenic Carboniferous is well developed in both areas. It is risky to infer what happens under the Mesozoic cover between Tarragona and Oviedo. However, it is possible that we could find here the first expression of a positive area, the so-called "Ebro-Massif", which played an important part in the geological history of the Mesozoic and Tertiary. Coarse clastics deposited in the Ordovician and in the Silurian of the Cordillera Iberica, Catalan Coastal Ranges, and Pyrenees can be proved to originate from the area now occupied by the Ebro Valley. The sediments are here also very strongly folded, with slight to no metamorphism.
5. A basinal area appears to have developed around Barcelona, with deeper marine facies in the Devonian than in the Cordillera Iberica and Cordillera Cantabrica.
6. Carbonates and clastics of pre-Silurian (?) to Lower Carboniferous age crop out in the Betic nappes. The rocks on the highest nappe (known as the Betico de Malaga) display only slight metamorphism, and have well-preserved fossils. Epimetamorphic or mesometamorphic rocks belonging to the lower nappes (Alpujarrides and Sierra Nevada) have been interpreted as being of Palaeozoic age and time equivalents of the sediments of the Betico de Malaga.

The above discussion covers the distribution of Cambrian to Lower Carboniferous (Namurian) rocks. Strong folding and faulting took place over the entire peninsula during the Middle and Upper Carboniferous (Hercynian Orogeny). The post-orogenic Carboniferous, mainly composed of clastics with coal common, developed irregularly all over the peninsula. The thickest coal-bearing sections are in the Westphalian and Stephanian of the Cordillera Cantabrica, where most of the coal produced in Spain is mined.

Triassic (including the Permian where present). Over most of Spain, the Triassic exhibits the typical continental-marine-evaporitic Germanic facies. A completely marine Triassic section (Alpine facies) is present only in the Betic nappes (Alpujarrides).

The Lower Triassic exhibits over most of Spain the typical Buntsandstein facies of conglomerates, red sandstones, and red claystones. These sediments resulted from erosion of post-Hercynian relief and were deposited in troughs caused by post-orogenic, tension-relieving faulting. The thickness of the Buntsandstein varies considerably, from less than 100 m to more than 1000 m. The maximum thickness of the Buntsandstein is reached in the Pyrenees and in the Cantabrian Mountains, where it includes the Permian.

In Middle Triassic time, the Iberian Peninsula was more or less a peneplain. A transgression of the sea from the east and southeast followed, depositing the marine sediments of the Muschelkalk. The sea never covered central Spain, for only Buntsandstein type classics occur in the Triassic along the rims of the Spanish Meseta (Fig 5). In the Cordillera Iberica and the Pyrenees, the Muschelkalk is typically composed of very thin-bedded carbonates, dark shales, and vuggy dolomites, with some marine fossils. In the Catalan Coastal Ranges, the Maestrazgo, and the central Ebro Valley (east of Zaragoza) the Muschelkalk carbonates are well developed. A unit of gypsiferous red marls is sandwiched there between two carbonate units. This so-called Middle Muschelkalk (with the same facies as the typical Keuper) has also been found by drilling in the Maestrazgo and the Ebro Valley, where it includes much salt. Marine fossils are common in the outcrops of the Catalan Coastal Ranges. The total thickness of the Muschelkalk in all these areas is several hundred metres.

The Upper Unit of the Germanic Triassic, the Keuper, is composed of gypsiferous vari-coloured clays, containing much salt in the subsurface. It overlies the Muschelkalk, except along the rims of the Meseta. The Keuper is characterized by its high mobility, being easily extruded through the younger sediments along fault planes and salt domes. The structure of the Mesozoic and Tertiary in Spain has been complicated by the salt tectonics of the Keuper (and in places by the lithologically similar Middle Muschelkalk). The thickness of the Keuper in undisturbed areas is only 100–200 m, but it must have been much thicker in the areas where flowage and tectonic complications make any thickness estimate impossible. The areas of thick evaporite deposits in the Triassic are shown in Fig 5.

It appears that the Buntsandstein, Muschelkalk, and Keuper are rock rather than time units. It is easy

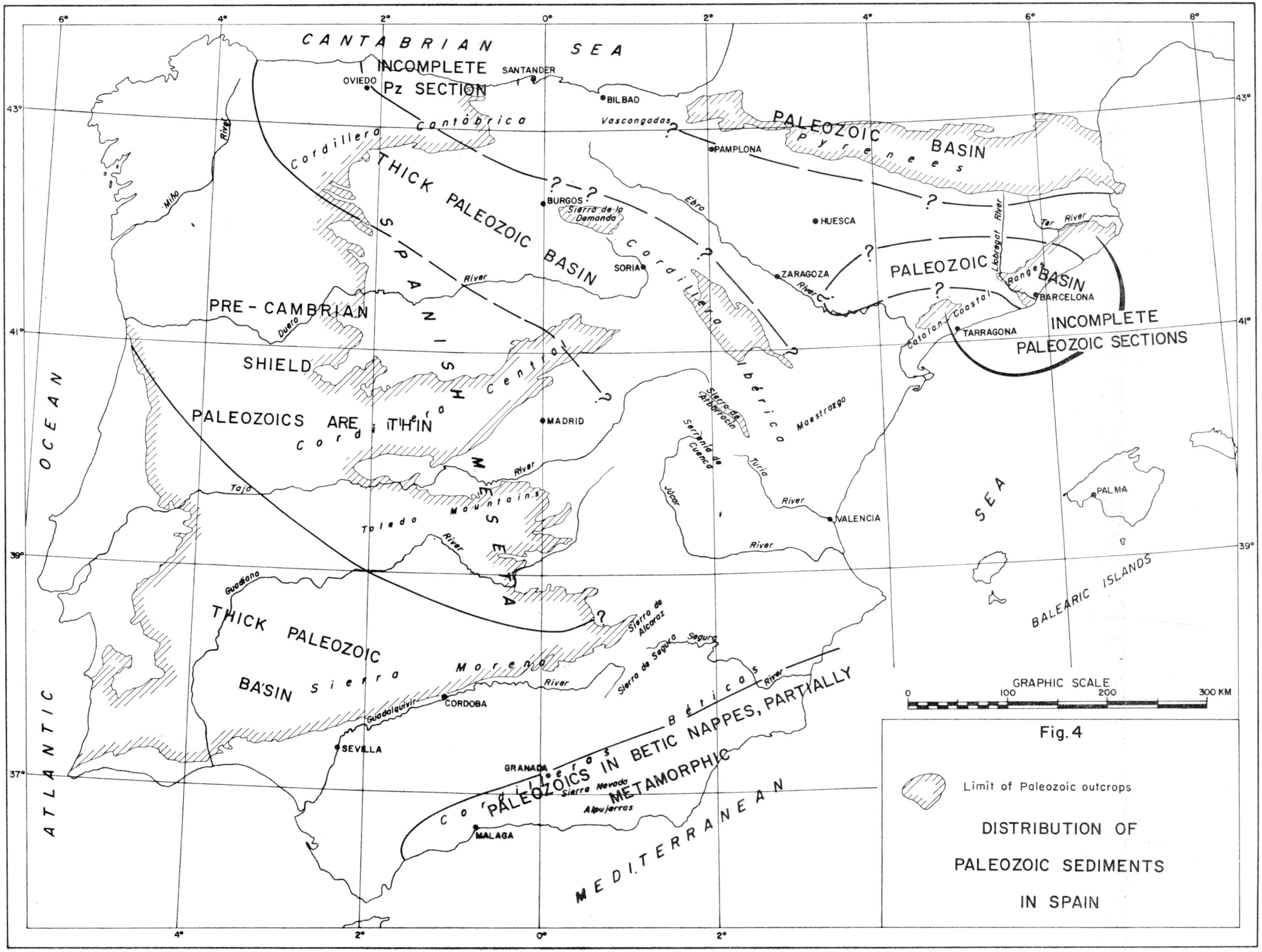

Fig. 4

DISTRIBUTION OF PALEOZOIC SEDIMENTS IN SPAIN

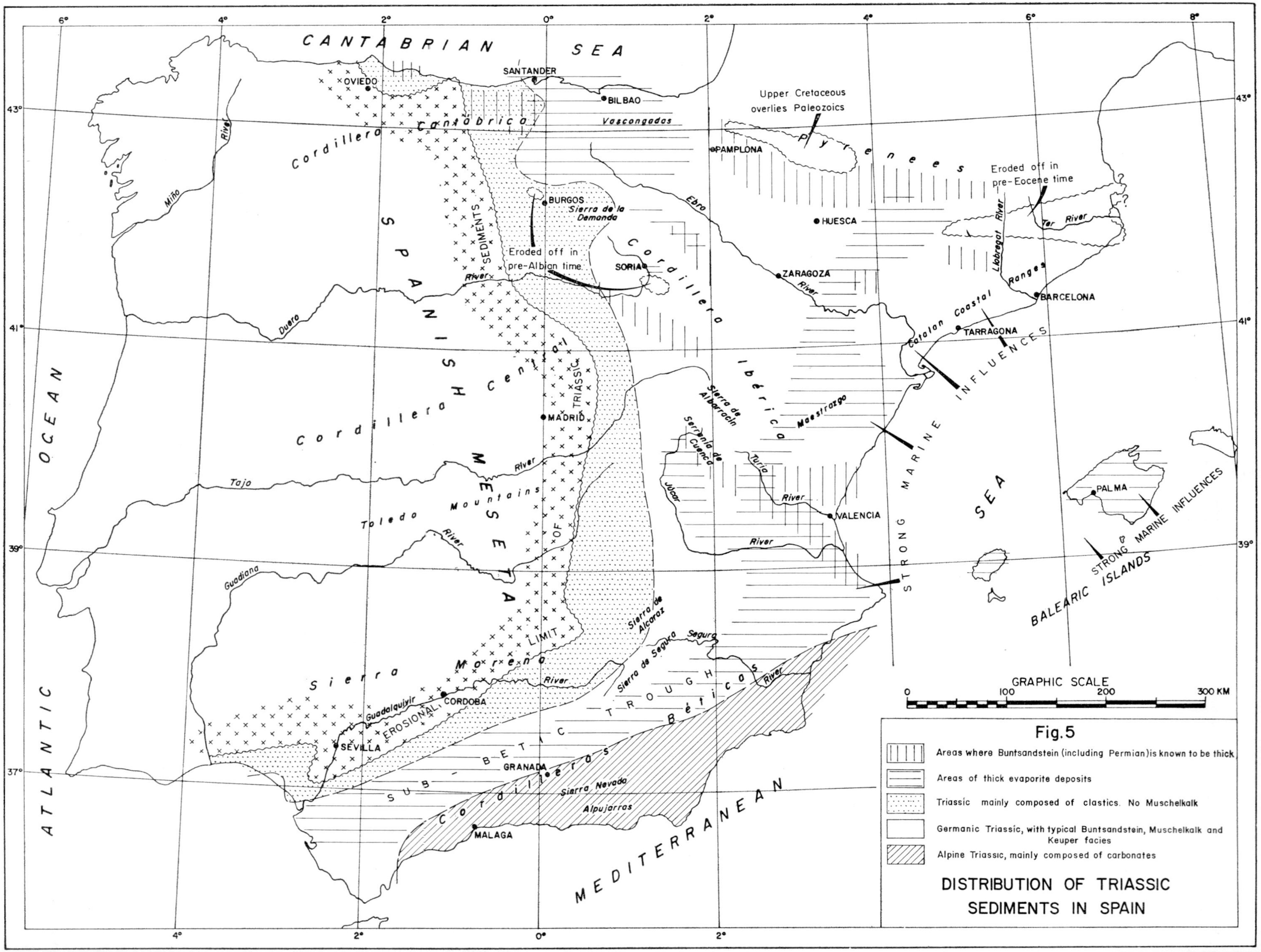

CANTABRIAN SEA
OCEAN
ATLANTIC
MEDITERRANEAN
SEA
OVIEDO
SANTANDER
BILBAO
Vascongadas
Cordillera Cantábrica
Upper Cretaceous overlies Paleozoics
Pyrenees
PAMPLONA
Eroded off in pre-Eocene time
Ter River
Llobregat River
HUESCA
Ebro
BURGOS
Sierra de la Demanda
SEDIMENTS
Eroded off in pre-Albian time
SORIA
Cordillera
ZARAGOZA
River
Catalan Coastal Ranges
BARCELONA
TARRAGONA
INFLUENCES
STRONG MARINE
Duero
River
SPANISH MESETA
Cordillera Central
TRIASSIC
OF
LIMIT
EROSIONAL
MADRID
Ibérica
Maestrazgo
Sierra de Albarracín
Serranía de Cuenca
Turia
River
Júcar
VALENCIA
PALMA
STRONG MARINE INFLUENCES
BALEARIC ISLANDS
Tajo
Toledo Mountains
Guadiana
Sierra de Alcaraz
Segura
Sierra de Segura
SUB-BETIC TROUGH
Béticas
Cordilleras
Sierra Morena
Guadalquivir
CORDOBA
SEVILLA
GRANADA
Sierra Nevada
Alpujarras
MALAGA
Miño
6°
4°
2°
0°
8°
43°
41°
39°
37°
GRAPHIC SCALE
0
100
200
300 KM
Fig.5
Areas where Buntsandstein (including Permian) is known to be thick
Areas of thick evaporite deposits
Triassic mainly composed of clastics. No Muschelkalk
Germanic Triassic, with typical Buntsandstein, Muschelkalk and Keuper facies
Alpine Triassic, mainly composed of carbonates
DISTRIBUTION OF TRIASSIC SEDIMENTS IN SPAIN

to follow the facies changes in the Triassic from the rims of the Meseta (where all of it is of clastic Buntsandstein facies) to the Cordillera Iberica, where the three facies co-exist, and to the Catalan Coastal Ranges and Balearic Islands, where the carbonates are thick.

The facies changes in the Triassic are much more spectacular in the south of Spain. In the outcrops of the Sierra de Alcaraz, only Buntsandstein type sediments are found between the Palaeozoics and the Jurassic. Farther south along the sub-Betic trough, the Triassic is exposed as a very thick and chaotic mass of variegated marls, with abundant gypsum; "Muschelkalk" type carbonates often float on this Triassic. The base of the Triassic is nowhere exposed in the sub-Betic trough and, accordingly, it is not known how much Muschelkalk and Buntsandstein are present there; however, the enormous volume of "Keuper" facies which can be seen in the outcrops is quite striking.

The lowest structural unit cropping out in the area of the Betic nappes is the Sierra Nevada complex, which is composed of metamorphics, including carbonates of possible Permo-Triassic age. The Sierra Nevada complex is overridden by the Alpujarrides complex, presumably overthrust from the south. The Alpujarrides are a metamorphic complex of probable Palaeozoic age, overlain by a complete Triassic section, several hundred metres thick and nearly entirely composed of carbonates with marine fossils (Alpine-type Triassic). The transition from the Palaeozoics to the Triassic carbonates is composed of very typical violet or purple schists, in places containing gypsum and marbles, which, on the basis of similarities with the section in the Alps, is believed to represent the Werfenian (lowermost Triassic). The Alpujarrides are in turn tectonically overlain by the Betico de Malaga complex of fossiliferous, slightly metamorphic Palaeozoics, capped by very thin Mesozoic and Eocene. The Triassic of the Betico de Malaga is usually referred to as the Permo-Triassic and is composed of a generally thin section of red coarse clastics.

Marine Jurassic (Fig 6). A widespread transgression from the east took place in Spain during the Jurassic. The transition from the Triassic is composed of anhydrites and dolomites (Carniolas) which grade upward into well-bedded open-sea fossiliferous limestones and marls. In a peneplained country, as most of the peninsula was at that time, the Carniolas could constitute the near-shore facies of the marine Jurassic, which might have been deposited over most of the peninsula. The areas of maximum deposition of the marine Jurassic (including the Carniolas), with thicknesses of 500 to 1000 m, are to the northeast and southeast of the Meseta, *i.e.* along the eastern Cordillera Cantabrica, Cordillera Iberica, and sub-Betic trough. The Jurassic appears to have been eroded off in large areas of the Ebro Valley, Pyrenees, and the Meseta. South of the sub-Betic trough, a very thin Jurassic section amounting to a few tens of metres can be observed in the Betico de Malaga complex, while no Jurassic has been reported on the Sierra Nevada and Alpujarrides nappes.

The marine sedimentation in the Jurassic was interrupted by the first terrestrial deposits of "Wealden" facies, the result of epeirogenic uplifts in the old Precambrian and Palaeozoic areas in the western half of the Peninsula. The "Wealden" deposits begin at different ages within the Jurassic. Fig 6, showing the age of the youngest marine rocks immediately underlying the "Wealden", strongly suggests that the sea retreated toward the east during the Middle and Upper Jurassic. It seems that, as during the Triassic, the Jurassic sea was connected to the Mediterranean sea rather than to the Atlantic ocean. One possible explanation for this fact is that the Palaeozoic shield area of the Meseta was connected to the Armorican Massif in Brittany and constituted an effective barrier to the sea during the Triassic and Jurassic. The barrier was broken in late Jurassic time, as proved by marine Kimmeridgian (overlying the Wealden) east of Oviedo, which had to come from the north. Also the Purbeckian and Neocomian deposits in the eastern Cordillera Cantabrica and in Vascongadas represent marine transgressions from the north. These transgressions could have originated from tectonic movements depressing the Palaeozoic fold belts, more or less contemporaneously with Wealden deposition. Proponents of continental drift theories could visualize the late Jurassic as the time when Spain drifted away from western France.

The tectonic movements during the Jurassic could also be detected in southern Spain, in the sub-Betic trough, where the first deep-water deposits, limestones and marls, with radiolarites and a pelagic fauna of ammonites, make their appearance.

Jurassic-Cretaceous transition (Fig 7). In the Mesozoic trough bordering the Meseta to the northeast, the transition from the Jurassic to the Cretaceous is known as the "Wealden" and consists of terrigeneous materials, chiefly red shales, siltstones, and sandstones, with some fresh-water limestones. It is now well known that the term "Wealden" should be used as indicating facies, because the age of the "Wealden" varies from uppermost Lias (?) to Neocomian. The nature of the Wealden deposits reflects tectonic activity, and uplift of large areas, like the Meseta. Incipient movement of Triassic salt and growth of salt structures can also be proved in these periods.

The westernmost of the Wealden basins exists east of Oviedo. Terrestrial deposits overlie the Liassic and contain in the upper part marine beds with Kimmeridgian fossils.

Between Burgos and Santander, the lowermost "Wealden" beds overlie the Callovian and consist of red shales and sandstones with some fresh-water limestones. They are overlain by brackish to marine

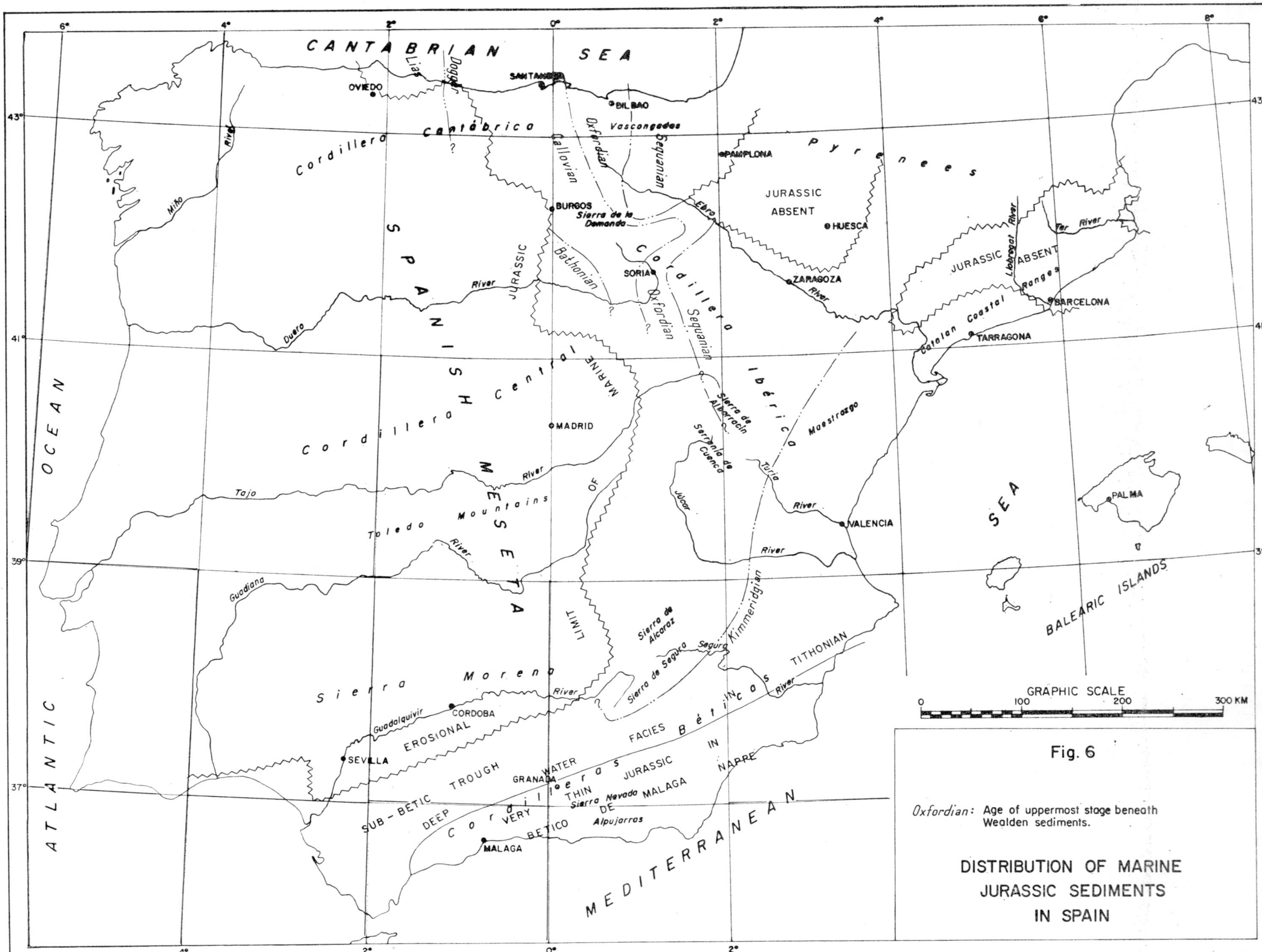

Fig. 6

Oxfordian: Age of uppermost stage beneath Wealden sediments.

DISTRIBUTION OF MARINE JURASSIC SEDIMENTS IN SPAIN

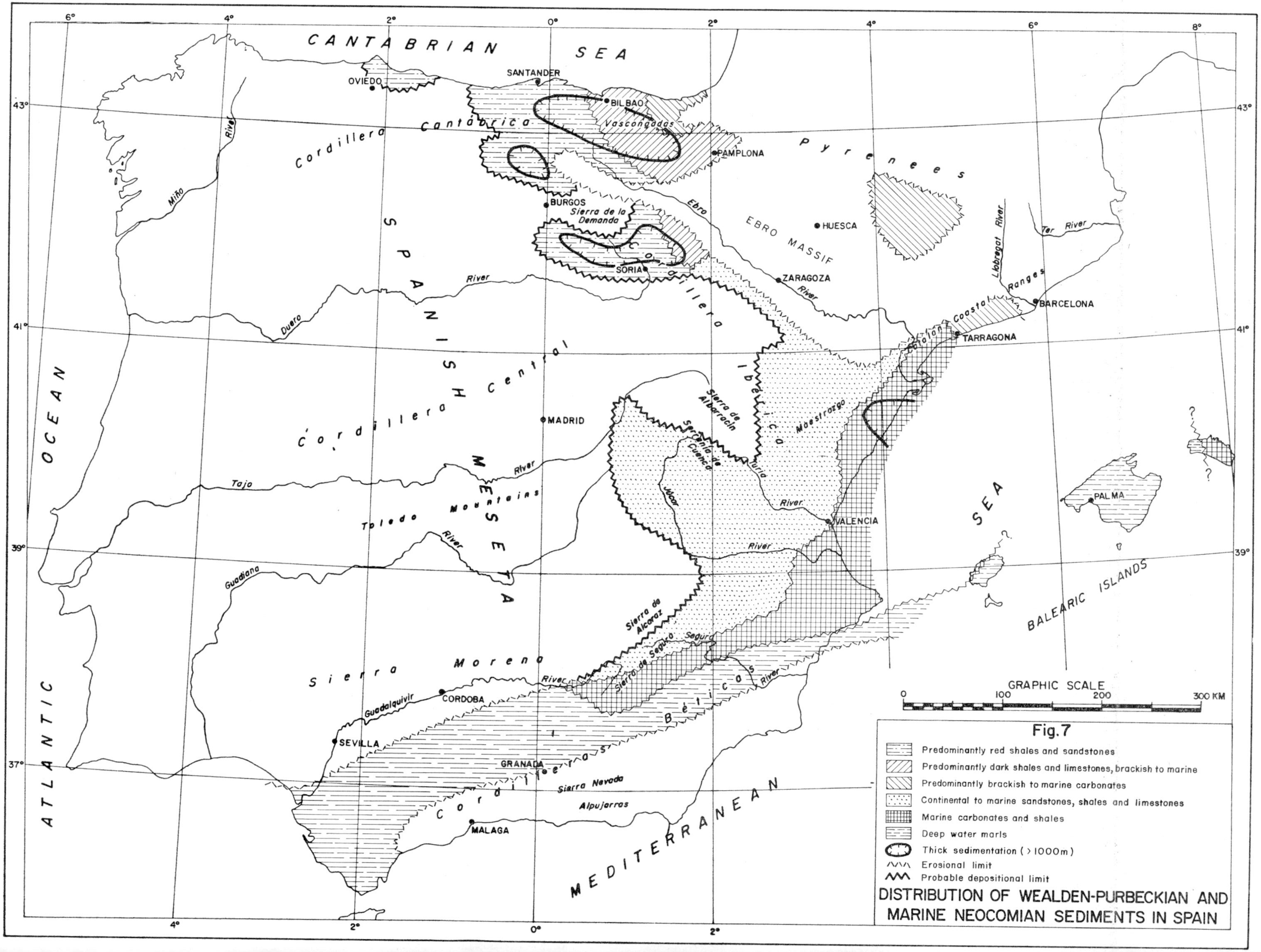

Fig.7

DISTRIBUTION OF WEALDEN-PURBECKIAN AND MARINE NEOCOMIAN SEDIMENTS IN SPAIN beds with Purbeckian to Neocomian fossils, followed in turn by a very thick sequence of red and green clays and sandstones, which are covered near the northern coast by fossiliferous Aptian and near the Meseta by coarse sands of Utrillas facies.

Marine influences from a sea to the north are observed in Vascongadas, where the uppermost marine Jurassic is overlain by black limestones and black shales with brackish to marine Neocomian fossils. Anhydrite and salt have also been found by drilling. The Neocomian of Vascongadas is quite thick and it is overlain by the marine Aptian.

In another deep basin north of Soria the Wealden was deposited on marine Jurassic (Dogger to Oxfordian) and consists of shales, sandstones, and brackish-water limestones. Ostracods in the "Wealden" limestones prove a Jurassic age. These rocks are overlain by coarse sands of Utrillas facies, of presumably Albian age.

Typical red clays and sandstones of Wealden facies can also be found in the eastern areas of Spain. This Wealden overlies the uppermost Jurassic and contains common marine tongues of Neocomian age. Marine Aptian overlies the red beds. Near the eastern coast, a thick and entirely marine Neocomian is present above the Jurassic and below the Aptian.

In the sub-Betic trough and the Balearic Islands, the Neocomian is also entirely marine and, at least in large part, of deep-water facies. Intraformational slumping and breccias can be observed in the Neocomian north of Granada. This proves that the tectonic activity which resulted in deposition of the "Wealden" clastics was also responsible for early movements in the Betic areas.

Some of the areas now covered by the Tertiary of the Ebro Valley appear to have been uplifted during the late Jurassic or early Cretaceous. Some of the terrestrial deposits in the "Wealden" Basin north of Soria originated from the north. The uplifted area is known in the literature as the "Ebro Massif". The Wealden deposits preserved in some areas of the Ebro Massif (*e.g.* east of Huesca and north of Tarragona) are brackish to marine carbonates, with a shallower environment than the marine sediments cropping out elsewhere along the eastern coast of Spain.

Aptian-Albian (Fig 8). Important epeirogenic movements in Aptian and Albian time are reflected by an important hiatus in the sedimentary section and also by the continental or near-shore nature of many Aptian-Albian deposits. Movement of the Triassic salt continued along many structures. The Aptian is generally transgressive, while the Albian represents a widespread regression.

Near the northern coast of Spain, the Aptian-Albian is very thick, and predominantly composed of fine clastics. Limestone reefs of Aptian to lower Albian age are interbedded.

Along the eastern coast of Spain, the Aptian-Albian is thinner, although it exceeds 1000 m halfway between Tarragona and Valencia. Bioclastic limestones with Aptian fossils predominate in the lower part, and the Albian is generally sandy.

The Aptian-Albian in central Spain is represented by unfossiliferous white or reddish, coarse, loose sands, in places with coal. They are overlain by fossiliferous Cenomanian, and for that reason they are referred to as "Albian". An alternate and better name than "Albian" sands is Utrillas facies. The Utrillas sands unconformably overlie older rocks, from the Palaeozoics to the Aptian.

The Aptian-Albian in the Pyrenees is found only in the central part and near the eastern coast. It is composed of marine limestones and marls, quite thick in the Central Pyrenees. Very little Lower Cretaceous has been found by drilling in the Ebro Valley. This and the thick Utrillas sands south of the Ebro Massif suggest that much of it was above sea level at this time.

Marine facies generally indicative of deep-water environment are found in the sub-Betic trough and the Balearic Islands. Tectonic activity can also be proven in this area, with evidence of gravity sliding of the sediments from the south.

Upper Cretaceous (Fig 9). In the Upper Cretaceous there was a widespread transgression with deposition of marine limestones and marls as far west as the central part of the stable Meseta. Anhydrites at the Cretaceous-Eocene transition represent the shallowest facies and are common in central Spain.

The thickest sedimentation, more than 4000 m, occurred in Vascongadas, with two distinct facies. To the south, limestones and marls similar to the Upper Cretaceous of the Cordillera Iberica and the Meseta were deposited, and near the coast the Upper Cretaceous is of flysch facies with interbedded volcanics.

The Upper Cretaceous is also thick in the Central Pyrenees, where it is composed of marine marls and limestones. Near-shore facies, with common sand, indicate the existence of a land to the south (Ebro Massif). Exploratory drilling in the Ebro Valley has demonstrated the absence of Upper Cretaceous sediments in this area.

In the sub-Betic trough, the Upper Cretaceous is represented by reddish and white chalky limestones, very characteristic in the field, which, on the basis of the predominance of pelagic foraminifera, are thought to represent deep-water deposits.

Eocene (Fig 10). The transition from the Cretaceous to the Eocene contains continental or evaporitic beds in central and eastern Spain. They reflect new tectonic movements that uplifted most of the Iberian Peninsula. Marine sediments are restricted to the areas adjacent to the Pyrenees and the Cordilleras Beticas.

The Eocene sediments are thick in the southern foothills of the Pyrenees (exceeding 2000 m in places). In the eastern areas, red marls and fresh-water limestones represent the Lower Eocene, while marine

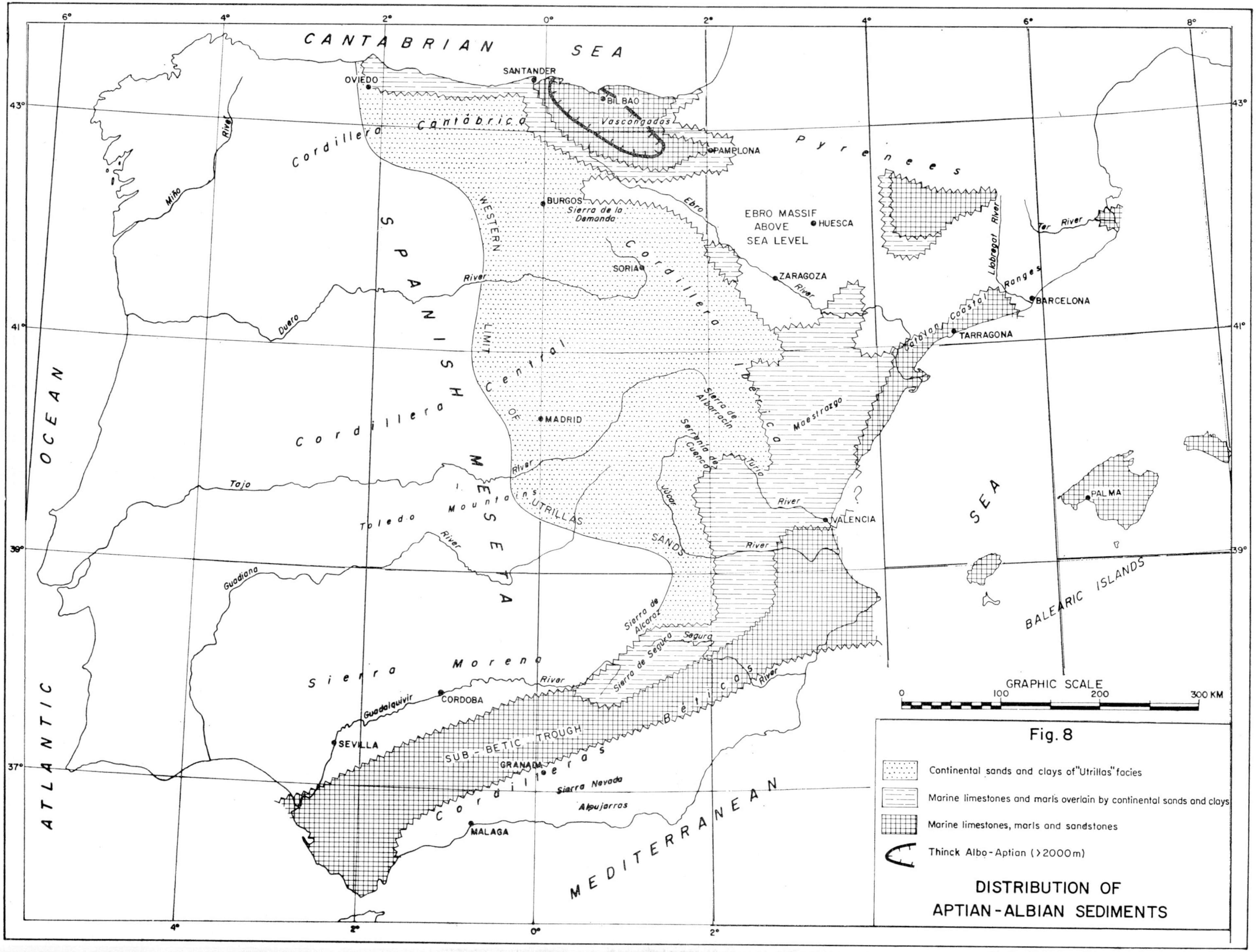

Fig. 8

DISTRIBUTION OF
APTIAN-ALBIAN SEDIMENTS

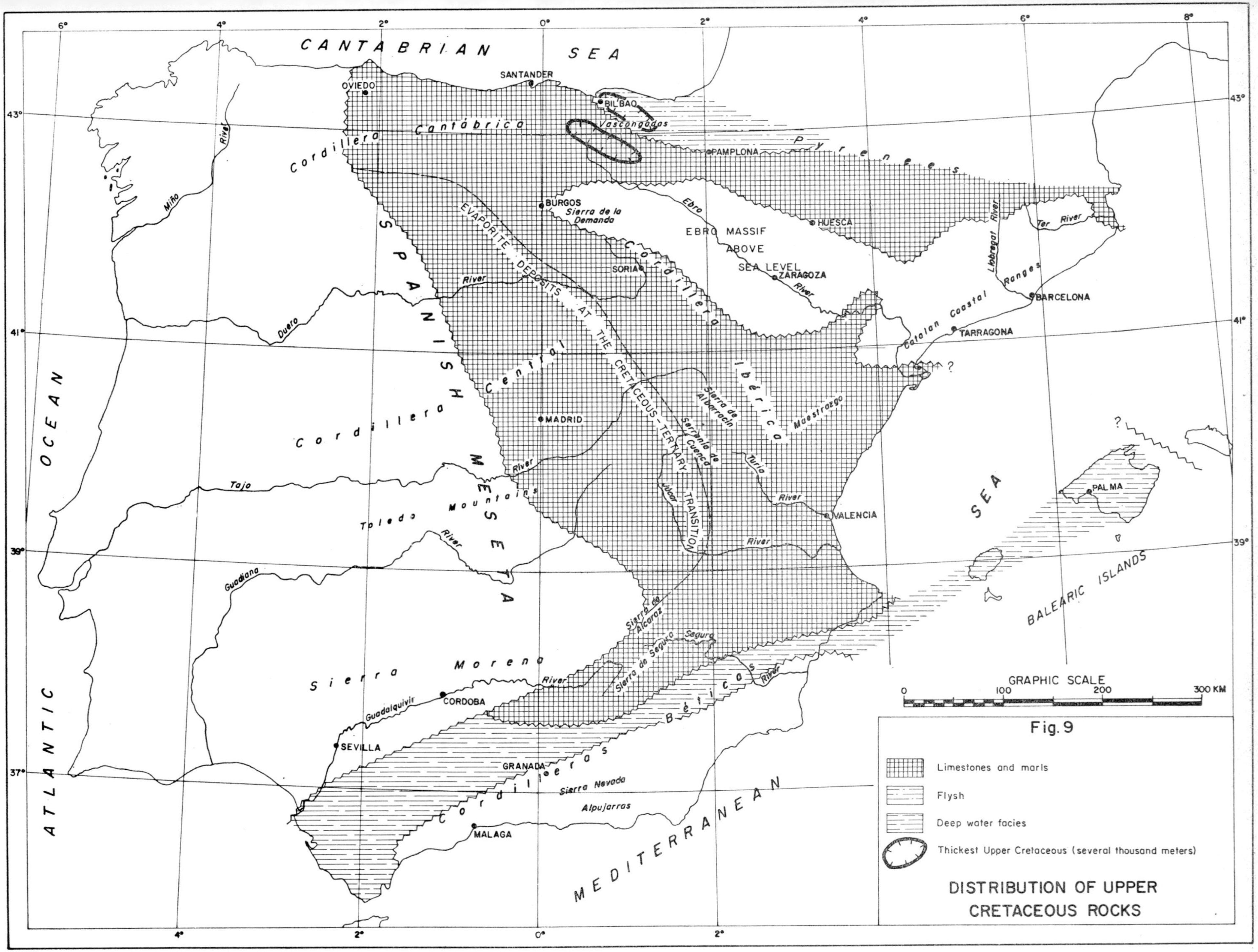

Fig. 9
DISTRIBUTION OF UPPER CRETACEOUS ROCKS

sediments were deposited to the west. Marine limestones, marls, and "flysch" represent the younger Eocene from the Cantabrian to the Mediterranean sea. These sediments are the result of early Alpine movements.

Marine sediments are thinner along the trough north of the Cordilleras Beticas. Marine limestones and marls predominate in the eastern areas, and "flysch" facies represent much of the Eocene in the west.

Oligocene-Miocene-Pliocene (Fig 11). The main Alpine movements in Spain took place between the Upper Eocene and the Lower Miocene. The Alpine mountains of Spain were raised above sea level, and continental sediments accumulated in a number of deep internal basins, all of them with thicknesses of the order of thousands of metres (Ebro, Castilla la Vieja, and Castilla la Nueva Basins). Large-scale diapirism of the Triassic salt complicated the structure of the Mesozoic sediments.

In northern and central Spain continental clastics were deposited with thick conglomerates near the edges of the rising mountains. Evaporitic conditions are common in the centre of these basins.

Marine sedimentation prevailed during the Miocene in southern Spain and along the eastern coast. The uplift of the Cordilleras Beticas is reflected in the deposition of olistostromes interbedded with fine clastics of authtocthonous nature. Marine clays with subordinate limestones and sandstones constitute the post-orogenic sediments of the Guadalquivir, Levante, Murcia, and Almeria Basins.

Pliocene vertical movements have been important. They have raised the mountains to their present altitudes above sea level. Also the whole of the Iberian Peninsula was tilted towards the west. As a result, the interior basins of the Meseta were drained to the west by the Duero, Tajo, and Guadiana rivers. Small consequent streams developed in the mountains bordering the Mediterranean Sea, and one of them drained the Ebro lake after eroding its way through the Catalan Coastal Ranges.

PETROLEUM EXPLORATION

Table III summarizes petroleum exploration activities in Spain.

Scientific methods had little or no application before World War II; generally, shallow wells were

TABLE III

Exploration activities in Spain

Year	Seismic	Exploratory wells		Development wells	
	Party months	No. of wells spudded	Metres drilled	No. of wells spudded	Metres drilled
		15 wells drilled before 1941	10,500 m before 1941		
1941		1	445		
1942		1	602		
1943	35 party-months	—	—		
1944		1	820		
1945		1	918		
1946		—	—		
1947	before 1951	2	2,970		
1948		1	2,211		
1949		1	1,264		
1950		1	1,610		
1951	4	1	1,221		
1952	6	4	6,523		
1953	6	5	5,628		
1954	20	4	6,914		
1955	10	3	7,295		
1956	56	14	24,308		
1957	36	13	20,299		
1958	48	12	23,576		
1959	36	18	35,834		
1960	43	12	24,817		
1961	101	15	38,509		
1962	102	19	40,673		
1963	64	20	41,956		
1964	50	11	25,684	3	4,377
1965	73	15	31,404	15	23,772
1966	41	18	35,711	12	17,677
1967	33	18*	35,489	1	2,312
1968 (1st half)	16	8†	15,217		
Total to 30 June 1968	780	234	442,398	31	48,138

* Three still drilling on 30 June 1968.

† Five still drilling on 30 June 1968.

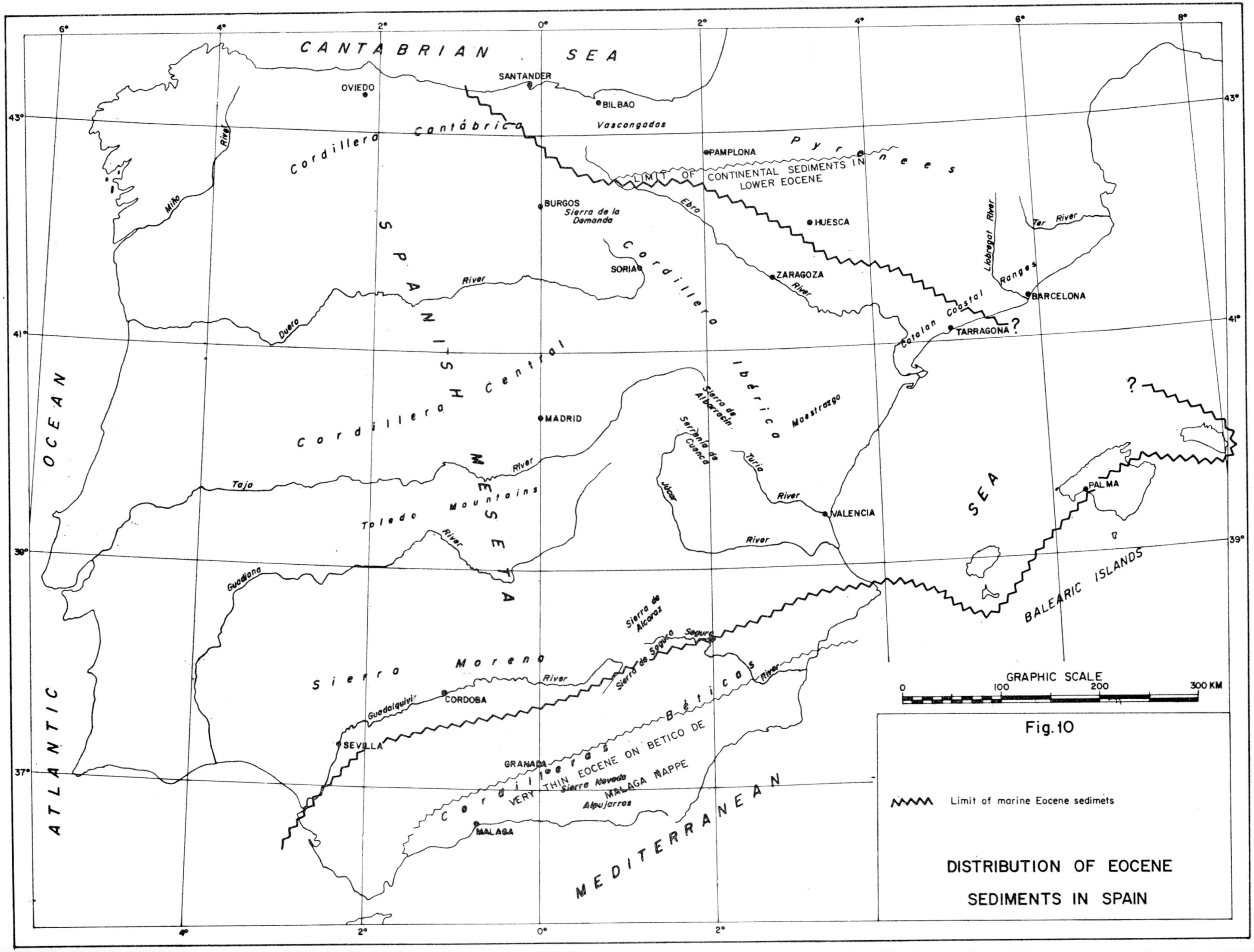

Fig.10

DISTRIBUTION OF EOCENE SEDIMENTS IN SPAIN

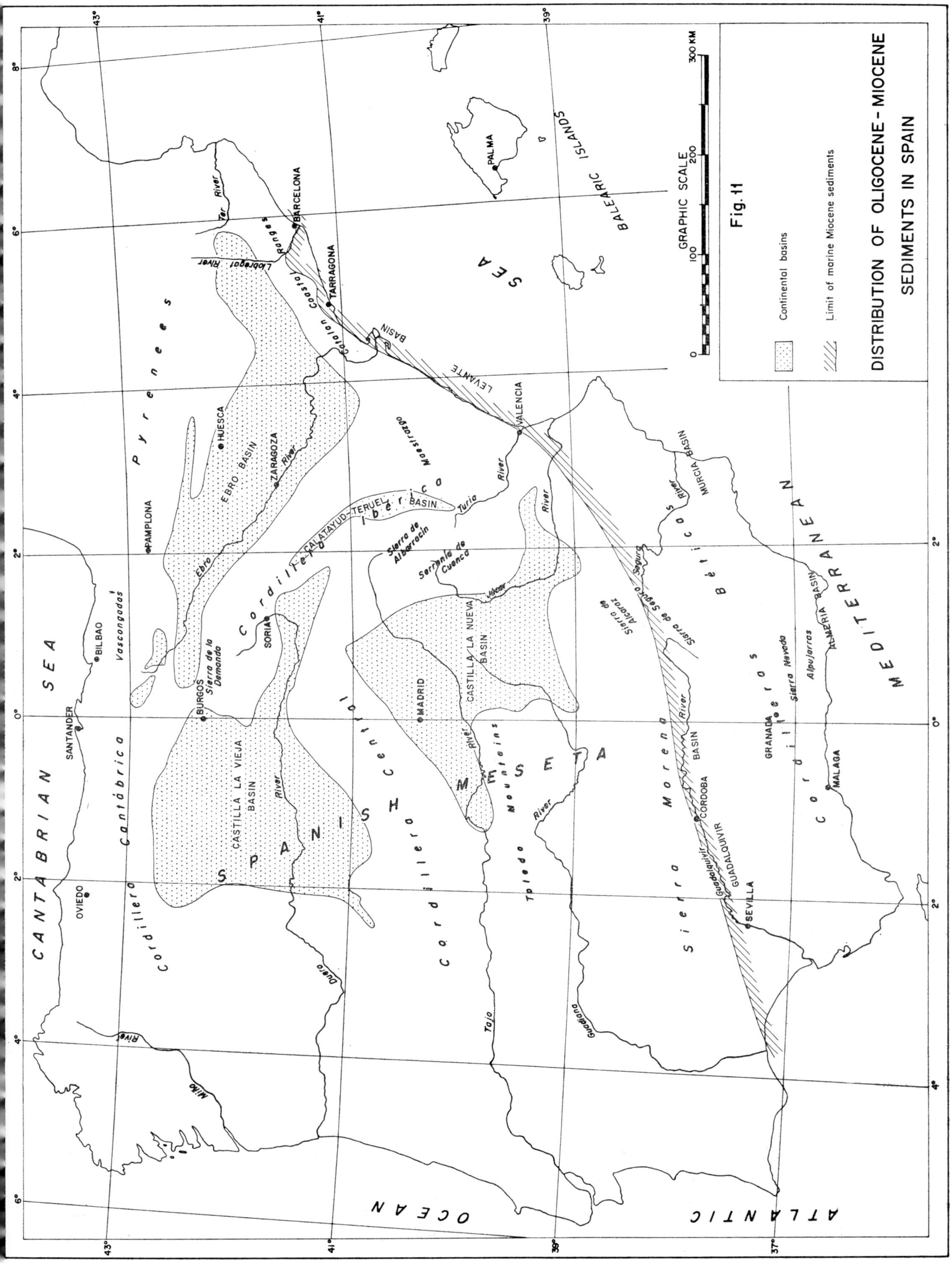

Fig. 11

DISTRIBUTION OF OLIGOCENE - MIOCENE SEDIMENTS IN SPAIN

drilled near oil-sand outcrops or seeps. Petroleum exploration rights were granted under the mining law. Between 1940 and 1959 CAMPSA, CIEPSA (with Mobil as partner, later Deilmann), and INI (associated to General American in 1952) were the sole explorers.

The Hydrocarbon Act of 1958 provided for reservation of large areas to INI, with French companies participating as operators of the exploration effort.

The remainder of the country was opened to other companies, resulting in major- to intermediate-sized international oil companies obtaining exploration permits. The law provided for granting of permits of up to 40,000 ha for a six-year term, with fixed financial obligations. After the initial six-year period the permits were renewable for three more years, with 25 per cent reduction of the area. Additional extensions of two years (with another 25 per cent reduction of area) and three years were left to the judgment of the Administration. Exploitation concessions could be granted for up to 50 per cent of the exploration permit area.

Intensive exploration was started by Spanish, American, and French companies, chiefly in northern Spain. Lack of success during the initial six-year period of exploration resulted in relinquishment of most of the permits at or before expiration of their initial term. The results in the Ebro Tertiary Basin and on the south border of the Pyrenees were particularly disappointing. The best hydrocarbon indications were found north of Burgos, where the Ayoluengo field was discovered in 1964.

Continued work since 1964 has been largely confined to north-central Spain, with extension of efforts to southeastern areas and the Guadalquivir valley, involving associations of CIEPSA and INI with Petrofina and French companies. The most significant trend of recent years was the acquisition of offshore blocks by Shell-CAMPSA and INI-Coparex-Continental, and the drilling of the first two offshore wells.

Table IV shows the distribution of exploratory wells according to their total depths.

TABLE IV

Total depth of exploratory wells in Spain as of 30 June 1968

Total depth in metres	No. of wells
0– 500	23
500–1000	39
1000–1500	28
1500–2000	45
2000–2500	35
2500–3000	19
3000–3500	23
3500–4000	9
4000–4500	7
4500–5000	4
more than 5000	2
Total	234

The only producing field in Spain is Ayoluengo (see description below). Non-commercial oil production has been obtained in the Wealden of the Tozo and Huidobro wells. The recent Hontomin No. 2 discovery well produces oil from Jurassic carbonates. Sub-commercial quantities of gas are produced in the Upper Cretaceous of two Castillo wells (Fig 1).

The Ayoluengo field (see location on Fig 1) was discovered by CAMPSA-Amospain in 1964. The section drilled by the discovery well, Ayoluengo No. 1, was as follows:

0–248 m. Upper Cretaceous marine limestones and marls. The oldest fossils are Cenomanian.
248–610 m. Albian (Utrillas facies), predominantly composed of loose sands, fresh-water bearing. This formation overlies unconformably the section below.
610–924 m. Wealden, continental, azoic, consisting of vari-coloured clays with some sandstones and dolomites. Pyrite and coal are common. The formation contains fresh to brackish water. Sandstones are common at the base of this unit, and an unconformable relationship to the underlying formation is suspected.
924–1115 m. Wealden, with a brackish fauna of ostracods of Purbeckian age. The section consists of dark grey shales with some sandstones and limestones. A resistive limestone at the base of this unit constitutes an excellent marker for correlation (base of Unit A of Fig 13). Water salinities are of 15,000–20,000 ppm. Some oil is produced in the field from the sandstones and limestones of this unit.
1115–1580 m. Wealden, azoic vari-coloured clays, with anhydrite and some dolomites, of continental or deltaic origin. Irregular sand bodies in this unit contain most of the oil in the field. Water salinities are in the order of 50,000 ppm. A calcarenite with silt to pebble-size grains of limestone in an argillaceous or calcareous matrix is found at the base of this sequence and strongly suggests an unconformable relationship with the underlying marine Jurassic. This section constitutes Units B and C of Ayoluengo geologists (scc Fig 13).
1580–2375 m. Jurassic, marine carbonates and marls. Some light oil has been recovered in a calcarenite within this unit. Anhydrites and dolomites (Carniolas) constitute the basal part of this unit. The contact with the Keuper is gradational.
2375–2397 m. Keuper anhydrite, salt, and dolomites.

The reservoirs at Ayoluengo are mainly sands, with very variable porosities (average 18 per cent). Permeabilities are also very variable. A limestone in the Purbeckian produces where fractured. The field is divided in a number of reservoirs by faulting and the discontinuous nature of the sand bodies.

The Ayoluengo field is located on the culmination of a large domal uplift (Figs 12 and 13). It is cut by several faults, giving a collapse graben in the central part. Growth of the Ayoluengo anticline can be proved during the Wealden, the collapse features and faulting being probably of Tertiary age. It is quite possible that the field is underlain by a Triassic salt dome.

The gravities of the Ayoluengo oil become lighter with depth, with a range of 20° to 39° API. The oil

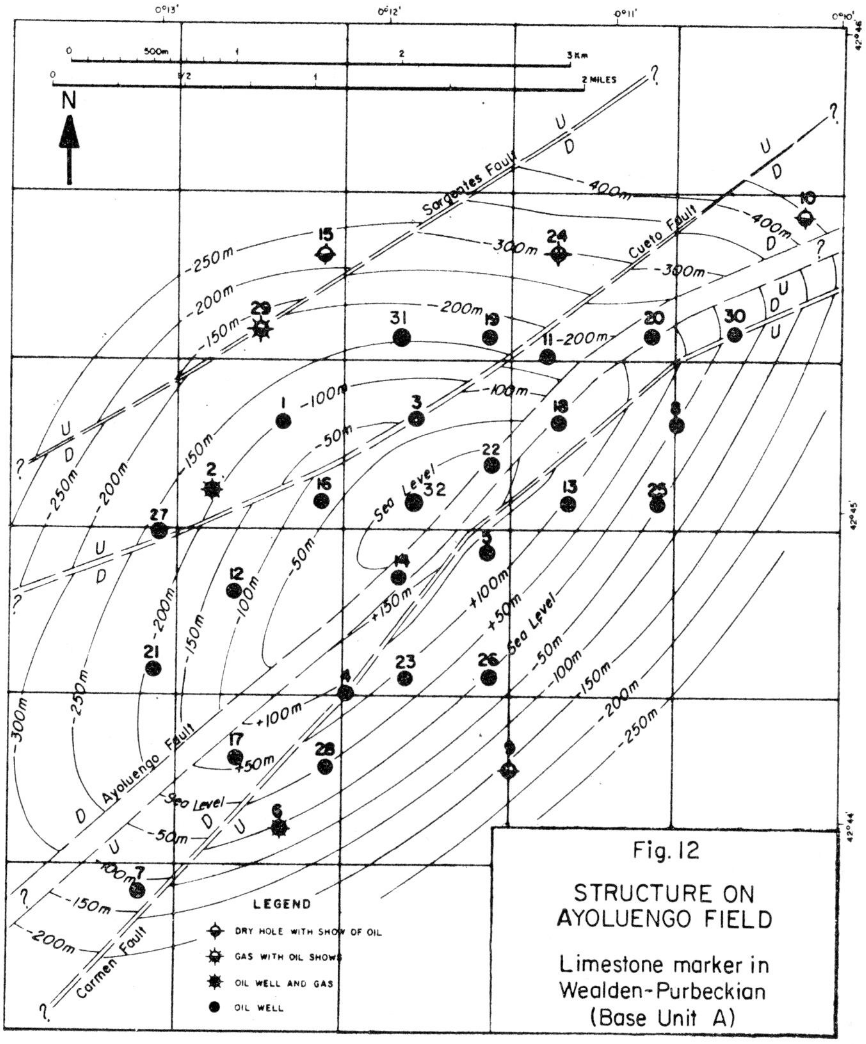

Fig. 12
STRUCTURE ON AYOLUENGO FIELD
Limestone marker in Wealden-Purbeckian (Base Unit A)

has a high wax content, as is the case with much petroleum found in continental sediments. The sulphur content is low, about 0·17 per cent. On the other hand, the arsenic content is high (22–24 ppm) and a desalting plant is needed to reduce it. Total recoverable reserves have been officially estimated at 2 million tons. Cumulative production at different dates is indicated below:

1 Jan. 1966	51,955 brl	(Approx. 6900 tonnes)
1 Jan. 1967	249,208 brl	(Approx. 33,100 tonnes)
1 Jan. 1968	846,791 brl	(Approx. 112,500 tonnes)
1 July 1968	1,364,154 brl	(Approx. 181,500 tonnes)

PETROLEUM PROSPECTS

The oldest formations with possibilities of commercial accumulations of hydrocarbons are in the Carboniferous. Coal-bearing sand-shale sequences of this age are known for their gas shows. At least one well has been drilled near Oviedo to explore the Carboniferous possibilities.

The Triassic has good prospects for accumulated hydrocarbons. Oil shows are known in the Muschelkalk, which probably constitutes a good source rock, but generally lacks porosity. The clastic Buntsandstein could provide an excellent reservoir either for Muschelkalk oil or for Carboniferous gas. Salt from the Upper Triassic constitutes an excellent cap rock. It should be remarked that structure on the Lower or Middle Triassic should be considerably more simple than in the younger sediments, since movement of the Keuper salt provides for additional complications above it. It is hoped that the new seismic methods will soon be able to map below the Keuper and that pre-salt structures will be drilled in the future.

The best shows of oil and gas in Spain have been found in the Jurassic and in the Lower Cretaceous. Good source rocks are found in the marine Jurassic and in the Purbeckian-Neocomian, while there are reservoirs in the Jurassic carbonates and in the Wealden and Albian sands. Several wells in northern Spain had small productions or good shows in these sediments. Asphalt shows are found in outcrops of the Wealden or in the Albian in a number of localities in north-central Spain. It appears most likely that new

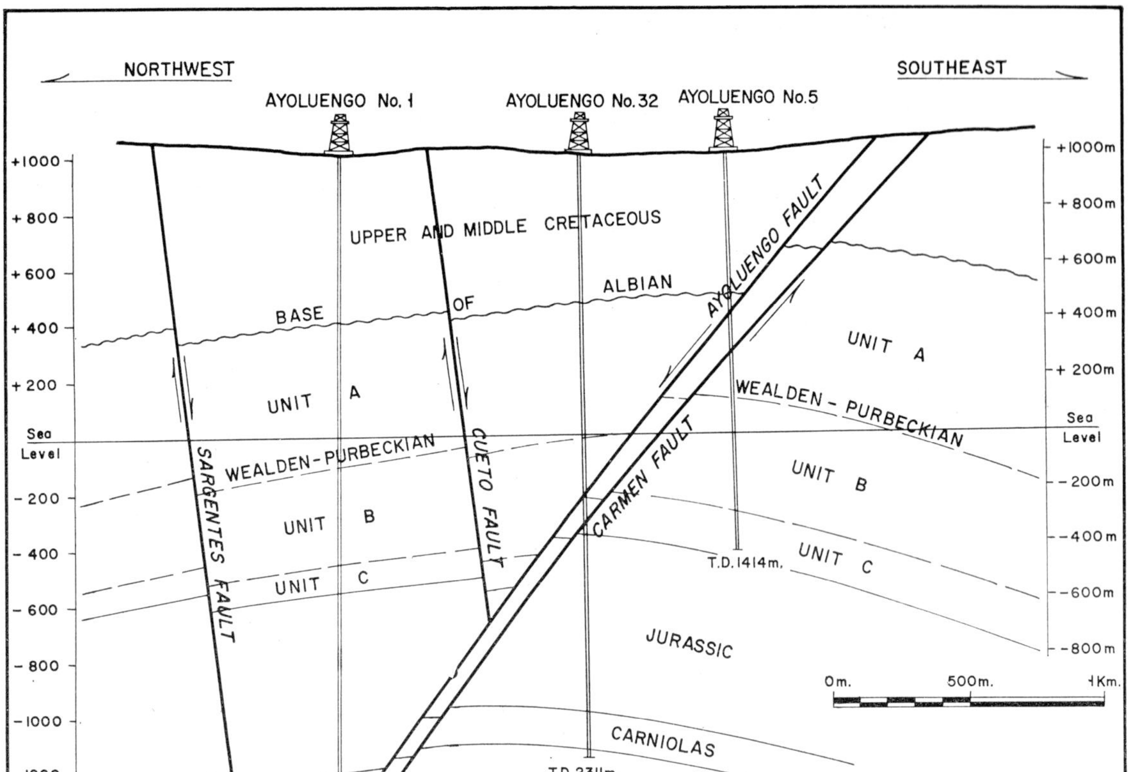

Fig. 13
AYOLUENGO FIELD
NW - SE SECTION

discoveries of small or moderate size will be made in the Jurassic and Lower Cretaceous of Spain.

Gas shows are common in the Upper Cretaceous and Eocene of northern Spain, where source rocks appear to be present. If adequate porosity is found in these formations, gas fields of economic value could be found.

Little exploration has been carried out in offshore areas of eastern Spain, or in onshore and offshore areas of southern Spain. The marine Miocene has gas shows in several localities. Further exploration might be able to discover hydrocarbons either in the Miocene or in the underlying Mesozoic.

Exploration activities in Spain have declined from a peak in 1961–63, in spite of the discovery of the Ayoluengo field in 1964. This is the result of the poor ratio of recoverable reserves to exploration investments (over $100 million). In spite of that, petroleum exploration in Spain should be continued actively, since hydrocarbon consumption represents nearly half of the energy requirements of the country and about 7 per cent of the value of all imports. Adequate incentives in the form of tax exemptions or direct subsidies would help considerably to maintain an active exploration effort.

It should be remarked that gas fields, even if found in deep traps, would easily be commercial in Spain. Thus gas exploration is still a very attractive venture. It appears likely that future activities in Spain will be directed towards areas inaccessible to older exploratory techniques, such as deep structures, offshore plays, areas of difficult geophysical mapping, and stratigraphic traps.

SUPPLY OF FUTURE HYDROCARBON REQUIREMENTS

Growth of consumption cannot be met by expectable indigenous supplies. Petroleum exploration in the Spanish Sahara and Equatorial Guinea has not been successful so far. As a result, the government has encouraged Spanish companies to engage in exploration for oil in North Africa and the Middle East. Esso has contracted to supply LNG from Libya.

ACKNOWLEDGMENTS

The writer was a geologist with American Overseas Petroleum (Spain) Ltd from 1959 to 1968. He is grateful for the opportunity provided by his work during that period to familiarize himself with the aspects of geology of the Iberian Peninsula related to petroleum exploration.

BIBLIOGRAPHY

Alastrué, E., Almela, A., and Ríos, J. M. "Mapa Geológico de la Provincia de Huesca. Escala 1 : 200.000." Madrid, Inst. geol. y min. España, 1957.

Almela, A., and Ríos, J. M. "Mapa Geológico de la Provincia de Lérida. Escala 1 : 200.000." Madrid, Inst. geol. y min. España, 1947.

Almela, A., and Ríos, J. M. "Investigación del Hullero bajo los Terrenos Mesozoicos de la Costa Cantábrica (Zona de Oviedo-Gijón-Villaviciosa- Infiesto)". Madrid, Empr. nac. Adaro, 1962.

Ashauer, H. "Die östlische Endingung der Pyrenäen." Abh. Ges. Wis. Got., Mat. Ph. Kl., vol. III, h. 10, 115 p., 1934.

Ashauer, H., and Teichmüller, R. "Die Variscische und Alpidische Gebirgsbildung Kataloniens." Abh. Ges. Got. Mat. Ph. K 1., vol III, h. 17, 78 p., 1935.

Bard, J. P. "Introduction a la Géologie de la Chaîne Hercynienne dans la Sierra Morena Occidental (Espagne). Hypothéses sur les caractéres de l'évolution géotectorique de cette chaîne" Paris, Revue de Géographie Physique et de Géologie Dynamique, vol. VII. Fasc 4, 1965.

Beuther, A., Dahm, H., Kneuper, F., Mensink, H., and Tischer G. "Der Jura und Wealden in Nordost-Spanien." Beih. Geol. Jb., h. 44, 1966.

Blumenthal, M. *Bol. Inst. geol. y min. España*, 1949, **LXII**, 17–203.

Brinkmann, R. "Betikum und Keltiberikum in Sudostspanien." Beitr. geol. westl. Mediterr., n°. 6, 1931.

Brinkmann, R., and Gallwitz, H. "El-borde externo de las cadenas Béticas en el Sureste de España." Madrid, Publicaciones extranjeras sobre geología de España Inst. "Lucas Mallada", Cons. Sup. Inv. Cient., vol. V, pp. 171–290, 1933.

Busnardo, R. "Regards sur la géologie de la Région de Jaen," *in* "Livre a la mémoire du Professeur Paul Fallott." Paris, Soc. géol. France, t.I, pp. 189–98, 1962.

Canerot, J. "Découverte de l'Albien marin et paléogéographie des Cretacé dans le Maestrazgo nord-oriental (Espagne)." C. R. somm. S.G.F., pp. 182–83, 1967.

Capdevilla, R. "Sur la géologie du Précambrien et du Paléozoique dans la région de Lugo et la question des plisements assyntiques et sardes en Espagne." Notas y Com. Inst. geol. y min. España, n°. 80, pp. 157–75, 1965.

Chauve, P. "Structure géologique de la région centrale de la Province de Cadix, Espagne," *in* "Livre a la mémoire du Professeur Paul Fallot." Paris, Soc. geol. France, t. I, pp. 257–64, 1962.

Ciry, R. *Bull. Soc. Hist. nat. Toulouse*, 1940, **74**, 519.

Colchen, M. *Bull. Soc. Géol. France*, 1966, **8** (7), 87–97.

Colom, G., and Escandell, B. "L'évolution du Géosynclinal Baléare," *in* "Livre a la mémoire du Professeur Paul Fallot." Paris, Soc. géol. France, t. I, pp. 125–30, 1962.

Compte, P. "Recherches sur les terrains anciens de la Cordillére Cantabrique." Mem. Inst. geol. y min España, t. LX, 1946.

Crusafont, M., Truyols, J., and Riba, O. "Contribución al conocimiento de la estratigrafía del Terciario Continental de Navarra y Rioja." Notas y Com. Inst. geol. y min. España, n°. 90, pp. 53–77, 1966.

Darder Pericás, B. *Bol. Inst. geol. y min. Espana*, 1945, **LVII**, 69–362.

Didon, J. "Les unités ultra-bétiques de la zone du Flysch gaditan au Nord et Nord-Est d'Algesiras," *in* "Livre a la mémoire du Professeur Paul Fallot." Paris, Soc. géol. France, t. I, pp. 265–72, 1962.

Didon, J. *Bull. Soc. géol. France* 1966, **8** (4), 521–27.

Dupuy de Lome, E. *Bol. Inst. geol. y min. España*, 1965, **LXXVI**, 23–76.

Dürr, S., Hoeppener, R., Hoppe, P., and Kockel, F. "Geologie du Montagnes entre le rio Guadalhorce et le Campo de Gibraltar. (Espagne Meridionale)," *in* "Livre a la mémoire du Professeur Paul Fallot." Paris, Soc. géol. France, t. I, pp. 209–27, 1962.

Egeler, C., and Booy, T. "Signification tectonique de la présence d'elements du Bétique de Malaga dans le sudest des Cordillères Bétiques," *in* "Livre a la mémoire du Professeur Paul Fallot." Paris, Soc. géol. France, t. I, pp. 155–63, 1962.

Evers, H. T. "Geology of the Leonides between the Bernesga and Porma rivers, Cantabrian Mountains, NW Spain." Leidse geol. Med., d. 41, pp. 83–151, 1967.

Fallot, P. "Estudios geológicos en la zona Subbética entre Alicante y el río Guadiana Menor." Madrid, Inst. "Lucas Mallada", Cons. Sup. Inv. Cient, 1948.

Fallot, P. "Les Cordillères bétiques." Madrid, Est. geol., n°. 8, 1948.

Fallot, P., and Darder, B. "L'ille de Majorque." XIV Cong. Geol. Int. Liv-Guide, 113 p., 1926.

Felgueroso, C., and Coma, J. "Estudio Hidrogeológico de una zona de la provincia de Guadalajara." Notas y Com. Inst. geol. y min. España, n°. 71, pp. 213–41, 1963.

Felgueroso, C., and Coma, J. *Bol. Inst. geol. y min. España*, 1964, **LXXV**, 115–209.

Fontboté, J. M., and García Dueñas, V. "Essai de systématisation des unités subbétiques allochtones dans le tiers central des chaînes bétiques." C.R. Ac. Sc, t. 266, pp. 186–89, 1968.

Fontboté, J. M., and Guitard, G. *Bull. Soc. géol. France*, 6° serie, **VIII**, 884–90.

Foucault, A. "Problemes paléogéographiques et Tectoniques dans le Prébétique et le Subbétique sur la transversale de la Sierra Sagra (prov. de Granade, Espagne)," *in* "Livre a la Mémoire du Professeur Paul Fallot." Paris, Soc. géol. France, t. I, 175–83, 1962.

Foucault, A. "Relación entre las zonas Prebética y Subbética entre Cazorla (Jaen) y Huéscar (Granada)." Notas y Com. Inst. geol. y min. España, n°. 83, pp. 71–77, 1966.

García Fuente, S. *Bol. Inst. geol y min. España*, 1952, **LXIV**, 345–63.

García Fuente, S. *Bol. Inst. geol. y min. España*, 1953, **LXV**, 271–324.

García Rodrigo, B. "Sur la structure du pré-bétique au nord d'Alicante (Espagne)," *in* "Livre a la Mémoire du Professeur Paul Fallot." Paris, Soc. géol. France, t. I, pp. 137–43, 1962.

Gavala, J. "Mapa geológico de la Provincia de Cadiz." Madrid, Inst. geol. y. min. España.

Guerin-Desjardins, B., and Latreille, M. *Bol. Inst. geol. y min España*, 1962, **LXXIII**, 329–70.

Helmig, H. M. "The Geology of the Valderueda, Tejerina, Ocejo, and Sabero Coal Basins (Cantabrian Mountains, Spain)," Thesis, Leiden, 1966.

Hollister, J. "La posición de las Baleares en las orogenias varisca y alpina." Madrid, Publicaciones alemanas sobre geología de España, Inst. "Lucas Mallada," Cons. Sup. Inv. Cient., Vol I, pp. 71–102, 1942.

Instituto Nacional de Colonización. "Mapa geológico general de la Cuenca del Duero," 1966.

Julivert, N. *Bol. Inst. geol. y min. España*, 1960, **LXXI**, 1–347. *Bol. Inst. Geol. y min. España*, 1963, **LXXIV**, 87–170.

Kanis, J. "Geology of the Eastern zone of the Sierra del Brezo (Palencia, Spain)." Leidse geol. Med., d. 21, Af 1.2, 377–445, 1956.

Karrenberg, H. "Die Postvariszische Entwicklung des Kantabroasturischen Gebirges (Nord-West Spanien)." Abh. Ges. Wis. Got., Mat. Ph. Kl., vol. III, h. 11, 104 p., 1934.

Koopmans, B. "The Sedimentary and Structural History of the Valsurvio Dome." Leidse geol. Med., d. 26, 121–232, 1962.

Lamare, P. "Recherches géologiques dans les Pyrénées basques d'Espagne." Mém. Soc. géol. France, nouv. sér, t. 12, n°. 27, 465 p., 1936.

Llopis Lladó, N. "Sur la paléogéographie du Devonien du Nord de l'Espagne." C.R. somm. S.G.F., 290–92, 1965.

Lotze, F. "Observaciones respecto a la división de las Variscides de la Meseta Ibérica." Madrid, Publicaciones extranjeras sobre geología de España, Inst "Lucas Mallada", Cons. Sup. Inv. Cient., t. V, 149–66, 1950.

Lotze, F. "Estratigrafía y tectónica de las cadenas Paleozoicas Celtibéricas." Madrid, Publicaciones extranjeras sobre geología de España, Inst. "Lucas Mallada ", Cons. Sup. Inv. Cient., t. VIII, 1955.

Lotze, F. "Mapa Geológico de la zona W de los Pirineos y E de la Cordillera Cantábrica." Compañía Petrolífera Iberica, S.A., 1958.

Lotze, F. "Das Kambrium Spaniens." Wiesbaden, Akademie der Wissenschaften und des Literatur in Mainz, 1961.

Mangin, J. P. *Bull. Soc. géol. France*, 1959, **1** (7), pp. 16–30.

Mangin, J. P. "Le Nummulitique sud-pyrénéen á l'Ouest de l'Aragon." Zaragoza, Pirineos, n°. 51–58, Cons, Sup. Inv. Cient., 1960.

Martínez Alvarez, J. A. "Estudio Geológico del reborde oriental de la Cuenca Carbonífera central de Asturias." Oviedo, Instituto de Estudios Asturianos, 1962.

Martínez Alvarez, J. A. "Rasgos geológicos de la zona oriental de Asturias." Oviedo, Instituto de Estudios Asturianos, 1965.

Misch, P. "Die Bau der Mittleren Südpyrenaen," Abh. Ges. Wis Got., Mat. Ph. Kl., vol. III, h. 12, 1597–764, 1934.

Navarro, A., and Trigueros, E. "Mapa geológico de la provincia de Murcia (España)." Madrid, Inst. geol. y min. España, 1966.

Navarro, A., Trigueros, E., Villalón, C., and Ríos, J. M. "Derniers progres dans la connaissance de l'extrémité Nord-Est du Chaînes subbétiques (Région d'Altea-Benisa prov. d'Alicante, Espagne)," *in* "Livre a la Mémoire du Professeur Paul Fallot." Paris, Soc. géol. France, t. I, 143–55, 1962.

Nederlof, M. "Structure and sedimentology of the Upper Carboniferous of the Upper Pisuerga Valley, Cantabrian Mountains, Spain." Leidse geol. Med., d. 24, 604–703, 1959.

Paquet, J. "Nota preliminar sobre la Prebética y la Subbética al N. de la Sierra Espuña (Murcia, España)." Notas y Com. Inst. geol. y min. España, n°. 74, 185–95, 1966.

Parga Pondal, I., Matte, P., and Capdevila, R. "Introduction a la géologie de 'l'Ollo de Sapo,' formation porphyroide antesilurienne du nord-ouest de l'Espagne." Notas y Com. Inst. geol. min. España, n°. 76, 119–55, 1964.

Pastor Gomez, V. *Bol. Inst. geol. y min. España*, 1965, **LXXVI**, 1–22.

Perconig, E. "Sur la constitution géologique de l'Andalusie occidentale, en particulier du bassin du Guadalquivir," *in* "Livre a la Mémoire du Professeur Paul Fallot." Paris, Soc. géol. France, t. I, 229–56, 1962.

Peyre, Y. "Etat actuel de nos connaissances sur la structure des Cordillères Bétiques sur la transversale de Malaga," *in* "Livre a la Mémoire du Professeur Fallot." Paris, Soc. géol. France, t. I, 199–209, 1962.

Radig, F. "Stratigraphie und tektonik der Asturichen Kuste zwischen San Esteban de Pravia und dem Cabo Torres (Prov. Oviedo, Nord Spanien)." Thesis, Münster, 1958.

Rat, P. "Les pays Crétacés Basco-Cantabriques (Espagne)." Publ. Univ. Dijon (Presses Universitaires de France), t. 18, 525 p., 1959.

Riba, O. "Estudio geológico de la Sierra de Albarracín." Madrid, Monogr n°. 16, Inst. "Lucas Mallada", Cons. Sup. Inv. Cient, 1959.

Richter, G. "Die Iberischen Ketten zwischen Jalon und Demanda." Beitr. geol. Mediterr., vol. 5, h. 3, 1930.

Richter, G., and Teichmüller, R. "Die Entwicklung der Keltiberischen Ketten." Abh. Ges. Got., Mat. Ph. Kl., Vol. III, h. 7, 1067–186, 1933.

Ríos, J. M. "Relación de los principales sondeos para investigación llevados a cabo en España desde 1939." Notas y com. Inst. geol. y min. España, n°. 50, 1st fasc, 47–73, 1958.

Ríos, J. M. "Relación de los principales sondeos para investigación de petroleos llevados a cabo desde 1939 hasta Mayo de 1960." Notas y Com. Inst. geol. y min. España, n°. 59, 133–66, 1960.

Ríos, J. M. "Relación de las principales actividades para investigación de hidrocarburos llevadas a cabo en España durante 1960." Notas y Com. Inst. geol. y min. España, n°. 63, 67–102, 1961.

Ríos, J. M. "Relación de las principales actividades para investigación de hidrocarburos llevadas a cabo en España durante 1961." Notas y Com. Inst. geol. y min. España, n°. 65, 143–88, 1962.

Ríos, J. M. "Relación de las principales actividades para investigación de hidrocarburos llevadas a cabo en España durante 1962." Notas y Com. Inst. geol. y min. España, n°. 70, 147–200, 1963.

Ríos, J. M. "Relación de las principales actividades para investigación de hidrocarburos llevadas a cabo en España durante 1963." Notas y Com. Inst. geol. y min. España, n°. 75, 183–234, 1964.

Ríos, J. M. "Relación de las principales actividades para investigación de hidrocarburos llevadas a cabo en España durante 1964." Notas y Com. Inst. geol. y min. España, n°. 79, 89–148, 1965.

Ríos, J. M. "Relación de las principales actividades para investigación de hidrocarburos llevadas a cabo en España durante 1965." Notas y Com. Inst. geol. y min. España, n°. 87, 7–58, 1966.

Ríos, J. M. "Relación de las principales actividades para investigación de hidrocarburos llevadas a cabo en España durante 1966." Notas y Com. Inst. geol, y min. España, n°. 93, 7–68, 1967.

Ríos, J. M. "Relación de las principales actividades para investigación de hidrocarburos llevadas a cabo en España durante 1967." Madrid, Industria Minera (Boletin de Información del Consejo Superior de Minas de España), n°. 94, 5–75, 1968.

Ríos, J. M., and Almela, A. "Estudios sobre el Mesozoico del Borde Meridional de la cuenca del Ebro," *in* "Libro Jubilar." Inst. geol. y min. España, t. II, 247–300, 1951.

Ríos, J. M., Almela, A., and Garrido, J. *Bol. Inst. geol. y min. España*, 1945, **LVIII**, 45–228.

Ríos, J. M., García Fuente, S., and Martínez Peña, I., "Contribución al conocimiento de la geología de la region Manchega." Madrid, Mem. Empr. nac. Adaro, 1945.

Ríos., J. M., Garrido, J., and Almela, A. *Bol. Real. Soc. esp. Hist. nat.*, 1944, **42**.

Rupke, J. "The Esla Nappe, Cantabrian Mountains, Spain." Thesis, Leiden, 1966.

Saeftel, H. "Paleogeografía del Albense en las cadenas Celtibéricas de España." Notas y Com. Inst. geol. y min. España, n°. 63. p. 163–91, 1961.

Sanz, R. "Ayoluengo Field, Southwest Cantabrian Basin, North-Central Spain," *in* Proceedings Seventh World Petroleum Congress (Mexico, 1967): New York, Elsevier, Vol 2, pp. 251–8.

Schröder, E. "Das Grenzgebiete von Guadarrama und Hesperischen Ketten (Zentralspanien)" Abh. Ges. Wis. Got., Mat. Ph. Kl., vol 16, h. 3, pp. 687–748, 1930.

Simon, O. J. "Geological investigations in the Sierra de Almagro, South-eastern Spain." Thesis, Amsterdam, 1963.

Sitter, L. U. *Bol. Inst. geol. y min. España*, 1963, **LXXIV**, 393–451.

Sitter, L. U., and Boschma, D. "Explanation geological map of the Paleozoic of the Southern Cantabrian Mountains." 1 : 50.000, sheet 1, Pisuerga: Leidse geol. Med., d. 31, pp. 191–238, 1966.

Sitter, L. U., and Zwart, H. J. "Geological Map of the Palaeozoic of the Central Pyrenees," 1 : 50,000. Leidse geol. Med., d. 22, pp. 351–412, 1959.

Sitter, L. U. "Excursion to the Central Pyrenees, Sept. 1959." Leidse geol. Med., d. 27, pp. 191–236, 1961.

Sterp, N., "The Geology of the San Isidro-Porma Area (Cantabrian Mountains, Spain);" Leidse geol. Med. d. 39, pp. 55–128, 1966.

Viallard, P. "Sur le Crétacé de la Chaîne Ibérique Castillane entre le rio Turia et la haute vallée du rio Jucar (provinces de Valencia et Cuenca, Espagne)." C.R. somm. S.G.F., pp. 366–68, 1966

Virgili, C. *Bol. Inst. geol. y min. España*, 1958, **LXIX**, 856 p., 1958.

Wagner, R. H. *Bol. Inst. geol. y min. España*, 1963, **LXXIV**, 171–324.

Walter, R. "Resultados de las investigaciones geológicas en el NO de la provincia de Lugo (NO España)." Notas y Com. Inst. geol y min. España, nº. 89, pp. 7–17, 1966.

Wiedmann, J. "Contribution a la Paléogéographie du Crétacé vasco-gotique et celtibérique septentrional (Espagne)" *in* "Livre a la mémoire du professeur Paul Fallot." Paris, Soc. géol. France, t. I, pp. 351–67, 1962.

MS received 6 February 1969

Discussion

J. D. Moody (Mobil): What are the relations of time of folding of Majorca, the Betic Cordillera, and the Alps?

Ramon Querol: I think that the geological history of the Balearic Islands, especially in Majorca and Ibiza, is very similar to all the Cordilleras Beticas. Probably the early development of the S nappes is late Eocene in time, but the movement in the outer, more N zones is younger, so the latest movements are probably of Lower Miocene age.

Hydrocarbon Distribution in the Aquitaine Basin of SW France

By H. COUSTAU, J. GAUTHIER, G. KULBICKI, and E. WINNOCK

(*Société Nationale des Pétroles d'Aquitaine*)

SUMMARY

The only significant petroleum province in France is the Aquitaine Basin. Here various numerous important fields have been discovered. Their distribution does not seem to be erratic.

Stratigraphically, most of these fields are situated close to the Cretaceous/Jurassic unconformity. Geographically, they are restricted to two narrow areas of Lower Cretaceous sedimentation—the area of Parentis and the Lacq–St. Marcet area along the N Pyronean border.

A geochemical study of organic matter and trace elements indicates three main horizons variously favourable to the generation of hydrocarbons. Their importance is variable inside the basin. They are mostly Jurassic and to a lesser degree Lower Cretaceous. They seem to be at the origin of the known fields.

The exclusive occurrence of oil deposits in the area of Parentis, contrasting with the gas fields in the area of Lacq–St. Marcet, is a feature which has not been definitively explained. The degree of evolution of the organic matter, estimated on the base of the carbon ratio, and the diagenesis of the clay minerals are much higher in the Upper Jurassic of the Pyrenean border. This seems to be in connection with the occurrence of gas in this area.

INTRODUCTION

Up to the present, the Aquitaine Basin is the only significant oil-producing basin of France, judging by the number and the importance of the discovered pools. The distribution of these pools points out three important facts.

Geographically, the oil pools and the most important shows are all located in the zones of deepest subsidence of the basin, the Parentis Basin and the Lacq–St. Marcet Basin (Fig 3). These are also the only portions of the Aquitaine Basin which are covered by Lower Cretaceous sediments.

Stratigraphically, the majority of the fields and all of the most important shows are localized near the Cretaceous/Jurassic unconformity (Fig 4).

The exclusive occurrence of oil deposits in the Parentis Basin contrasts with the large predominance of gas fields in the Lacq and St. Marcet Basin.

The results of geochemistry studies, particularly the original work done in the SNPA laboratory at Pau, explain the relationships existing between the major geological facts and the distribution of the most important accumulations of hydrocarbons which have been discovered up to now in the basin.

GENERAL GEOLOGICAL SETTING

The Aquitaine Basin has a strong asymmetry, shallow and rising progressively to the N, and deepening strongly towards the S, with more than 10,000 m of sediments in front of the overthrust of the Pyrenean front.

North of a line from Arcachon to Toulouse, the Mesozoic series are reduced and formed of uniform platform-type sediments where Lower Cretaceous formations are absent.

On the contrary, S of this line (Fig 2), the Mesozoic series shows in time and space rapid and enormous variations of thickness.* This is controlled by changes in subsidence and important structural differentiations, largely due to the salt tectonics of the Triassic. Also their lithology becomes diversified. The series becomes complete with the deposition of the lower Cretaceous in two separated areas:

The Parentis Basin, S of Bordeaux, contains 117×10^6 tons of oil reserves in place, of which 39×10^6 are recoverable. In this basin, the Parentis field is by far the most important, with 27×10^6 tons of recoverable reserves and a production of 3600 tons/day in 1968.

The Lacq Basin, on the N flank of the Pyrenees, contains 300×10^9 cu m of gas, two-thirds of it in the Lacq field. This field, of which the recoverable reserves amount to 200×10^9 cu m and which produces 23×10 cu m/day, is remarkable by its high H_2S content, corresponding to 50×10^6 tons of sulphur reserves.

Concerning the stratigraphic segregation of hydrocarbons, an emphasis has been put in the past on the role of the Cretaceous:

"Cenomanian" breccia at St. Marcet, the first field discovered in Aquitaine, in 1939.
Neocomian at Lacq (1951).
Neocomian at Parentis (1954).
Albian at Cazaux (1959).

* This is shown particularly well on Fig 11. 4000 m of Lower Cretaceous of the Arzacq syncline pinch out northward and southward. In this last direction, beyond the Meillon ridge, the Lower Cretaceous is completely absent; on the other hand, the Upper Cretaceous, very thin N of the ridge, thickens beyond 5000 m to the S of the ridge.

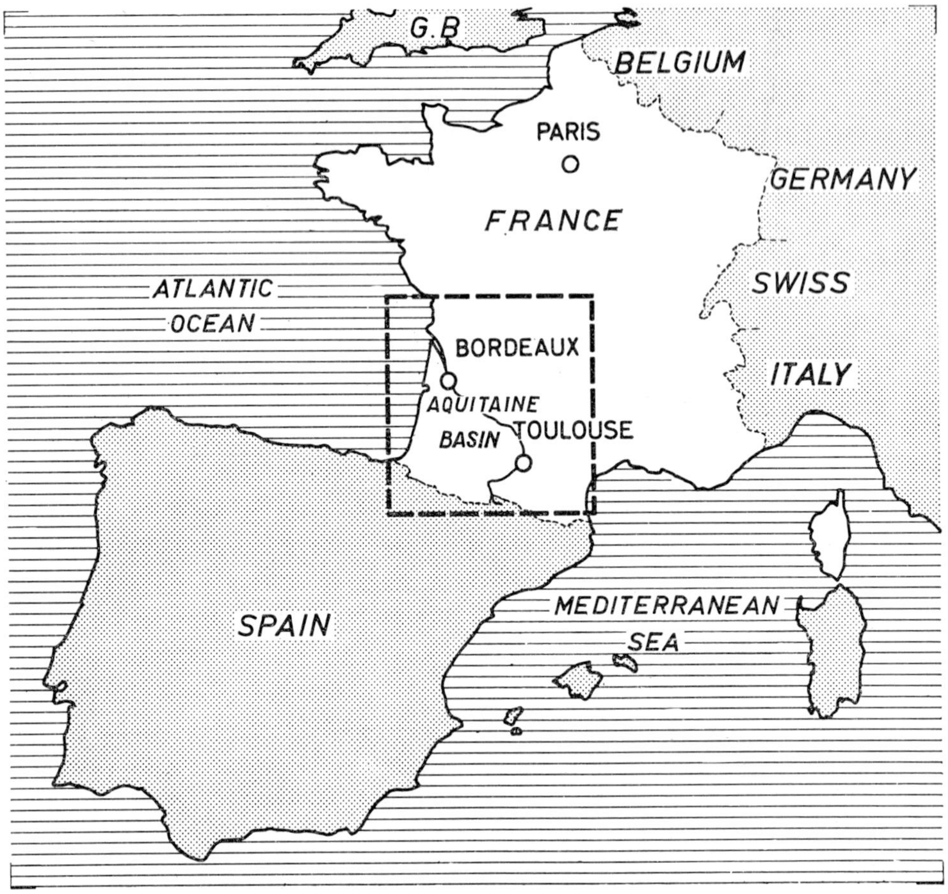

FIG 1. Key map.

In fact (Fig 4) at St. Marcet the most important part of the gas is in the Jurassic, the breccia of St. Marcet containing only the small part of the gas trapped at the base of the Upper Cretaceous cap. In the same way, in Lacq, the main part of gas is in the Jurassic and the sandy-dolomitic Purbeckian on top of it. In the new fields discovered near Pau*, only the Jurassic is productive. In the Parentis Basin, the oil is more likely at the base of the Lower Cretaceous, where several horizons of reservoir rocks exist but in the main fields the oil-water level goes down into the Jurassic.

Because of the surprisingly complex history of the basin, the vertical distribution of hydrocarbons is very variable from one field to another. It essentially depends on the cap rock qualities of the Cretaceous overlying the erosional unconformity. But altogether 90 to 95 per cent of the liquid or gaseous hydrocarbons discovered in Aquitaine are in direct connection with this Cretaceous/Jurassic unconformity†.

* Meillon–Rousse (1965).

† Only two fields are an exception to this rule:

Upper Lacq: oil has been found in the Lower Senonian, under the unconformity of Ypresian blue marls. The origin of this oil is either *in situ*, or from the sub-reefal Lower Cretaceous known on the anticline of Lacq. Hydrocarbon shows have been found also at these levels in the area of Pau.

Upper Cazaux: the relatively important oil reserves known in the Albian of Cazaux are not directly connected with the Cretaceous/Jurassic unconformity. Anyway, the oil of the Albian is of the same family as those of the Aptian-Neocomian on which the Albian is laying unconformably.

SOURCE ROCK CONCEPT

It seems useful to briefly recall the conditions which a rock must satisfy to be considered as a source rock for hydrocarbons.

The first essential feature, held by the authors, is evidently that the rock must contain a certain amount of residual organic matter. This means getting from a simple analysis a clearly quantitative criterion (content of organic carbon) to characterize the petroligenous value of a formation. But this criterion is not sufficient to solve the difficult problem of recognition of the source rocks. In fact, the quantity of hydrocarbons formed is a function of the quantity of organic matter deposited, but it also depends upon other parameters like the nature and the origin of this organic matter, its state of oxidation and evolution at the time of deposition, and the intensity of the diagenetic processes to which it has been submitted during its burial.

Numerous studies undertaken in our geochemistry laboratories have shown that a close relationship often exists, in the rocks, between the organic matter content and some trace metals, such as molybdenum,

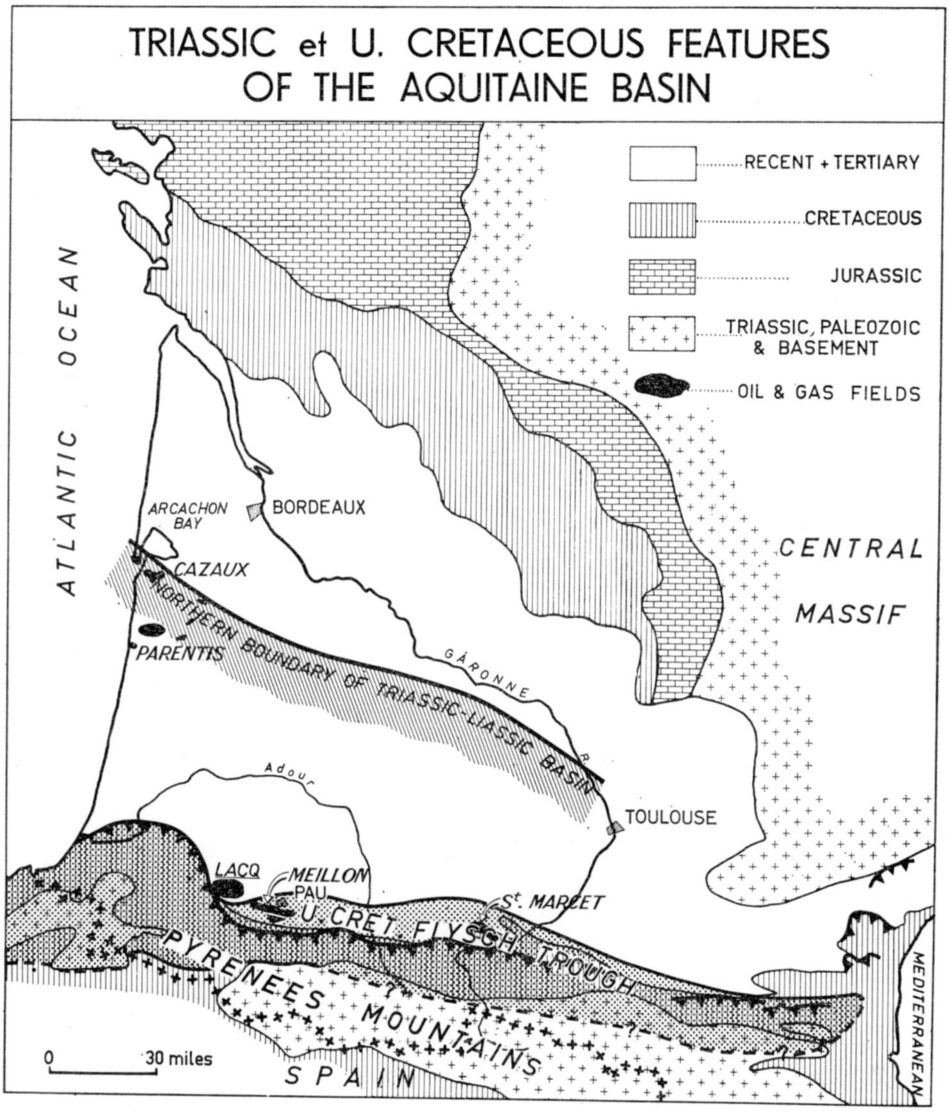

FIG 2. Triassic and Upper Cretaceous features of the Aquitaine Basin. Triassic salt is the main cause of mobility of subsidence and tectonic in S Aquitaine. Flysch trough provides closure of gas fields in the S part of the Lacq Basin.

nickel, copper, zinc, and lead. These metals can be integrated into the structure of the living organic matter or, for the most part, absorbed by the dead organic matter in reducing environments. Also these metals occur proportionally in small amounts in terrestrial plant organic matter and are abundant in the deposits of planktonic marine organic matter. Thus the correlations between organic matter and associated metals seem to us one of the most essential criteria for the estimation of the nature of the organic matter and the physico-chemical conditions prevailing in the sediments at the time of deposition.

Temperature and pressure increase, determined either by the burial, or by the regional metamorphism inside the basin, constitute another important factor in the evolution of the organic matter. The evaluation of the diagenetic processes to which the rocks have been submitted is thought necessary to estimate the hydrocarbon yield of a specific formation.

MAIN RESULTS OF GEOCHEMICAL STUDIES

Our work was essentially based on subsurface and outcrop samples situated in the S of the Lacq Basin. The formations from Lias to Upper Cretaceous have been studied with a special emphasis on the Jurassic and Lower Cretaceous.

In the series studied, the content of organic carbon is low and does not exceed the average values found in the sedimentary rocks of comparable lithology*. There are no horizons with high concentration of organic matter.

THE CORRELATIONS: ORGANIC MATTER—TRACE ELEMENTS

Statistical studies show that a good correlation exists between the organic carbon contained in the fine fraction of the insoluble residue and some metals

* Ronov, 1958; Hunt, 1961.

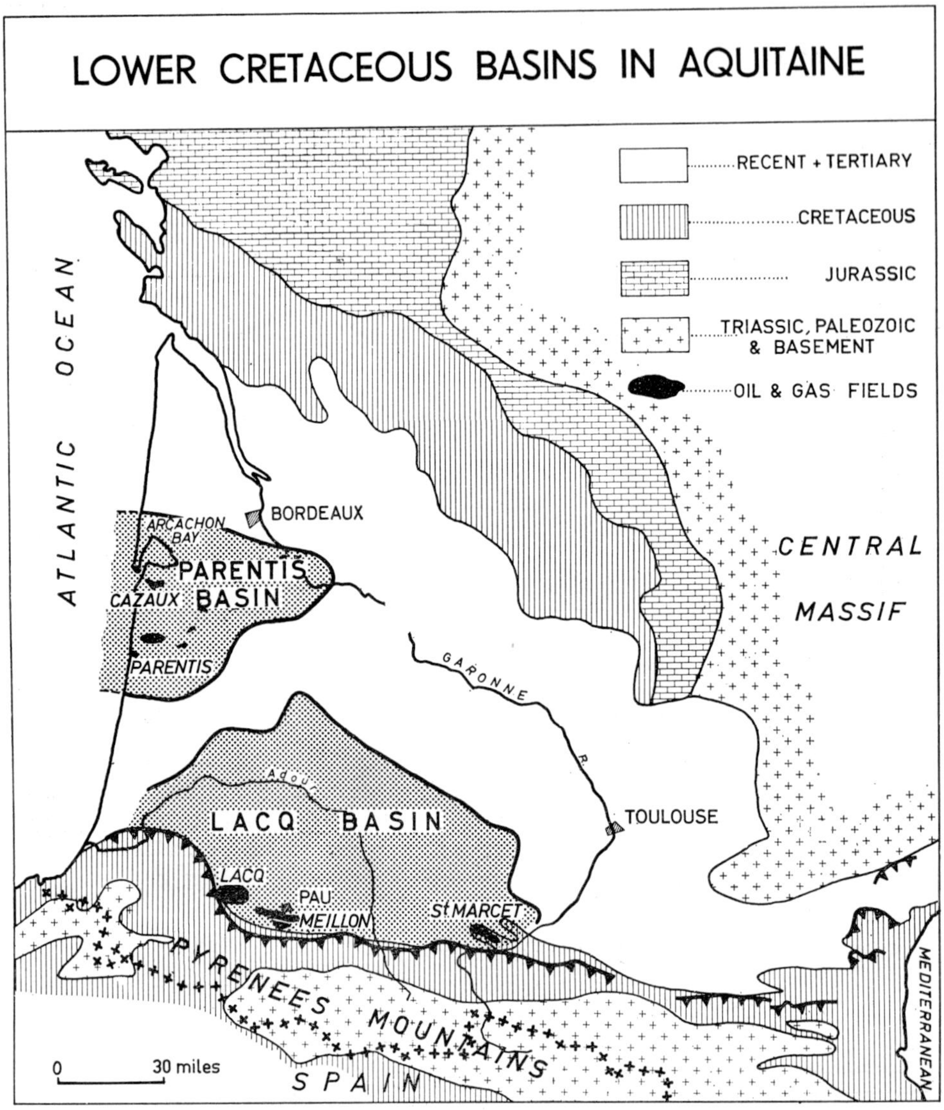

FIG 3. Lower Cretaceous basins in Aquitaine. No hydrocarbon accumulations outside the Lower Cretaceous basins; only oilfields in Parentis Basin, only gas fields (and minor oilfields) in Lacq Basin.

such as molybdenum, copper, zinc, nickel, and lead. In order to neglect the role of the detrital contribution in the process of concentration of these metals, we have chosen molybdenum and copper which are, in a first approximation, weakly associated with the silicate phase.

On Fig 5 are plotted the values of organic carbon, against molybdenum, both determined in the fine fraction of the insoluble residue. The figure shows that the average values of carbon and moy bdenum are different according to the different formations.

They are low in the Upper Cretaceous and in most of the Lower Cretaceous (Zone I, Mo between 2 and 20 ppm, carbon between 1 and 4 per cent).

They are higher in the Gaujacq formation (Lower Cretaceous) and in the Jurassic (middle part of Ossun formation = Oxfordian and Lons and Mano formation = Kimmeridgian and Portlandian Arkell)—(Zone II, Mo between 10 and 60 ppm, carbon between 3 and 20 per cent).

They are very high in the most part of the Ossun formation (Dogger and Lower Kimmeridgian Arkell)—(Zone III, Mo between 60 and 800 ppm, carbon between 10 and 80 per cent).

The enrichment of molybdenum can be a result of either the process of biological concentrations or of adsorption on the organic matter (Krauskoff, 1956). Our studies lead us to think that, for this metal, the biological processes are responsible only for low concentrations, but that the probable mechanism of enrichment in molybdenum is adsorption, favoured by the existence of reducing conditions which lower its solubility. Very high values of the Mo/organic carbon ratio, associated with the existence of a correlation between these two elements, would be an indication of reducing environment (Rumeau, 1969).

On Figs 6(*a*) and 6(*b*) it can be seen that in all

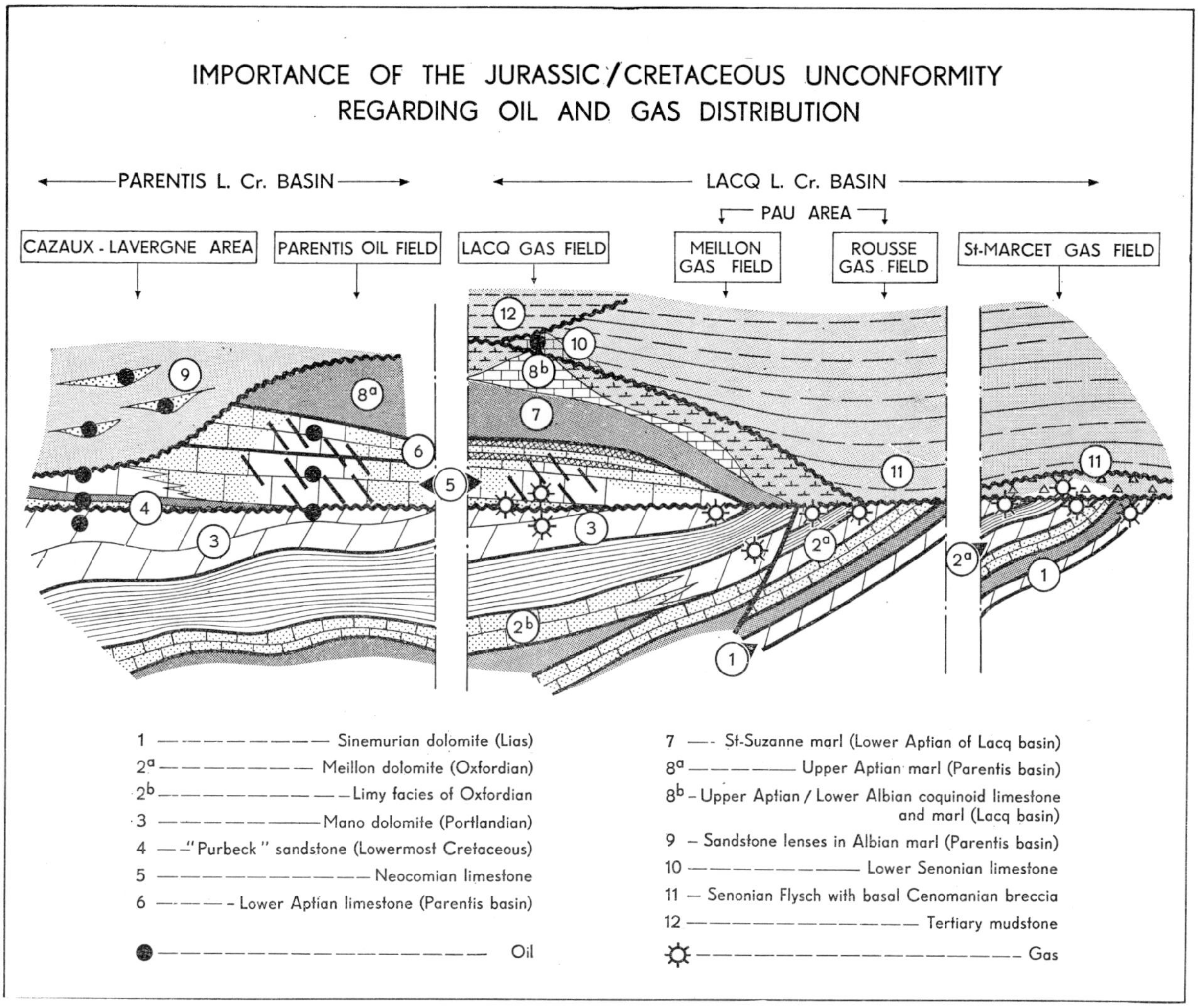

FIG 4. Importance of the Jurassic/Cretaceous unconformity regarding oil and gas distribution. Jurassic is always present in connection with major hydrocarbon accumulations. Lower Cretaceous provides impervious cover and additional reservoir rocks. Upper Cretaceous flysch provides impervious cover in front of the Pyrenean Range when Lower Cretaceous is lacking.

Cretaceous formations, as well as in some Jurassic formations (Lower Ossun, Lons, and Mano) these ratios are very low (low slope of least square line). On the contrary, they are higher in the dolomites of Ossun formation, indicating a more strongly reducing environment.

The study of the ratio Cu/Mo permits another approach to the understanding of the conditions of deposition of a sediment. When the redox potential near the water-sediment interface is close to 0, there is enough oxygen to allow the development of benthic life, which can produce an enrichment of copper in the sediment. If the redox potential decreases, the benthic life tends to disappear, whereas the adsorption of molybdenum on planktonic organic matter in the superficial layers of the sediments is increased: the ratio Cu/Mo decreases. A low value of the ratio Cu/Mo would indicate a reducing environment, a higher value, a more oxidizing environment, all other conditions being the same (Rumeau, 1967, 1968).

Fig 7 shows that three formations have very low values of the Cu/Mo ratio.

1. The extreme base and the upper part of the Ossun formation.
2. Some levels of the Neocomian.
3. The Gaujacq formation (Lower Cretaceous).

Of these, the Ossun formation has the lowest values, and was therefore deposited in the most reducing environment. In this formation, the values of the Mo/organic carbon ratio are widely scattered. In fact, there is a relation between the values of molybdenum, copper, and organic carbon. The high values of the Mo/organic carbon ratio correspond to the low values of the Cu/Mo ratio (Fig 8), indicating the most reducing environments. This dispersion in this formation can therefore be explained by important variations of the depositional conditions. These variations are to be compared to the great variety of the facies found within the formation (oolithic dolomites, or coral, crinoïds, or unfossiliferous dolomites, etc.).

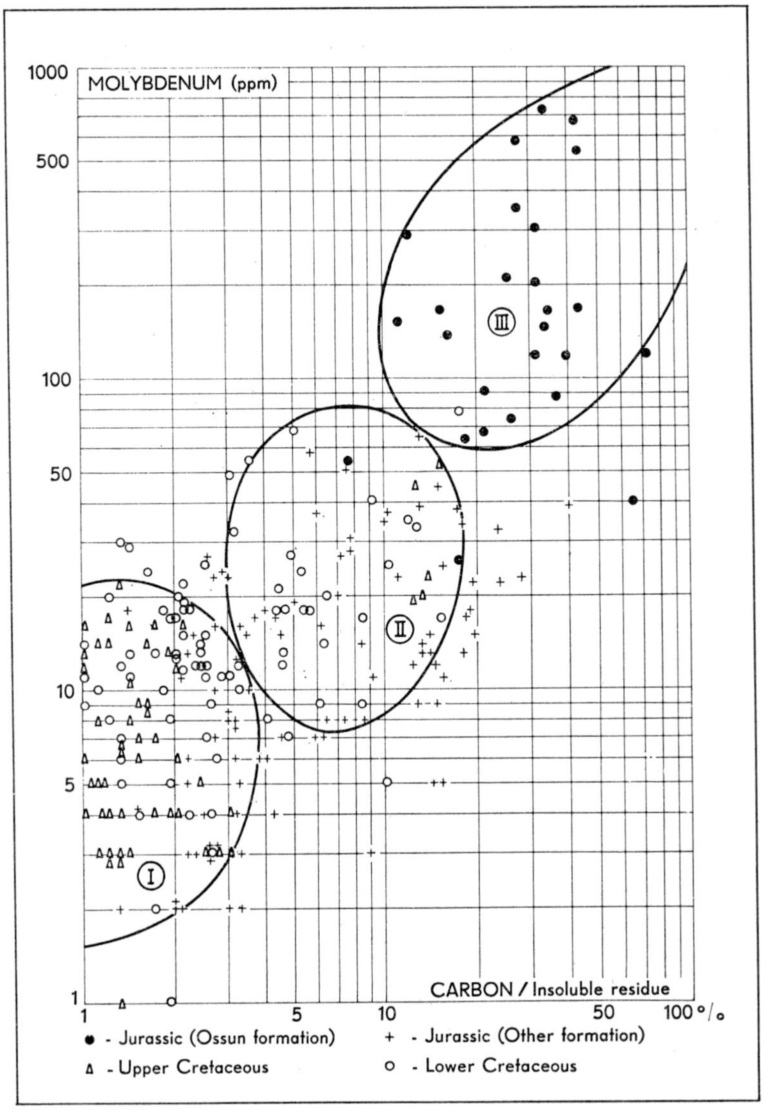

FIG 5. Plot of molybdenum *v.* organic carbon in S Aquitaine Mesozoic formations.

Thus there is in the Jurassic, and to a lesser degree in the Lower Cretaceous, organic matter of marine origin, deposited in an environment which favoured its preservation. This organic matter is presumed to have been, at least qualitatively, favourable to the generation of hydrocarbons.

TENTATIVE SYNTHESIS

First let us note that the studies mentioned above deal essentially with the S of Aquitaine. They indicate the particularly favourable features of the Upper Jurassic facies as known on the longitude of Pau and to the E of it. Outside the studied area, the abundance of oil shows encountered in drilling in the marly or limy formations and limestones of the Malm Basin or Parentis leads us to think that the Upper Jurassic may also play an important role in the supply of hydrocarbons. This conclusion can only be reinforced by the distribution of traps along the Jurassic unconformity, a general feature in Aquitaine (Fig 4). We will see now how the post-Jurassic evolution has been unfavourable in some regions of the basin and very favourable in others.

Importance of Burial; Precocity of Burial and of Reservoir Trapping

Some recent studies have shown the effect of temperature and pressure in the process of maturation of source rocks (Philippi, 1965; Louis and Tissot, 1967). In the case of the Liassic shales of the Parisian Basin, a minimum depth of burial of 1200 m was indicated as a threshold for the production of hydrocarbons (Louis, Tissot, 1967). This helps to explain why in the northern part of Aquitaine, where the Jurassic sediments have never been buried deep enough, no oilfields exist, not even significant shows.

In the part of the basin situated S of the line Arcachon–Toulouse, where hydrocarbon shows are abundant, minimum conditions of burial favourable to the genesis of hydrocarbons have often been

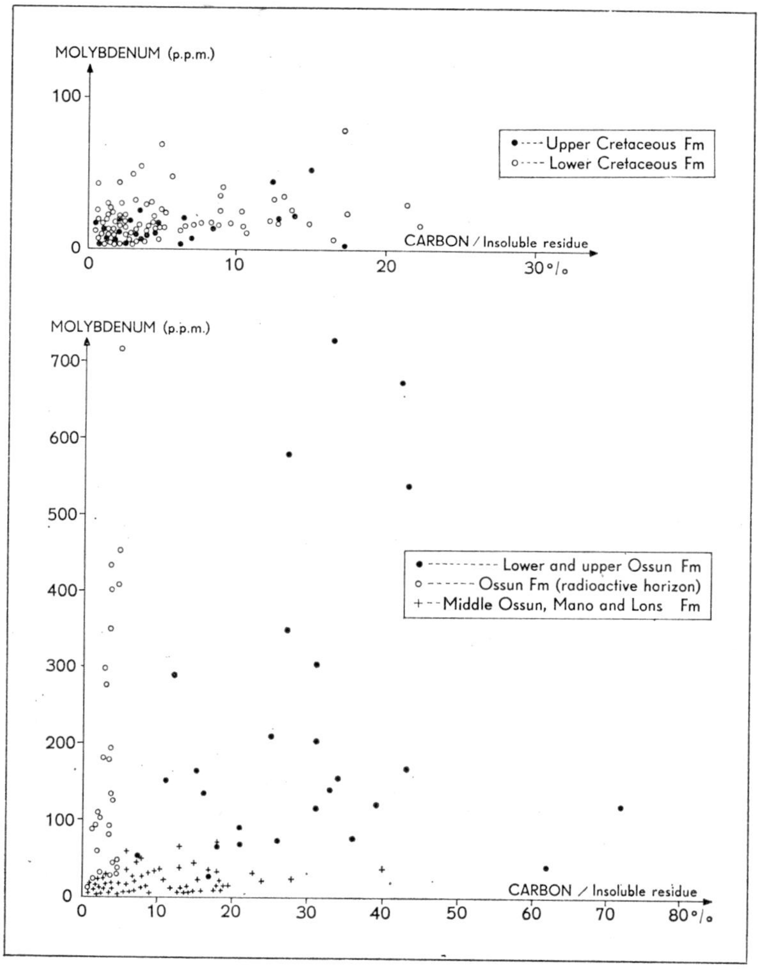

FIG 6. Plot of molybdenum *v.* organic carbon.

reached or exceeded as soon as Lower Aptian time.

Furthermore, as soon as the hydrocarbons were liberated they were finding not only reservoirs inside the basins of the Lower Cretaceous (Purbeckian–Valanginian in Lacq; Purbeckian–Neocomian in the Parentis Basin), which were able to extend the capacity of the reservoirs of the Jurassic but also the impermeable rocks necessary for their protection: evaporites of the Upper Neocomian at Lacq followed by the thick deposits of the Ste. Suzanne marls; Wealdian clays in the Parentis Basin completed by the "marly series" going from the Upper Aptian to the Lower Cenomanian.

Finally, and this is not the least important feature of the fields of Aquitaine, the trapping of the hydrocarbons has been favoured by a very important structural fact—the very early origin of the traps closely related to movements of the Triassic salt bodies, which are well developed in the southern part of the basin. Thus the anticline of Parentis can already be identified by the end of the Neocomian (Fig 9). The trap of Meillon was formed as far back as the Lower Senonian and it is probable that the structures of Lacq and St. Marcet became established at the same time.

The role of hydrodynamic conditions in the distribution of the fields inside the basins of the Lower Cretaceous is also very important. As a matter of fact, in these privileged basins, we can note a tendency of migration towards the borders. In these peripheral areas, the most favourable factors for the concentration and trapping of hydrocarbons formed inside the basin are contrasted by the increased possibility of leakage when the cap rocks thin out and disappear. The Parentis Basin encloses in its central area sodium chloride waters (20,000 to 65,000 ppm), iodized and reduced, indicating an effective protection. The fields are distributed on structures along the border of the protected zone. Fresh water appears beyond them as a result of the direct superposition of the Upper Cretaceous calcareous formations on the Jurassic floor.

The situation is the same in the Lower Cretaceous basin of Lacq and this is in fact what led to the neglect of the northern border of this basin. Already, on the ridge of Audignon, which is a high zone in the middle of the basin, the contrast is sudden between the northern flank (Audignon 3) invaded by fresh water, and the southern flank (Audignon 4) where the water is salty (115,000 ppm), chlorinated, and reduced (Fig 10). The 3000 m of calcareous and argillaceous Lower Cretaceous rocks on the southern flank

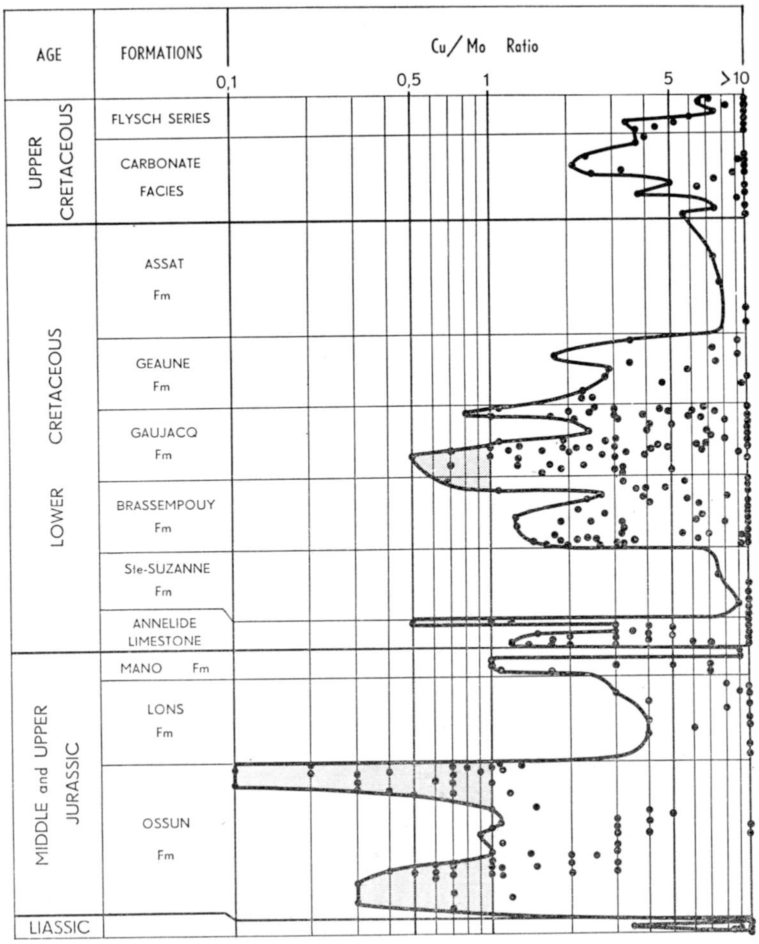

FIG 7. Cu/Mo ratio variations in Mesozoic formations of S Aquitaine.

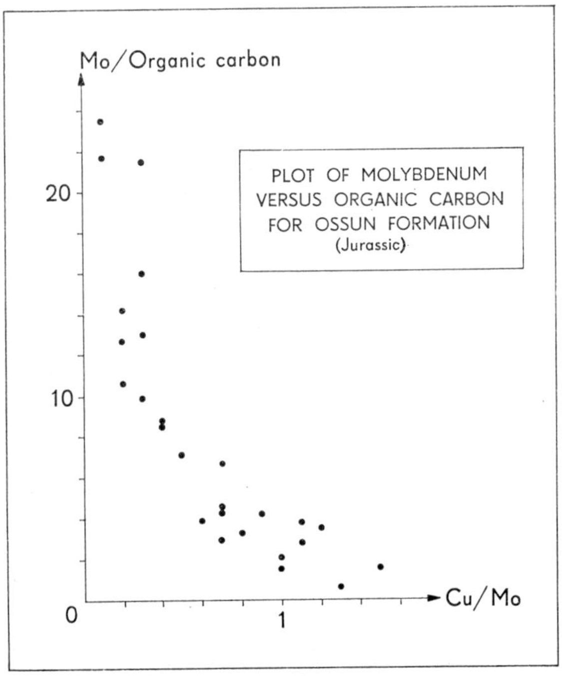

FIG 8. Plot of molybdenum *v.* organic carbon for Ossun formation (Jurassic).

isolating the Jurassic reservoir from the superficial aquifer are opposed to the several hundreds of metres of limy and dolomitic formations of the Lower Cretaceous on the N flank.

The S flank of the Lacq Basin presented simultaneously the favourable conditions for the migration of the hydrocarbons and the increased dangers of leakage due to the importance of erosion which denuded all of the Jurassic series of an area, extending from the S of Lacq to the S of St. Marcet. But in contrast to what was happening on the northern flank of the basin, thick deposits of impermeable flysch were deposited in the vicinity of the Pyrenean chain, closing the southern flank of the structure in Meillon–Rousse (Fig 11) and isolating completely St. Marcet structure.

The example of Meillon–Rousse merits an examination of the composition of the gases produced on either side of the geological accident separating these two accumulations.

	Composition			
Field	% CO^2	% SH^2	% C^1	Gasoline g/cu m
Lacq	9·6	15	69	25
Meillon–	8·6	6	78	40 to 60
Rousse	4·6	0·8	76	470
St. Marcet			89	79

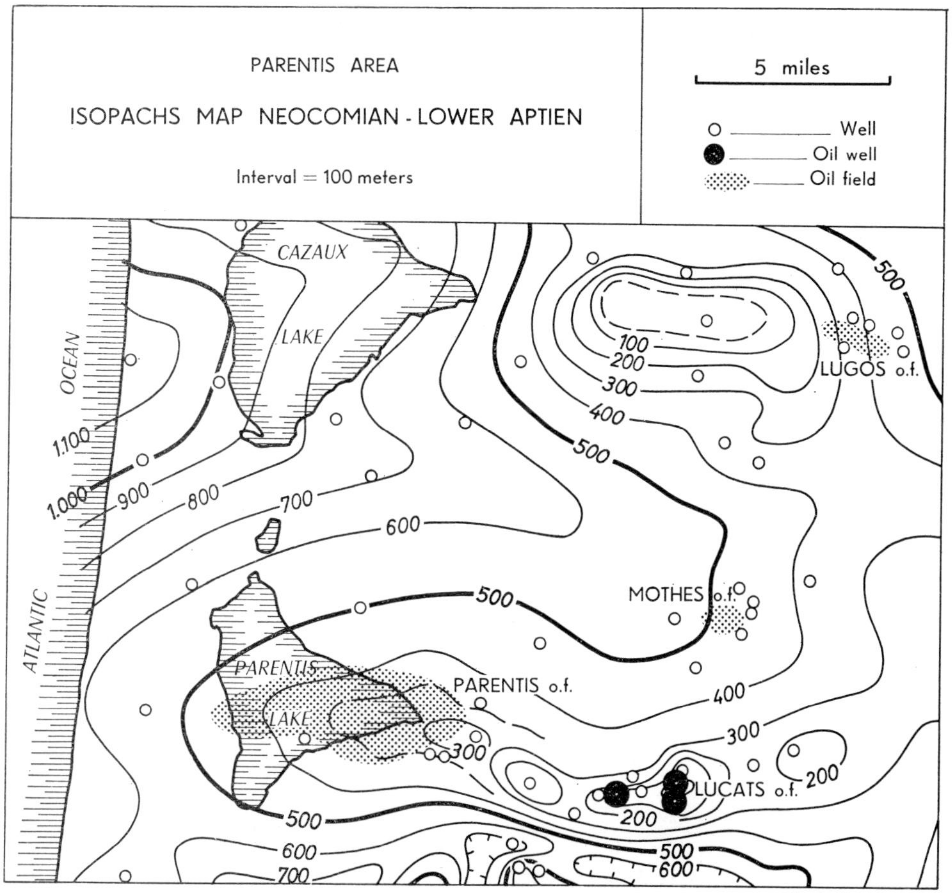

FIG 9. Parentis oilfield area—Neocomian + Lower Aptian isopach map. Parentis anticline trend appears clearly at the end of Lower Aptian.

The gas of Meillon is rich in H_2S, which seems to be related to the anhydrite evaporites present at the base of the huge wedge of Lower Cretaceous covering the northern monocline. This interpretation is even more plausible because at Lacq, where the pool is completely covered by the same Neocomian anhydrite, the H_2S content of the gas is even higher.

The characteristic of the southern pool (Rousse) which appears as a cliff of a Jurassic mass directly buried under the Upper Cretaceous, is the abundance of condensates (470 g/cu m). It is necessary to point out that shows of wet gas of similar composition or light oil were found in the Upper Cretaceous of other wells in the same area.

The different thermal history of the two portions of the field furthermore seem to furnish a remarkable element for the explanation of the lack of condensates in the Meillon pool. This aspect will be treated later.

Oil Province and Gas Province

Temperature and pressure increase, either due to the burial or to the regional metamorphism inside a basin, is an important factor for the evolution of the organic matter, and consequently for the nature (and quantity) of the formed products. This parameter can be studied from the analysis of the organic matter itself (carbon ratio*) or from measurements of the mineral fraction of the sediment (estimation of the crystallinity of illite).

1. *Study of the carbon ratio*

In the Lacq Basin and in all the studied series (Lias to Upper Cretaceous) the carbon ratio is very high and at least equal or superior to 0·7. This indicates that in all these formations the maturation of organic matter is advanced. However, it is in the Jurassic, and especially in the well of Pont d'As, that the measured values are the highest (0·9). They express a very high degree of carbonization of the organic matter.

On the other hand, in the Parentis Basin these high

* The carbon ratio is computed from the "fixed-carbon", which is the organic carbon left in the rock after heating for one hour at 1000° C under inert atmosphere. The carbon ratio may vary from 0 (theoretical case where all organic matter is volatilized by heat) to 1 (case of the graphites which give no volatile product upon heating). This is an indirect way of estimating the H/C ratio of the organic matter, which is practically impossible to measure in a sediment.

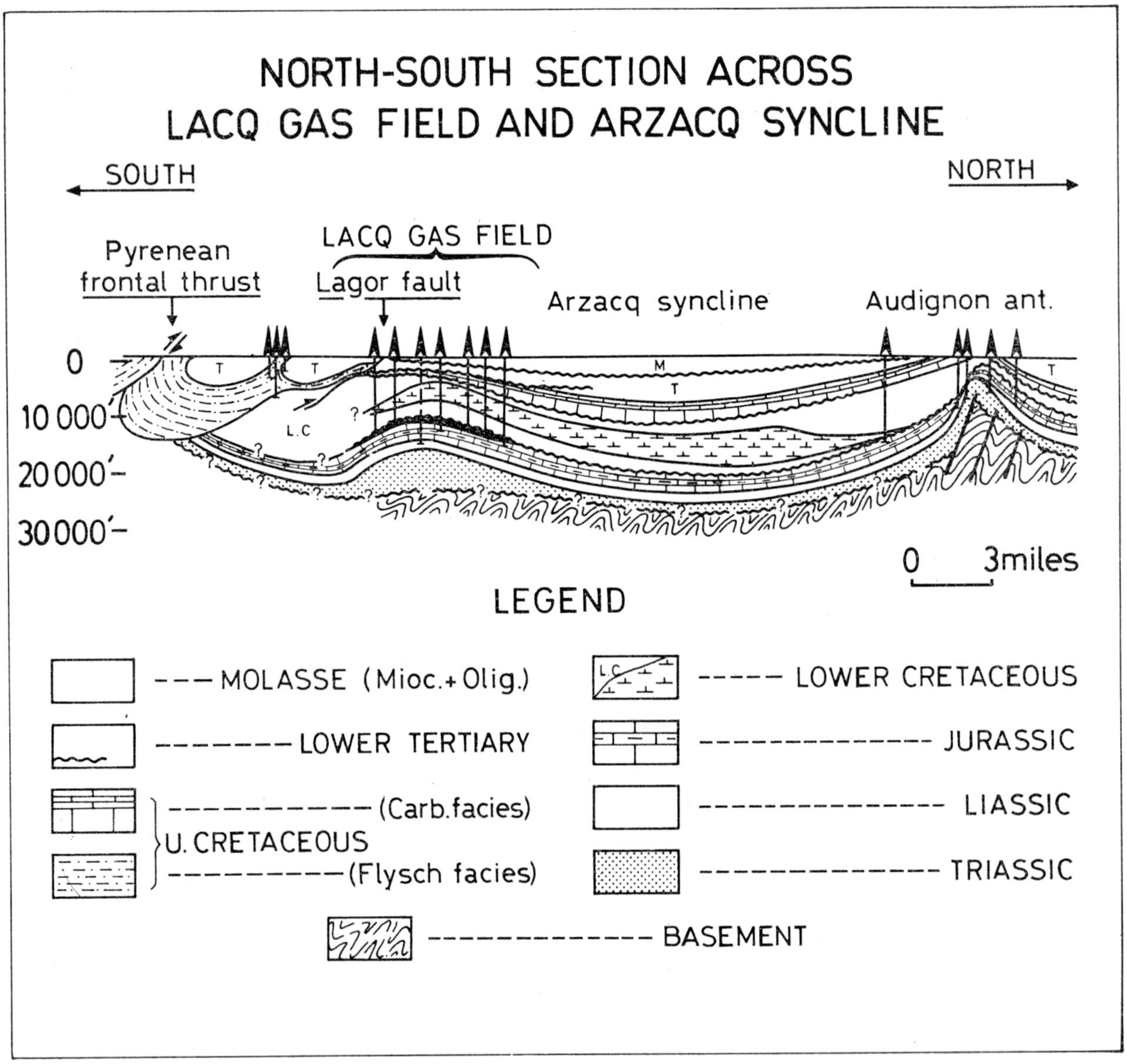

FIG 10. N–S section across Lacq gas field and Arzacq syncline. Lacq gas field deeply buried below Lower Cretaceous rocks. This cover of impervious rock is also present on the S flank of Audignon anticline but is absent on its N flank.

values are never reached. The carbon ratio never exceeds 0·8 on the deepest portions.

2. *Study of the crystallinity of illite*

In the Lacq Basin, the study has shown that the different formations of the Cretaceous are all located in a non-metamorphic zone. For the Jurassic, on the contrary, the different drillings are distributed in three areas: non-metamorphic (Orthez 102, Lacq, Tournay), anchimetamorphic (Meillon 2, Bordes, Rousse, Ossun 1), epimetamorphic (Baysere, Pont d'As, St. Faust, Mazères, Meillon) (Esquevin, 1969).

The effects of this metamorphism of the epizone are maximal along an axis Baysère—Pont d'As—St. Faust—Meillon 1 (Fig 12). On the NW of this axis (Lacq), as well as on the SE (Meillon 2—Bordes) the intensity of the metamorphism decreases progressively.

On the other hand, in the Parentis Basin all sediments are located outside of the metamorphic zone in the zone of diagenesis.

It can then be understood why the evolution of the organic matter may be less advanced in the Parentis Basin, where the Jurassic sediments have been buried only rarely beyond 4000 m (in the fields, the top of the Jurassic is at 2500–3000 m) while they have been buried beyond 7000 m in the deeper parts of the southern basin. Already, at the end of the Lower Cretaceous, the thermic evolution was really different (2000 m of burial on the average against 4000 m). This difference of natural thermic evolution, as early as the end of the Lower Cretaceous (75° C against 150° C) and after the subsidence of the Upper Cretaceous and of the Tertiary (100° to 150° C against 150° to 300° C) seems, therefore, to be a very important element.

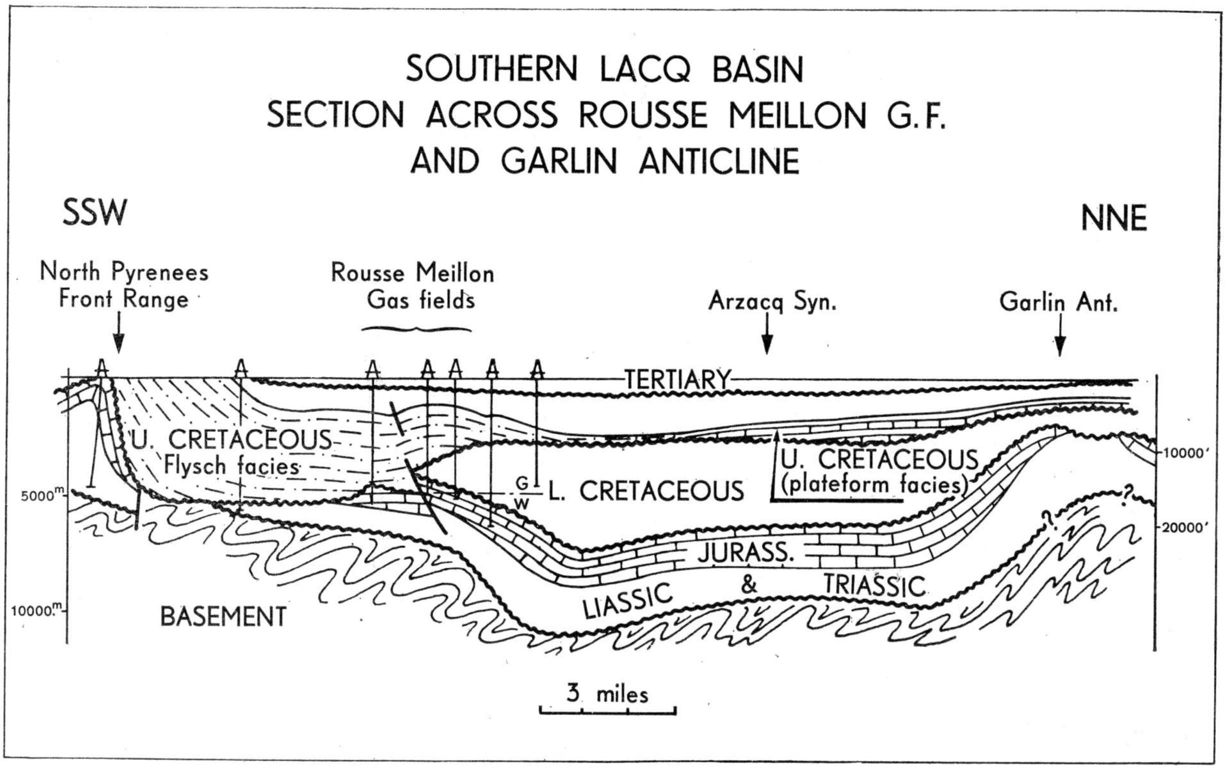

FIG 11. S Lacq Basin section across Rousse–Meillon gas field and Garlin anticline. These flysch deposits do not exist on the N flank of Arzacq syncline. Development of Upper Cretaceous flysch deposits provides hydrocarbons from dismigration in Rousse–Meillon erosional edge.

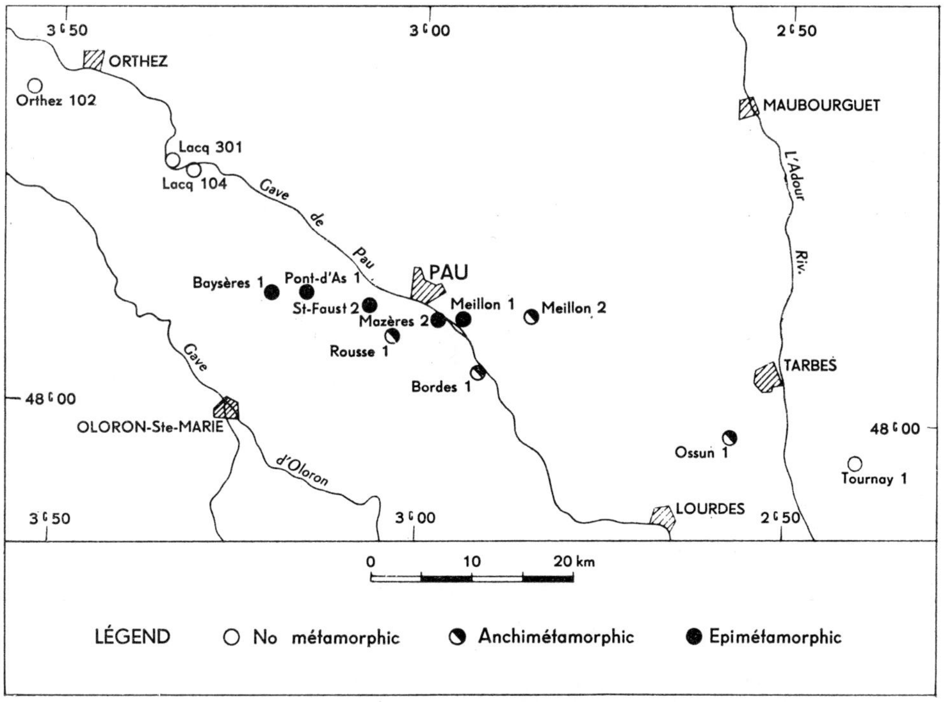

FIG 12. Occurrence of metamorphism in Jurassic formation of Pau area.

However, it has been pointed out that this metamorphism is not proportional to the maximal depth of burial reached by the sediments (J. Henry, 1966). Even when taking into account the time factor (time and duration of burial), the distribution of the phenomena does not appear quite clear. So we wonder if thermal phenomena linked to the Pyrenean orogeny should not be considered. As a matter of fact, in the front range the metamorphism, more advanced than in the basin, shows a very particular geographical distribution. A connection with the abundance of igneous activity which occurred near the Lower Cretaceous–Upper Cretaceous limit seems to appear. Thus we could admit that at this time an additional thermic effect related to the Pyrenean orogeny could have been added to the normal temperature gradient in some sections of the basin.

This early and very accentuated thermic evolution is, in our opinion, the most determining factor of the exclusive presence of gaseous hydrocarbons in the deep fields of the S of the Lacq Basin, in front of the Pyrenean chain.

CONCLUSIONS

In Aquitaine, the carbonate formations of Upper Jurassic, although not very rich in organic matter, were deposited in environments favourable to the preservation of this organic matter. On the condition that possible additional contributions of some horizons of Lower Cretaceous and perhaps Upper Cretaceous are not neglected, the contribution of the Jurassic could have been predominant, even outside the studied area.

In the northernmost part of Aquitaine the shallow burial of these formations has not allowed a sufficient thermal evolution. In any case, the lack of cap rock horizons, as well as the flushing by meteoric waters, would alone explain the meagre interest of this area.

On the other hand, inside the two large zones of deposition of Lower Cretaceous the important burial, together with the early availability of the traps, were favourable to the formation and accumulation of hydrocarbons.

In the southern portion of the basin the stronger thermal and diagenetic evolution, due to a more active subsidence and the vicinity of the Pyrenean chain, can explain the presence of gas and condensate, as opposed to oil, in the Parentis Basin.

ACKNOWLEDGMENTS

The authors are deeply indebted to all of their oil geologist and geophysicist colleagues from ESSOREP, ERAP, CFP, and SNPA, whose work in Aquitaine made this paper possible. They also want to thank Dr G. Colo for reading and correcting this manuscript.

BIBLIOGRAPHY

De Chevilly, F., Colo, G., Henry, G., and Winnock, E. "Le champ de gaz de Meillon-Rousse", to be published in *Rev. Inst. Franç. Pétrole*, 1969.

Dupuy, J. P., Oswaldt, G., and Sens, J. "Champ de Cazaux, Géologie et Production" *Proc. 6th World Petrol. Congr.*, 1963.

Esquevin, J. "Influence de la composition chimique des illites sur leur cristallinité." *Bull. Centre Recherches de Pau*, vol. 3, 1.

Garreau, B. "Le gisement de Parentis." 20th International Geological Congress, Mexico, 1956.

Henry, J. "Le problème des étages tectoniques dans la zone nord-pyrénéenne occidentale (France)." Colloquium on tectonics, Neuchâtel 18–21 April 1966, pp. 253–68.

Hunt, J. M. *Geochem. Cosmochim. acta*, 1961, **22**, 37–49.

Krauskopf, K. B. *Geochim. Cosmochim. acta*, 1956, **9**, 1–32.

Louis, M. C., and Tissot, B. P. "Influence de la température et de la pression sur la formation des hydrocarbures dans les argiles à kérogène." *7th World Petrol. Congr.*, 1967.

Michel, P. "Le gisement de gaz de St. Marcet." 20th International Geological Congress, Mexico, 1956, Vol. V.

Philippi, G. T. *Geochim. Cosmochim. acta*, 1965, 1021–49.

Ronov, A. B. "Organic carbon in sedimentary rocks, in relation to the presence of petroleum." Geochemistry No. 5, 1958.

Rumeau, J. L. "Caractérisation géochimique des conditions d'oxydo-réduction dans une sédimentation marine carbonatée. Exemple de la Keg River formation (Alberta-Canada)." *Bull. Centre Recherches de Pau*, vol. 3, 1.

Winnock, E., and Pontalier, Y. "Geology of Giant Fields." AAPG, to be published.

MS received 18 February 1969

Discussion

Dr Donald P. McGookey (Texaco Inc.): A question about the origin of the Pyrenees Mountains. Are they the result of compression, as suggested by some of the continental drift school, or are they the result of vertical uplift of a core area accompanied by gravity sliding to the N and S to account for the imbricate thrusting?

E. Winnock: The history of the Pyrenees mountain building is very intricate and, as yet, not well known. It seems probable that both mechanisms suggested were active.

J. D. Moody (Mobil): Would you comment on sulphur in the Lacq gas and its relation to your comments on oil *v.* gas? Also, what can you tell us of the reservoir rock at Lacq?

G. Kulbicki: Although we have not studied this problem very thoroughly, we can say that the hydrogen sulphide has always had close relationships with anhydrite in the reservoir or in the adjacent formations.

Concerning the reservoir rock, it is mainly a fractured, low porosity (5 per cent) dolomite.

Dr P. E. Kent (British Petroleum Co. Ltd): First, do the authors support the view that the Bay of Biscay owes its origin to the anti-clockwise rotation of Spain, and secondly, if so, where would they look for the continuation of the Lacq gas field belt? Which side of the gap?

E. Winnock: Many offshore studies are under way in the Bay of Biscay. We have no final opinion about a possible opening of the bay by continental drifting. If this really happened, it should have occurred very recently, probably during late Tertiary.

Concerning the second question, the Lacq gas field belt already has a direction indicated by the line extending from Meillon to Lacq. The trend of Lacq may be on the S part of the opening, but because of the proximity of the mountains the problem is highly complex.

Dr Anthony Reso (Tenneco Oil Co.): In Fig 4 of your paper you show an unconformity at the base of the Cenomanian in the Lacq Basin. The unconformity is quite severe because it is suggested from the legend that the Upper

Albian is absent and in the Lacq gas field Lower Senonian limestones rest on Lower Albian beds with Upper Albian and Cenomanian missing. In the Parentis Basin (left side of Fig 4), however, there is an angular unconformity at the base of the Albian. Do you know if there is also an unconformity at the base of the Cenomanian in the Parentis Basin as there is in the Lacq Basin, which might provide similar conditions for the production of oil and gas above the unconformity as there is at Lacq gas field and in the St. Marcet gas field?

Is there Cenomanian oil and/or gas production in the Parentis Basin, in the Cazaux–Laverone area, for example?

E. Winnock: In the Parentis Basin there is also an unconformity at the base of the Cenomanian, but the formations are tight and no reservoir exists in the vicinity of this unconformity.

Neither gas nor oil production is known in the Cenomanian. The lack of reservoir can explain this fact.

Rudolf Blaser (Bataafse Internationale Petroleum Maatschapij): Have the authors of this paper constructed any maps representing the present-day geothermal gradients (as established from BHT measurements)? If not, could they provide information on BHT in the oil and gas fields?

E. Winnock: A map has been constructed representing the present-day geothermal gradients. The mean BHT in the Lacq gas field is 266° F at 13,120 ft; in the Meillon gas field it is 320° F at 16,400 ft.

Abu Amr (Sonatrach): In the Aquitaine Basin do enough data exist to map potentiometric surfaces of the hydrodynamic condition mentioned?

E. Winnock: Enough data now exist to map potentiometric surfaces. For each reservoir, a potentiometric map has been drawn, but the Lower Cretaceous and Jurassic reservoirs inside the Lacq Basin are "confined type" and there is no gravity flow. In the Lacq gas field area the pressure is abnormally high, whereas in the Meillon–St Marcet area the pressure is nearly hydrostatic in the Upper Jurassic reservoir.

Exploration for Oil and Gas in Italy

By MARCO PIERI

(*Regional Exploration Manager*, *AAPG*, *Soc. Geol. Ital.*)

SUMMARY

Exploration and production of hydrocarbons in Italy began to take an increasing importance following the end of the Second World War. With discovery of natural gas in the Po Plain by the state-owned petroleum exploration and production company, AGIP, Italy began to enjoy the fruits of a domestic resource of energy fuel.

The positive effect of this new availability upon the economic development of the country was additionally accentuated by concentrating the distribution and sale of this gas in the hands of public enterprise through the constitution of ENI, within the framework of a national energy policy.

An outline of the regional geology of Italy is given on the basis of recent developments in geological studies and petroleum exploration. The Mesozoic, Palaeogene, and partly Miocene sediments were deposited in basins following, in general, a NNW–SSE trend on the mainland and a W–E trend in Sicily. The main tectonic phases in late Palaeogene and Miocene caused the thrusting of the "inner" (*i.e.* Tyrrhenian) units over the "outer" ones (Adriatic and S Sicily). The Upper Miocene to Quaternary sediments were deposited in small post-tectonic basins along the W side of the Apennines and in a deeper and wider foretrough following their E side from the Po Plain to the Gulf of Taranto. Pliocene and Quaternary tectonics caused remarkable structural complexity for the Apenninic margin of the Pliocene foretrough.

Principal gas-producing areas on land are, at present, the Pliocene basins of the Po Plain and of Central-Southern Italy, and the Lower Miocene basin of N Sicily. Up to now commercial quantities of oil have been discovered only in Triassic carbonates at Ragusa and Gela, SE Sicily.

A recent law (July 1967) regulates exploration and production activities in territorial waters and the continental platform, and amends the earlier (1957) law concerning onshore activities. Regulations favourable to exploration companies have given fresh impetus to prospecting activities.

The new law gives ENI a status of leadership in offshore exploration activity without conceding unrestrained privilege to the company. At the end of December 1968, 44 exploration permits had been granted and 160 applications were pending for the Adriatic Sea, N of 42nd parallel. Recent important gas discoveries confirmed the prospectivity of this area.

According to the law, exploration permits would be granted in the S Adriatic, Ionian, and Tyrrhenian seas at the end of 1968 and in the Sicilian offshore at the end of 1969.

The actuation by ENI of a nationwide transmission system will provide an important instrument for rapidly increasing the Italian consumption of methane and for promoting a more articulated spread of its uses.

THE DEVELOPMENT OF PETROLEUM EXPLORATION ACTIVITY AND POLICY IN ITALY

ITALY is one of the Western European nations where petroleum exploration activity has been most intense during the past two decades. The important discoveries made during this period have encouraged the intensification of activities in an effort to increase the effective incidence of domestic hydrocarbon resources on the energy fuels supply situation of the country.

The first positive results of petroleum prospecting were felt in 1944, when a large gas field was discovered at Caviaga in Lombardy by AGIP, the state-owned company for petroleum exploration and production. This discovery initiated intense exploration and development activity, which in following years led to exploitation of large natural gas reserves found in the Po Valley. Accessibility of this new fuel, particularly for industrial use, was highly instrumental in post-war reconstruction and subsequent development of the Italian economy due to both a broadening of the range of energy fuel supplies available in the country, and to the reduced price at which gas offerings were made.

In Italy, which is traditionally poor in coal reserves, the appearance on the market of methane fuel supplies helped to accelerate the substitution of the gas and oil for coal, which later became general throughout Europe. With consumers rapidly re-converting for for utilization of the new fuels and in view of the necessity for the country to exploit its gas reserves efficiently, it became essential for the government to adopt a national policy on hydrocarbons which would take into account the fundamental exigencies of the nation's future economic development and would place the important new sources of fuel under some form of public control. As a result, in 1953, the Italian Parliament voted to constitute a state-owned enterprise, ENI, giving the new entity the objective of promoting and realizing, in the interest of the nation, enterprises concerning the exploration, production, distribution, and uses of hydrocarbons.

In the Po Plain, ENI has exclusive rights to explore for and to produce hydrocarbons, and to build and operate a pipeline network.

ENI immediately initiated a policy based upon the principle that the use of natural gas in the country would thereafter continually increase. A first step

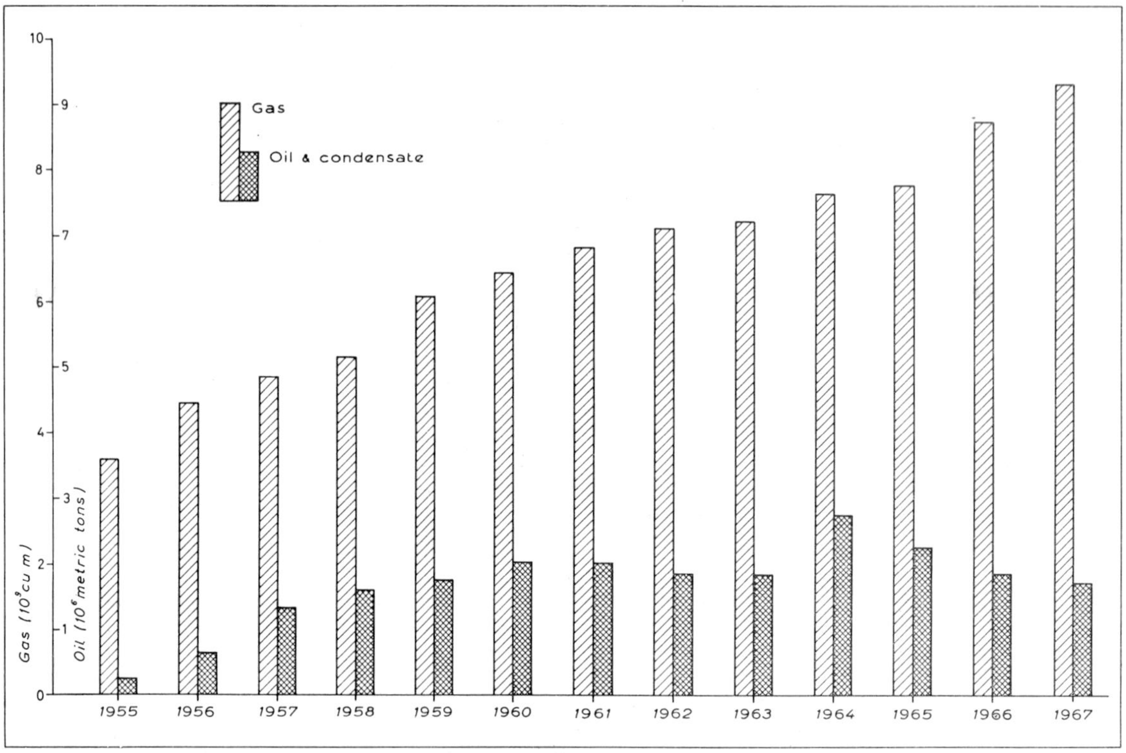

FIG 1. Production of natural gas (in billion cu m), and oil (in metric tons) in Italy, 1955–1967.

was necessarily the construction of a transmission system to transport gas from the Po Valley fields to all the N Italian industrial zones. ENI also began working towards an extension of petroleum exploration throughout domestic territory under rules decreed in a 1957 law.

Gas was subsequently made available as quickly as possible to all potential industrial consumers at a price lower than the quivalent cost of other commonly available fuels (particularly fuel oil). Under ENI policy and activity, the consumption of natural gas in Italy rose from 63 million cu m in 1946 to 3500 million in 1955 (Fig 1). This meant that from practically a zero incidence of natural gas in the Italian energy balance in 1946, the use of this fuel rose to a position covering 11·3 per cent of the country's energy needs in 1955. At that time the major part of available gas supplies went for industrial heating and for steam electric station use.

But the desirability of creating a modern petrochemical industry in Italy, capable of competing with the foreign chemical industry, induced ENI in 1954 to begin offering natural gas for chemical uses at very convenient prices. Thus stimulated, certain chemical facilities were erected to use a gas feedstock, and since that time a steady increase of demand has raised gas consumption in N Italy alone to the present total of some 1500 million cu m/year.

Characteristic of the Italian energy structure are the high incidence of hydrocarbon fuel consumption (in 1967, domestic natural gas accounted for 8·8 per cent and imported crude for 74·9 per cent of total demand) and the fact that there is only one fully integrated national company with dimensional, technical, and financial capabilities for undertaking the onerous domestic and foreign ventures necessary for supplying both hydrocarbon and nuclear fuels at prices and in quantities sufficient effectively to keep the country's energy costs reasonably low.

The increasing need for hydrocarbon fuels in Italy (where ENI sells more than 97 per cent of gas consumed and has a market share of around 25 per cent of oil products) is, of course, caused by the constantly larger share of oil and gas required to cover needs in the gross energy consumption picture (Fig 2). This has made imperative ENI's increased effort to accentuate domestic and foreign petroleum exploration activities with the double objective of finding more oil abroad and more gas in Italy.

Two factors have now become favourable for reaching these objectives. Abroad, during 1967 and 1968, ENI was able to acquire exploration rights to 381,559 sq km of onshore and offshore areas (299,200 sq km of which were net of other participants). In Italy the new law of 1967, which has favourably modified the 1957 law for what concerns onshore exploration and introduced regulations for offshore areas, has fostered vastly extended exploration (Figs 3, 9) and has already resulted in large new gas discoveries.

The new Italian law has also attributed to ENI, as producer and distributor of practically all Italian natural gas, certain rights and obligations different than those of private enterprises, committing the state-owned enterprise to taking on the role of guide in domestic exploration activities. ENI has been

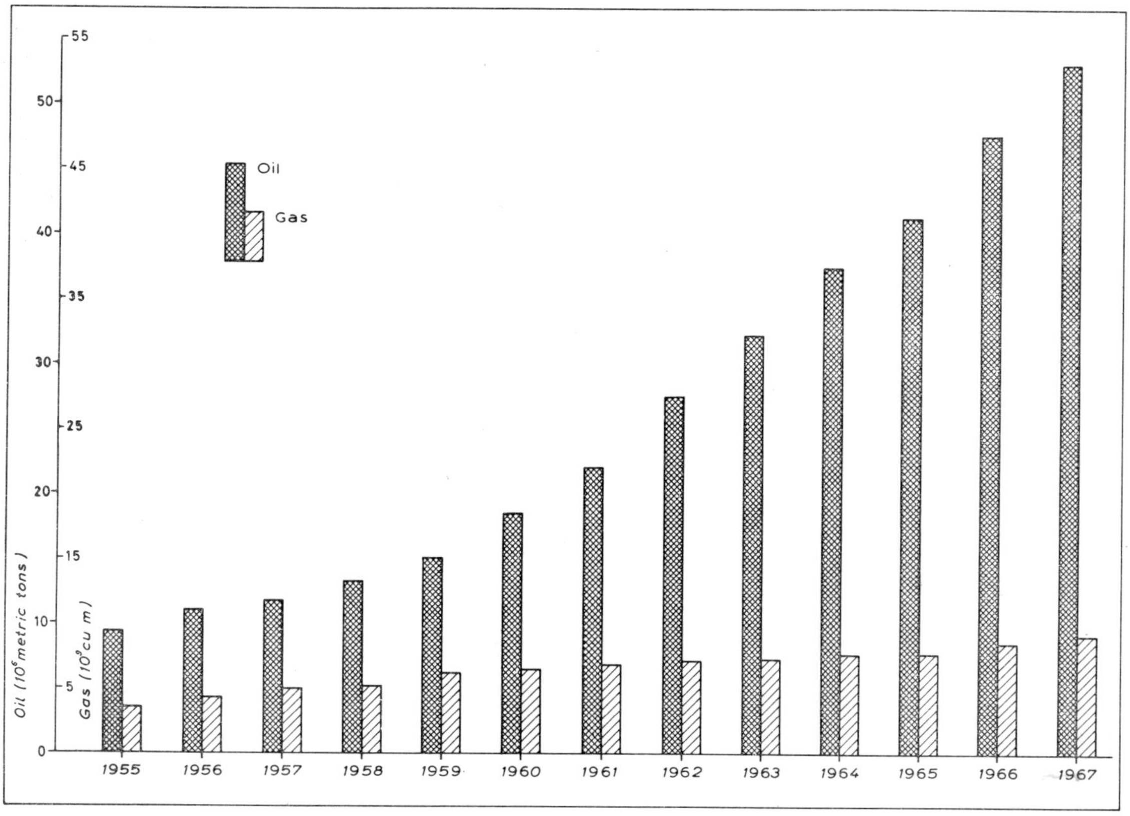

FIG 2. Consumption of natural gas (in billion cu m) and oil (in metric tons) in Italy, 1955–1967.

given priority in the selection of up to 25 per cent of any offshore areas designated for exploration by the government, and has the privilege of option, like the British Gas Council, to purchase gas discovered by others; ENI, however, is obliged to make all preliminary geophysical exploration of the designated areas, the results of which will be available to other companies as well. Another important point in the 1967 law is the possibility given to ENI for associating with private companies for exploration joint ventures.

GEOLOGICAL SUMMARY

The geological history of Italy reflects the evolution of a part of the Mediterranean geosyncline, whose first stage can be recognized in the Permian and Triassic sedimentary section of the southern calcareous Alps. The successive development of the geosyncline led to the formation of distinctive isopic zones or geological provinces characterized by different sedimentary series (Fig 4).

Whether starting in Permian, Triassic, or Lias, the sedimentary succession in different geological provinces is generally characterized by clastic basal deposits, overlying a folded and eroded Palaeozoic basement.

These clastic deposits are mostly followed by evaporites and carbonates of Triassic–Jurassic age. The Jurassic and Cretaceous sequences usually consist of either shelf carbonates or pelagic limestone and chert formations. Then nearly all geological provinces show a flysch cover of different ages, varying from Upper Cretaceous to Middle Miocene.

The Upper Miocene, characterized by the deposition of evaporitic, paralic, and continental sediments, was a period of significant change in the palaeogeography of Italy. The landscape began to evolve into its present shape, while the sea occupied relatively smaller basins in former sedimentary provinces (yet having no direct relationship to them). During the Pliocene and early Quaternary these basins were then filled with a remarkable volume of sand and shale deposits. Facies variations were controlled mainly by tectonic movements contemporaneous to sedimentation.

During the latest stages of the geosynclinal development (Lower Miocene to Upper Pliocene), the main elements of the tectonic framework took form, prevailingly but not necessarily following the pattern of the former isopic zones (Fig 5).

All along the Apennine range and in Sicily a common structural feature is the thrusting of the "inner" units (those located nearer to the Tyrrhenian massif which was the main source of clastic sediments) over "outer" ones. The resulting allochthonous units of various kinds (nappes, orogenetic landslides, olistostromes) become thinner and show less defined structural features the further they are displaced.

Recent and still active tectonic deformation, which masks older structural and stratigraphical features, is responsible for the present physiography of Italy.

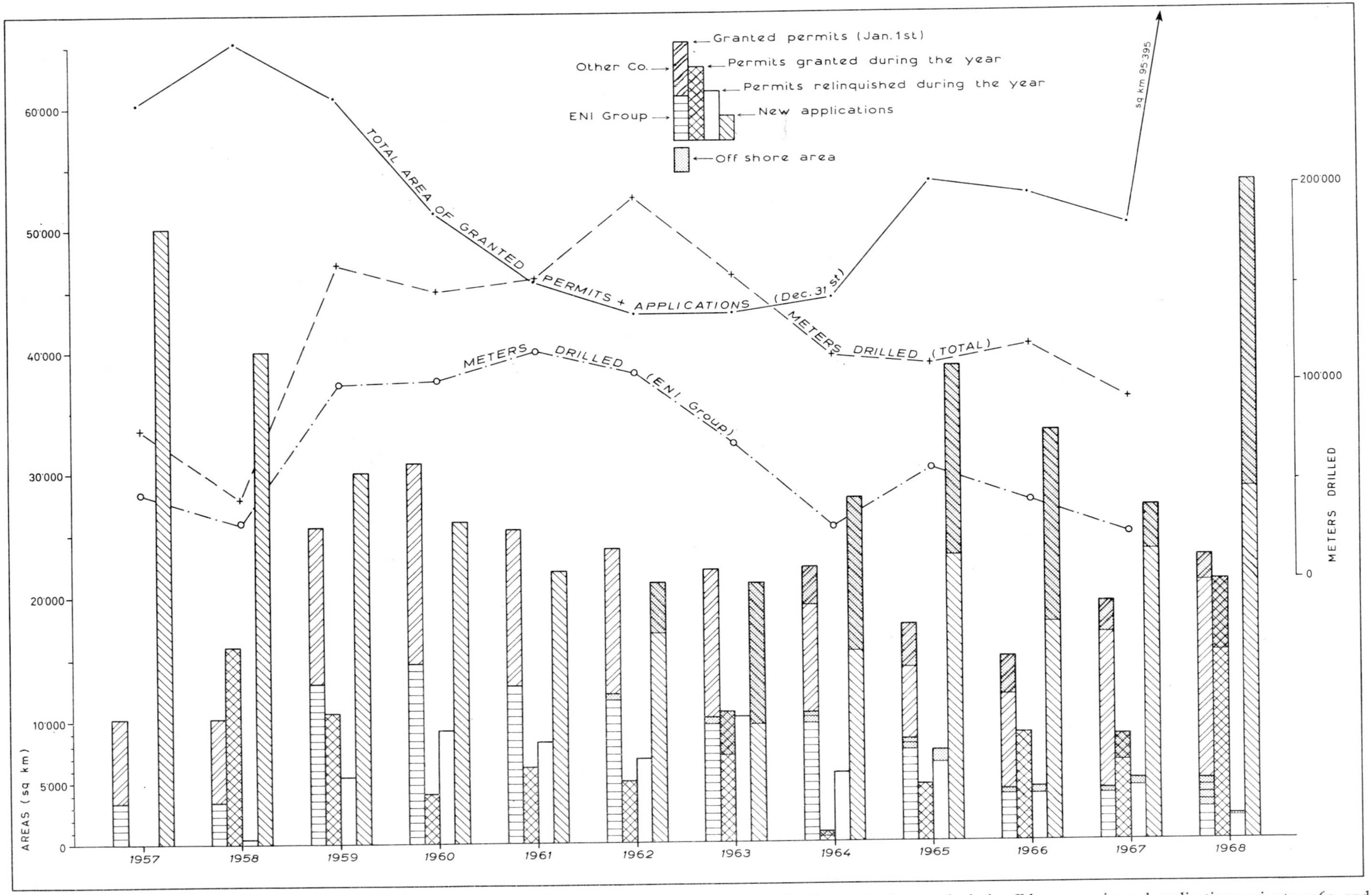

FIG 3. Area of permits granted and applied for according to the Italian laws of 11 Jan 1957 and 21 July 1967 (the graph does not include offshore permits and applications prior to 1962, and permits and applications in the areas subject to the Regional laws of Sicily (20 March 1950), Trentino-Alto Adige (21 Nov 1958), Sardinia (19 Dec 1959), and Friuli–Venezia Giulia (not yet ratified). Data from *Bollettino Ufficiali degli Idrocarburi*, 1968, data up to vol. 12, No. 12.

Total of metres drilled (exploratory, development, stratigraphic, and structural wells); data from *Bull. Amer. Ass. Petrol. Geol.*; this total does not include drilling in ENI area and Sicily and Sardinia.

FIG 4. Simplified geological map of Italy.

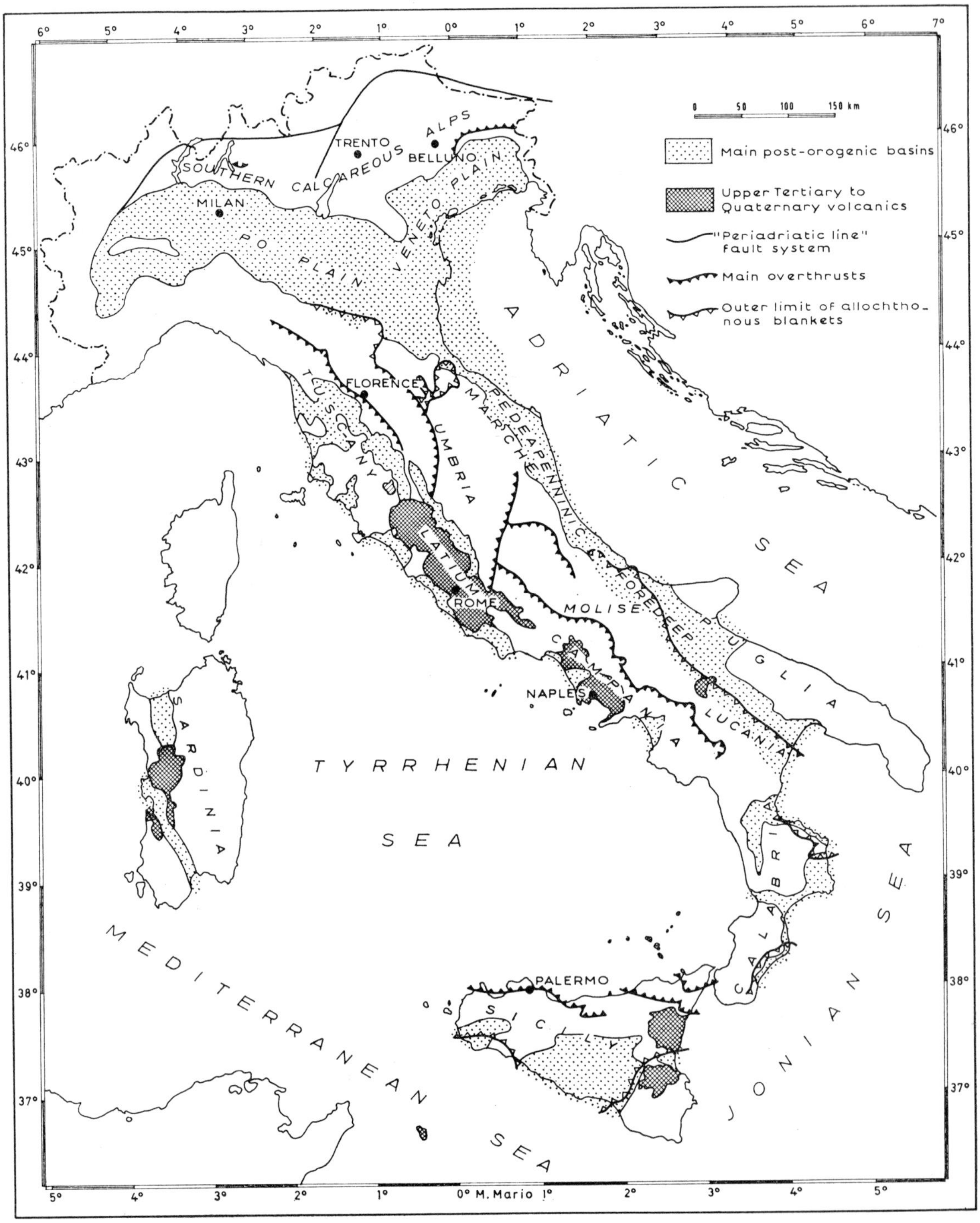

FIG 5. Tectonic sketch of Italy.

Stratigraphic and structural features of peninsular Italy are fairly symmetrical to those of the Dinarides and Hellenides which form the E coast of the Adriatic sea (Aubouin, 1965; Pieri, 1966). The S part of Apennines, in its general outline, may also be considered symmetrical to Sicily.

Alps

Sedimentary series are developed only in the area of the southern calcareous Alps, along a belt varying from a few km to 100 km or so wide, extending from Lago Maggiore to the E Italian state boundary.

The most significant facies variations occur along a W–E trend (Fig 6), and due to this the southern Alps may be sub-divided into provinces in which stratigraphic characteristics were controlled by the depth of the sea, the rate of subsidence, and the distance from the clastic sources (Aubouin, 1955; Bosellini, 1965).

The latest orogenic movements, some of which

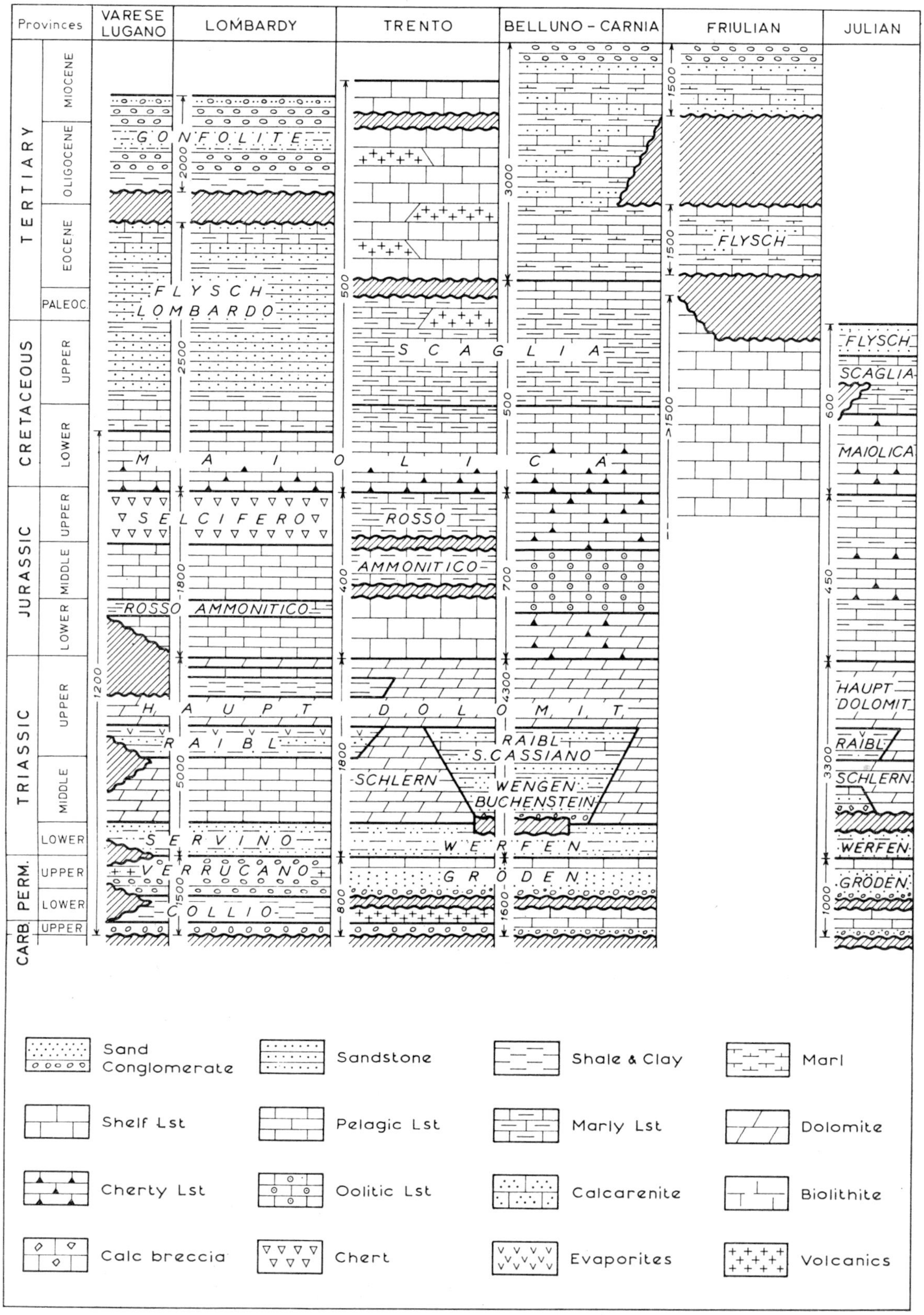

FIG 6. Stratigraphic chart of the S calcareous Alps: type sections, indicative thicknesses (in metres), and names of the more important formations.

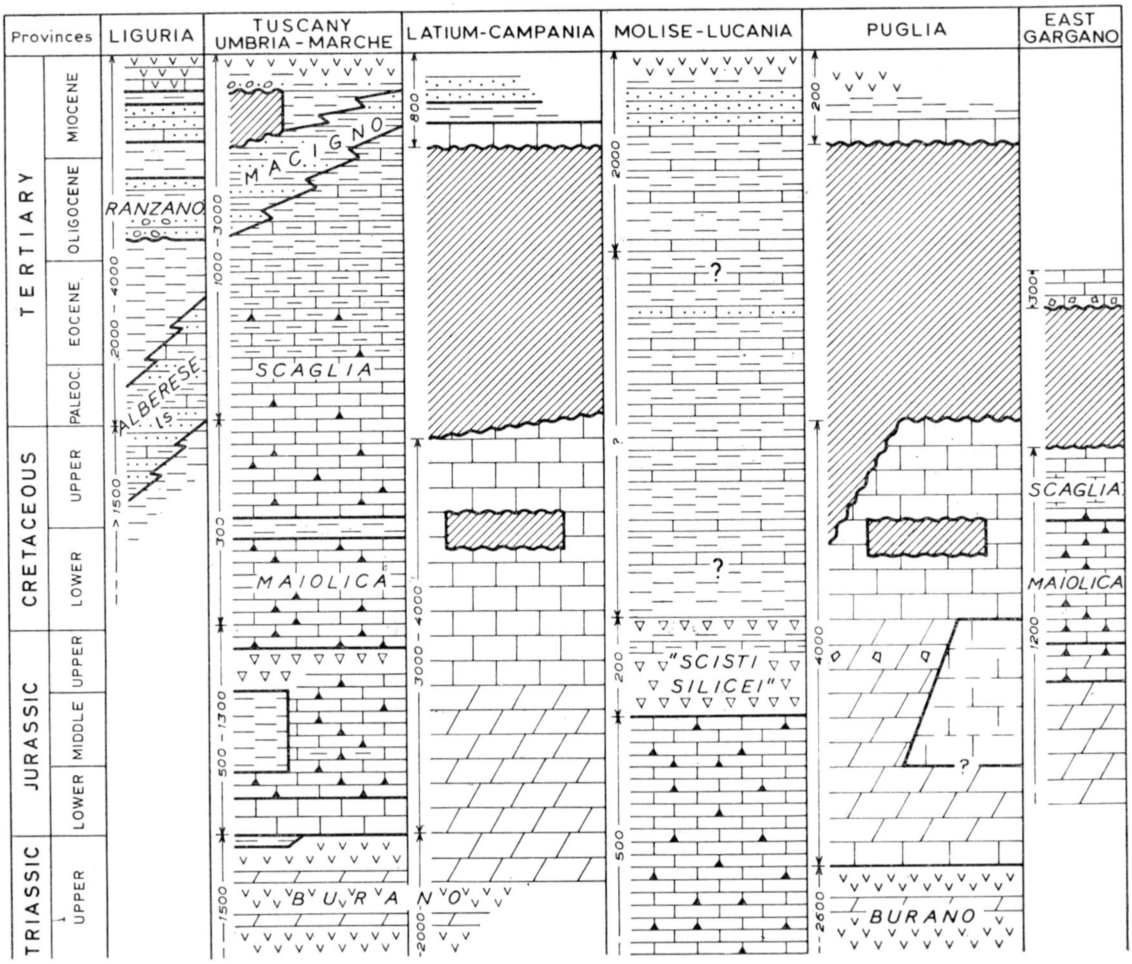

FIG 7. Stratigraphic chart of peninsular Italy: type sections, indicative thicknesses (in metres), and names of the more important formations. Occurrence and lithology of allochthonous material is omitted.

certainly occurred after the Miocene, caused systems of asymmetrical folds and thrust faults. These, in general, are more important in the N areas, while they are comparatively less outstanding along the foothill border. Their general trend is broadly parallel to the southern Alpine border, with some exceptions, as in the Trento province being separated from the Lombardy province by a system of SSW striking faults known as the Giudicarie fault system.

Po and Veneto Plain

This area, bordered by the Alps, the N Apennines, and the Adriatic Sea, broadly corresponds to a Pliocene-Quaternary sedimentary basin now covered by recent alluvial deposits (AGIP, 1969).

The thickness of the sedimentary section varies considerably, ranging from a few hundred metres to the thousands of metres that fill the synclines of the central and SE part of the plain. The deposits consist of shales and sands; their distribution was mainly controlled by synsedimentary tectonics.

Large anticlines are located to the S of a line drawn from Milan to the Po Delta area, while only minor folds are to be found to the W of Milan, along the Alpine border and in the Veneto plain.

Information about the geology of the pre-Pliocene units is scarce, especially in areas where a great thickness of Quaternary and Pliocene sediments is piled up. Yet sufficient data are available to assume that Tertiary and Mesozoic under the unconformity which normally marks the base of the Pliocene–Quaternary cycle are very similar to those outcropping in the Alps and in the Apennines.

Peninsular Italy

The main stratigraphic variations occur in a W–E, direction (Fig 7); based on stratigraphic criteria Peninsular Italy may be sub-divided into seven provinces: (1) Liguria; (2) Tuscany, Umbria and Marche; (3) Latium and Campania; (4) Molise and Lucania; (5) Puglia; (6) E Gargano; (7) Calabria.

Their typical stratigraphic sections may be summarized as follows (AGI, 1964; Crescenti, 1966; Merla, 1959; Pieri, 1961; Selli, 1962):

1. *Liguria*: Upper Triassic shelf carbonates; Jurassic and lowermost Cretaceous pelagic cherty limestones; Lower–Upper Cretaceous and Eocene flysch; unconformable Oligocene coarse clastic neritic sediments; Lower–Middle Miocene marls and sandstones; Upper Miocene evaporites.

2. *Tuscany, Umbria, and Marche:* Upper Triassic transgressive coarse clastics, evaporites, carbonates, and marls; Lower Liassic shelf limestones; Jurassic and lowermost Cretaceous pelagic cherty limestones and cherts; Cretaceous–Eocene pelagic limestones and marls; Oligocene, Lower–Middle Miocene flysch. The age of these flysch deposits ranges from Oligocene in W Tuscany to Middle Miocene in Umbria. In E Marche the flysch units are absent. Upper Miocene clastics and evaporites mark the beginning of a new sedimentary cycle in small post-orogenic basins on the Tyrrhenian side of the Apennines and in a foredeep as continuation to the Po Plain Basin on the Adriatic side. Sediments of Pliocene to Quaternary age consist of shale and sand.

3. *Latium and Campania:* Upper Triassic evaporites and dolomites; Jurassic and Cretaceous shelf carbonates; Palaeogene detrital limestones (locally missing); Lower Miocene transgressive limestones; Middle Miocene shales and sandstones.

4. *Molise and Lucania:* The stratigraphy of this province is quite controversial. At present we adhere to the following interpretation: Upper Triassic and Jurassic pelagic cherty limestones and cherts; Cretaceous and Palaeogene flysch (?); Lower and Middle Miocene flysch; Upper Miocene clastics and evaporites; Pliocene transgressive conglomerates, sand, and shales (deposited in post-orogenic basins originally connected with the Apennine foredeep).

5. *Puglia:* Upper Triassic dolomites and evaporites; Jurassic and Cretaceous shelf carbonates; Palaeogene detrital limestones (locally missing); Lower Miocene transgressive detrital limestones; Middle Miocene shales; Upper Miocene evaporites. The W part of the Puglia area is covered by the Apennine foredeep, locally known as Abruzzo–Bradano trough, filled with thick Pliocene–Quaternary shale and sand deposits.

6. *Eastern Gargano:* Upper Jurassic and Cretaceous pelagic cherty limestones and marls; Eocene detrital limestones. The sedimentary features of this zone are analogous to those of the Ionian province in the Albanian and Greek coasts of the Adriatic.

7. *Calabria:* The crystalline basement and the scanty Mesozoic and Palaeogene outcrops are intensely tectonized. Some coastal post-orogenic basins are known. They are filled with a considerable thickness of conglomerates, shales, and sands of Miocene to Quaternary age.

The oldest record of a tectonic phase (late Eocene) in peninsular Italy is found in the Ligurian area and in Calabria; progressively younger phases affected the peninsula shifting from W to E. The age of tectonics is dated as Lower Miocene in W Tuscany and Upper Miocene to Lower Pliocene in Marche and in N Latium–Campania provinces.

Following the uplifting of inner areas, tectonic gravity sliding was favoured by lubricating rocks, as Triassic evaporites and shale beds layers at the base of flysch type series of various ages. Huge rock masses were thus moved from their original location towards outer areas and came to rest on them as allochthonous covers. Thereby Ligurian series are now found to overlie Tuscany-Umbria series in a large part of N Apennines and Tuscany; Mesozoic and Tertiary flysch series of inferred Tyrrhenian origin overlie Latium–Campania series; Latium-Campania series are thrusted over the Molise–Lucania province, while a part of the Molise–Lucania section is found to cover Pliocene sediments in the S Apennine foredeep of the Puglia province.

The only areas which remained unaffected by these complex structural features are the seaward margin of the Apennine foredeep in the Puglia province and E Gargano. These show a relatively regular structure, mainly controlled by vertical displacement during the latest orogenic phases.

Sicily

Here the most significant facies variations occur in a N–S direction. According to Schmidt di Friedberg (1962, 1964–65) Sicily can be subdivided into five geologic provinces: (1) north-eastern; (2) northern; (3) central; (4) south-eastern; (5) south-western. A sixth province, the Tyrrhenian one, must be added as the original area of the allochthonous formations known in N Sicily. The typical stratigraphic sections of these provinces can be summarized as follows (Fig 8):

1. *North-eastern province:* Upper Triassic (in places Lower Lias); coarse clastic deposits lying over metamorphic basement; Upper Triassic to Lower and Middle Liassic carbonates; Middle and Upper Liassic pelagic limestones and marls; Middle and Upper Jurassic pelagic limestones, marls, and cherts; Upper Jurassic to Eocene pelagic limestones and marls; Oligocene transgressive conglomerates and flysch; Lower and Middle Miocene shallow marine detrital limestones; Middle and Upper Miocene clastics and evaporites; Pliocene conglomerates, sands, and shales.

2. *Northern province:* Upper Triassic (possibly Carnian) shales and limestones; Upper Triassic to Lower Liassic carbonates; Jurassic and Cretaceous pelagic limestones, shales, and cherts; Middle and Upper Eocene transgressive marls and limestones; Oligocene and Lower Miocene flysch; Middle Miocene conglomerates, sandstones, and marls; Upper Miocene evaporites; Pliocene marls, clays, sandstones, and conglomerates.

3. *Central province*: Allochthonous complex formed by formations with heterogeneous lithology and age; unconformable Middle Miocene conglomerates, sandstones, and shales; Upper Miocene evaporites; Pliocene and Quaternary shales, sands, and calcarenites.

4. *South-eastern province:* Middle–Upper Triassic shelf carbonates; Upper Triassic to Lower Jurassic shales and carbonates; Lower Jurassic pelagic limestones chert and marls; Upper Jurassic, Cretaceous, and Eocene cherty pelagic limestones; Oligocene, Lower and Middle Miocene calcarenites and marls; Upper Miocene evaporites.

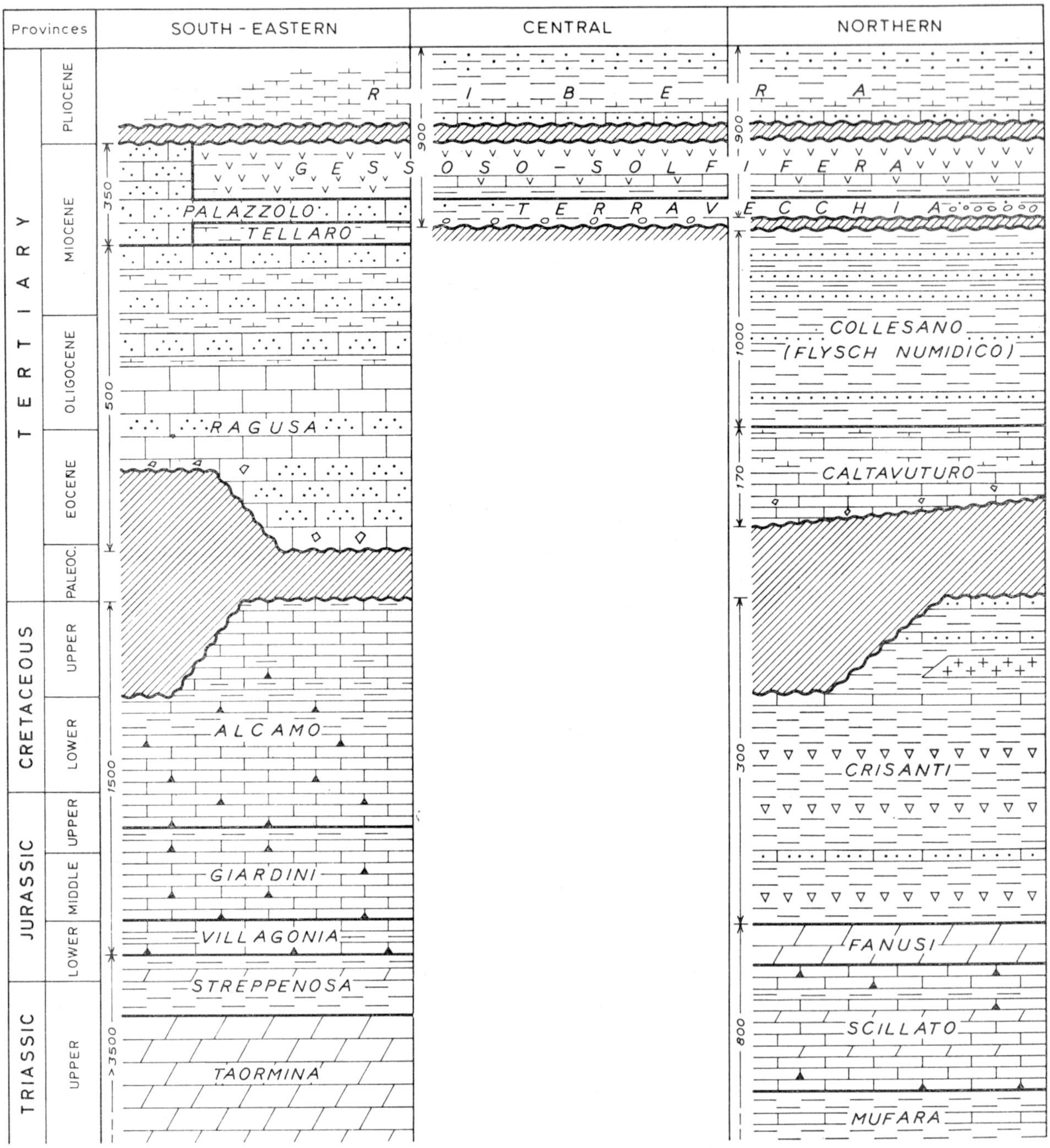

FIG 8. Stratigraphic chart of Sicily (SE, central, and N provinces): type sections, indicative thicknesses (in metres), and names of the more important formations. Occurrence and lithology of allochthonous material is omitted.

5. *South-western province:* Triassic to Liassic shelf carbonates; Middle and Upper Jurassic to Cretaceous pelagic limestones and marls; Eocene and Oligocene limestones and marls; Miocene and Pliocene as in the central province.

6. *Tyrrhenian province:* Upper Triassic (possibly Carnian) shales and limestones; Upper Triassic to Upper Cretaceous reef and shelf carbonates; Middle and Upper Eocene transgressive marls and limestones; Oligocene and Lower Miocene flysch.

The earliest tectonic phase may be dated to the end of the Oligocene as recorded in the Tyrrhenian and northern provinces. Other tectonic phases affected the central and S provinces during Miocene, Pliocene, and Quaternary times. As in peninsular Italy, in Sicily uplifting of "inner" areas (in this case N areas) caused the sliding of allochthonous rock masses southward over the "outer" (southern) areas (Ogniben, 1960).

Allochthonous blankets are missing only in the SE province and in a part of SW Sicily.

Sardinia

Diastrophism, metamorphism, and erosion of Palaeozoic and Mesozoic rocks eliminate their prospectivity; Palaeogene series are poorly developed.

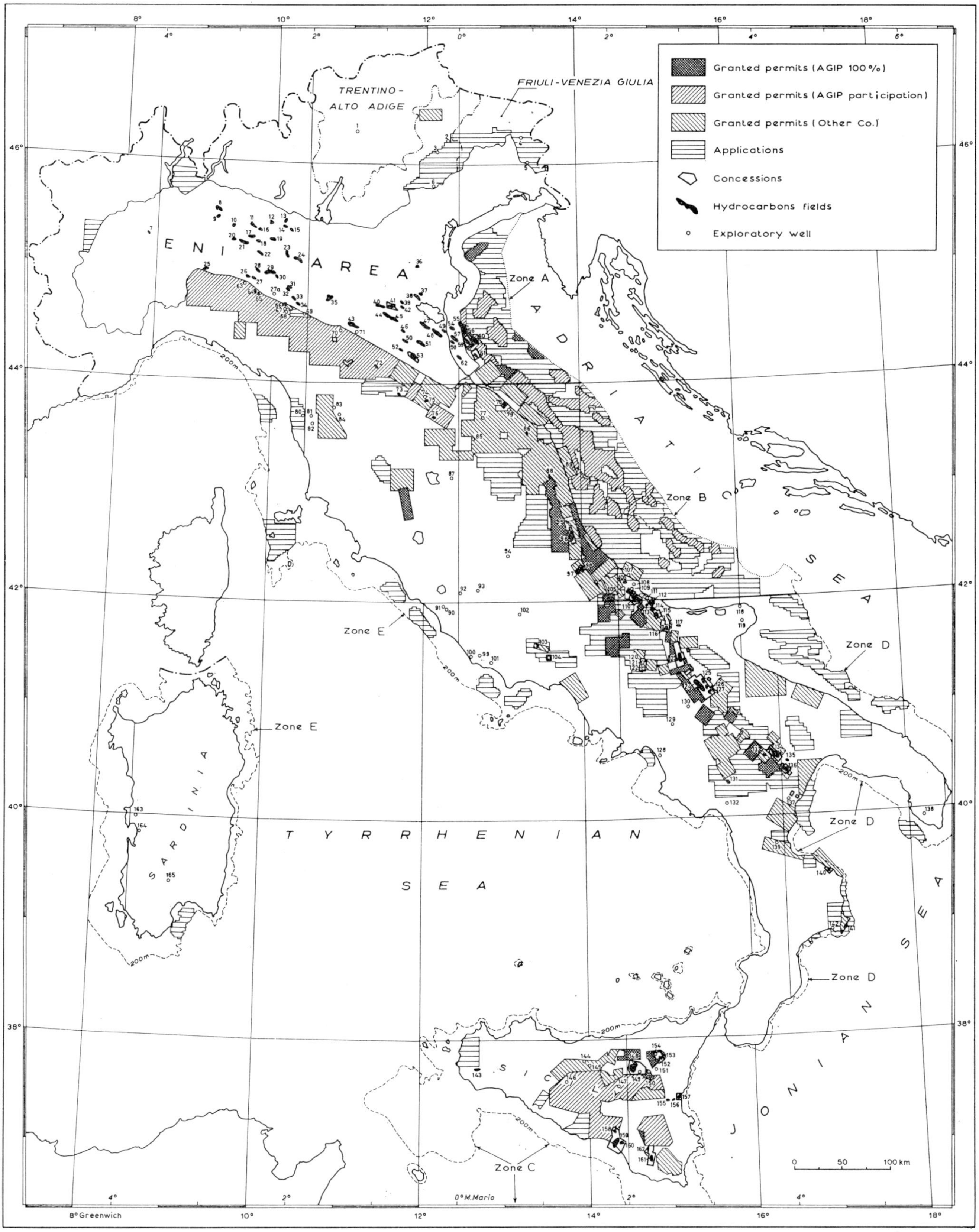

Fig 9. State of permits and applications at the end of 1968, and index map of fields and wildcats mentioned in the text; numbers refer to the list in Appendix 1.

A considerable thickness of Miocene and Pliocene clastics, interbedded with volcanic rocks fills up the post-orogenic basin of Campidano and Sassari.

The thickness and structural features of the Mio-Pliocene formations were controlled by faulting which started in the Upper Oligocene and continued through the Pliocene. The main elements in the structural pattern are normal faults with regional strikes NNW–SSE, N–S, and E–W bordering post-orogenic basins.

HISTORY AND RESULTS OF EXPLORATION FOR OIL AND GAS

An extensive survey on exploration activities prior to 1953 is given by Reeves (1953).

Detailed information on the Po Valley, the Abruzzi–Bradano and the Calabria Basins, and on the SE Sicily province was reported in relatively recent literature (AGIP, 1959; Bronzini, 1959; Carissimo, D'Agostino, Loddo, and Pieri, 1963; Jaboli and Roger, 1952; Kafka and Kirkbride, 1959; Rocco, 1959). In the following paragraphs emphasis will be given to unpublished or lesser known results of recent exploration. The location of the main fields and wildcats is shown in Fig 9; analysis data of some gas and crudes are given in Tables I and II.

Alps

Geological and geophysical surveys were carried out in the southern Alps, from Lombardy to the Yugoslav border, from 1956 to 1962.

The complex structural pattern and the lack of definite stratigraphic prospects led to the discontinuation of exploration after a first reconnaissance survey in the Lombardy province. In the Trento province the Coredo 1 wildcat (T.D. 2067 m) was drilled by AGIP in 1958; some oil and bitumen shows were found in Lower Lias and Upper Triassic carbonates (Pavan, 1968).

In the Belluno province the AGIP wildcat Belluno 1 (1960, T.D. 2613 m) drilled Triassic–Jurassic carbonates but was abandoned because of the complexity of the subsurface structure (Martinis, 1966). Sedico 1, drilled by the same company (1961, T.D. 2297 m) tested the Eocene–Middle Jurassic section, however, only fresh water levels were found in the Mesozoic carbonates. The Volpago 1 wildcat (T.D. 3442 m) was drilled in 1960 by Idrocarburi Piave near the margin of the Veneto Plain, on an outcropping Upper Miocene anticline, and was abandoned in Lower Eocene carbonates. Gas shows were reported from Miocene porous levels.

The Friuli province was explored by two AGIP wildcats. Buttrio 1 (1955, T.D. 1442 m) penetrated the lowest members of Eocene flysch and 600 m of Upper Cretaceous limestones with gas and oil shows. Bernadia 1 (1959, T.D. 2570 m) was drilled on an asymmetrical Cretaceous anticline. The well had the objective of testing the Lower Mesozoic section but has to be abandoned because of evidence of multiple overthrusting of Mesozoic and Tertiary levels in the lower part of the drilled section (Martinis, 1966).

Po and Veneto Plain

The history and results of exploration carried out by AGIP in the Po and Veneto Plain have been illustrated in the Proceedings of the Symposium on Western European Gas Fields (1959). At that date the main gas fields of Caviaga, Cortemaggiore (gas and oil), Ripalta, Bordolano, Ravenna, Selva, Minerbio, Alfonsine, and some 25 minor ones were already discovered.

During the last ten years exploration has actively continued, resulting in the discovery of the gas fields of Brugherio (Fig 10), Sabbioncello, and over 15 minor ones. The original recoverable reserves which were assumed to amount to 93 billion cu m in 1958 have increased to about 200 billion cu m (comprising the Po Plain offshore area) by the end of 1968. The yearly output was contemporaneously raised from 4·8 to more than 6 billion cu m.

TABLE I

Molecular composition of the gas produced by some Italian fields

Moles, per cent \ Field	Corte wells (38–70)	Caviaga	Ravenna	S. Salvo (sand reservoir)	S. Salvo (lst reservoir)	Ferrandina (sand reservoir)	Ferrandina (lst reservoir)	Gagliano	Mazara (well 5)	Catania (well 5)
Methane	91·21	97·03	99·54	98·34	60·18	98·98	88·22	89·27	99·15	42·55
Ethane	5·85	1·66	0·06	0·54	5·49	0·06	3·14	4·99	0·12	0·12
Propane	1·80	0·56	0·01	0·42	2·78	0·04	0·84	2·35	0·03	0·08
iso-Butane	0·27	0·19	—	0·02	0·31	0·01	0·07	0·61	Traces	0·04
n-Butane	0·32			Traces	0·63	0·01	0·11	0·69	,,	
iso-Pentane	0·08	0·04	—	—	0·16	Traces	0·03	0·34	,,	0·02
n-Pentane	0·05		—	—	0·17	,,	0·01	0·21	,,	
hexanes	0·07	Traces	—	—	0·20	—	—	0·48	—	0·01
Heptanes plus	—	—	—	—	—	—	—	—	—	—
Helium	—	—	—	—	—	—	—	—	—	—
Carbon dioxide	0·03	0·07	Traces	0·02	8·40	0·16	4·13	0·69	0·02	48·83
Nitrogen	0·32	0·45	0·39	0·66	21·62	0·74	2·19	0·37	0·68	8·35
Hydrogen sulphide	Zero	Zero	Zero	Zero	0·06	Zero	1·26	Zero	Zero	Zero
Total combined sulphur (except H_2S) g/st cu m	0·0041	0·0142	0·0025	0·004	0·0198	0·0061	0·0313	0·0005	0·003	0·006

TABLE II

Composition of some Italian crude oils

Field	Cortemaggiore (Lower Pliocene)	Vallecupa ("Tocco" limestone)	Torrente Tona (Well 13)	Pisticci (Well 5)	Ragusa (Oil gathering station)	Gela (Well 20)
Density at 60° F	0·801	0·967	0·855	1·005	0·938	0·990
API Gravity	45·1	14·8	34·0	9·3	19·3	11·4
Engler viscosity at 20° C	1·08	48·6	1·37	—	124·9	—
Engler viscosity at 50° C	—	—	—	66·0	—	312·8
Sulphur, % by weight	0·08	4·91	2·09	6·23	2·41	6·15
Hydrogen sulphide	zero	zero	zero	zero	zero	zero
Composition of the 1st key fraction						
Aromatic carbon, %	30·5	25·0	18·0	19·5	14·0	13·0
Naphthenic carbon, %	24·0	31·0	25·5	37·0	30·0	40·0
Paraffinic carbon, %	45·5	44·0	56·5	43·5	56·0	47·0
Composition of the 2nd key fraction						
Aromatic carbon, %	26·0	36·0	22·0	35·0	24·5	19·0
Naphthenic carbon, %	20·0	9·5	22·5	22·0	15·0	35·0
Paraffinic carbon, %	54·0	54·5	55·5	43·0	60·5	46·0
Base of the crude	Naphthenic-intermediate	Naphthenic	Intermediate	Naphthenic	Intermediate	Naphthenic

The Pliocene sediments, where the main gas reserves were found, lie unconformably on the pre-Pliocene formations; another main unconformity separates the Lower Pliocene from the Middle-Upper Pliocene section. Lower Pliocene accumulations are generally found at a depth varying from 450 m to 3200 m in anticlinal traps located along a belt 25–50 km wide running from SE of Milan to the Ravenna–Rimini area ("pedeapenninic folds"). In the area between Milan and Brescia, where Lower Pliocene sediments overlie a deeply eroded Miocene section, at the depth of 900–1200 m gas fields were discovered in stratigraphic traps controlled by the distribution of sand and conglomerate bodies at the base of the transgressive sediments (Sergnano, Leno, Brugherio, etc.).

Middle–Upper Pliocene fields are located in the SE part of the Po Plain; the gas accumulations are mainly found in combination traps, controlled by the pinch-out of sand wedges that occur upwards on the flanks of the anticlines.

Minor accumulations of hydrocarbons (gas, condensate, and oil), were discovered in the Miocene (Cortemaggiore, Desana, Vigatto, etc.).

No hydrocarbon indications were found in the pre-Miocene section that was tested by a number of deep wildcats.

Adriatic offshore

Exploration in the Adriatic was carried out in the ENI area (that includes a 15 km wide offshore belt) and in a few exploration permits granted under the 1957 Hydrocarbon Law.

In the SE Po Plain offshore the gas fields of Ravenna Mare (Fig 11), Cesenatico Mare, Cervia Mare, South Ravenna Mare, Porto Corsini Mare, and West Porto Corsini Mare were discovered between 1960 and 1967 by AGIP. In 1968, after the granting of new permits according to the 1967 Hydrocarbon Law, three more fields were discovered by AGIP (Porto Garibaldi) and AGIP-Shell (Agostino and Amelia).

The main geological features of these fields are the same as for the Po Plain ones; the gas is found in Pliocene sand reservoirs in structural and combination traps at depths of 1500–4000 m.

Further S, along the Marche and Abruzzi coast, two discoveries were reported: S. Stefano Mare (Elf, 1967, T.D. 1890 m), where gas was found in the Pliocene sands, and S. Giorgio Mare (Elf, 1968, T.D. 1774 m), where the reservoirs seem to be in pre-Pliocene limestones.

Other possible traps were tested by 18 new field wildcats which proved to be dry.

At the end of 1968 three deep wildcats were drilling in the N Adriatic offshore.

Tuscany, Umbria, Marche province

(a) *Northern Apennine*

Several small gas and oilfields were discovered during the past century in areas where surface evidence for hydrocarbons had been known since historical times. Most of those fields are now exhausted; their maximum comprehensive yearly oil output was about 20,000 metric tons in 1931–32, at present it is less than 3000 metric tons. The reservoir rocks are either porous units of the allochthonous blanket belonging to the Ligurian series, as at Montechino, Velleia, Vizzola, Vallezza, Pietramala, Viamaggio, etc., or sandstones of Miocene autochthonous series, as in the fields of Salsomaggiore, Selvapiana, and Castagno. New data and theories on the structure of the Apennine led to a different approach to exploration from 1956, when exploration permits were granted to AGIP, MPI, and SNIA. Gravity and reflection seismic surveys were carried out with the main aim of exploring the autochthonous section under the thick allochthonous cover.

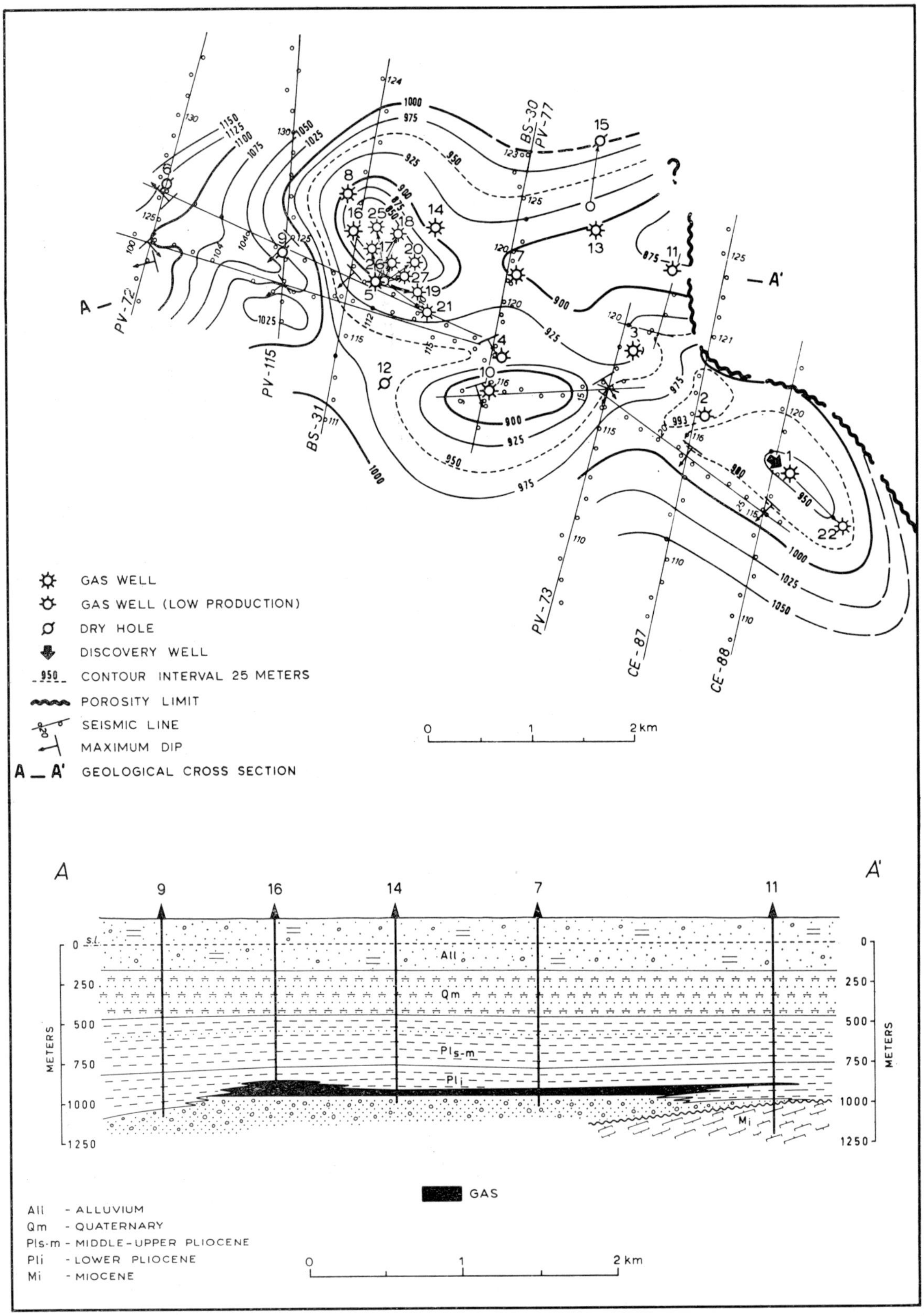

FIG 10. Brugherio field: top of Lower Pliocene gravels and geological cross-section.

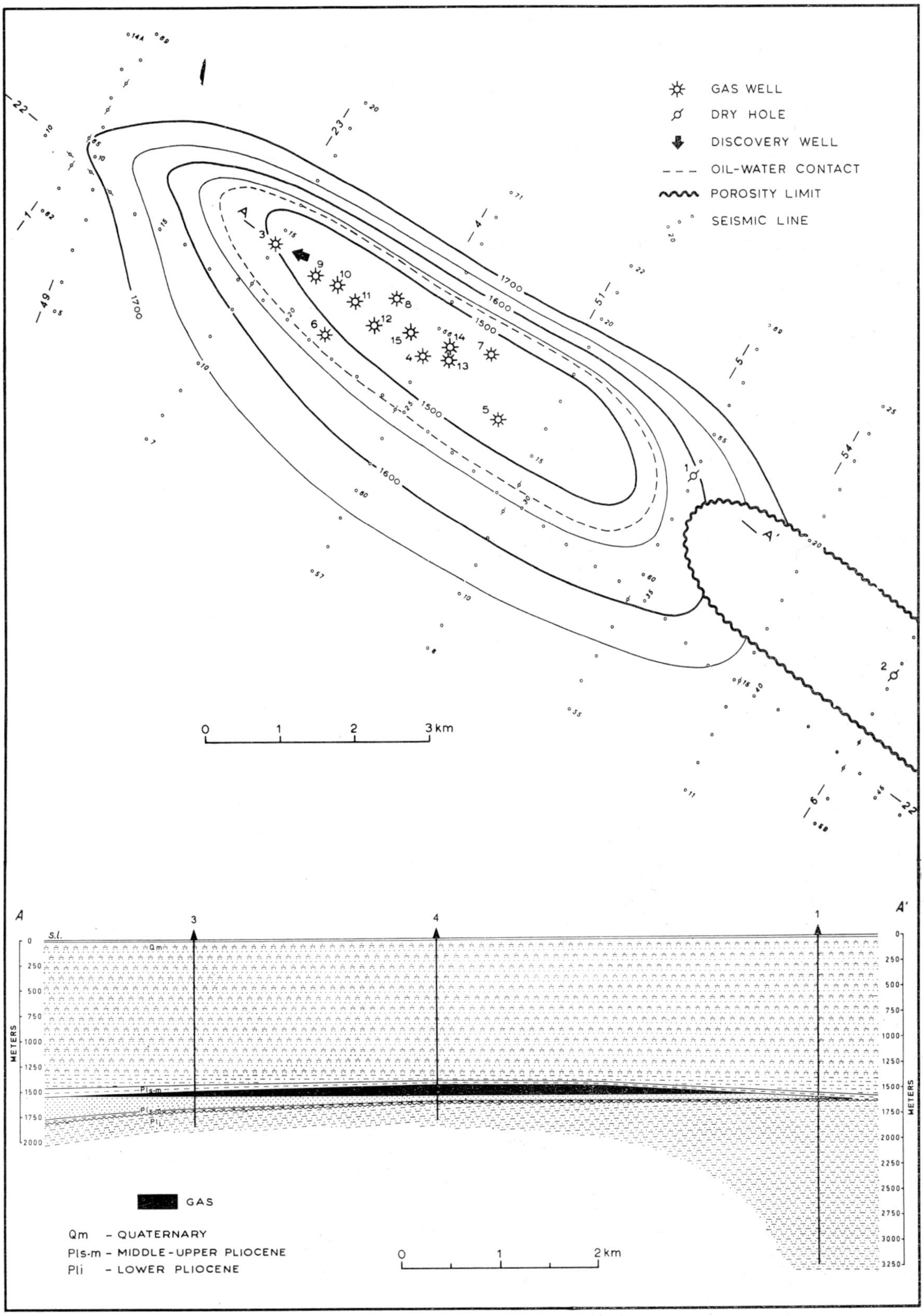

FIG 11. Ravenna Mare field: top of Middle Upper Pliocene sands and geological cross-section.

Only a few wells reached the autochthonous substratum, finding either Miocene (Levizzano 1, Vallezza 178 and 179, Ponte dell'Olio 1, Sevizzola 1, S. Polo d'Enza 1) or Pliocene formations (Monteveglio 1). Other tests did not reach beyond the allochthonous cover, which in places has a thickness of more than 3000 m.

Two dry wildcats, Salsomaggiore 1 (AGIP, T.D. 3326 m) and Monte Pietra 1 (SNIA, T.D. 3619 m), were located on outcropping Miocene autochthonous structures to test the deeper Miocene horizons.

Besides the small gas field of Santerno (AGIP), the only significant production from autochthonous series is the one discovered in 1955 by SPI at Marzeno. Cumulative production to 1967 adds up to 234 million cu m of gas from Pliocene and Middle Miocene sands.

(b) Umbria and Western Marche

Subsurface exploration of the Mesozoic section in this area was carried out from 1956 to 1961; the wildcats Burano 1, Fossombrone 1 and 2, Perugia 1 and 2 were drilled by AGIP on outcropping anticlinal structures. The Cretaceous and Jurassic section did not show any presence of hydrocarbons; the Upper Triassic evaporites and dolomites (Burano Formation; Martinis and Pieri, 1963) yielded several faint oil shows. Only the Perugia 2 well, located on the W part of the area, reached the lowest part of the Triassic section, with impervious quartzites, clays, and micaschists.

(c) Neogene Tyrrhenian basins

The sediments in these basins may reach a thickness of about 2000 m. In the Upper Miocene and Pliocene section porous sand beds and lenses interbedded with shales are present. A first exploration programme was carried out from 1950 to 1960 by Dalmine with the drilling of four dry wildcats: Ponsacco 1, Poggio 1, Zanone 1, and Cerbaie 1.

Recent drilling activity by AGIP consisted of the completion of four wildcats, of which three gave some gas shows (Certaldo 1, 2, 3, T.D. 1738, 1860, 1668 m).

(d) Apennine foredeep in the Marche and N Abruzzo region

Exploration in this region started before World War II and is still going on. Approximately 40 wells have been drilled. From 1952 till now the small gas fields of Iesi (AGIP), Rapagnano (AGIP), and Cellino (Petrosud) have been discovered, producing from Pliocene sand reservoirs. Cumulative production from these fields up to 1968 is estimated at 33, 53, and 470 million cu m respectively; smaller production of gas is also obtained from Fano (Metauro) and Marotta (Corime) wells.

Latium–Campania province

(a) Valle Latina

Exploration carried out before World War II resulted in the discovery of the two small oilfields of Ripi and S. Giovanni Incarico. The production of low API gravity oil from small accumulations in Miocene sandstones never reached an economic rate and the fields are at present abandoned. Further exploration was carried out on the same prospect without results.

(b) The Mesozoic prospect

Numerous surface evidence of bitumen connected with faults, affecting Mesozoic carbonate rocks, encouraged exploration of the Mesozoic subsurface series. However, the results of two AGIP wildcats, Antrodoco 1 (1959, T.D. 2875 m) and Trevi 1 (1965 T.D. 3549 m), only pointed to the complex structure of the Mesozoic section, which in places is thrusted over plastic Miocene formations (Pieri, 1966). Shearing off of the Mesozoic sediments possibly occurred at the level of Triassic evaporites (Burano formation).

(c) Neogene post-orogenic sedimentary basins

Based on gravity and seismic data some wells (Roma 1 and 2, Ladispoli 1, Moricone 1, Fogliano 1, Latina 1 and 2) were drilled between 1953 and 1965 by Pontina Metano and AGIP in the Tevere Valley basin and in the Pontina Plain. The wells penetrated the whole section of Quaternary and Pliocene sediments whose thickness averages 1000 m and reached underlying Miocene, Palaeogene, and Mesozoic formations. As no evidence of hydrocarbons was found in these wells, this prospect was abandoned.

Molise–Lucania province

Production of oil and gas has been obtained from the small field of Tramutola since 1936; the reservoirs are in a Palaeogene flysch series overlying Mesozoic cherty limestones. From 1948 to 1959 the cumulative production of the field reached 6000 cu m of oil.

During the last ten years exploration was directed mainly toward the following prospects.

(a) Lowest Mesozoic section

The Mesozoic section in pelagic-calcareous facies is known only for the interval from Upper Triassic to Cretaceous, but no data are known about older formations.

The Tramutola 45 (T.D. 2000 m) and Lagonegro 1 (T.D. 1447 m) wildcats were drilled by AGIP in 1959 and 1961 respectively on anticlinal structures, but they were abandoned after finding evidence of reversals in the stratigraphic sequence, apparently due to overthrusting.

(b) Flysch complexes and shelf carbonates section

The complex and controversial structural features of the S Apennines have been mentioned previously.

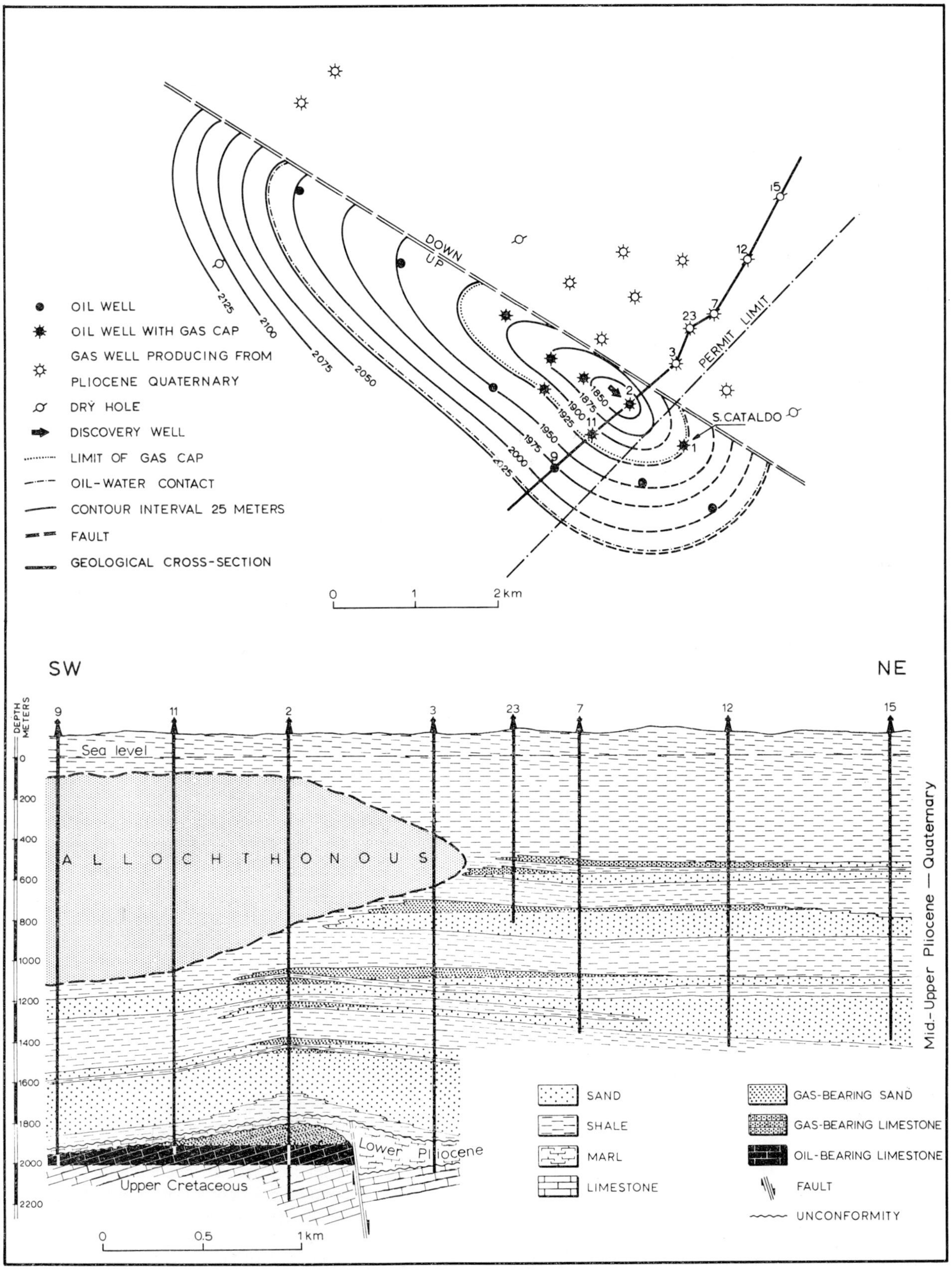

FIG 12. Pisticci field: top of Cretaceous limestone and geological cross-section (from Carissimo *et al.*, 1963, modified).

The thick Cretaceous–Miocene sequences are interpreted as more or less laterally displaced units. Even the shelf carbonate formations of essentially Mesozoic age forming the main orographic elements are believed to be allochthonous. Nearly 20 wildcat were drilled to test these series; Serroni 1 and 2 (1958, Somicem, T.D. 2486 m) found gas shows in the flysch; Nusco 1 and 2 (1955–56, Fondedile, T.D. 1542 m) found oil and gas shows in Mesozoic carbonate horizons.

A group of wells, S. Croce (1962, Idroc. Castelgrande, T.D. 3195 m), Cercemaggiore 1 (1963, Idroc. Ariano, T.D. 3215 m), and Jelsi 1 (1966, Idroc. Castelgrande, T.D. 2607 m) were completed as oil producers from small accumulations in Mesozoic carbonates.

(c) Post-orogenic Neogene basins on the Tyrrhenian coast

Under geological conditions similar to those described for post-orogenic Neogene basins in Tuscany and Latium, geophysical and geological surveys and drilling of wildcats were carried out in the alluvial plains of the Garigliano, Volturno, and Sele rivers. The first two plains were explored by Samet, AGIP, and Ripemi with ten wells down to a maximum depth of 3006 m; the Sele plain was explored by the wildcat Sele 1 (1954, Samet, T.D. 1497 m). The penetrated section consists of Quaternary to Upper Miocene shales, sands, and conglomerates. Tuffs and basalts were encountered in the Volturno area. The bottom of the basin was not reached, nor was any evidence of hydrocarbons found.

Puglia Province

(a) Abruzzi–Bradano trough

The history of exploration carried out by AGIP in this area was reported by Carissimo *et al.* in 1963. Fields already discovered to that date included Cupello–S. Salvo (gas), Grottole–Ferrandina (gas), Pisticci (oil and gas, Fig 12), Cigno–Vallecupa (oil), and Ascoli–Satriano (gas); total reserves of the first three fields were evaluated to more than 21 billion cu m of gas and one million metric tons of oil. During the past five years exploration has been very active both by AGIP and by other companies, as Montecatini, Idroc. Castelgrande, SNIA, AMI. Reserves of known fields were extended and new discoveries were obtained, namely, the gas fields of Montecilfone (AGIP), Guglionesi–Montecapraro–Portocannone (AGIP–AMI), Monte Stillo–Roseto (AGIP), Candela (SNIA), Bomba (IM), Accettura (AMI), Colle Scalella (SORI), and the oil and gas field of Torrente Tona (AGIP).

Recoverable gas reserves of ENI fields in this area are evaluated at about 40 billion (10^9) cu m. Some additional 10–15 billion cu m may be estimated to represent the reserves of fields owned by the other companies.

The main reservoirs were found in the top part of Cretaceous and Miocene limestone horst blocks and in combination traps of sand beds lying within the Lower Pliocene and Middle–Upper Pliocene to Quaternary. The trapping of hydrocarbons is younger than Lower Pliocene. In several fields, as at the SW margin of the trough, allochthonous material is found within the Pliocene to Quaternary section. The allochthonous masses are mainly composed of Miocene and Pliocene sediments originally deposited in areas situated further inward.

Their displacement took place between late Lower Pliocene and Quaternary. The presence of this allochthonous material, which in places reaches a thickness of 1000 to 2000 m, seriously hampers the use and the interpretation of seismic surveys.

A number of wells, namely, Cupello 19, Rotondella 4, Casalbordino 1 and 2, Casaborselli, Villalfonsina 1, Vallecupa 45, Petacciato 1, penetrated under the Pliocene basin floor into a considerable thickness of shelf carbonates (3566 m from Upper Cretaceous to Middle–Upper Lias in the Cupello 19 well section).

Interesting oil shows in the Cretaceous were found in some of these wells.

(b) Gargano, Murge, and Salento

Two deep wildcats were drilled in this area with the objective of testing the Mesozoic. Foresta Umbra 1 (1961, AGIP, T.D. 5912 m) on the Gargano Peninsula drilled through Jurassic–Upper Triassic dolomites and dolomitic limestones and, from 3291 m to T.D. into the anhydritic-dolomitic Burano formation, where several slight evidences of oil were found. Ugento 1 (1959, Somicem, T.D. 4535 m) was drilled in the southern-most part of the Salento area in shelf dolomites and limestone of Upper Cretaceous–Upper Jurassic age. Hydrocarbon indications were encountered.

Eastern Gargano Province

In this area only the wildcat Peschici 1 (1961, AGIP, T.D. 1275 m) was drilled in Cretaceous–Jurassic carbonates, but without any hydrocarbon shows.

Calabria

The history of exploration carried out by Montecatini in the post-orogenic Neogene Calabria Basins has been summarized in a paper by Bronzini (1959). In the Crati Basin the dry wildcat Doria 1 (T.D. 1919) was drilled in 1956. In the Crotone Basin a small gas accumulation in Pliocene sands was discovered in 1953 by the wildcat Capo Cimiti 1 (T.D. 2972 m) and gas shows were reported in 1953 by the wildcat Capo Cimiti 1 (T.D. 2972 m) and gas shows were reported in the S. Leonardo 1 well (T.D. 1270 m). Gas production from shallow accumulations in Upper Pliocene–Quaternary sands was found in 1964 in the W margin of a seaward developed basin in the Cirò area.

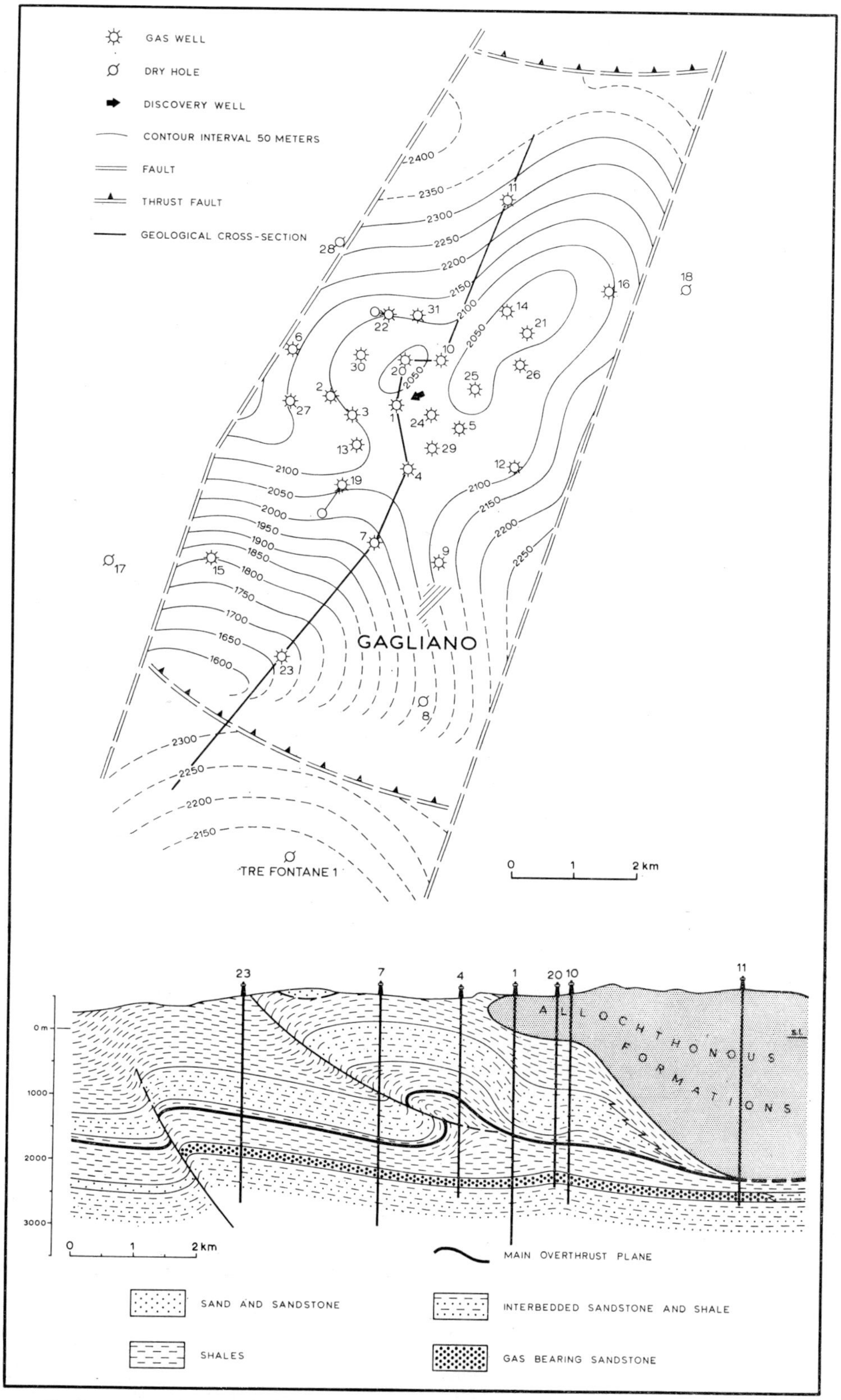

FIG 13. Gagliano field: top of the main producing sands and geological cross-section; the structure of the overthrusted units is greatly simplified.

Sicily

North-eastern Province

Stratigraphic and structural features in this province do not encourage exploration for hydrocarbon.

Northern province

Intense exploration has been carried out in this province since 1955 to test the Mesozoic section and the Tertiary flysch series.

The Mesozoic carbonate section, tested by more than ten wildcats, yielded oil shows in the wells Avanella 1 (T.D. 3051 m) and Colla (T.D. 2205 m), drilled by Montecatini group companies in 1956 and 1966 (Schmidt di Friedberg, 1966); the other tests were completely dry. The exploration of the Oligocene–Miocene flysch prospect (Collesano formation = Numidian flysch), where oil and gas production was obtained at Gioitto and Bronte in 1940–43, led to the discovery of the "class A" gas field of Gagliano (Fig 13) by AGIP in 1960.

Data from the wells show that the main accumulations found at an average depth of 2500–3200 m belong to a faulted deep anticline covered by upper tectonic units that, at places, are 2000 m thick. These upper units consist mainly of flysch which is part of the same formation as the one of the lower unit (*i.e.* Collesano formation).

In spite of further exploration carried out with the drilling of more than 30 wildcats by AGIP and other companies, no other important discoveries were made. Only small, most uneconomic, gas productions were obtained in: Feudo Grande 1, 2 and 4, Castelluzzo 2, Miraglia 1, Randazzo 1 and 2, Casalini 1 and 2.

Central Province

This province consists of a basin filled with post-orogenic clastics and evaporites deposited during Middle–Upper Miocene and Pliocene over a thick blanket of allochthonous rocks displaced from their original location in northern provinces. The nature of the underlying autochthonous formations is known only in the SE part of this province, where it consists of a sequence of Upper Triassic to Quaternary sediments similar to the ones of the type section of the SE province.

The discoveries in the central province are so far limited to the oilfields of Gela, Ponte Dirillo, and Cammarata.

Gela field (Fig 14), discovered by AGIP in 1956, produces low gravity oil from faulted anticlinal traps at 3000–3500 m in Upper Triassic dolomites (Taormina formation) and the overlying alternating shales and carbonates (Streppenosa formation).

Details on this field are reported in a previous paper (Rocco, 1959); cumulative production up to the end of 1968 was of 8 million cu m.

Smaller quantities of oil were found in the same reservoirs in Ponte Dirillo field (5 km E of Gela); in the Cammarata–Pozzillo field (7 km N of Gela) oil is produced from porous carbonates, possibly the lateral equivalent of the Liassic Villagonia formation covered by dense limestones of Middle–Upper Jurassic age.

Along the N border and in the central part of the province some 13 wildcats were drilled to test the Tertiary section either above or under the allochthonous blanket.

No commercial results were obtained. Gas shows were reported from Tortonian sands (Terravecchia formation) in the well Platani 1 (Platani Petr., 1956, T.D. 701 m) and from Lower–Middle Miocene sandstones (probably belonging to the allochthonous complex) in the Enna 1 and 2 wildcats (SNIA–AGIP 1955–1958, T.D. 736–2077 m).

South-eastern Province

Petroleum geology and exploration in this province have been discussed in earlier papers (Kafka and Kirkbride, 1959; Rigo and Cortesini, 1961). The Ragusa oilfield, discovered in 1953 by Gulf, is still the only producing one of the province (though the Gela oilfield, in the central province, produces from the same horizon); its cumulative production, up to 1968, amounts to 14 million cu m.

In most of the 45 wildcats drilled to an average depth of 2000–3000 m in this province by several companies (AGIP, CISDA, Edison, Arpe, Gulf, Petroli dell'Isola, etc.) oil and/or gas shows were reported, but no commercial production has been obtained so far, with the possible exception of Comiso.

The origin and distribution of oil in the Triassic–Liassic carbonates of the SE province of Sicily is controlled by the areal extension of the Streppenosa formation. This consists of an alternation of black shales and dark dolomites of Liassic age overlying the shelf dolomites of the Triassic Taormina formation; the oil is trapped at the top of the Taormina formation and in permeable carbonate levels of the Streppenosa, or, as in the above-mentioned Cammarata–Pozzillo field in the SE part of the central province, in porous carbonates related to the reef type sediments bordering the N edge of the Streppenosa Basin.

In the northern most part of the province gas accumulations were discovered by Miso in the small post-orogenic basin of the Catania plain. The gas was found in sand lenses enclosed in Quaternary clays at the maximum depth of 500 m; the cumulative production of the Fontana Rossa, Cisina, and Rizzo field up to 1968 amounts to about 120 million cu m.

South-western province

In this province exploration was carried out by AGIP and SOIS in 1955–1962. Twelve wildcats were drilled to test both the Tertiary and Mesozoic sections down to 2583 m (Lippone 1).

The "class D" field of Lippone–Mazzara was discovered in 1957; gas was found in Tortonian sands (Terravecchia formation) belonging to the Neogene section filling a post-orogenic basin.

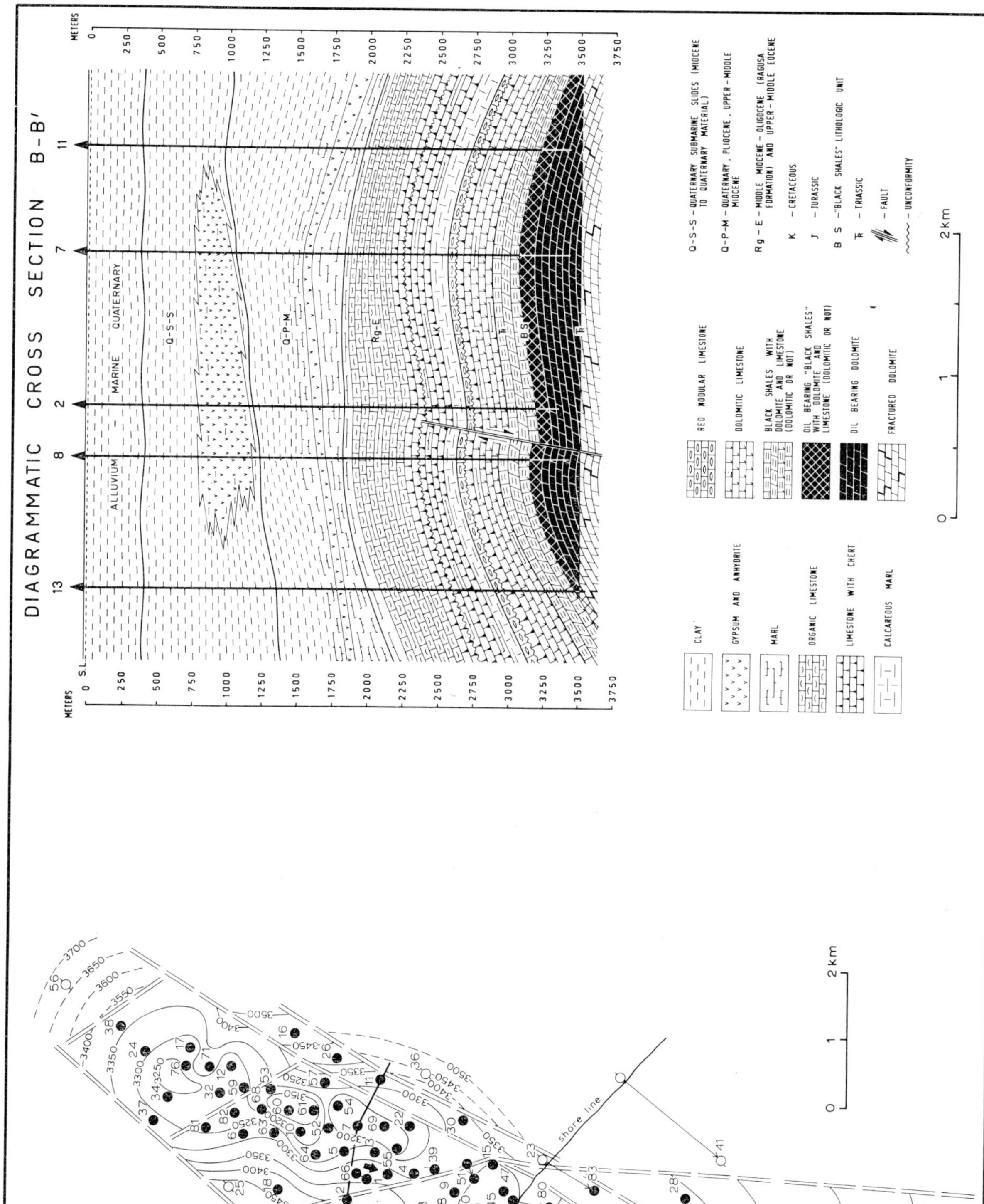

FIG 14. Gela field: top of Taormina formation and geological cross-section (from Rocco, 1959, modified).

Sardinia

From 1957 to 1965 SAIS (a joint venture of Wintershall AG and the Sardinian Administrative Government) carried out the exploration of Neogene post-orogenic basins in the Campidano and Sassari areas. Three widlcats (Oristano 1, T.D. 1802 m; Oristano 2, T.D. 1700 m; Campidano 1 T.D. 1700 m) were drilled in Miocene and Pliocene sediments and reportedly reached the bottom of the basins. All the wells were abandoned as dry holes.

FUTURE PROSPECTS FOR NEW OIL AND GAS RESERVES

The present distribution of permits and applications (Fig 9) shows the main trends of exploration for the next years.

In the E Alps, in spite of disappointing results of earlier exploration, SNIA applied for a number of permits of which the main objective seems to be the testing of the Mesozoic.

For the Po and Veneto Plain (where ENI holds exclusive rights of exploration) the recent improvements of seismic techniques promoted a new extensive survey aimed at ascertaining both the interesting remaining possibilities of the Pliocene sediments and the structural situation of the mostly untested pre-Pliocene section.

A number of permits were granted in the N Apennines to an association of several companies (AGIP, Montedison, BP, Fina, Union Oil, Italsignal, API). The failure of the exploration carried out between 1956 and 1964 was due mainly to the very poor results of geophysical surveys. New techniques will possibly overcome the handicaps of this otherwise attractive prospect.

In the Umbria-Marche Apennines the exploration of the unknown Lower Mesozoic section underlying the Burano formation is at present the main objective of the permits granted to or applied for by AGIP, SNIA, Gulf, Union Oil, and Westates. The conjectural tectonic unconformity between shallow and deeper structures will require geophysical experimtation in order to define locations for wildcats.

The pedeapenninic Pliocene foredeep, from Marche to the Gulf of Taranto, is still one of the more attractive areas in Peninsular Italy. In the Marche–Northern Abruzzi part of the foredeep the improved quality of seismic surveys and a better understanding of the somewhat complex structural features led to the revaluation of the prospects of the thick Pliocene section.

The Abruzzi–Bradano trough, notwithstanding the active exploration carried out during the last ten years, is regarded as an area where new discoveries of some importance are still possible. Actually the entrapment of hydrocarbons is often controlled by combination of stratigraphic and tectonic factors; modern and detailed seismic reflection surveys and a better knowledge of the evolution of the basin will doubtlessly contribute to the development of its remaining resources.

The tectonic complexity of the S Apennines in the Molise–Lucania province and the controversial structure and stratigraphy of its subsurface sedimentary section are a serious handicap to exploration for oil and gas. On the other hand, the hydrocarbon shows and small productions known in the area and the actual possibility of important discoveries in areas where the general geological conditions are fairly analogous (*i.e.* the Gagliano field in Sicily) justify a futher effort.

In the Puglia province the main objective is the testing of the Lower Mesozoic section. This will actually require the drilling of deep and expensive wildcats in order to penetrate the thick Cretaceous and Jurassic section.

Important discoveries in the Miocene–Pliocene post-orogenic basins of Tuscany, Latium, Campania, and Calabria seem improbable.

Exploration in Sicily is at present lingering, compared with the rest of Italy. In fact both the disappointing results of the exploration carried out during the last years and the onerous conditions of the Sicilian Hydrocarbon Law do not encourage explorative investments. Exploratory activity is at present aimed at two main objectives: (1) the testing of remaining possibilities in the Mesozoic of SE Sicily; (2) Tertiary gas prospects in the N and central provinces.

The results of the activities carried out in Sardinia during past years led to an almost complete abandonment of exploration.

In the Italian offshore the geophysical surveys were started by AGIP in 1958; the first offshore well in Italian and European waters was completed as an outpost of the Gela field in 1959, and the first offshore new field was discovered in 1960 (Ravenna Mare). Larger exploration started following the application of the 1967 law; an appraisal of the future prospects is therefore premature, particularly for the areas where the selection of the permits is not yet completed (S Adriatic, Ionian and Tyrrhenian seas) or will be undertaken at the end of 1969 (Sicilian offshore).

The N Adriatic ("Zones A and B") is at present covered almost totally by exploration permits granted to AGIP–Shell and AGIP (for 25 per cent of the area) and by applications of other companies. In this area the prospective interest of the Pliocene basins has been enhanced by the recent important gas discoveries in the Po Plain offshore and along the Marche and Abruzzi coast. The actual possibilities of the lower sedimentary section are at present less clearly defined. Hydrocarbon indications in the Mesozoic carbonate section on both sides of the Adriatic and the inferred lack of intense tectonization will certainly lead to thorough testing of the pre-Pliocene sediments.

In the S Adriatic and in the Ionian seas ("Zone D") the 200 m outer limit of the platform binds the prospecting to a relatively narrow strip. The main objective of the exploration will probably be the Mesozoic section in the offshore of Puglia and the seaward extension of the Pliocene basins in the Lucania and Calabria offshore.

The prospects of the Tyrrhenian Sea may be inferred to be in general similar to those of the post-orogenic Miocene–Pliocene basins of Peninsular Italy and Sardinia.

The Sicilian offshore ("Zone C") is mainly developed in the area facing SW and Sicily and between Lampedusa island and the Tunisian border. The sedimentary sections and the structural features of the geological provinces framing the Sicilian offshore show a wide range of conditions. It would therefore be premature to suggest any definit hypothesis on its prospects; however, the inferred presence of a thick sedimentary section and the oil and gas discoveries and/or shows in the subsurface of S Sicily and E Tunisia justify an optimistic appraisal of the area.

THE IMPORTANCE OF HYDROCARBONS TO ECONOMIC DEVELOPMENT OF ITALY

The new reserves discovered recently in the N Adriatic, plus those already known, mainly in the Po Plain, S Italy, and Sicily, will permit ENI to operate more securely and efficiently in promoting and undertaking industrial initiatives as it has already done in the under-developed regions of Italy. Ravenna, Gela, and Pisticci are three locations showing examples of ENI's ability to exploit the natural resources it has found, where the company has built petrochemical works not only industrially bold but also economically and socially important for certain regions of the country.

Since the Italian economic structure is based upon industry mainly engaged in manufacturing which depends, for efficiency and competitiveness, upon the availability of low cost energy, Italian energy policy is necessarily always inclined towards insuring such availability. In practice the success of this policy depends upon maintaining a certain degree of competition among the various energy fuel sources. For hydrocarbon fuel sources the competition between gas and oil has caused a division of the market into somewhat well-defined sectors of consumption.

An optimum exploitation of domestic reserves and the satisfying of those fuel needs which may be more appropriately supplied with greater economic advantage by gas than any other fuel, now render necessary the importation of gas to integrate the country's availabilities. In consequence, ENI is planning to import natural gas beginning in 1969. At the same time, the national gas transmission system is being extended to cover the whole country eventually, not only to permit distribution to localities never before served, but also to distribute methane more efficiently for non-industrial uses (home heating, etc.), allowing municipalities to substitute undesirable manufactured gas facilities.

ACKNOWLEDGMENTS

The author wishes to express his sincere appreciation to AGIP, Direzione Mineraria, for granting the use and release of information contained in this paper. He is also indebted to G. Marruzzo for his contribution to the first and last paragraphs, and to F. La Torraca and W. O. Gigon for their co-operation in preparing the illustrations and in revising the English version.

BIBLIOGRAPHY

AGI Intern. Field Inst., Italy, Guidebook: Am. Geol. Inst., Washington, 129 p, 1964.

AGIP. "Relazione geologica e mineraria sulla Pianura Padana" *in* "Giacimenti gassiferi dell'Europa Occidentale". Roma, Acc. Naz. Lincei, vol. 2, pp. 63–497, 1959.

Aubouin, J. "Geosynclines." Elsevier, Amsterdam, 335 p, 1965.

Bosellini, A. "Lineamenti strutturali delle Alpi Meridionali durante il Permo–Trias e alcune considerazioni sui possibili rapporti con la tettonica alpidica." Mem. Museo St. Nat. Ven. Trident., vol. 15, f. 3, pp. 1–72, 1965.

Bronzini, S. "Note sulle ricerche di gas in alcune zone del litorale ionico" *in* "Giacimenti gassiferi dell' Europa occidentale." Roma, Acc. Naz. Lincei, vol. 1, pp. 399–405, 1959.

Caflisch, L., and Schmidt di Friedberg, P. *Mem. Soc. Geol. Ital.*, 1967, **6** (4) 449–74.

Carissimo, L., D'Agostino, O., Loddo, C., and Pieri, M. *Proc. 6th World Petrol. Congr.*, 1963, **1**, 267–92.

Crescenti, U. *Boll. Soc. Geol. Ital.*, 1966, **85**, 541–79.

Jaboli, D., and Roger, A. "Esquisse structurale de la Fosse Bradanique (Italie meridionale)" *in* Int. Geol. Congr. Proc., Sect. 9, pp. 305–24, 1954.

Kafka, F. T., and Kirkbride, R. K. *Proc. 5th World Petrol. Congr.*, 1959, **1**, 233–57.

Martinis, B. "Prove di ampi sovrascorrimenti nelle prealpi fruilane e venete." Mem. Ist. Geol. e Min. Univ. Padova, vol. 25, 31 p, 1966.

Martinis, B., and Pieri, M. "Alcune notizie sulle formazioni evaporitiche del Triassico superiore nell'Italia centrale e meridionale." *Mem. Soc. Geol. Ital.*, vol. 4, pp. 649–78, 1963.

Merla, G. "Essay on the geology of the Northern Apennines" *in* "Giacimenti gassiferi dell'Europa occidentale." Roma, Acc. Naz. Lincei, vol. 2, pp. 629–51, 1959.

Ogniben, L. *Riv. Min. Siciliana*, 1960, **11** (64–65), 183–212.

Pavan, G. "Le ricerche petrolifere condotte dall'AGIP nel Trentino–Alto Adige" *in* Synposium Int. Giacimenti Minerari delle Alpi. Trento, vol. 4, pp. 973–85, 1968.

Pieri, M. *Boll. Soc. Geol. Ital.*, 1961, **80** (1), 1–34.

Pieri, M. *Geologica Romana*, 1966, **5**, 407–24.

Reeves, F. *Bull. Amer. Ass. Petrol. Geol.*, 1953, **37** (4), 601–53.

Rigo, M., and Cortesini, A. *Boll. Serv. Geol. d'Italia*, 1961, **81**, (2–3) 361–9.

Rocco, T. *Proc. 5th World Petrol. Congr.* 1959, **1**, 207–32.

Rocco, T., D'Agostino, O. "The Sergnano gas field, a typical stratigraphic trap in the Po basin (Italy)" *in* "Stratigraphic Case History." Soc. of Exploration Geophys., Tulsa (in print), 1969.

Schmidt di Friedberg, P. *Rev. Inst. Franç. Pétrole*, 1962, **17**, 635–68.

Schmidt di Friedberg, P. *Riv. Min. Siciliana*, 1954–65, **15** (88–90), 198–217; **16** (91–93), 50–71.

Schmidt di Friedberg, P. *Boll. Soc. Geol. Ital.*, 1966, **85**, 687–704.

Selli, R. "Il Paleogene nel quadro della geologia dell'Italia meridiobale." *Mem. Soc. Geol. Ital.*, v. 3, pp. 737–89, 1962.

Appendix 1

List of Fields and Exploratory Wells (Fig. 9)

1. Coredo 1
2. Sedico 1
3. Belluno 1
4. Bernadia 1
5. Buttrio 1
6. Volpago 1
7. Desana
8. Brugherio
9. Lambrate
10. Pandino
11. Sergnano
12. Orzivecchi
13. Maclodio
14. Bagnolo Mella
15. Leno
16. Romanengo
17. Ripalta
18. Soresina
19. Bordolano
20. Cornegliano
21. Caviaga
22. Cremona S
23. Piadena W
24. Piadena E
25 Caddogio
26. Podenzano
27. San Giorgio
27a. Salsomaggiore
28. Ponte Nure
29. Cortemaggiore
30. Busseto
31. Bellena
32. Fontevivo
33. Collecchio
34. Vigatto
35. Correggio
36. Villadose
37. Tresigallo
38. Sabbioncello
39. Monestirolo
40. S. Pietro C.
41. Malalbergo
42. Bova
43. Spilamberto
44. Minerbio
45. Selva
46. Budrio E
47. Alfonsine
48. Ravenna
49. Porto Corsini
50. Imola
51. Cotignola
52. Santerno
53. Marzeno
54. Porto Corsini mare W
55. Agostino
56. Porto Garibaldi
57. Ravenna mare
58. Ravenna mare S
59. Porto Corsini mare
60. Amelia
61. Cervia Mare
62. Cesenatico Mare
63. Ponte dell'Olio
64. Montechino
65. Velleia
66. Vizzola
67. Vallezza
68. Sivizzola 1
69. S. Polo 1
70. Levizzano
71. Monteveglio
72. Pietramala
73. Castagno
74. M. Pietra 1
75. Selvapiana
76. Viamaggio
77. Fossombrone 1
78. Fano
79. Marotta
80. Ponsaccol
81. Poggio 1
82. Zanone 1
83. Cerbaie 1
84. Certaldo 1
85. Burano 1
86. Iesi
87. Perugia 1–2
88. Rapagnano
89. S. Giorgio mare 1
90. Roma 1
91. Roma 2
92. Ladispoli 1
93. Moricone 1
94. Antrodoco 1
95. Bellante
96. Cellino
97. Vallecupa
98. Cigno
99. Latina 1
100. Latina 2
101. Fogliano 1
102. Trevi 1
103. Ripi
104. S. Giovanni Incarico 1
105. Bomba
106. Villalfonsra
107. S. Stefano Mare
108. Casa Borselli 1
109. Casalbordino 1
110. S. Salvo–Cupello
111. Petacciato 1
112. Colle Scalella
113. Montecilfone
114. Guglionesi
115. Portocannone
116. Torrente Tona
117. Chieuti
118. Peschici 1
119. Foresta Umbra 1
120. Cercemaggiore 1
121. S. Croce 1
122. Jelsi 1
123. M. Stillo
124. Candela
125. Ascoli Satriano
126. Carapelle
127. Serra del Riposo
128. Sele 1
129. Nusco 1
130. Serroni 1–2
131. Tramutola
132. Lagonegro 1
133. Accettura
134. Grottole–Ferrandina
135. Pomarico
136. Pisticci
137. Rotondella 4
138. Ugento 1
139. Doria 1
140. Cirò
141. Capo Cimiti
142. S. Leonardo 1–3
143. Mazara–Lippone
144. Colla 1
145. Avanella 1
146. Platani 1
147. Enna 1
148. Gagliano
149. Feudo Grande 1
150. Miraglia 1
151. Castelluzzo 2
152. Bronte
153. S. Nicola
154. Randazzo 1–2
155. Cisina
156. Rizzo
157. Fontana Rossa
158. Cammarata
159. Gela
160. Ponte Dirillo
161. Ragusa
162. Comiso 1
163. Oristano 2
164. Oristano 1
165. Campidano 1

MS received 25 February 1969

Discussion

Dr Donald P. McGookey (Texaco Inc.): What is the distribution of control on the thickness and extent of Triassic evaporites? Are there any indications of salt diapiric or piercement structures on the Italian peninsula?

Dr Marco Pieri: Evaporites of Upper Triassic age (Burano Formation) outcrop in the NW Apennines and in W Tuscany; in most of the outcrops solution and re-cementation by ground water transformed the original anhydrite–dolomite sequence into de-dolomitized collapse breccia ("calcare cavernoso"). In Central Italy the Burano Formation was drilled in the wells Perugia 1 and 2, Fossombrone 1 and 2, Antrodoco 1; Trevi 1, Foresta Umbra 1 (Martinis and Pieri, 1963; Pieri, 1966). Information about the distribution of Upper Triassic evaporites in W Yugoslavia and W Greece is reported by W. D. Gill ("The Mediterranean Basin", in "Salt Basins around Africa", Institute of Petroleum, London, 1965).

The thickness of the Burano Formation in the W Tuscany outcrops may be estimated as a few hundred metres; it rapidly increases towards the E (1400 m in the Perugia area, more than 2600 m in the Foresta Umbra 1 well, where the base of the formation was not reached).

In the wells Antrodoco 1 and Trevi 1 a thrust plane separates the Triassic evaporites from underlying younger formations.

In my opinion there are no definite indications of real salt diapiric or piercement structures in the Italian peninsula. The Triassic evaporites of the Burano Formation do probably fill the core of the anticlines in the N and Central Apennines, but this phenomenon seems to be related to a regional décollement, *i.e.* the upper part of the section slid on the pre-evaporitic section, causing at places the thickening of the incompetent Burano Formation.

In a few other cases, small outcrops of Triassic gypsum may be related to local intrusions along recent normal faults; however, this phenomenon is to be considered as secondary and of very limited importance.

Ramon Querol (Kuwait Spanish Petroleum Co.): Could Dr Pieri tell us something about the palaeogeography of Italy in Triassic time? In which areas was there extensive evaporite deposition and where did the carbonate facies predominate?

Dr Marco Pieri: The complete Triassic section is known

only in the S Alps, where it consists of Lower Triassic clastics sedimented in continental-shallow marine environments, followed by Middle Triassic carbonate-clastic and/or volcanic sequences and by the Upper Triassic Hauptdolomit, sedimented in shallow marine-intertidal–supratidal carbonate shelf environments.

Extensive evaporite deposition seems to be limited to the area S of the Po Plain, from the N Apennines to Central Italy, and in time to the Upper Triassic. The age of the evaporites is inferred from the age of the overlying Lower Liassic or Rhaetian Formations.

In the southernmost part of the Italian peninsula and in Sicily no extensive Triassic evaporites are known; Calabria and NE Sicilian province probably emerged during most of the Triassic time. In Sardinia scanty Upper Triassic outcrops referred to the German facies (Keuper) and, including anhydritic levels, are reported from the NW part of the island.

J. Vercellino (Rigo and Vercellino): Would it be possible for you to give us some industry figures on Italy's present gas reserves; what percentage of these reserves have been discovered since passage of the offshore petroleum law; and the present available market demand for gas in Italy?

Dr Marco Pieri: The amount of gas reserves is considered confidential data by the Italian companies. Some general data were recently published referring to the situation at the end of 1968.

Original recoverable reserves of the ENI group were then estimated to be 260 billion cu m. 100 billion cu m were produced in the last 15 years; residual recoverable reserves therefore amount to 160 billion cu m. Of these reserves, 75 billion cu m were found in the N Adriatic sea since the passage of the 1967 Hydrocarbon Law; of this, 56 billion cu m were discovered in 1968.

The programme of the ENI Group foresees the drilling of 20 exploratory wells and 22 development wells in the Adriatic Sea.

The market demand will certainly increase in the next few years; ENI is therefore planning to import natural gas beginning in 1969, and the gas transmission system is being extended to cover the whole country.

Dr Rudolf Martin: What is the relationship, geographically speaking, between the evaporites of the Triassic and the reef facies of the Dolomites? How are the corresponding environments related to each other?

Dr Marco Pieri: The reef facies of the Dolomites are classically developed in the Middle Triassic section, while the evaporites belong to the Upper Triassic.

The lateral equivalent of the Burano Formation in the Dolomites is represented by the Hauptdolomit, that according to the more recent studies and interpretations was deposited in a non-reefoidal carbonate shelf.

The dolomites interbedded in the evaporites of the Burano Formation are not reefoidal, but finely crystalline, often stromatolitic.

Reef carbonates are developed in the Jurassic section that overlies the Burano Formation, *e.g.* in the upper 2000–3000 m of the section drilled by the Foresta Umbra 1 well.

Rustum Byramjee (CFP): How do you place the Apennine Range in the general framework of the Alpine orogeny with regard to the rotation of the Italian peninsula and opening or shortening of the Tyrrhenian Sea?

Dr Marco Pieri: This is a rather complex question. In my opinion the palaeogeographical and structural model of the Italo–Dinaric system and the Alpine arc, as illustrated in the well-known recent works of J. Aubouin, is a useful scheme and clearly outlines the situation of the Apenninic range in the general framework of the Alpine orogeny. Some criticism may be posed as far as the palaeogeography of S Italy is concerned, but the main concept that an axis of divergence runs through the Tyrrhenian Sea (from the Genoa area to possibly the Calabria peninsula), is quite convincing; in the N Tyrrhenian area this same concept was clearly outlined by the Italian specialists on N Apennine tectonics.

The Tyrrhenian Sea should therefore be considered as an area of upheaving and shortening; in fact, it is in this area that the erosion feeding the clastic material to the Apenninic Cretaceous, Palaeogene, and Miocene flysch sequences took place. From this same area the allochthonous material slid over the outer Apenninic geosynclinal elements.

It is quite possible that the shortening of the Tyrrhenian Sea was in the same way compensated by an opening in the W Mediterranean area.

Geology of Petroleum and Natural Gas from the Neocene Complex and its Basement in the Southern Part of the Pannonian Basin, Yugoslavia

By R. FILJAK, Ž. PLETIKAPIĆ,
(*Ina-Naftaplin, Zagreb, Yugoslavia*),
D. NIKOLIĆ, and V. AKSIN
(*Naftagas-Naftna Industrÿa, Novi Sad, Yugoslavia*)

SUMMARY

The petroleum and natural gas industry in areas now under Yugoslav jurisdiction has a fairly long tradition. Its development has always been characterized by higher capacities of local refineries completed to serve the local market than the capacities of domestic crude oil producers. Consumption of natural gas, however, has steadily been below productive abilities of discovered reserves.

Up to the present time, quantities of crude oil and natural gas produced in Yugoslavia have been obtained from the portion of the Pannonian Basin in Yugoslavia.

The Pannonian Basin is a Neocene sedimentary basin framed by the orogenic belt of the Eastern and Dinaric Alps and the Carpathian Mountains. The deposition of Neocene sediments was ruled by rather strong movements of surrounding orogenic masses, as well as of the basin's floor.

The Yugoslav portion of the basin is divided into several local depressions: Mura, Drava, Sava, Slavonia-Srem, Banat, and Bačka. In the sedimentary sequences of each depression, regional reservoirs with many potential traps for hydrocarbons have been identified. While testing these traps by drilling, more than 65 oil and gas fields have been found, of which 23 are giving current domestic oil production.

The paper discusses general oil geological characteristics of the portion of the basin within the Yugoslav borders, as well as some aspects of future developments.

INTRODUCTION

THE authors of this study desire to acquaint those interested in oil and gas exploration problems with knowledge so far gained of oil-geological relations in the Yugoslav part of the Pannonian Basin; then, via results obtained through searching for hydrocarbon deposits, with the future possibilities of this work, and the significance of domestic oil and natural gas production and its role in the development of the Yugoslav oil industry.

The oil and natural gas industry in Yugoslavia started many years ago. It was in 1856 that oil, for the first time, was produced from the dug-out wells in Peklenica. Oil wells were first drilled in 1880, while crude oil processing started in 1882 at Rijeka. Gas production began in 1918 at Bujavica. However, the interval between World War I and World War II and the period of World War II were not periods when the oil industry played an important role in the country's economy. Researches for hydrocarbon deposits were restricted to small, isolated structures. By the end of World War II a single oilfield was in production, with a capacity of 32,000 tons/year, and one gas field, with a yearly output of 2,000,000 Nm^3. During the war all the refineries that had been processing up to 320,000 tons/year of oil before the war were demolished.

There has been a rapid development of the oil industry in Yugoslavia since World War II, after 1955 in particular. It expanded at a rate that has only rarely been recorded in the world oil industry. All the branches of the industry began to expand rapidly, while the processing of oil and its distribution were in the forefront of the expansion. These two branches were those that had been most directly stimulated by the general expansion of the Yugoslav economy, and by the specific advantages of hydrocarbons as an energy source. When in 1968 consumption reached 4,000,000 tons, there were in operation five refineries with a total capacity of 8,000,000 tons of processed oil per year. Two of them, those in Novi Sad and Pančevo, began operation recently, but all five are ready and able to satisfy the present market demands and to meet the expected demands that increase from day to day.

In spite of the fact that capacities in oil industry processing were always greater than those of oil production, Yugoslav oil production nevertheless achieved important results in the period from 1955 to 1960.

The companies for the production of oil established after the war again undertook search work for hydrocarbon deposits. Exploration took place in promising areas of the country, and by the end of 1968 more than 60 oil and gas fields had been discovered, of which 23 are now under exploitation with an annual capacity of 2,360,000 tons of oil and 467,000,000 Nm^3 of natural gas (See Fig 1). All the hydrocarbon deposits

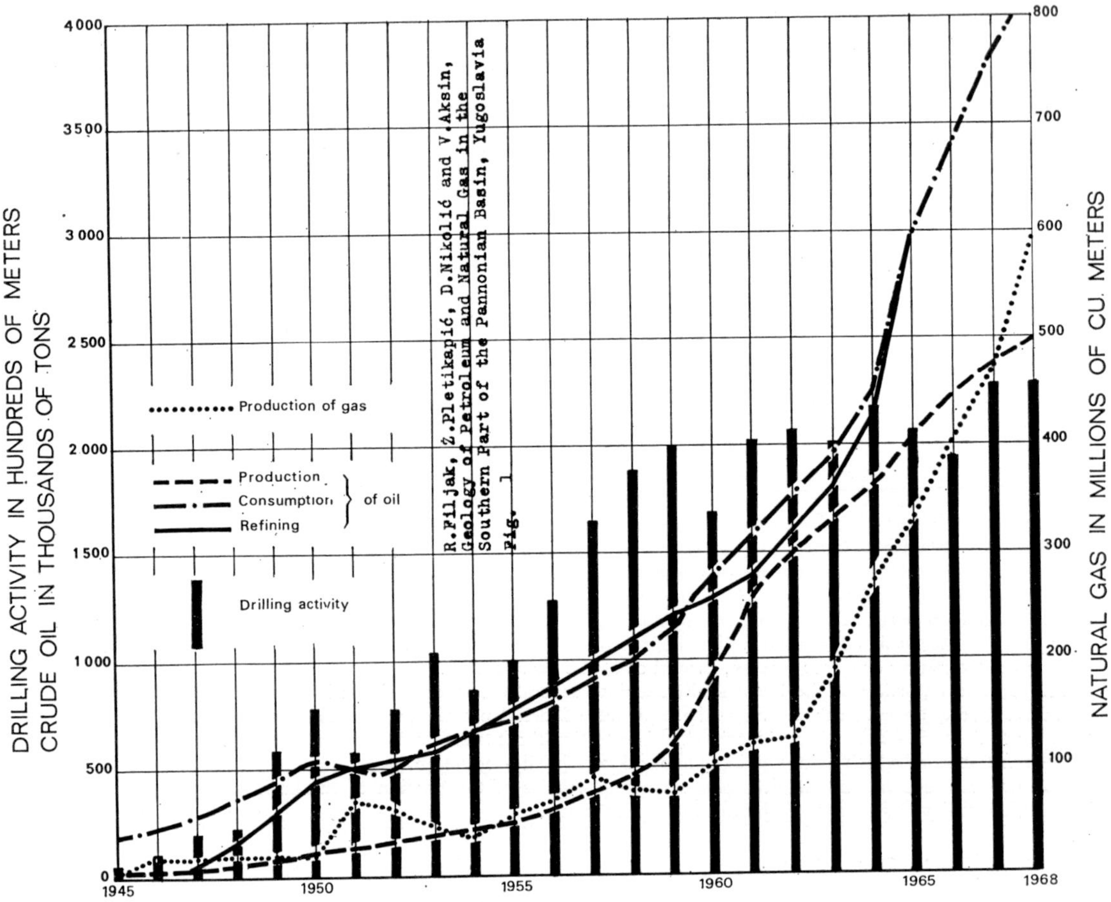

FIG 1. Annual crude oil and natural gas production in Yugoslavia compared with the annual consumption and refining of oil, as well as with drilling activity.

that have been discovered so far in Yugoslavia are in the zone of the Pannonian oil and gas bearing basin. As seen on the map (Fig 2) these deposits are in the closest vicinity of processing plants and within the country's main consumption area.

This is a very important fact as far as it concerns the prospect of domestic oil production. This fact compensates to a certain extent for the unfavourable effect of natural conditions from an economical point of view. The unfavourable effect of natural conditions means that the oil and gas reserves are dispersed in numerous small deposits and the average yield of production wells is rather low.

LOCATION OF THE BASIN

The Pannonian intermontane structural depression is in the very centre of Europe. It includes the Pannonian oil and gas bearing basin. The depression covers the territories of Hungary, Yugoslavia, Rumania, Austria, Czechoslovakia, and the U.S.S.R. Its greater part, *i.e.* the N part of the Pannonian Basin, is in Hungarian territory, while its S part is in Yugoslavia (Fig 3).

Together with the Caspian Basin, Black Sea Basin, Getic Basin, and some other smaller ones, the Pannonian sedimentary basin developed as a part of Paratethys, the internal sea that appeared after the splitting of Tethys in consequence of the emergence of the Alpine orogenic belt in S Europe. Paratethys extended from France to Turkestan in S Russia, and during the process of development it split to several separated palaeogeographic units, each of which had its own specific conditions and sedimentation pattern.

The Pannonian Basin, surrounded by the mountain ridges of the Eastern Alps, the Carpathian Mountains, and the Dinarides, reaches more than 500 km in

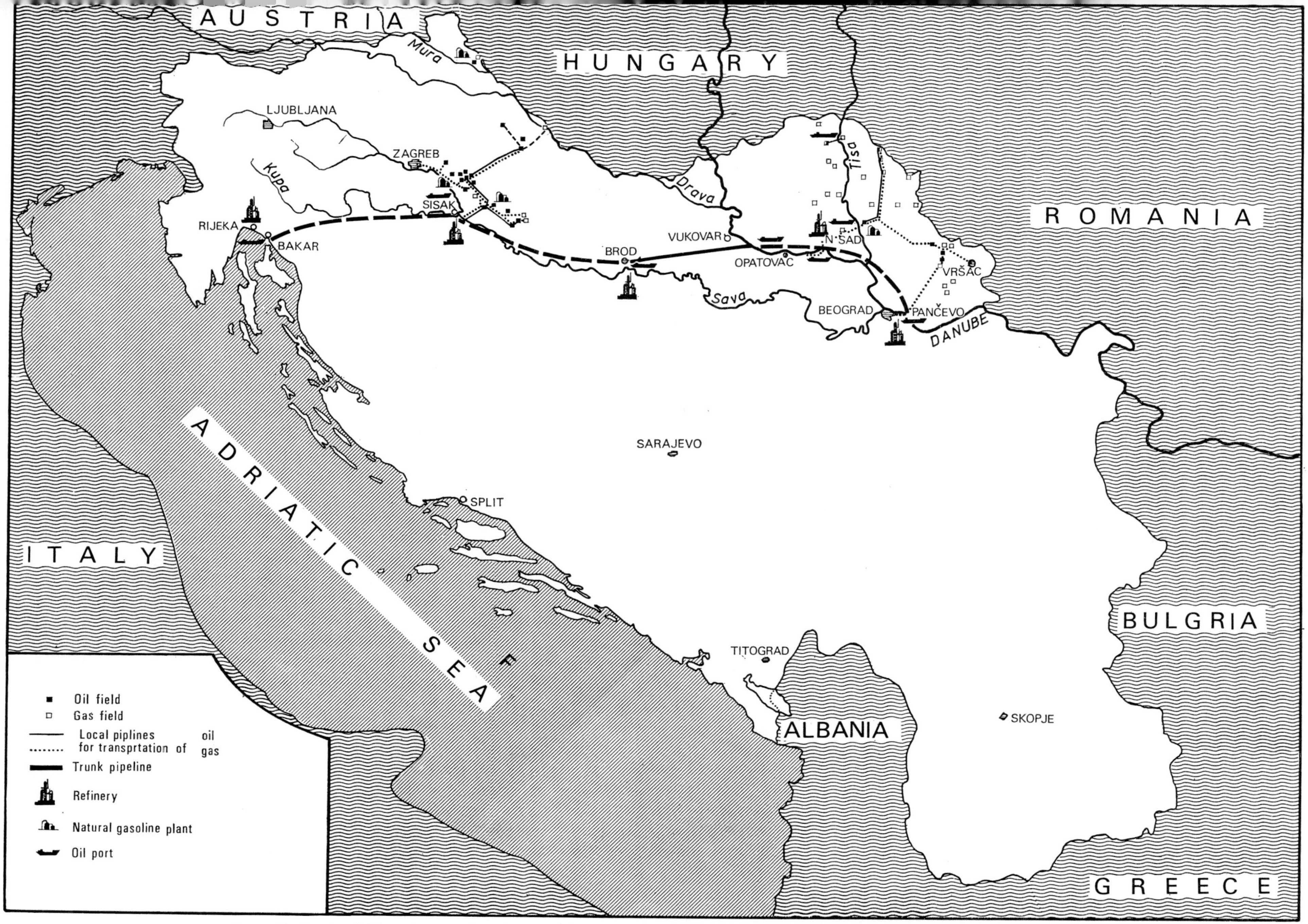

FIG 2. Production fields and processing plants with transportation facilities of petroleum and natural gas industry in Yugoslavia.

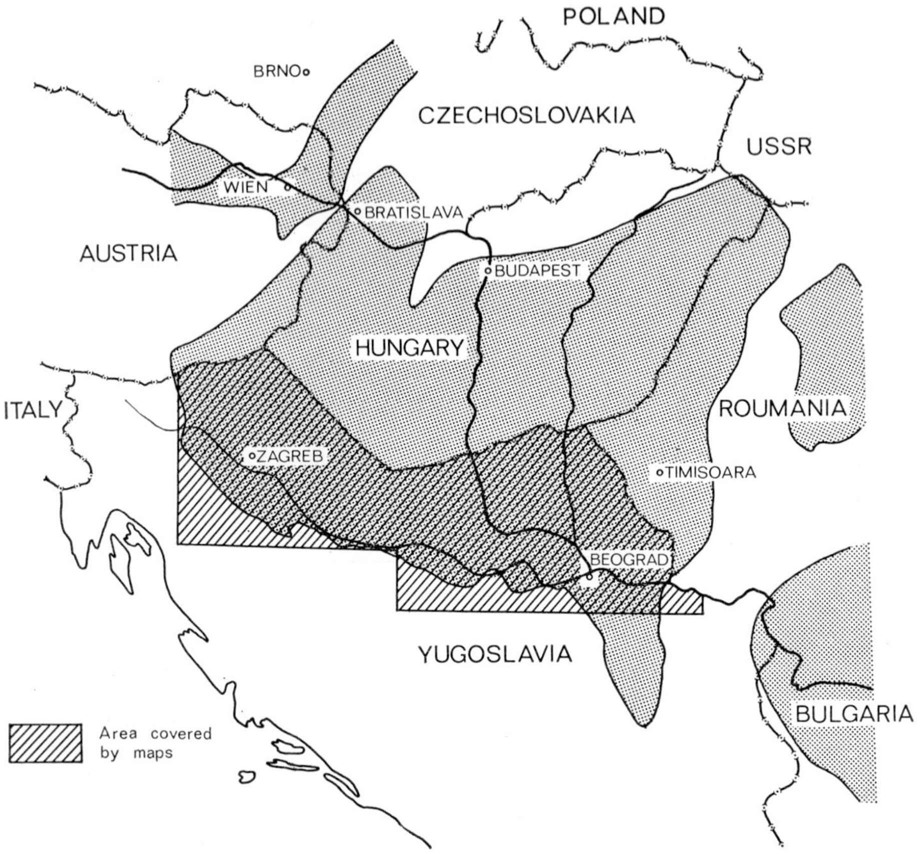

FIG 3. Pannonian and neighbouring Neocene Basins in Central Europe—Vienna, Transylvanian, and Walachian Basin (redrawn from V. Aksin, "Geologija nafte." Novi Sad, 1967, p. 690).

diameter. It is one of the world's greatest intermontane structural depressions. The part of the Pannonian Basin in Yugoslavia covers some 50,000 sq km.

GEOLOGICAL HISTORY AND STRUCTURE OF THE BASIN

The origin and evolution of the Pannonian Basin are closely related to the evolution and structure of the Alpine–Carpathian tectogene, and the explanation of the origin of the former is in close connection with that of the latter. However, in spite of the fact that these problems had been coped with by many authors, it cannot be said that any definite solution has ever been reached in that respect so far.

L. Kober, the author known for his theory about the bilateral orogene, dealt with this subject in a study published in 1931 which considered the problem as a whole. According to his hypothesis, the Pannonian Basin represents the consolidated remnants of an old platform that because of the folding of the neighbouring geosynclines became yoked and started to subside, making place for the deposition of the youngest, the Neocene complex of sediments. These remnants of platforms have been named by Kober "Intermediary massif" ("Zwischengebirge"). According to him, the intermediary massif acted upon the directions in which the surrounding mountain ridges developed. Its structure conditioned the forming of a depression where fault tectonics played the most important role.

The opinion of more recent authors, such as M. V. Muratov (1949), is that within the Pannonian Basin huge folded structures exist under the Neocene sedimentary cover—the anticlinoria, having in their cores crystalline schists and Palaeozoic sediments, and the synclinoria, with thick and very complexly arranged Mesozoic and Palaeocene sediments.

Searches for oil and gas deposits have shown that the lithologic composition and the tectonic structure of the basement of Neocene are rather complex. There are crystalline schists, gneisses, granites, and the so-called "green schist" belonging to the Pre-Cambrian, Cambrian, and Lower Palaeozoic, then argillaceous schists and limestones belonging to the Carboniferous, sandstones, limestones, and dolomites of the Permian, Triassic, Jurassic, and Cretaceous. Some portions of the basin, those marginal ones in particular, had suffered subsidence of a type such as that which gave origin to grabens. These parts are usually built up of thick flysch sediments. Sediments of Palaeogene have also been found in marginal parts of the Pannonian Basin. They are flysch-like and their thickness is rather high. (See Fig 4.)

The basin is of a very complex structure. First, the rocks of the basement are highly broken, often

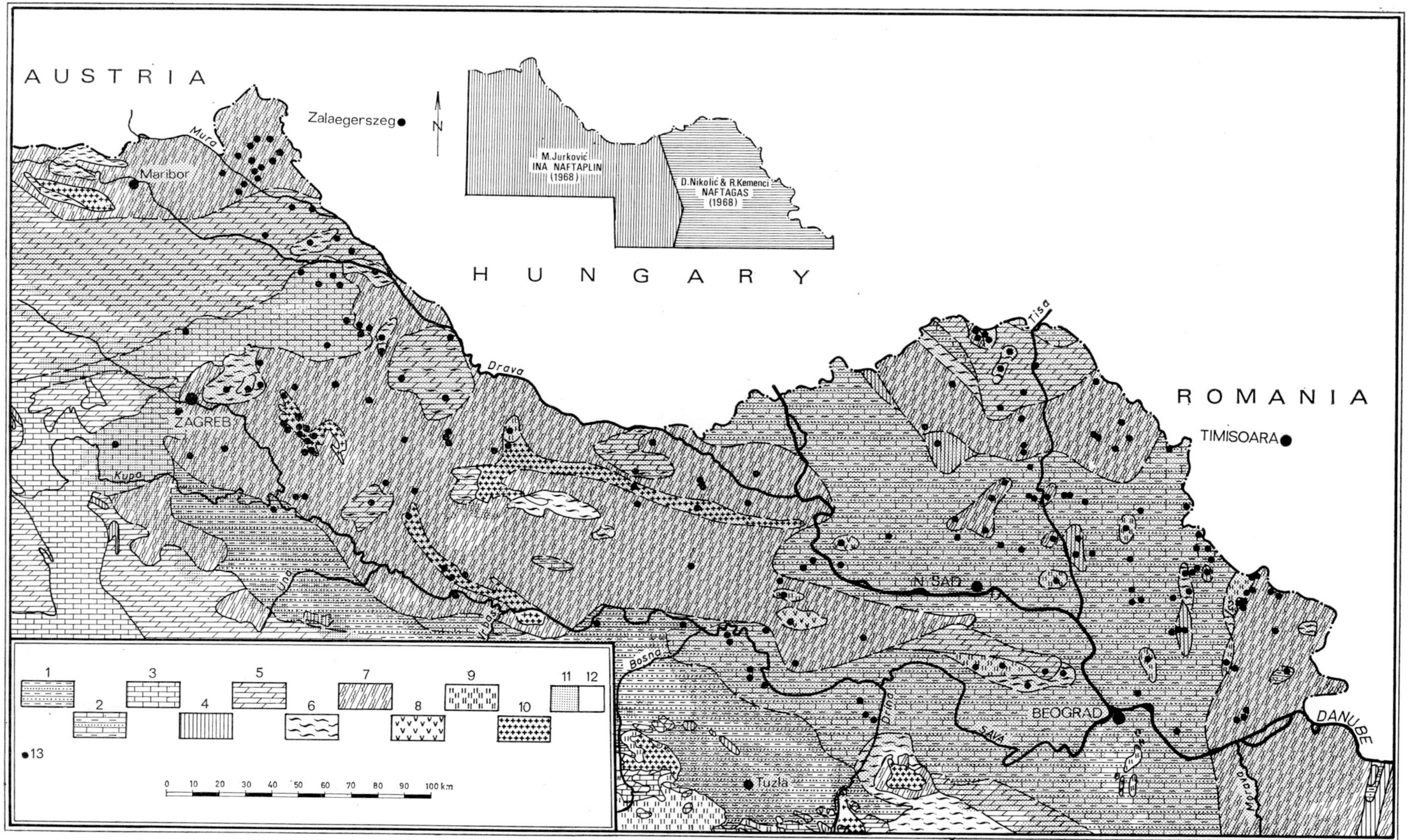

FIG 4. Palaeogeologic map. Basement of the Neocene Complex (compiled by Ž. Pletikapić, 1968). 1. Alternation of sandstone, shale, and marl (Eocene). 2. Alternation of sandstone, shale and limestone (Cretaceous). 3. Dominantly limestone and dolomite with some intercalated clastics (Cretaceous). 4. Dolomite, limestone, rarely fine clastics (Jurassic). 5. Dolomite, limestone, shale, and sandstone (Triassic). 6. Argillaceous schist, sandstone, rarely carbonate (Upper Palaeozoic). 7. Highly metamorphosed schist, gneiss, chloritic and amphibolitic schist (Lower Palaeozoic, Cambrian, and Pre-Cambrian). 8. Diabase and melaphyre. 9. Gabbro and serpentine. 10. Granitic rocks. 11. Covered by Neocene complex. 12. Appearing on the surface. 13. Subsurface control point.

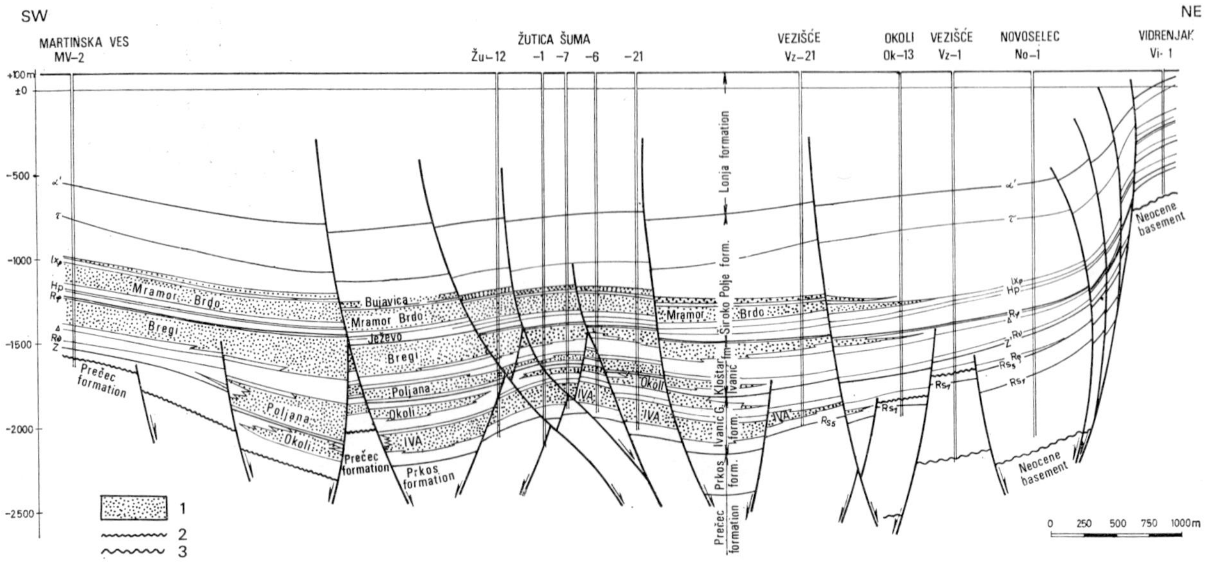

FIG 5. Section across Sava Depression (after J. Simon, 1967). 1. Sandstone members within the middle part of Neocene complex. 2 Intraformation unconformities. 3. Regional surface of unconformity at the base of Neocene complex.

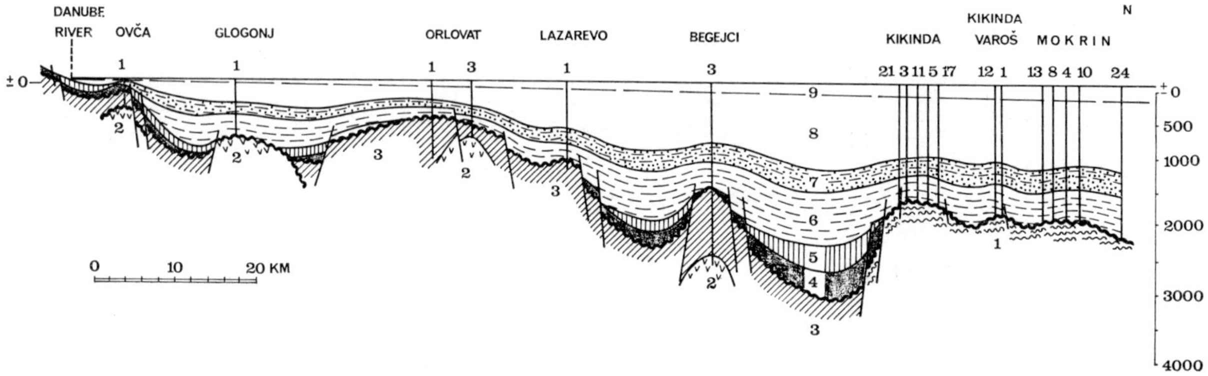

FIG 6. Section along Banat depression (after D. Nikolić, 1968). 1. Schists. 2. Serpentinites. 3. Flysch or flyschlike sediments (Cretaceous). 4. Conglomerate, shale, and algal limestone (Pre-Tortonian, Tortonian, and Sarmatian). 5. "White marl" (Pannonian). 6. Shaly marls (Pontian). 7. Sandy marls and sand (Pontian). 8. Sand, clay, and lignite (Dacian and Levantian). 9. Loess, clay, and gravel (Quaternary).

folded. Older rocks have undergone metamorphosis. No folding features characterize the structure of sediments belonging to the Neocene. However, folding, if any, was the consequence of vertical movements, on which occasion some rocks blocks became yoked. Variable movements of blocks limited by faults and compactions of sediments, deposited over protruding features caused by such block tectonics on the bottom of the basin, formed the structure of the Neocene complex. (See Fig 5).

In most cases blocks continued to emerge or subside throughout the entire period of the Neocene deposits. However, later tectonic movements crushed those blocks and some of those which had previously emerged now subsided. Due to regional tilting of the basin's bottom in Slavonia, Bačka, and Banat, groups of islands went down and became overlain by younger members of the Neocene sediments (See Fig 6).

In general, it may be said that throughout the geological evolution of the Pannonian Basin the movements that were the most important had been those rhythmical, vertical ones, of regional significance. It happened that some blocks emerged while others at the same time subsided, and this phenomenon gave rise to many a horst massif and local depression, where variable compensatory sedimentation developed.

Deep boreholes that were drilled to discover oil and gas deposits have proved that, except on horsts where they outcrop, the basement is overlain by the thinnest Neocene cover in the central parts of the basin at a depth of 1000–1500 m, while in the basin's marginal parts the rocks belonging to the Pre-Neocene are at much greater depths, down to more than 6000m. Such position of depressions with thick sequences of Neocene sediments deposited on the margins of the basin are clearly noticeable in the S part of the basin, the Yugoslav portion (Fig 7). As seen on the map, the following marginal depressions exist in the S part of the basin: Mura, Drava, Slavonia-Srem,

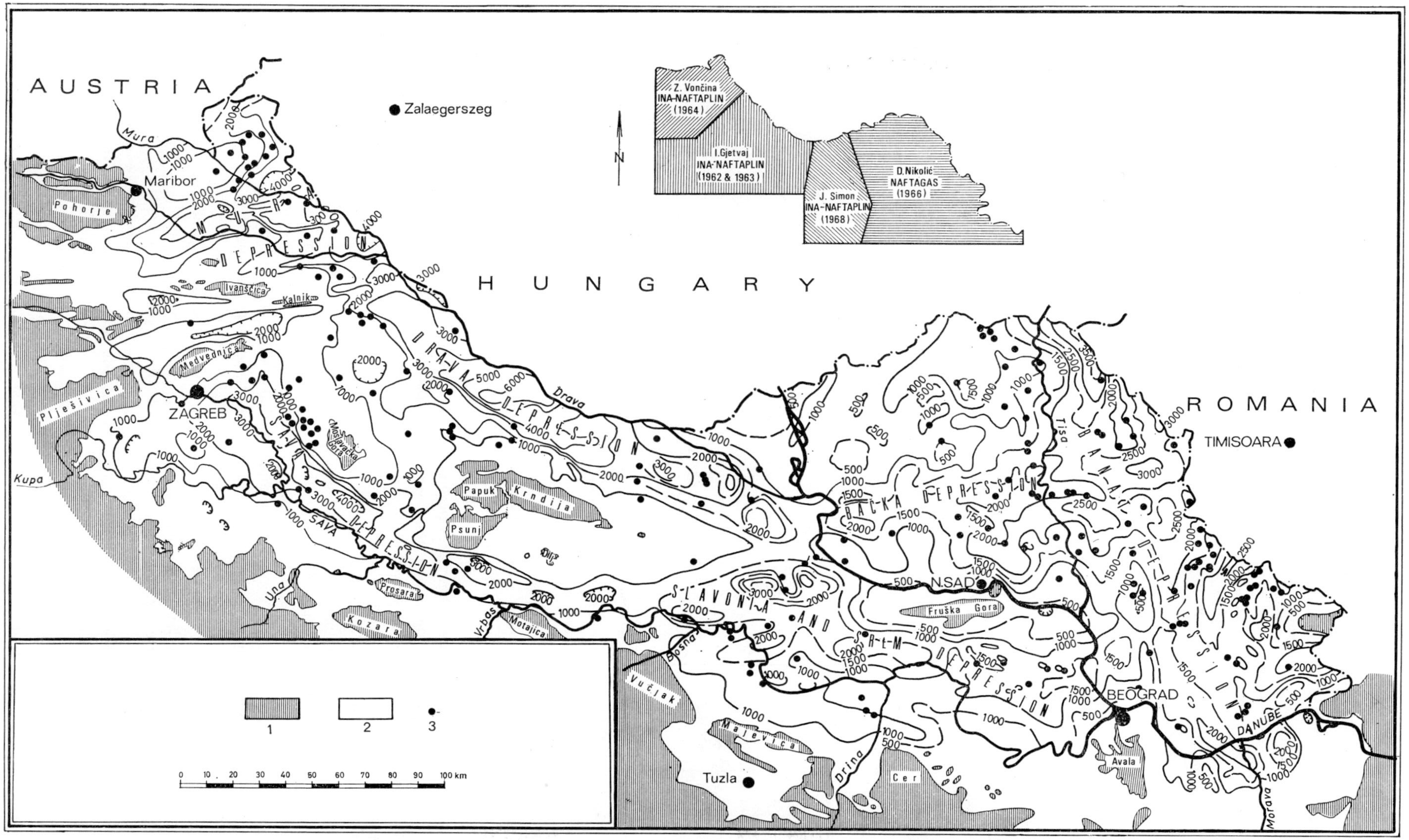

FIG 7. Thickness of the Neocene complex (compiled by Ž. Pletikapić, 1968). 1. Outcrop of the basement rocks. 2. Area covered by the Neocene sedimentary complex. 3. Subsurface control points.

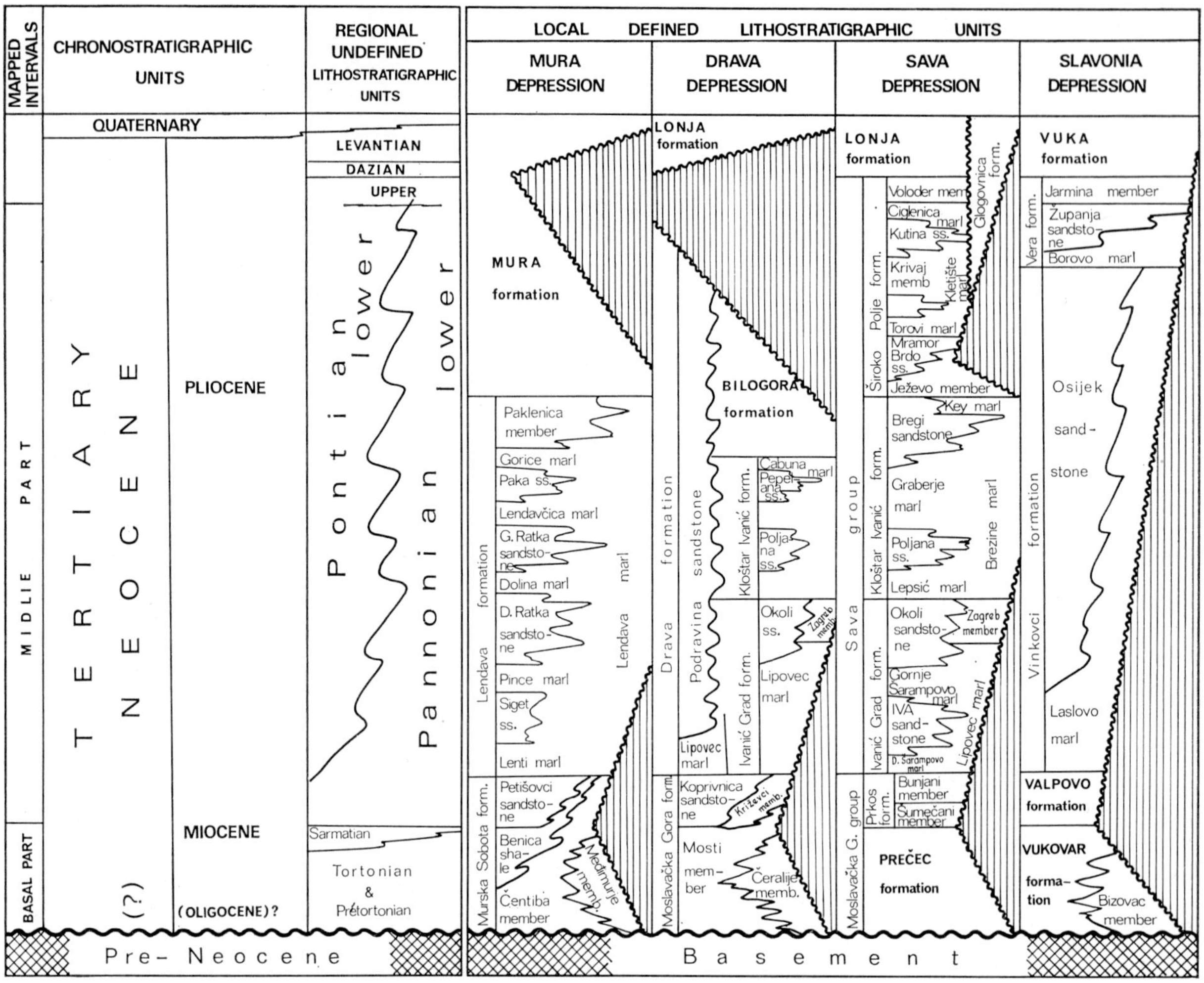

FIG 8. Stratigraphic correlation chart (after J. Simon and Z. Pletikapić, 1968).

Bačka, and Banat. The marginal depressions are separated from one another by horsts or subsided swells. In the course of deposition which developed in the Neocene, these depressions sometimes became linked with each other and sometimes they were individual, separate sedimentary environments. As a consequence, in spite of the fact that in general characteristics of the Neocene sediments resemble one another, they differ from depression to depression in some specific characteristics. Moreover, the marginal depressions differ in size, thickness of Neocene rock complex, basement composition, and the direction of cordilleras against which they lean. So the trends of the Sava, Drava, Slavonia-Srem, and Bačka depressions display the Dinaric Alps direction, whereas the Banat depression extends in the same way as does the corresponding part of the Carpathians (Fig 7).

The Neocene complex represents a sedimentary macrocyclus that can be sub-divided into three smaller parts.

The first and the oldest, the basal part of the complex, is made up of coarse-grained clastics (breccia, conglomerates, sandstones) clays, marls, and biogenic limestones deposited in fresh water and marine environments. According to their mineralogical composition, the clastics belong to arkoses and quartzites. In one part of the complex also intercalations of extrusive rocks and basaltic magma tuffs have been found.

According to the stratigraphic nomenclature being applied in the preparation of the general geological map of Yugoslavia, this portion of the complex comprises Pre-Tortonian, Tortonian, and Sarmatian sediments. Recently, the lithostratigraphic units have also been determined in the basal part of the complex but the study of the same is still in the initial phase (Fig 8).

The maps (Figs 9 and 10) show the thicknesses and lithofacies relationships within the basal portion of the Neocene complex. Thicknesses change abruptly. On the map can be seen numerous local maxima and minima within the same depression. These strong changes are attributed to a high degree of vertical diversity of the basin's bottom and to rapid filling of unevennesses with sediments whose material was derived from adjacent intensively eroded island masses that had been considerably

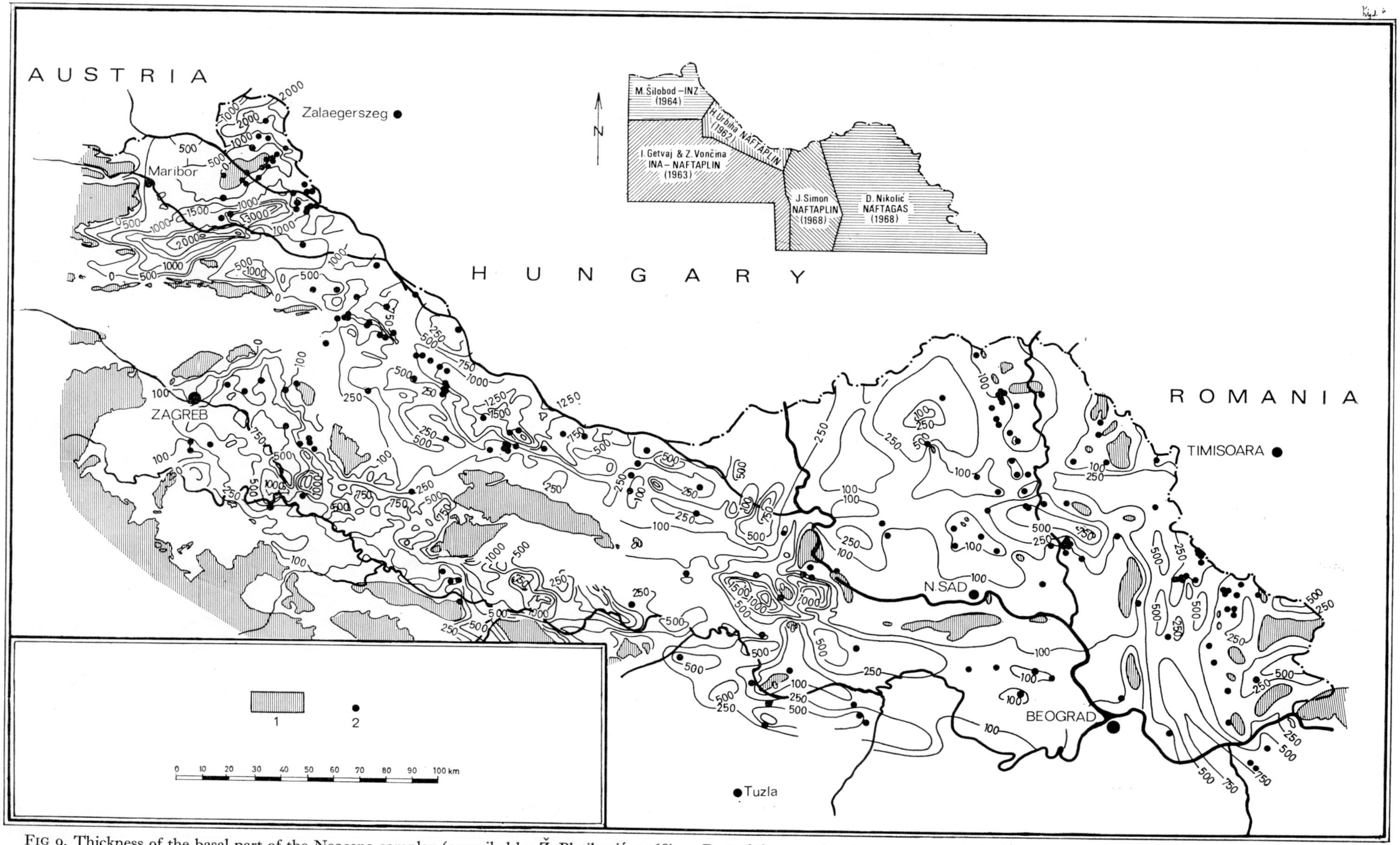

FIG 9. Thickness of the basal part of the Neocene complex (compiled by Ž. Pletikapić, 1968). 1. Part of the complex absent—non-deposited or completely eroded. 2. Subsurface control point.

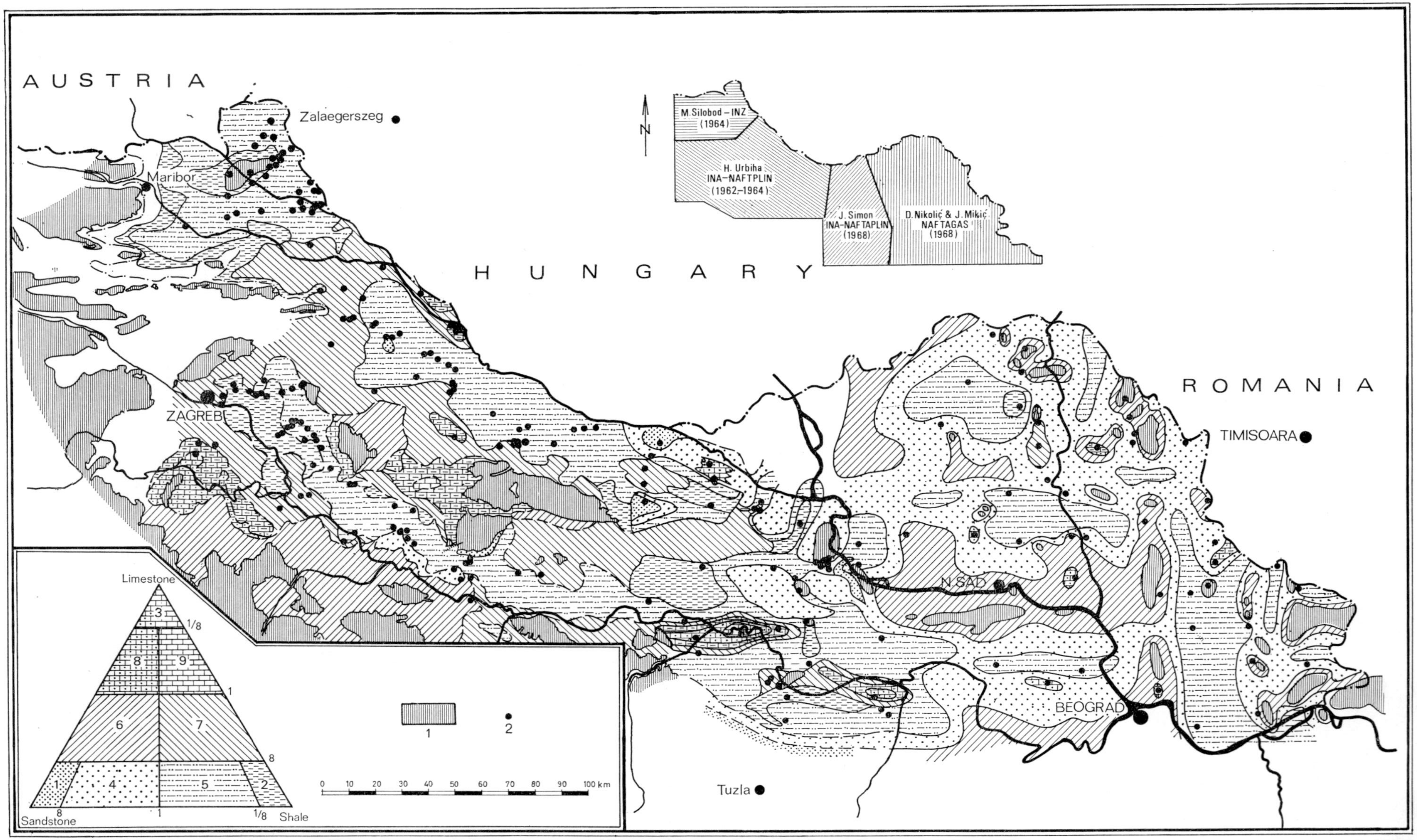

FIG 10. Lithofacies map of the basal part of the Neocene Complex (compiled by Ž. Pletikapić, 1968). 1. Basal part of the complex absent. 2. Subsurface control point.

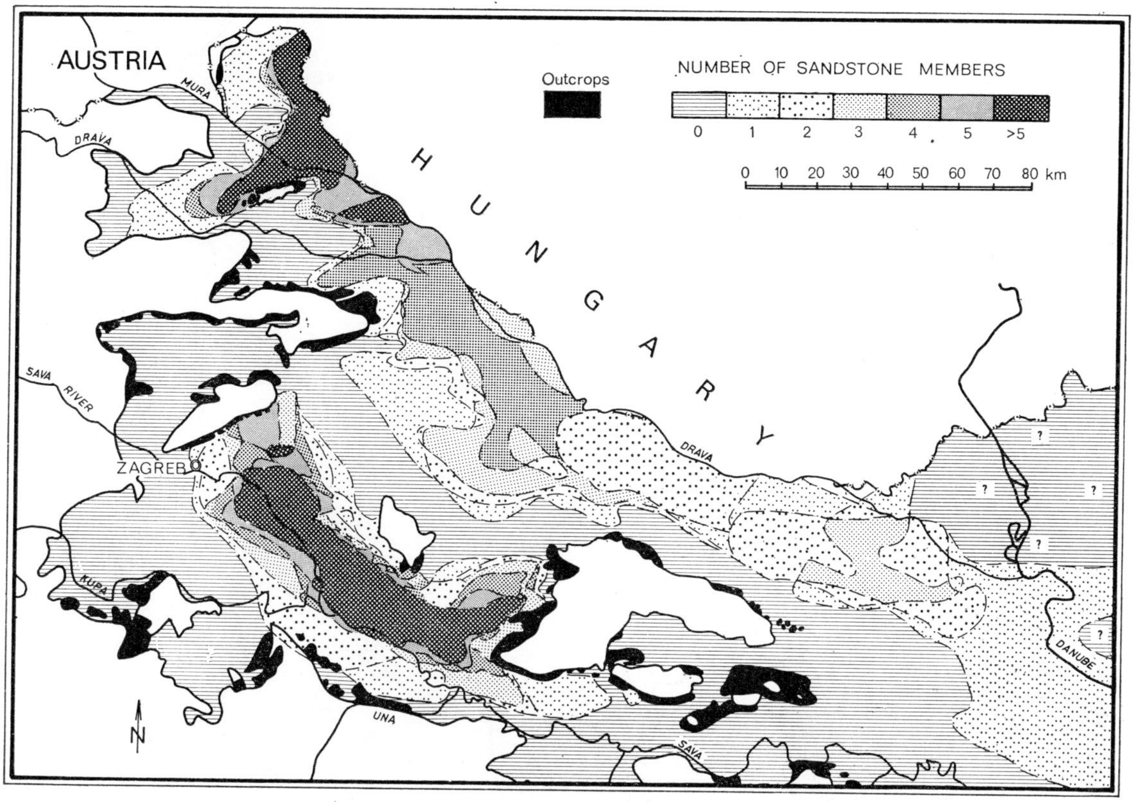

FIG 11. Distribution and number of sandstone members in the middle part of the Neocene complex (after J. Simon, 1968).

weathered during the preceding continental phase.

The lithofacies map shows the share of the three final members in the composition of the basal part of the complex. Characteristic of the main zones of subsidence in depressions is the dominant content of fine clastics with rather frequent alternation of pelite and psammite sediments. In marginal and shallow parts of the basin either coarse clastics or biogenic limestones prevail.

The middle part of the complex consists of marls and sandstones belonging to the type of calcarenite sub-graywackes. It was deposited in the Caspi-brackish environment. According to the stratigraphic nomenclature given in the standards of preparation of the general geological map of Yugoslavia (Fig 8), this part of the complex corresponds to Pannonian and Pontian deposits. Recently determined in it, in the W part of Yugoslavia, were also the lithostratigraphic units, whose extent, thickness, and facies characteristics are currently being examined. As the sandstones of this part of the complex represent the main oil and gas reservoirs, these investigations are of great importance.

Fig 11 is a synthesis representing outlines and number of sand members intercalated in marls of this part of the complex. It is seen from the map that the sandstone distribution is associated with the basin's main depression area. The material removed from the surrounding cordilleras was being brought into the basin by rivers and often carried in suspension by currents over great distances. It is evident that rhythmical tectonic movements in the source area had greatly influenced the amount and composition of material supplied, while the tectonic pattern of the depositional are a determined position and the shape of sand bodies.

The final portion of the complex is composed of a cyclic series of sands and gravels, clays and lignite, representing the result of work of rivers, and was deposited in extensive marshes along the river courses, the general direction of which corresponded to the trends of marginal depressions.

For the most part, the Neocene complex is covered with Quaternary sediments—gravels, clays, and loess, brought into the basin by action of rivers and winds.

OIL AND NATURAL GAS ACCUMULATIONS

The areal distribution of discovered oil and gas fields in the S part of the Pannonian Basin points to an unquestionable connection existing between hydrocarbon accumulations and depression zones in which the main sedimentary masses of the Neocene complex have been deposited. Their locations are shown on the map (Fig 12). The greatest number of fields has hitherto been discovered in the Sava and Banat depressions.

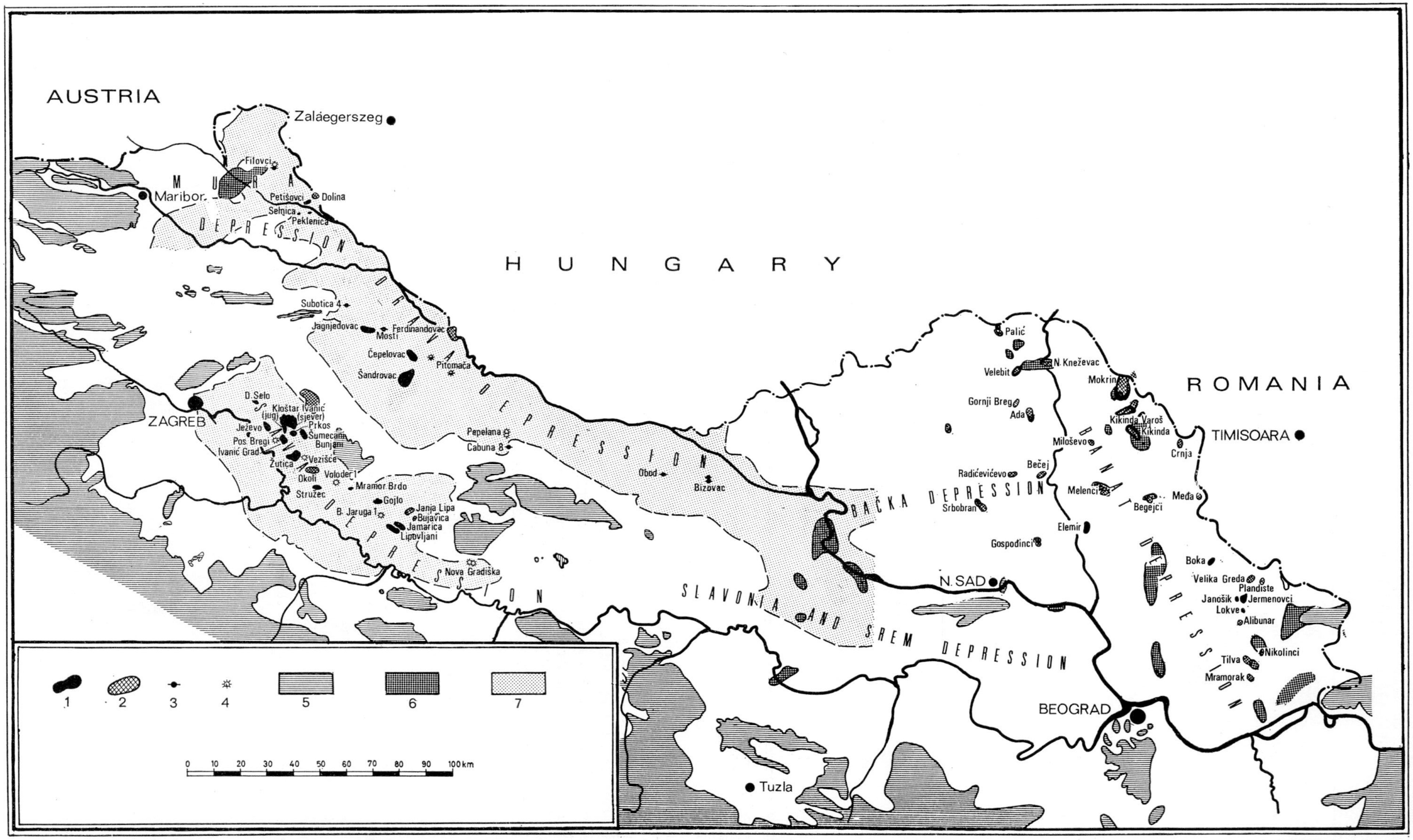

FIG 12. Oil geological map, S portion of Pannonian Basin (Yugoslavia). 1. Oilfield. 2. Gas field. 3. Oil discovery. 4. Gas discovery. 5. Outcrop of Pre-Neocene rocks. 6. Subcrop of Pre-Neocene rocks below Pannonian. 7. Areal distribution of Pannonian and Pontian sandstone members.

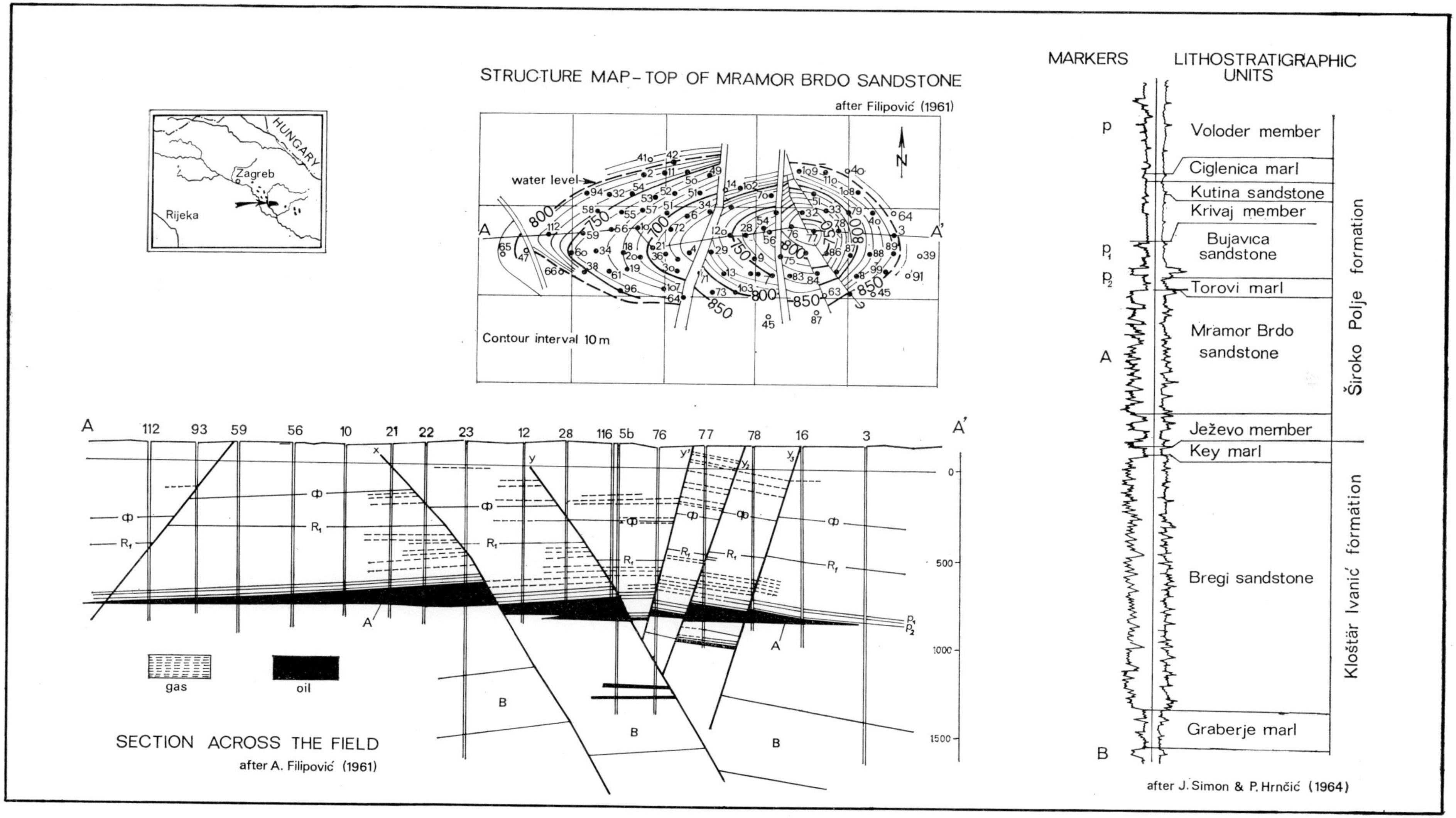

FIG 13. Strużec—oilfield, Sava depression (after A. Filipović, 1961).

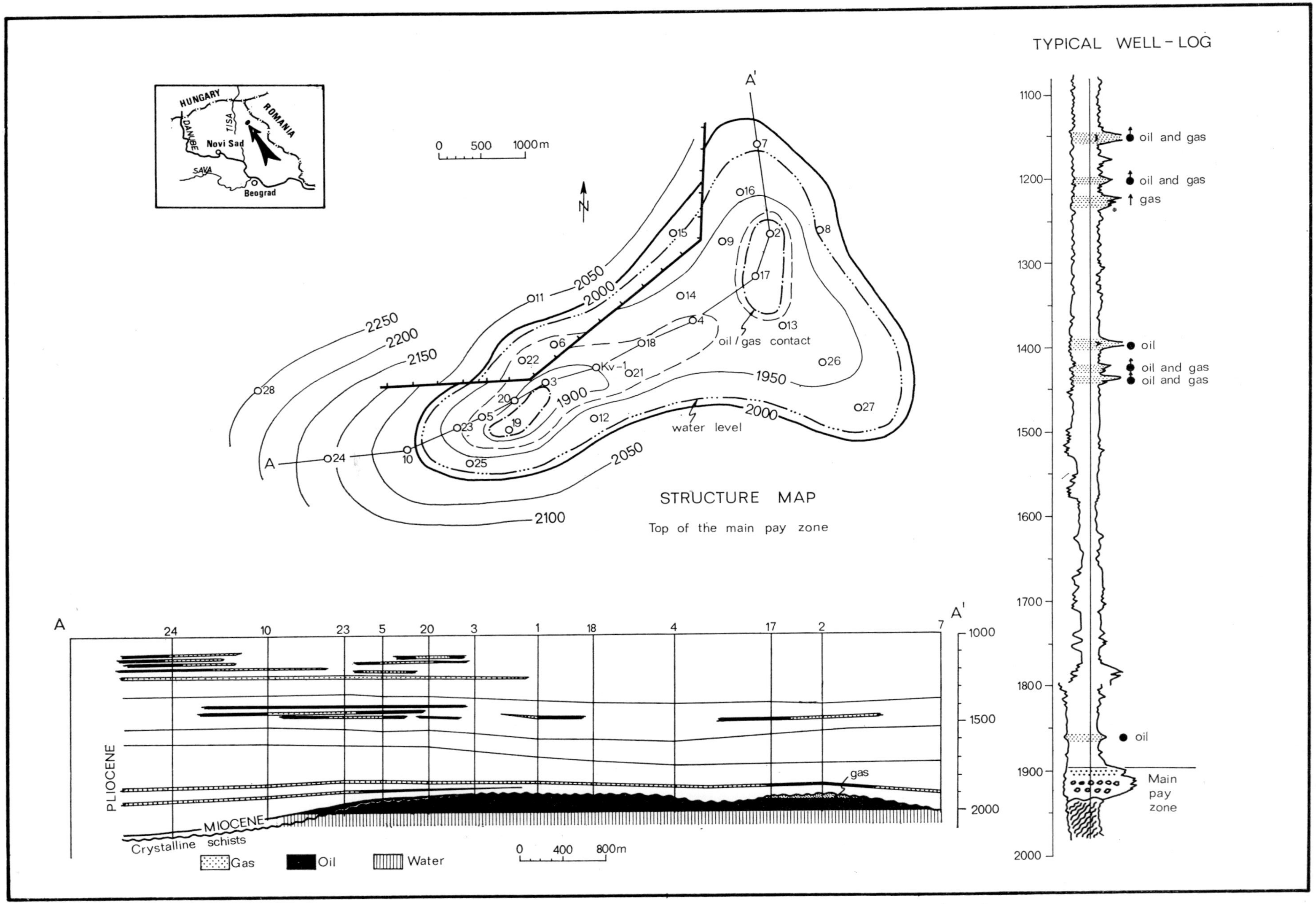

FIG 14. Kikinda (Varoš)—oil and gas field, Banat depression (after D. Nikolić, 1968).

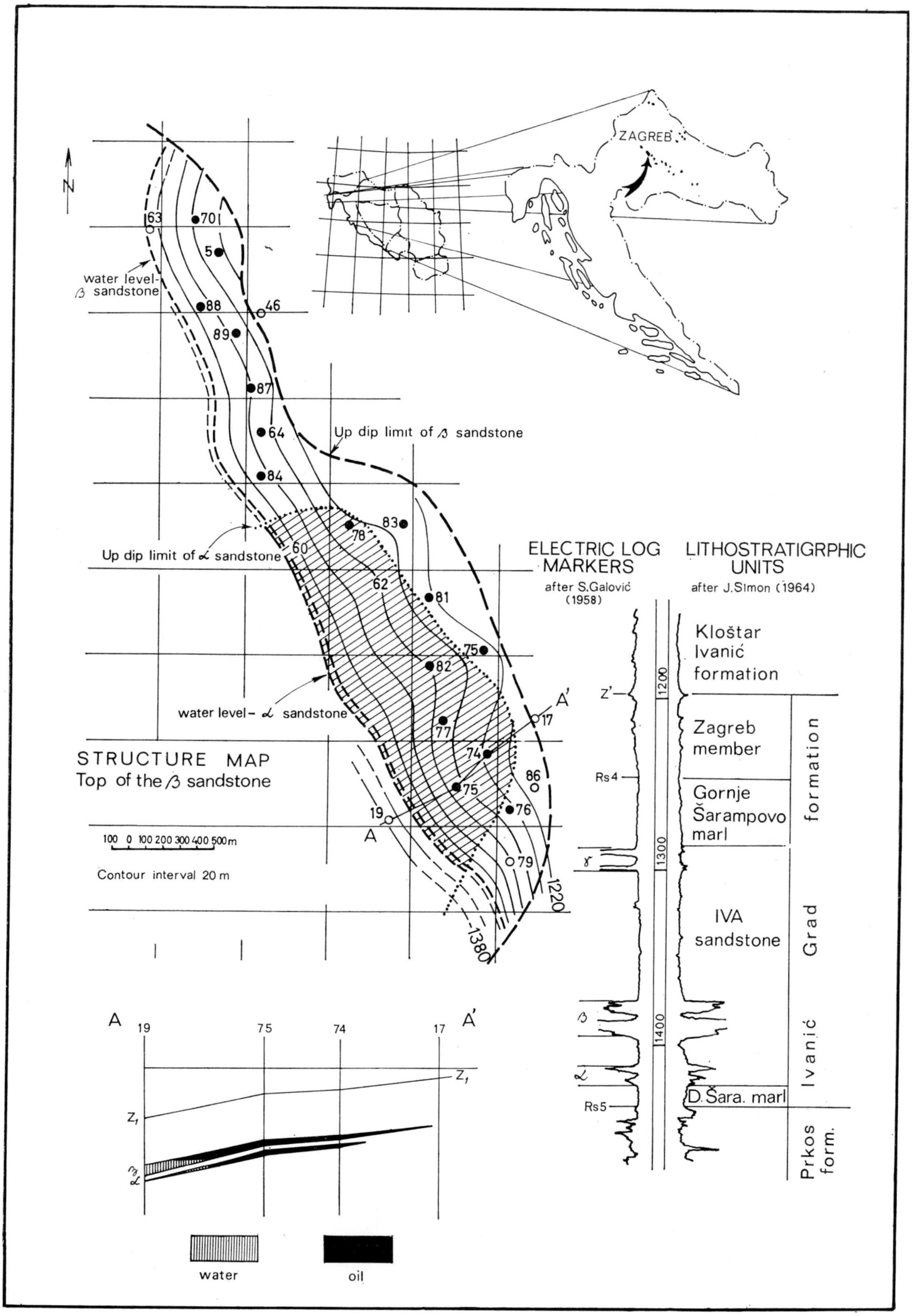

FIG 15. Kloštar (south)—oilfield, Sava depression (partly redrawn from Klcštar field, Core Lab. Inc., Eng. Consulting Department, 1959).

Oil and gas accumulations occur in the Neocene complex sediments, as well as in rocks underlying Neocene deposits. The reservoirs of the latter exhibit secondary porosity and are represented by carbonates, schists, and magmatic rocks.

Neocene reservoir rocks are either sandstones characterized by intergranular porosity or porous biogenic limestones. Clays and marls represent the cap rocks in all reservoirs.

The reservoirs are highly unhomogeneous, their characteristics varying over a wide range. Permeabilities are relatively low, so that an average oil production per well ranges about 12 cu m/day, and a maximum about 100 cu m/day.

Discovered pools occur at various depths. Production is obtained from depths ranging from 400 to 2700 m.

Natural traps in which the oil pools have been formed exhibit all Levorsen's classification types. The greatest number, however, may be classified as combined traps. With regard to the reserves they contain, the most significant are some structural traps of some reservoirs in the Sava depression.

The largest oil and gas fields discovered are either in production or are being prepared for it. A certain number of fields have been depleted, whereas the remaining discovered fields, for the time being, do not represent objects of greater commercial interest.

According to reserves and the production they yield, the most important fields are Stružec (Fig 13) and Žutica, both situated in the Sava depression. Next follow the Kikinda field (Fig 17), in the Banat depression, Kloštar (Fig 15), in the Sava depression, Šandrovac, in the Drava depression, and Elemir, in the Banat depression.

The most significant gas fields are Mokrin in the Banat and Okoli in the Sava depression.

The main oil properties from the majority of producing oilfields are shown in the table (Table I).

PROSPECTS OF FURTHER DISCOVERIES OF OIL AND GAS POOLS

By a simple comparison of the number of fields discovered by the end of 1955 and 1968 respectively (the state is shown on the comparison maps, Fig 16) it can be concluded that during that period great progress was made.

This progress shows that exploration staffs working in existing geonomic conditions have successfully overcome the risks of their job. There is every reason, therefore, to anticipate that by carrying out exploration activities in prospective areas through application of up-to date technology and working methods they will also be successful in discovering new fields in the future. The portion of the Pannonian Basin within Yugoslavia is completely covered by the network of regional geological and geophysical surveys. However, only the most prospective areas in the Sava and Banat depression have been explored in detail. It will be noticed that the main producing fields in these area are concentrated. The general oil-geological characteristics of other areas (in these and other depressions) do not essentially differ from characteristics encountered in areas being explored in detail in those two depressions. Objectively, it could therefore be expected that substantial oil and gas accumulation could be found there too. Such references are especially due to the N parts of the Banat depression and the zones of hydrocarbon accumulation in the Drava, Slavonia-Srem, and Mura depressions.

By detailed stratigraphic differentiation of the Pannonian—Pontian sandy—marly column within the Neocene complex, as well as by studying spatial position and oil geological characteristics of sand bodies—the regional reservoirs—within the above-mentioned column, prerequisites could be ensured for successful discoveries of pools in lithologic traps.

TABLE I

Analysis of Yugoslav crude oils (rewritten from V. Askin, "Geologija nafte." Novi Sad, p. 787)

	Stružec	Kloštar	Dugo Selo	Mramor Brdo	Bunjani	Lipovljani	Gojlo	Križ	Jagnjedovac	Lendava	Kikinda	Elemir	Jermenovci	Boka
Sp gr at −15° C	0·83	0·85	0·87	0·87	0·87	0·82	0·88	0·96	0·86	0·84	0·87	0·86	0·92	0·85
API degrees	38·98	35·3	30·8	30·9	30	40	28·6	15·3	33·8	37	30·92	32·2	22·1	32
Viscosity (cSt/20° C)	4·0	11·2	—	55	36·4	3·9	14·8	—	11·5	3·38	—	25	—	—
Viscosity (°E/20° C)	1·3	1·9	—	7·24	4·8	1·3	2·29	—	1·97	1·24	—	3·93	—	—
Viscosity (cSt/50° C)	—	—	11·2	—	—	—	—	18	—	—	30·20	—	262	25·1
Viscosity (°E/50° C)	—	—	1·94	—	—	—	—	2·65	—	—	4·12	—	34·6	3·45
Viscosity (°E 100° C)	—	—	—	—	—	—	—	—	—	—	—	—	2·55	—
Sulphur (%)	0·4	0·46	0·84	1·02	0·8	0·24	0·43	0·74	0·27	0·15	0·64	0·38	0·61	0·63
Ash (%)	0·02	0·01	0·03	0·23	—	0·02	—	0·04	0·01	tr.	tr.	tr.	—	tr.
Salt (ppm)	26	22	42	43	280	22	28	60	22	—	tr.	21	84	72
Water (%)	0·4	0·44	0·5	0·3	11·6	0·5	0·3	5	0·2	0·2	0·5	0·2	1·4	1 8
Pour point° C	+0	+11	+20	+14	+13	+2	−9	−6	+12	+8	+27	+25	+28	+35
Paraffin (%)	6·8	—	2·96	7	6·63	—	5·9	—	10·2	—	4·78	10	16·1	—
ASTM destillation (%)														
up to 100° C	10	9	6	2	7	12	2	—	5	2	6	6	—	1·6
up to 200° C	36	24	19	22	23	38	23	—	23	25	19·5	21	—	11
up to 300° C	56	43	34	42	43	60	49	12	48	67	35	36	—	25
up to 350° C	66	56	52	62	—	70	—	—	69	—	44	43	—	—

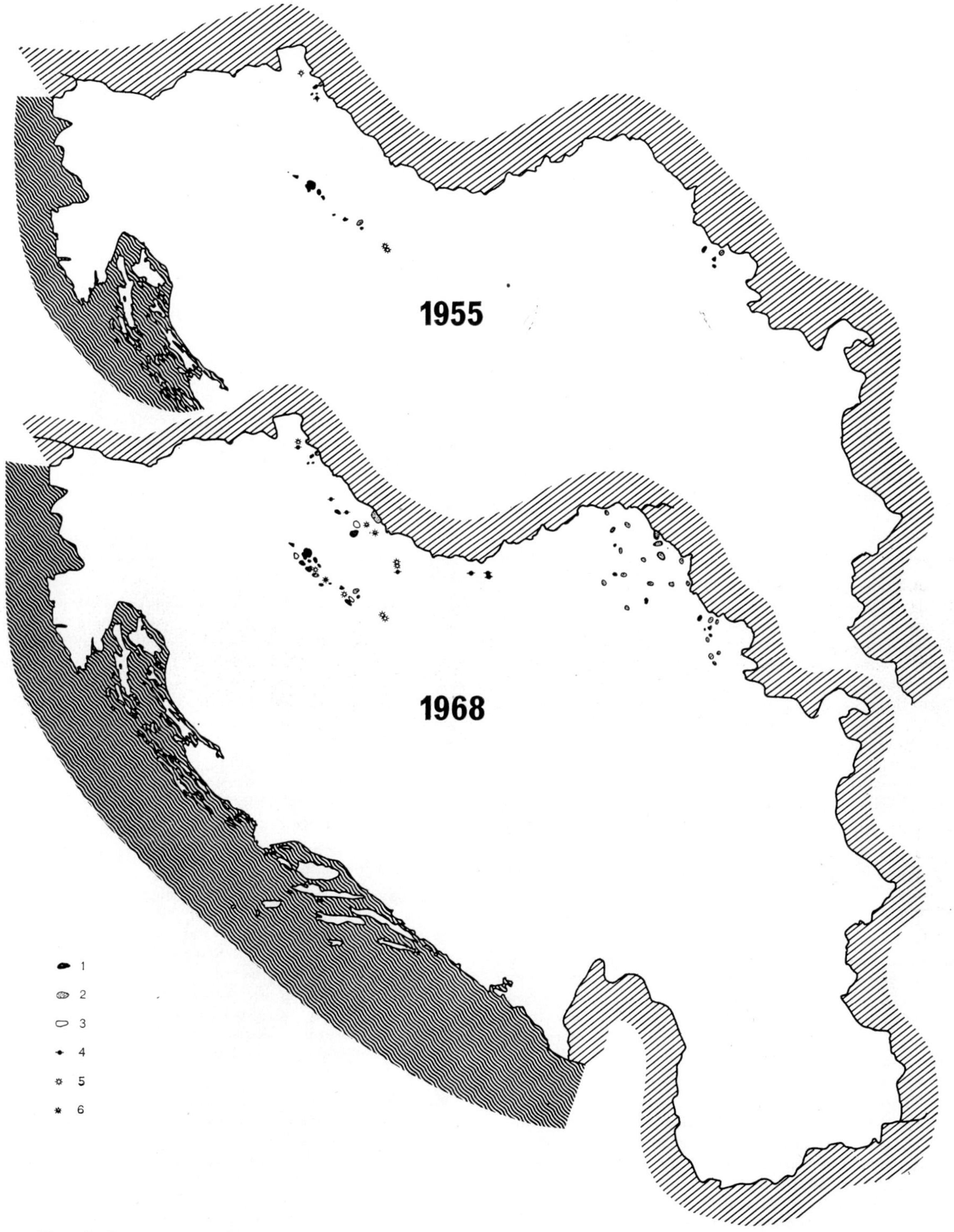

FIG 16. Comparison of the fields discovered in Yugoslavia by the end of 1955 and 1968 respectively. 1. Oilfield. 2. Gas field. 3. Developing field. 4. Oil discovery. 5. Gas discovery. 6. Oil and gas discovery.

Although exploration work has not hitherto been conducted in this direction, some of the discovered pools are closed lithologically (Fig 15). Such facts point out the possibility of finding new stratigraphic traps at the margins of sand bodies whose oil/gas bearing potentiality has an evidence pronounced by discovery of hydrocarbon pools in other trap types.

Similar expectations might be held with regard to the pools occurring in the base of the Neocene complex. Most of those traps already known have been discovered by testing gravimetric maxima. In addition, the thickness and composition of Mesozoic formations whose extent is associated with marginal zones of old Pannonian intermediary massif can also offer good possibilities of discovering new and probably significant fields.

Generally speaking, it could therefore be said that real possibilities of finding new fields do exist. It might, however, be expected that a future trend will continue to have its present characteristics.

ACKNOWLEDGMENTS

Thanks are due to the Director General of Kombinat Naftagas, Novi Sad, and to the Director General of Ina-Naftaplin, Zagreb, for permission to publish data in this paper, and in particular to Eng. J. Simon for helpful collaboration in preparing most of the illustrations in the section dealing with geologic structure of the basin.

BIBLIOGRAPHY

Aksin, V. "Geologija nafte." Novi Sad, Dnevnik, 1967.

Kober, L. "Das alpine Europa und sein Rahmen." Berlin, 1931.

Levorsen, A. I. "Geology of Petroleum." W. H. Freeman, San Francisco, 1954.

Muratov, M. V. "Tektonika i istorija razvitija Al'pskoj geosinklinal'noj oblasti juga Evropskoj časti SSSR i sopredelenih stran." Tektonika SSSR, Moskva, Akademija Nauk SSSR, 1949.

Vucković, J., Filjak, R., and Aksin, V. "Survey of Exploration and Production of Oil in Yugoslavia." *Proc. 5th World Petrol. Congr.*, 1959.

Discussion

Chairman: D. R. BROWN

Dr P. Bitterli (Bataafse Internationale Petroleum Maatschappij): Referring to your mentioning of oil from the Mesozoic and fractured basement below the Neocene cover, could you tell us something about actual oil production from such reservoirs?

Z. Pletikapic: Up to the present, we were not oriented towards targets in the fractured basement. All our discoveries in the reservoirs below the Neogene cover were made at random in boreholes, the targets of which were reservoirs within the Neogene complex.

Our production from the basement rocks in the Pannonian Basin is very low and is coming from fractured granites, gneisses, and schists. But in the Hungarian part of the Pannonian Basin big pools with fairly high production abilities were found in the Mesozoic formations, and I hope we shall have a chance to hear something about that from the following paper.

Georg W. Rippel (Deutsche Erdöl AG): You stated that oil has been found in the Pannonian Basin only so far. May I ask for the recent exception, the Dugi Otok oil discovery? What is the age of the reservoir, type of the trap, and your plans to develop it?

Z. Pletikapić: Our exploration activity outside of the Pannonian Basin is still in the initial phase and we could not give you much more information of our results there. In the outer zone of the Dinaric Alps we have selected some structures for testing. We have drilled very thick sequences of Mesozoic formations there with alternation of good cap rocks and poor fractured reservoirs. We did not succeed in finding commercial quantities of oil and gas but only very good shows.

Dr Rudolf Martin: Figs 7 and 9 give a very good picture of the palaeotopography of the buried landscape underlying the Neocene. They are isopach maps of the formations overlying this landscape, and as such represent a "cast" of this old surface. It would appear that most of the buried valleys are closed depressions, suggesting that the landscape was formed by underground drainage in a karst-type drainage system. The maps show very little evidence of any coherent surface drainage system. Have any systematic studies been carried out to determine the palaeogeomorphology of this buried landscape? What theoretical approach has been taken to resolve the composition of this old surface?

Z. Pletikapić: The map on Fig 7 was our first synthesis on the palaeotopography of the pre-Neocene formations. Recently, we obtained much more data for studying palaeogeomorphology, especially in the E part of the basin, in the Banat depression, where many fields are associated with buried hills. But up to now we did not undertake any systematic studies in the way as you consider, although we should be very interested in the results of such studies.

William Sax (Union Oil Co., Netherlands): How many seismic, gravity, and magnetic crews do you generally have employed and how many rigs?

Z. Pletikapic: We try to give some information about our activity in Fig 1. As you can see, our drilling activity is something more than 200,000 m annually. In the selection of the exploration well sites we have usually employed all surface exploration techniques but the seismic one predominantly, since detailed gravity and magnetic surveys have been nearly completed in our part of the Pannonian Basin. Recently, our activity in exploration is shifting towards employment of drilling and subsurface mapping only because the network of seismic lines became rather tight, offering a little chance to new lines in discovering any unknown features.

C. A. Doh (Schlumberger): What are the offshore possibilities of the Eastern Adriatic?

R. Filjak: We expect similar geological conditions and oil/gas possibilities in our part of the Adriatic Sea as they are in the neighbouring Italian part. We started the seismic survey there a year ago. The first results are promising, but we need at least a year more for the completion and interpretation of our survey and then we would be able to discuss the results. Today, our Adriatic coast is of more interest, even to petroleum geologists, as a holiday resort, and I could recommend you to visit it and spend your vacation there.

M. R. Warman (British Petroleum Co. Ltd): As the geology is relatively constant, it would be interesting to know if there is any systematic variation in gravity/oil quality relative to either depth or geographical position.

Z. Pletikapić: The scope of our paper was, of course, very limited but we have tried to display also the main properties of Yugoslav crude oils. Unfortunately, this table could not inform you about different properties and geological characteristics of reservoirs from which these crudes are produced. These data, had they been presented, would have shown not constant but relatively very varying geological conditions. If we tried to completely eliminate variations in geological conditions, by examining relationships between geographical position and gravity of oil within the same regional reservoir, we would find pools with gas-condensate and light oil in deeper, as well as dry gas and heavy oil pools in the shallower part of the reservoir. Variations, even then, would not be always systematic, since they are greatly influenced by hydrogeological anomalies.

Oil and Gas Exploration in Hungary

By V. DANK
(*Hungarian Oil and Gas Trust, Budapest*)

and J. KOKAI
(*Research Laboratory for Oil and Gas Industry, Budapest*)

SUMMARY

The pre-Tertiary basement belonging to the Hungarian portion of the Pannonian Basin is rather complex from the point of view of structure as well as that of stratigraphy. Palaeozoic metamorphic and sedimentary formations, as well as Triassic, Jurassic, and Cretaceous rocks, participate in the build-up. The major part of Tertiary sediments of a great thickness filling up the Pannonian Basin is formed by Pliocene (Lower and Upper Pannonian) sediments. The country's area extends over 93,000 sq km, of which 77,000 sq km are hopeful for hydrocarbon exploration. From the beginning of hydrocarbon prospection up to the end of 1968, 363 structures were proved, out of which 158 had been explored by drillings; 64 (40 per cent) from these proved to be successful. According to calculations up to now 50 per cent of the possible hydrocarbon reserves have been discovered. These hydrocarbon occurrences belong to a greater extent to the small group of the world category. Over 60 hydrocarbon occurrences are known. However, the major part of hydrocarbon reserves known till now are found in six fields only.

The overwhelming part of hydrocarbon occurrences is in structures formed by compaction above basement highs around the deep depressions filled up with thick Tertiary sediments. As regards further explorations, the most hopeful are the thick Tertiary sediment forming zones, the pinch-out zones, and the interfaces of the tectonic units. The importance of the Palaeozoic–Mesozoic formations under the Tertiary sediments are not yet known from the point of view of hydrocarbon accumulation, but their exploration is in course. Among these the fissured Triassic and the Upper Cretaceous limestones, as well as the flysch-like Upper Cretaceous formations, may be considered to be the most hopeful from the point of view of hydrocarbon prospection. At present oil production amounts to 1·8 million tons/year, gas production yields 2·8 milliard cu m/year. Hydrocarbon ratio amounted to 39 per cent in the country's energy consumption in 1968.

INTRODUCTION

THE territory of Hungary belongs to a greater extent to that young basin-like swale which developed in consequence of the tearing away of the Alps in the east and between the Carpathians and the Dinarides. The usual denomination covering the entire basin is known to be the Pannonian Basin. The area of Hungary extends over 93,000 sq km, 86,000 sq km of which are covered by sediments, and according to our present knowledge 77,000 sq km are suitable for hydrocarbon prospection.

The fundamental character of the geological structure is determined by the fact that the country's territory within the division of Europe into four big structure units falls on the area of Neo-Europe. The geological situation indicates also that this country is situated on a territory subject to rather intense movements which is a characteristic feature for the early periods of the Earth history. The geological circumstances are more complicated and more complex than those of many other basin areas.

The Hungarian part of the Pannonian Basin is constituted by the Great and the Little Plain as well as the Trans-Danubian Basin. All three basins extend beyond the country's borders. The basin of the Little and the Great Plain are separated by the Hungarian Central Mountains. The N boundary of the Great Plain is constituted by the Mátra and Bükk mountain ranges. In the SE and E direction the Great Plain Basin continues on Rumania's territory and is flanked by the Bihar (Apuseni) Mountains. In the S direction towards Yugoslavia the basin is open.

On the W border of the Little Plains Basin there are the Sopron–Kőszeg Mountains, which may be considered to be the outposts of the Alps. Towards the N the Little Plains Basin continues in the Nyitra river valley on Czechoslovakian territory.

In the Trans-Danubian Basin the Villányl and Mecsek mountain ranges emerge like island hills above the rolling topography.

In the S the Pannonian Basin extends beyond the Sava river into the Slavonian Basin (Fig 1).

The height above the sea level of these basins having a flat or hilly surface and appearing to be morphologically uniform varies between 80–130 m. Their sediments are identical or similar; their basement is, however, formed by rocks of rather different ages which likewise present a rather big variation with a view to their structure. In consequence of different geological factors they are partly sunken in the depth, partly folded; at some other places again they are settled in horst systems broken up into blocks under Tertiary sediments.

ROCKS AND STRUCTURE OF THE PALAEOZOIC AND MESOZOIC BASEMENT

The pre-Tertiary basement of the Pannonian Basin is very complex both with a view to structure

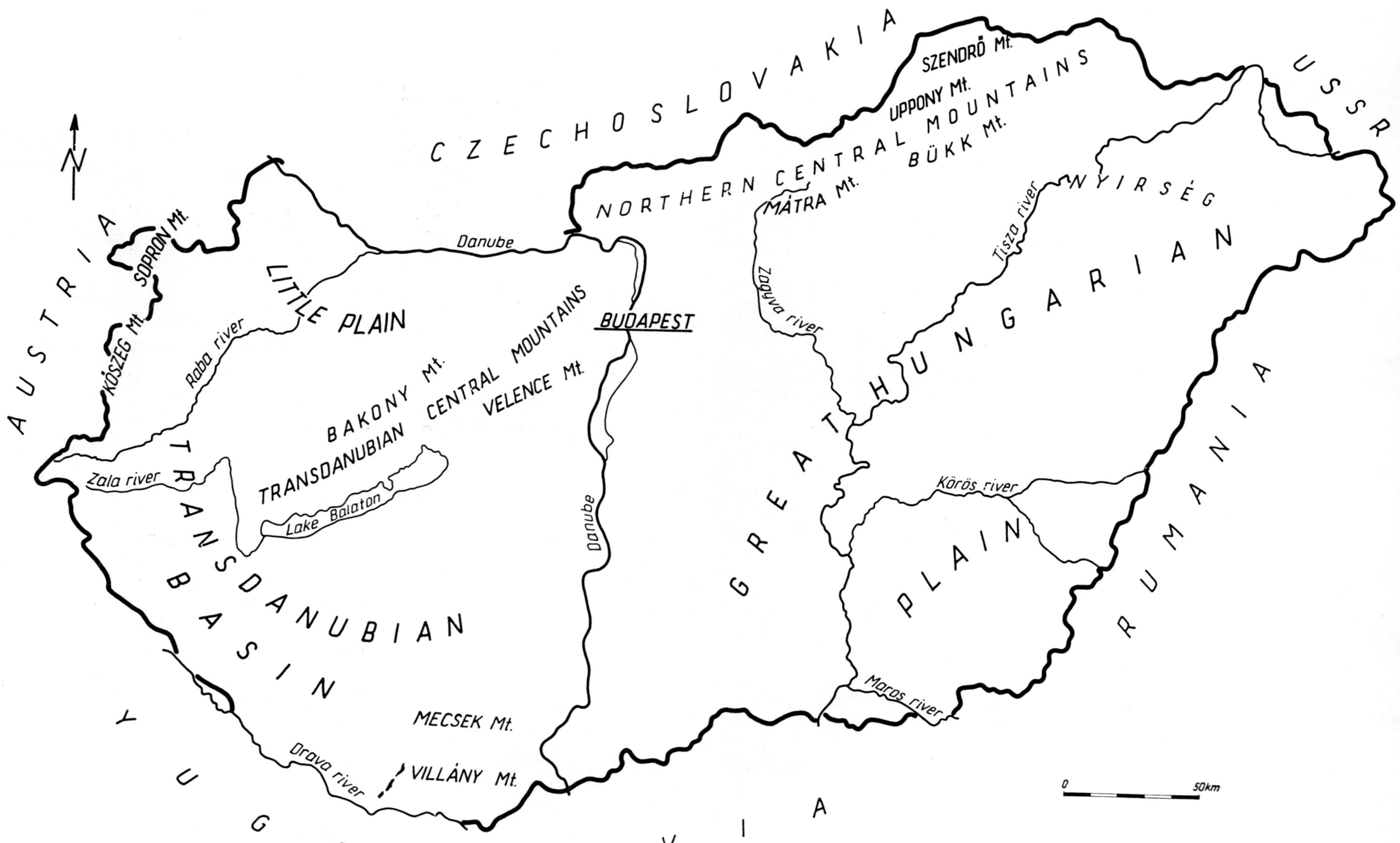
CZECHOSLOVAKIA
USSR
AUSTRIA
RUMANIA
YUGOSLAVIA
N
SZENDRŐ Mt.
UPPONY Mt.
NORTHERN CENTRAL MOUNTAINS
BÜKK Mt.
MÁTRA Mt.
NYIRSÉG
Tisza river
Zagyva river
GREAT HUNGARIAN PLAIN
Körös river
Maros river
SOPRON Mt.
KÖSZEG Mt.
Danube
LITTLE PLAIN
Raba river
BAKONY Mt.
TRANSDANUBIAN CENTRAL MOUNTAINS
VELENCE Mt.
BUDAPEST
Zala river
Lake Balaton
TRANSDANUBIAN BASIN
MECSEK Mt.
VILLÁNY Mt.
Drava river
0
50km

Fig 1.

and lithology. Mountain fractions connecting the pre-Tertiary folded-up formations of the Alps–Carpathians–Dinarides on the Hungarian part of the Pannonian Basin may be found to a greater extent under the Palaeogene–Miocene and Pliocene sediments having a thickness of many thousand metres here and there. Interpretation of their structural circumstances on the territories overlain with thick Tertiary sediments appears to be rather difficult, in spite of the fact that the basement had been discovered by a relatively high number of deep drillings. Clearing up of their geological structure is often intricate even on those places where they are on the surface (Bükk and Trans-Danubian Central Mountains, Mecsek Mountains) (Fig 2).

Extremely varied in structure, the pre-Tertiary basement includes the metamorphic formations of the Proterozoic, the metamorphic or sedimentary formations of the Early and Late Palaeozoic, and the sedimentary rocks of the Triassic, Jurassic, and Cretaceous systems.

Palaeozoic formations on the surface are known to exist on the territory of the Sopron–Köszeg Mountains, the Balaton Uplands, the Mecsek–Villány mountain ranges, the mountains of the Bükk and Szendrö, as well as on the NE part of the country at Vilyvitány. As to the connection among these superficial Palaeozoic formations, consequences may be drawn only on the basis of drilling penetrating up to the basement and serving mostly the purposes of hydrocarbon prospection (about 120 in Trans-Danubia and about 200 on the Great Plain) as well as from the results of geophysical measurements.

Palaeozoic metamorphites may be divided into three groups on the basis of the re-crystallization age:

1. Assyntian, predominantly epizonic metamorphosis partly buried by the Variscan epi-mezozonic and Alpine retrograde metamorphosis.
2. Variscan metamorphosis with a metamorphization grade being different by territories.
3. Alpine retrograde slightly epizonic metamorphosis only in the SE part of the country.

Metamorphites are settling in belts of a ENE–WSW direction. Starting from NW towards SE they are as follows:

(*a*) Variscan mezo-epizonic Early Palaeozoic (at Sopron, Pinnye).
(*b*) Variscan anchimetamorphous Early Palaeozoic (at Köszeg, Little Plain, uplands of Lake Balaton).
(*c*) Variscan granitoids (at Gelse, Buzsák, Ságvár, and Velence Mountains).
(*d*) Variscan plus pre-Variscan polymetamorphous migmatitized mezo-epizonic Proterozoic (S Trans-Danubia, region between the Danube and the Tisza rivers, Mid-Trans-Tisza) with a higher metamorphosis degree on the S edge of the Great Plain.
(*e*) Two polymetamorphous groups in the Vilyvitány block: Proterozoic plus Variscan mesozoic, as well as Caledonian plus Variscan epizonic group.

Palaeozoic magmatites are represented predominantly by granitoid rocks. In the SE foreground of Lake Balaton a Variscan range extends consisting of granite masses, NE closing members of which are the microcline-free orthoclase–oligoclase biotite granites of the Velence Mountains.

Variscan granite masses in the Danube–Tisza region are settling into two ranges parallel with one another. For the N range (at Mórágy, Miske, Soltvadkert, Kecskemét) the mainly red porphyrous microline granite is characteristic; for that of the S (at Sükösd, Jánoshalma, Szank, Jászszentlászló) light-grey granite of a finer grain. Chemically in both ranges the potassium is predominant and has a slightly alkaline character. Migmatic phenomena are frequent on the edges. In S Trans-Tisza (in the neighbourhood of Battonya) the Variscan granite of a similar feature is characterized by contact formations and by the postorogenous quartz–porphyry breaking through the granite mass.

The older representatives of the Palaeozoic vulcanism (acidic tuffs of the Ordovician, Devonian basic vulcanites) are known from Trans-Danubia and along with the enclosing sediments they have been slightly metamorphicized. Non-metamorphous Palaeozoic vulcanites are represented by the Permian quartz–porphyry (Mecsek Mountains, uplands of Lake Balaton and Battonya). The greater part of metamorphites are anchimetamorphites having not reached an equilibrium state. On the basis of their critical mineral association they may be predominantly classified into the epidotamphibolite facies and to a lesser degree into the green shale facies.

There is no doubt about the connection of granitization and regional metamorphosis which may be linked to the different phases of the Variscan orogenesis. Moreover, it is beyond any doubt that both granitization and the regional metamorphosis have been terminated with the various orogenous displacements. Permian–Cretaceous sediments have already settled on the eroded granite–metamorphite surface; all rock types of the crystalline basement can be traced in the gravel material. The greater part of anchimetamorphous rocks were destroyed by the erosion having occurred in the Permian period. As a result of Alpine displacements, crystalline formations were crushed to a greater extent along the faulted belts, and in addition they have brought about a slight retrograded recrystallization in the crystalline shales. In the greater part of the Trans-Tisza area the crystalline basement has sunken into a great depth, presumably partly under 5000 m, and therefore it is not duly known.

Due to the Variscan metamorphosis the Early Palaeozoic sediments were slightly re-crystallized. (Ordovician: sandy shale, quartzite, phyllite; Silurian: calcareous graptolite shale, graphite lime phyllite, sericite shale, conglomerate and breccia.) Anchimetamorph formations have come to the surface in the

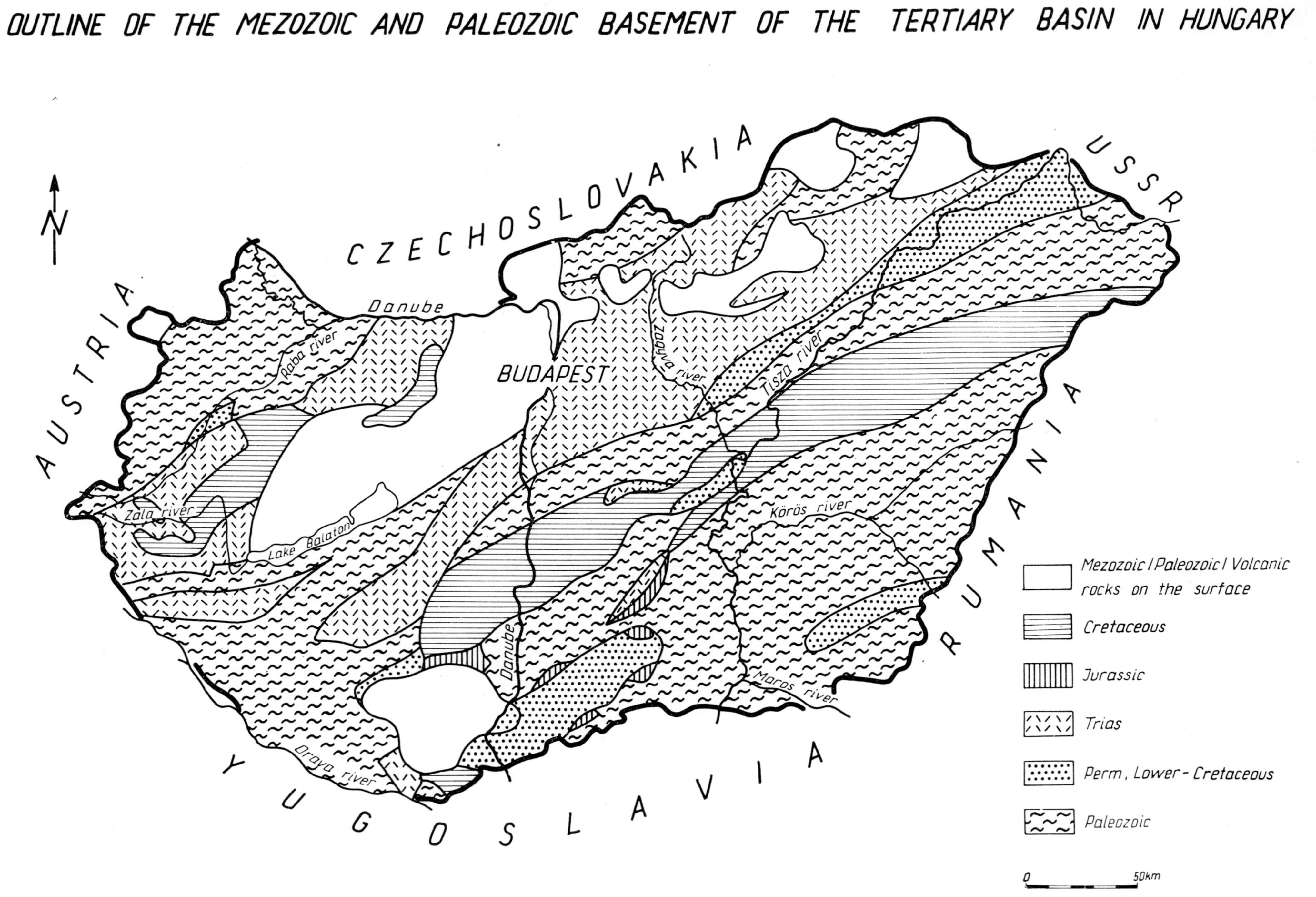

OUTLINE OF THE MEZOZOIC AND PALEOZOIC BASEMENT OF THE TERTIARY BASIN IN HUNGARY
N
CZECHOSLOVAKIA
USSR
AUSTRIA
RUMANIA
YUGOSLAVIA
BUDAPEST
Danube
Raba river
Zala river
Lake Balaton
Zagyva river
Tisza river
Körös river
Maros river
Drava river
Danube
Mezozoic / Paleozoic / Volcanic rocks on the surface
Cretaceous
Jurassic
Trias
Perm, Lower-Cretaceous
Paleozoic
0
50km

FIG 2.

Köszeg Mountain ranges, in the uplands of Lake Balaton, and in the Szendrö Mountains; subordinately also in the Mecsek Mountains. These are the prevailing formations characteristic for the basement of the Little Plain and the Tokaj Mountains but are also known in the basement of the Great Plain as the remnants of erosion.

Among the Early Palaeozoic formations the Carboniferous period is mainly represented by marine carbonate rocks. They are known in a not very frequent incidence in the Bükk Mountains, the Uppony Mountains, as well as in the drillings of Trans-Danubia (at Szabadbattyán).

Permian sediments represented to a greater extent by continental red sandstone, conglomerate, aleurolite, and clay are known to exist on a somewhat larger area on the surface in the Mecsek Mountains, the Balaton Uplands, and, according to the core analyses, here and there in the basement between the Danube and the Tisza rivers (at Nagykörös, Ásotthalom, Kunbaja) where they are settled on the highly eroded area of metamorphous shales. Marine Permian formations of carbonate structure are encountered on the surface of the Bükk Mountain area; in patches in the Danube–Tisza region (at Bugyl) and in Trans-Danubia (at Tabajd) basement.

The lagoon-like formation of Permian sediments indicates the transition towards the Triassic (Little Plain: at Tét, region between the Danube–Tisza rivers: at Nagykörös).

MESOZOIC FORMATIONS

Carbonate rocks are prevailing on the surface in considerable masses in the Trans-Danubian Central Mountains and in the Mecsek Mountains; in the N part of the country in the left-shore blocks of the Danube and the Bükk Mountains; they are settling in the basement in a corresponding range with the main strike direction of the mountains on the surface, Their role is rather considerable in the basement of Trans-Danubia, decidedly less important in the region between the Danube–Tisza rivers, while in the Trans-Tisza area it is subordinate.

The Lower Triassic of a clastic formation, as compared to the Middle and Upper Triassic of a carbonate formation, may be disregarded. The main area for the spreading of Triassic formations is the Central Mountain region of Trans-Danubia and a zone along the Bükk Mountains of a maximum width of about 100 km, covered to a greater extent by Tertiary sediments and on which carbonate masses of over 1000 m thickness may be found. The NW border of the Triassic zone is the line of the Raba river where it gets for the most part tectonically in touch with the early anchimetamorphites. Its SE border is the S front of the Lake Balaton, respectively in the Danube–Tisza region the line extending between Bugyi and Jászberény. In the basement of N Zala trough and the Little Plain Basin and in the Central Mountains of Trans-Danubia Triassic formations are covered here and there by Jurassic, Lower Cretaceous, and on a larger area in thicker layers by epicontinental Upper Cretaceous sediments.

Besides Triassic formations may be encountered S from Lake Balaton, mostly sunken into greater depth along the localities Budafa, Buzsák, Igal, as well as on the surface in the Mecsek and Villány Mountains. In the S part of the region between the Danube and the Tisza rivers, the Triassic basement is represented to a lesser extent by Low- and Mid-Triassic formations (at Nagykörös, Kerekegyháza, Lajosmizse, Tompa). In the greater part of the area of Trans-Tisza the Triassic, together with the other Upper Cretaceous older Mesozoic formations, are lacking probably in consequence of erosion (it is still encountered to a smaller extent next to Tótkomlós) (Fig 3).

Jurassic formations of a clastic facies may only be found in the Mecsek Mountains. In the NW part of the Central Mountains of Trans-Danubia they are encountered in patches on the surface and partly on their faulty edges in a carbonate facies. Jurassic sediments are known in the S part of the Danube–Tisza region (at Kiskörös, Rém, Jánoshalma, Eresztö, Madaras, Pusztamérges). Jurassic formations on the E part of the Great Plains basement have a subordinate importance according to our present knowledge.

Sedimentary formations of the Lower Cretaceous of a carbonate and clastic facies on the NW part of the Central Mountains of Trans-Danubia are to be found partly on the surface, but to a greater extent, however, on the faulty covered banks similar to the Jurassic formations. In the Villány Mountains, as well as in the middle of the Danube–Tisza range basement, they are present in relatively bigger independent masses on the surface (at Kerekegyháza, Lajosmizse, Nagykörös); in the drillings near to the areas of Tabd, Kiskörös, and Pusztamérges they may be traced in smaller patches of a flaky structure. They assume a subordinate role in the middle and the E part of the Great Plain and present a shaly folded character (at Pusztaföldvár).

Lower Cretaceous basic formations are to a greater extent products of sub-sea volcanic activity. They may be found in greater masses on the surface in the Mecsek Mountains (Trachydolerite, phonolite), as well as in the NE direction of the Danube–Tisza region basement (at Kiskörös: trachydolerite; at Szandaszöllös, Jászkarajenö, Nagykörös, Hangács: diabase tuff; at Tóalmás: diabase) and in the proximity of the Bükk Mountains along the Darnó line.

Sedimentary facies of the Upper Cretaceous may be divided into non-flysch-like and flysch-like types. The non-flysch-like carbonate sedimentary formations (Senonian Hyppuritas–Gryphean limestone and Inoceramus–Globotruncan marl) are limited in form of a broad zone to the N Zala trough, respectively to the S part of the Little Plain, as well as to the vicinity of localities Izsák, Kerekegyháza in the Danube–Tisza region. They are mainly settled on earlier Mesozoic, first of all Triassic sediments, elsewhere directly on the crystalline rocks.

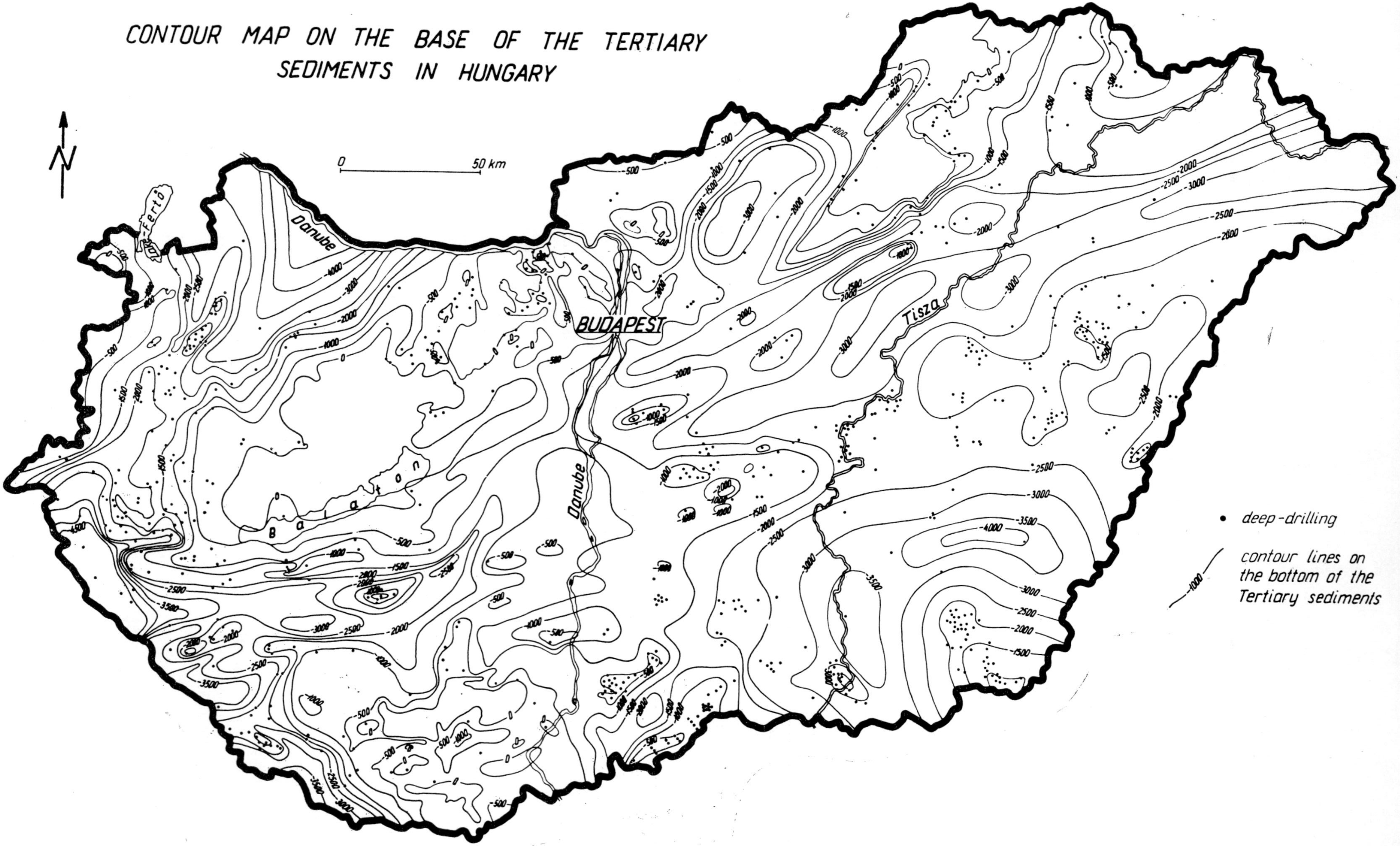
CONTOUR MAP ON THE BASE OF THE TERTIARY
SEDIMENTS IN HUNGARY
N
0
50 km
Danube
Fertő
BUDAPEST
Tisza
Balaton
deep-drilling
contour lines on
the bottom of the
Tertiary sediments

Fig 3.

The presently known belt of the flysch-like sediments (max width of 60 km) can be traced more or less continuously from the NE border line up to the Yugoslav border section of the Danube–Tisza region. Stratigraphically it can be divided into the Upper Cretaceous and Palaeogene formations. Based on the micro-fossils, Upper Cretaceous formations represent exclusively the Senonian (at Nádudvar, Kunmadaras, Szank, Csikéria, Madaras). Within the flysch-like zone sedimentary hiatus and erosion occurred between the Upper Cretaceous and the Palaeogene. Palaeogene sediments have a transgressive character and they exceed over the sedimentary area of the flysch-like Upper Cretaceous. The foraminiferous fossils of the Upper Cretaceous have often been washed over into the Palaeogene sediments. The flysch-like sedimentation in the Palaeogene is confined to the Palaeocene as well as to the Lower and Mid-Eocene. Upper Eocene sediments following above them and the sporadic Oligocene sediments no longer present a flysch-like character. In the vicinity of Hajduszoboszló, Balmazujváros, Józsa, Nádudvar, Szandaszöllös, and Törtel there are Palaeocene sediments of a flysch-like character having been proved by the micro-fossils.

The basement of the flysch-like zone is of a heterogeneous structure. The flysch-like formations are partly settled on Variscan crystalline rocks and partly on different Mesozoic sediments and even on Lower Cretaceous basic vulcanites. The basement of the Palaeogene flysch-like formations are formed to a greater extent by Upper Cretaceous flysch-like sediments and they are settling on a crystalline or earlier Mesozoic basement only at the edges.

The flysch-like sediments on the Great Plain are of a synorogene formation. Their settlement presents a tectonically faulted, slightly folded pressed and in general a gradient aspect. Sheared quartzite gravel can often be observed, which proves a very great power effect. However, up to now we were not successful in proving any covering structure. In any of the drillings could a reversed succession of layers be traced and not even the repetition of any of the stratigraphic levels.

The connection of the flysch-like zone in the NE direction towards the Transylvanian inner Carpathian formations can directly be proved: its S connection, however, still remains to be cleared. Most probably, however, there is a connection towards SE too, having been proved by results of drillings on the SE flanks of the Great Plain, on the territory of Rumania along the River Maros.

The present relatively low and high situation and prevalence of the basement formations have developed in consequence of tectonic, palaeographic, and palaeomorphologic circumstances. In any case, the effect of these circumstances can be separated but with difficulty.

At present we are not quite sure whether the hiatus between the Palaeozoic–Mesozoic and the Tertiary formations may be ascribed to erosion or not.

A sufficient knowledge of structural history of these formations may be hoped only on basis of research on lithofacies, biofacies as well as palaeogeography and geophysics, and last, but not least, of the synthesis of results of the deep drillings.

What appears to be, however, quite certain is that the pre-Tertiary basement was uneven. On some parts of the basement Palaeozoic metamorphic rocks are on a relatively high position, on other places again their subsidence is rather noteworthy.

The fact that the Mesozoic and particularly the Triassic sediments settled down in an open sea, far from the shorelines, suggests that these formations once covered most of the country and that their present lack in some areas is due to subsequent erosion in the first place, whereas structural and palaeogeographic factors probably played a secondary role.

Within the pre-Tertiary basement itself, various structural styles such as thrust-fault (region between the Danube–Tisza rivers), block-faulting (Trans-Danubian Central Mountains, Mecsek and Bükk Mountains), upwarping of folded subsurface ranges (flysch-like basement of the Tisza river region) can be recognized.

In Trans-Danubia Silurian sediments have also been proved in the pre-Tertiary basement.

Devonian and Carboniferous deposits were deformed by Hercynian foldings. The faulted and folded Triassic, Jurassic, and Cretaceous deposits are in connection with the Alpine orogenic cycle.

Investigation of the stratigraphic and structural circumstances of the pre-Tertiary basement as well as that of sediments filling up the Tertiary basin allows us to draw the consequence that the inner Carpathian area in general and within especially the area of Hungary cannot be identified with any of the well-known megatectonic types. This area is neither a platform, a folded belt, nor a foredeep. The most recent drilling results have conclusively refuted actions of a monolithic pre-Cretaceous block Tisia or even of a composite-zoned basement with zones running parallel to the main direction of faulting. In the development of the specific tectonic characteristics, a considerable role was assumed by the fact that on this territory crustal thickness is relatively thinner and is situated higher than on the adjacent areas. While elsewhere the mean crystal thickness is from 30 to 33 km, in the Pannonian Basin the Mohorovicic interface lies at a depth of 23 to 25 km.

TERTIARY BASIN SEDIMENTS

On the W and E borders of Trans-Danubia, as well as on the greater part of the Great Plain, the Pliocene (Lower and Upper Pannonian) sediments have directly settled on the pre-Tertiary basement. On the N border of the Great Plain, furthermore on that of the Central Mountains of Trans-Danubia and in the W direction of Lake Balaton, Tertiary sedimentation starts with nearshore and open-sea Eocene formations. The Tertiary basin N of the line

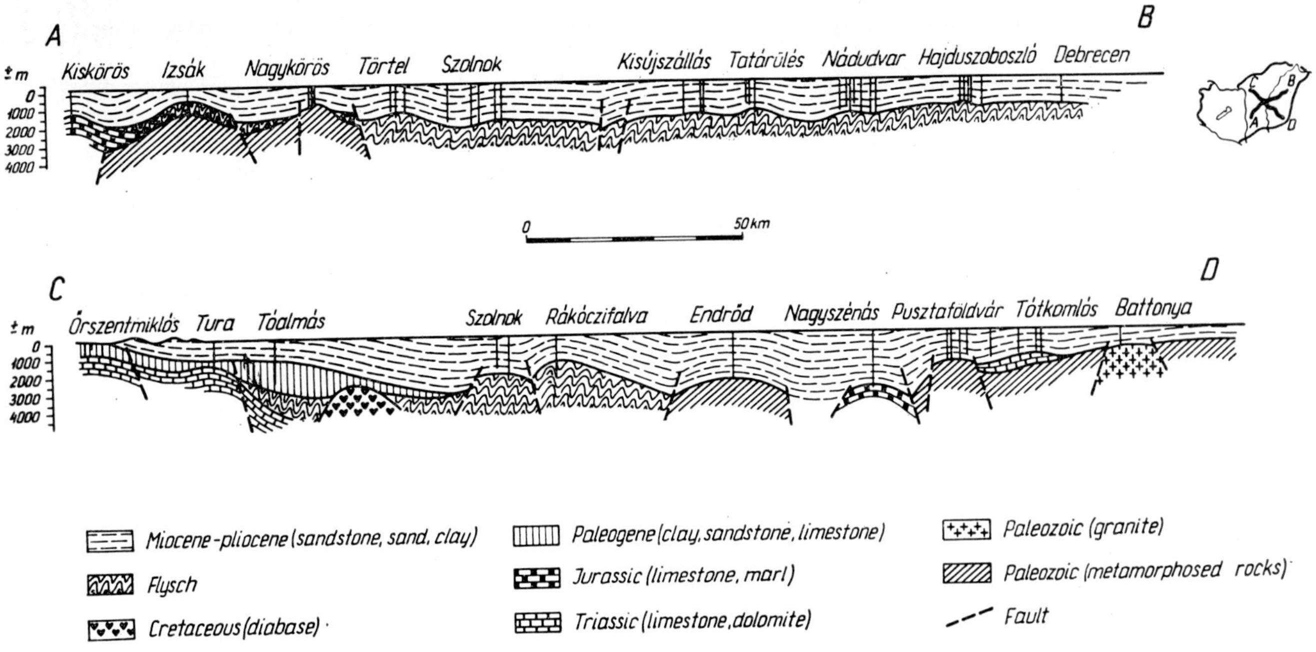

FIG 4.

Budapest–Miskolc essentially differs from the rest on the so-called Pannonian Basin. Pliocene sediments are thinner on this territory and beneath them thick Oligocene sandstone, clay-bearing sandstone conglomerates may be found having been penetrated in many places by vulcanic magmatic masses. On the Oligocene conglomerates on many places Miocene vulcanic (rhyolite, dacite) tuff layer have settled. The Oligocene conglomerate is densely faulted (at Demjén, Fedémes).

South from the above-mentioned line no Oligocene sediments are known to exist. On the territory of the Great Plain Miocene sediments are not continuous and are different in thickness and their spreading. Transition between the Mesozoic and Tertiary sediments on the Great Plain is represented by Palaeogene flysch-like formations. On the NE part of the Great Plain, in the territory of the Nyirség Miocene vulcanic tuff formations may be found under the Pliocene sediments, reaching a thickness of even 2000 m.

In the S part of Trans-Danubia, under the Pliocene sediments (Upper and Lower Pannonian) Upper Miocene, first of all Tortonian molasse-like sediments are known.

In the later period of the Alpine orogene the basement underwent a considerable but rather uneven swale which fundamentally determined the thickness of Tertiary formations.

Stratigraphy of the Pliocene sediments (Lower and Upper Pannonian), representing the bulk of the Tertiary sediments in Hungary, as well as their correlation with the adequate formations of the neighbouring basins, is still a matter of discussion among the experts. The comprehensive denomination of Pannonian applies to the formations having generally many hundred, in some places even 2000–3000 m thickness.

Neogene development of E Europe in the Upper Miocene began to differ from the W one.

In Hungary the Sarmatian (Upper Miocene) is overlain by a thick succession of widespread Congeria- and Limnocardium-bearing marls, clays, and sands. Their fauna comprises oligohaline and fresh water forms. Their stratigraphic correlations with the marine facies of similar age are still somewhat problematic. Given their position immediately above the Upper Miocene, their Pliocene age is hardly open to discussion, yet certain authors did place the whole succession into the Miocene. Sub-division of the sequence and correlation of particular horizons has been handicapped by the geographical isolation of the various facies of these Congeria beds. In fact, outcrops where the succession of horizons could be observed are very scarce. The name Pannonian has been generally adopted for the Congeria beds.

The most important single fact proved by drillings is that the open-sea facies of the Pannonian is much more uniform than the nearshore facies and that their thickness exceeds ten or more times that of the facies known from exposures. In Trans-Danubia, the Lower Pannonian deposits are sharply delimited against the Upper Pannonian, which differs little from the nearshore facies of the Lower Pannonian. In Central and E Hungary there is a gradual transition between the Lower and Upper Pannonian.

The various thicknesses of these sediments are determined first of all by the situation of the pre-Tertiary basement.

Lower Pannonian formations consist predominantly of clay and aleurite. Sandstones compared to the Upper Pannonian strata have a subordinate role.

The Upper Pannonian sediments consist mainly of mid- and fine-grain sandstones.

Detailed long-range correlation of the Pliocene layers, owing to their palaeogeographic situation, generally causes difficulties (especially in the case of the Upper Pannonian layers). However, identification of formations of a greater thickness belonging to the same sedimentation cycle can in general be done satisfactorily on the entire Hungarian part of the Pannonian Basin.

HISTORY OF PROSPECTING FOR OIL AND NATURAL GAS IN HUNGARY

In Hungary prospecting for oil and gas was started in 1910. H. Böckh and his co-workers applied first the torsion balance invented by Roland Eötvös for petroleum prospecting. In 1923 the Anglo-Persian Oil Co. carried out drilling in Trans-Danubia, but as it proved to be barren, prospecting was discontinued. In the following period large-scale hydrocarbon prospecting was at a standstill, although several wildcats have been drilled on the Great Plain and its borders (at Karcag, Örszentmiklós) and in the vicinity of Hajduszoboszló. These have yielded encouraging results and at Hajduszoboszló, along with hot water, considerable quantities of methane gas were obtained. That very fact that in the water of a great number of artesian wells having been drilled on the Great Plain hydrocarbon gases often appeared gave rise to further hopes on the part of the geologists for a successful outcome of the prospecting.

In 1935 a concession in Trans-Danubia was given to the European Gas and Electric Co. (Eurogasco) belonging into Standard Oil's sphere of interest. A Hungarian–German company carried out prospecting in the S part of the Great Plain in 1941. At the same time the Hungarian Government launched a gravity survey and deep drilling programme in the NE part of the country.

Eurogasco explored pools at Mihályl in 1935 with 94–96 per cent carbon dioxide content. Then at Budafa considerable natural gas and oil reservoirs were discovered in 1937. A less important oilfield was discovered at Bükkszék in the Palaeogene basin on the Great Plain's N border in 1937. The prospecting has led to ever-increasing results. In the period between 1939–41 new oil and natural gas fields were discovered at Lovászi, Hanót, and Ujfalu. The Hungarian–German company on the Great Plain has discovered pools of lesser importance at Körösszegapáti and Tótkomlós.

In the period between 1937–45 Hungarian oil and gas production was practically supplied by the Budafa, Lovászi, Ujfalu, and Pusztaszentlászló fields.

Between 1945–54 a Hungarian–Soviet company carried out prospecting in the E part of the Great Plain and at Biharnagybajom and Mezökeresztes discovered oil occurrences of minor importance.

In the history of petroleum prospecting in Hungary, the discovery of the Triassic–Cretaceous dolomite–limestone reservoirs at Nagylengyel in 1951 marked a noteworthy event.

In 1955 natural gas was discovered in the valley of the Dráva river next to Babócsa and oil was found in the proximity of Eger.

The Hungarian Oil and Gas Trust, which was called into life in the middle fifties, elaborated a hydrocarbon prospecting plan covering the entire country. Prospecting on the Great Plain, however, was always in the limelight as the results of geophysical and mainly a great number of seismic surveys have proved to be very hopeful.

Prospecting of the Great Plain has been successful since 1958. At Hajduszoboszló a natural gas field was discovered, which is significant from the point of view of the country's energy situation. In addition, smaller oil and more considerable natural gas reservoirs were explored on the SE part of the Great Plain at Kunmadaras, Battonya, and Pusztaföldvár (Fig 5).

The prospecting activity was later on concentrated in the Great Plain. Hydrocarbon prospecting activity was especially successful on the S part of the Great Plain. Oil and natural gas fields were discovered at Üllés (1962), Szank (1964), Algyö (1965), Asotthalom (1967), Tázlár (1968), and Kelebia (1968). Among these, at present the Algyö field represents the most important hydrocarbon occurrence of the country and contains nearly 50 per cent of the country's hydrocarbon reserves.

At the same time smaller oil and natural gas fields were found in Trans-Danubia next to Mezöcsokonya and Belezna.

The number of the country's hydrocarbon occurrences is over 60, but the major part of hydrocarbon reserves is concentrated in the pools of six fields, namely, Algyö, Hjduszoboszló, Nagylengyel, Pusztaföldvár, Kunmadaras, and Szank.

In the past decade the most significant result was the discovering of the Algyö field in the SE part of the Great Plain near Szeged.

A subsurface geological structure was detected by the seismic method and explored in 1965. The deeply buried Lower Palaeozoic metamorphic rocks were revealed by the seismic profiles to form an elongated ellipse of N–NW trend, 17 km axial length and 4 to 5 km width. The basement of the area consists of metamorphic rocks (chloritic mica-schist, micaceous quartzite, granite-gneiss, amphibolite) at 2500 to 3200 m depth. It is overlain by Palaeogene and Miocene formations in the W and by Miocene and Lower Pliocene conglomerates elsewhere. These are overlain in their turn by a Lower and Upper Pliocene marl and sandstone sequence.

According to present knowledge, sedimentation started in this area only in the Miocene. At the beginning of the Lower Pannonian territory began. The first phase of the transgression is represented by the basal-conglomerate containing the rock material of the metamorph basement. Hydrocarbons are contained in the Pliocene (Upper Pannonian and

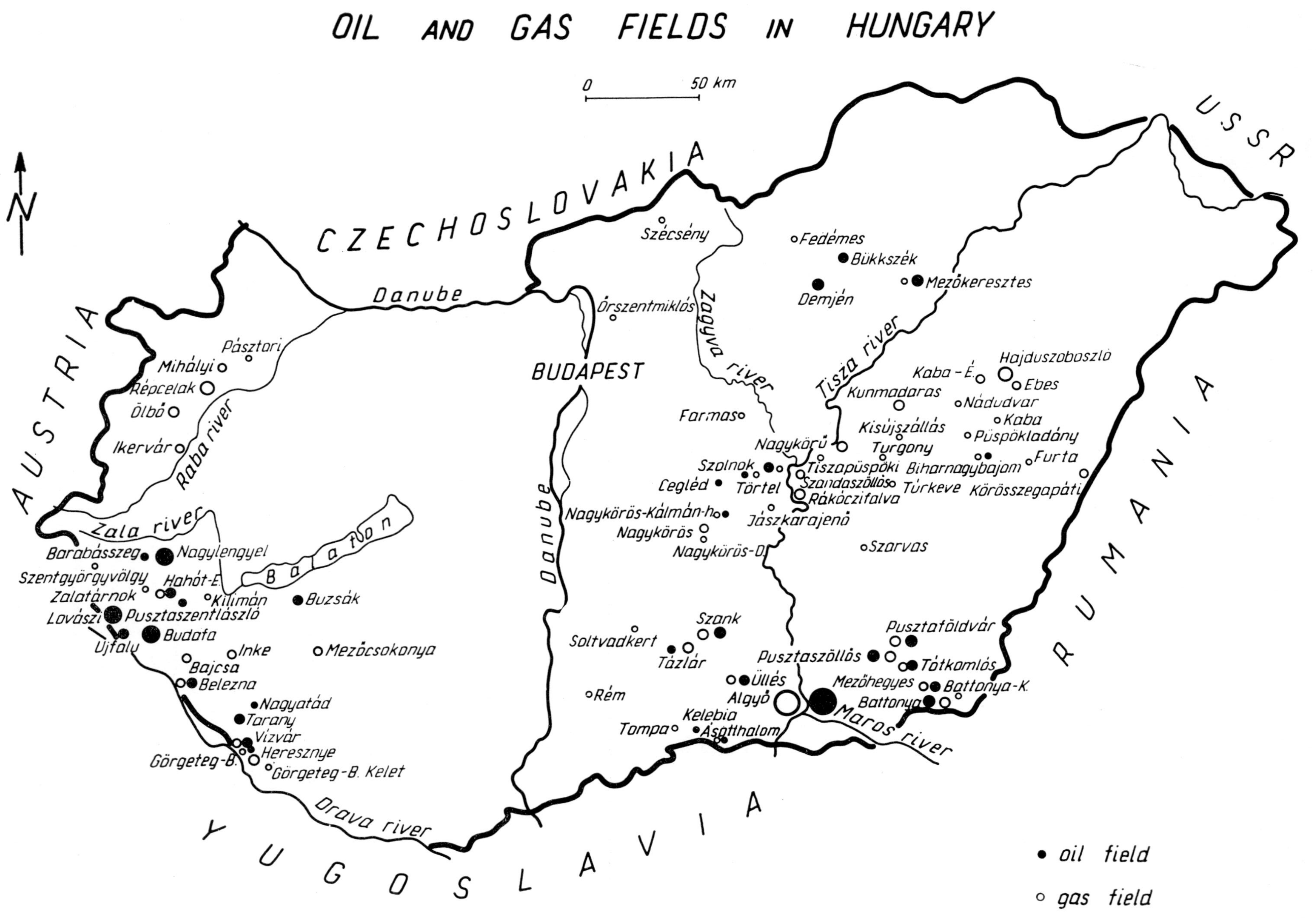
OIL AND GAS FIELDS IN HUNGARY
0 50 km
N
CZECHOSLOVAKIA
USSR
AUSTRIA
RUMANIA
YUGOSLAVIA
BUDAPEST
Danube
Zagyva river
Tisza river
Raba river
Zala river
Balaton
Drava river
Maros river
Szécsény
Örszentmiklós
Fedémes
Bükkszék
Demjén
Mezőkeresztes
Hajduszoboszló
Kaba-É.
Ebes
Kunmadaras
Nádudvar
Kaba
Kisújszállás
Püspökladány
Turgony
Biharnagybajom
Furta
Nagykörü
Tiszapüspöki
Szandaszőllős
Rákóczifalva
Túrkeve
Körösszegapáti
Farmos
Szolnok
Törtel
Cegléd
Nagykörös-Kálmán-h.
Jászkarajenő
Nagykörös
Nagykörös-D.
Szarvas
Pásztori
Mihályi
Répcelak
Ölbő
Ikervár
Barabásszeg
Nagylengyel
Szentgyörgyvölgy
Hahót-É.
Zalatárnok
Kilimán
Buzsák
Lovászi
Pusztaszentlászló
Budafa
Újfalu
Inke
Mezőcsokonya
Bajcsa
Belezna
Nagyatád
Tarany
Vizvár
Heresznye
Görgeteg-B.
Görgeteg-B. Kelet
Szank
Soltvadkert
Tázlár
Rém
Tompa
Kelebia
Ásotthalom
Üllés
Algyő
Pusztaszőllős
Pusztaföldvár
Tótkomlós
Mezőhegyes
Battonya-K.
Battonya
• oil field
○ gas field

FIG 5.

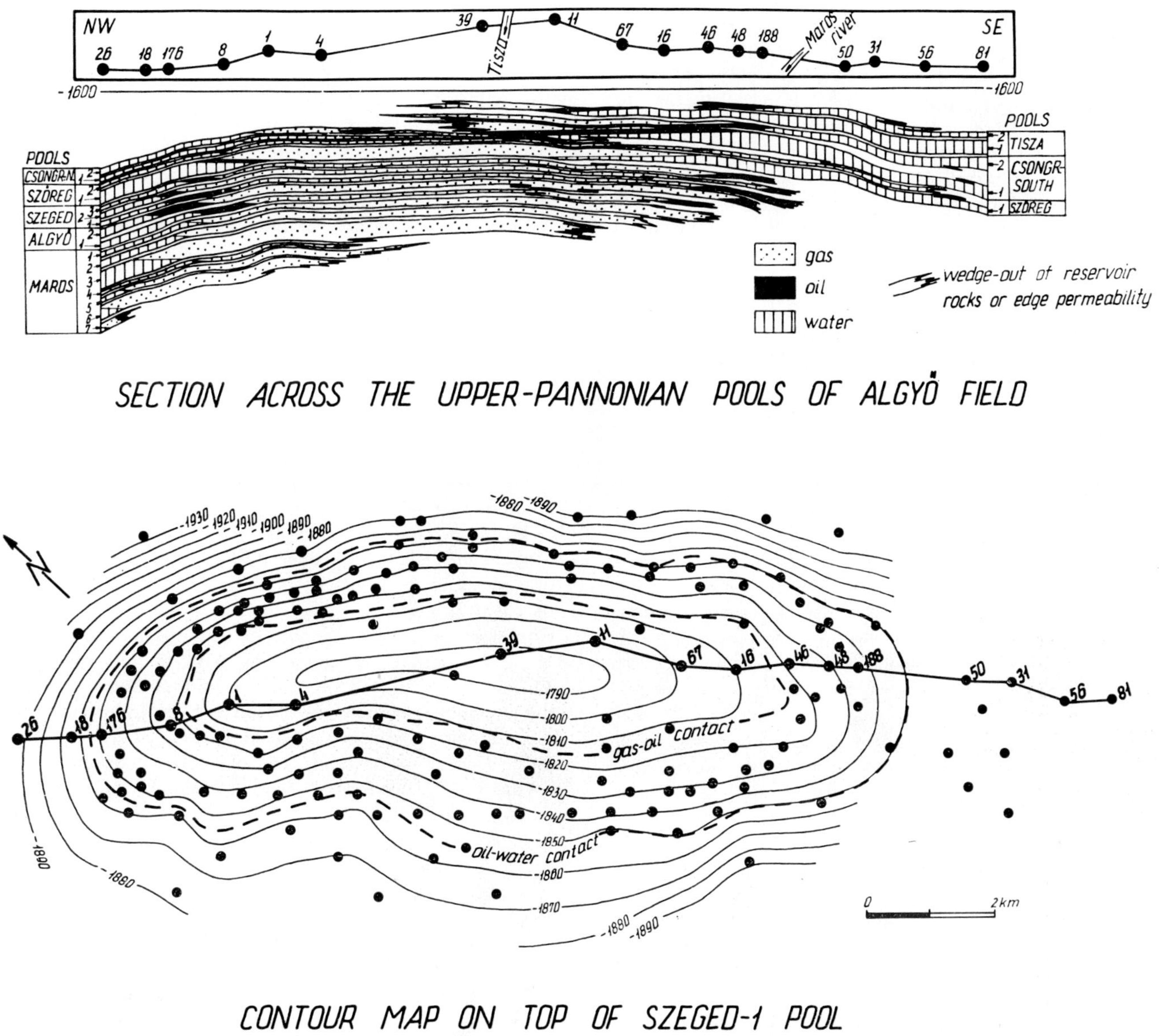

Fig 6.

Lower Pannonian) sandstones and in the Lower Pannonian to Miocene conglomerate sequence. The Upper Pannonian sediments are 1400 to 1800 m thick. The Lower Pannonian to Miocene conglomerates vary from 5 to 180 m.

The Upper Pannonian is characterized by frequent alternation of sandstones and marls, the former making up more than 50 per cent of the beds. From W towards SE the increase of the fine grain fraction in the sediments and the wedge-out of the reservoir rocks may be observed. The Lower Pannonian sediments consists mainly of marl layers and lenses.

In the Upper Pannonian in the depth between 1750–2200 m there are 19 oil and gas pools. In the Lower Pannonian only less important hydrocarbon pools are known, except the gas-bearing basal-conglomerate. Full-scale exploration of the area, including a thorough prospecting for oil and gas, is still in progress.

Taking into consideration the present technical means at disposal, 77,000 sq km of the country's total area of 93,000 sq km are suitable for hydrocarbon prospecting, from which 26,000 sq km fall to Trans-Danubia and 51,000 sq km to the Great Plain. In regard to the major part of the country forms a plain, first of all geophysical, mainly seismic methods are applied at the prospecting.

On the basis of geological considerations, hopeful territories are surveyed in detail by geophysical methods including gravity, magnetic, geoelectric, seismic measurements, and even parameter drilling in case of necessity and then they are prepared for exploration by drilling. Prospecting over the entire Hungarian territory is performed by the Geological Organization of the Hungarian Oil and Gas Trust. One geophysical and two drilling companies are working in the country under the Trust's direction. Preparatory work is carried out by ten geophysical groups (seven seismic ones); 34 deep-drilling rigs are working, of which seven are suitable for depths of 3500–5000 m and 27 for shallower holes. 350,000–400,000 m are drilled yearly. Equipment for the drilling of holes of medium depth (2000–3000 m) are drilling 13,000–16,000 m per year. The deepest hole having been drilled up to the present is at Kerkáskápolna in Trans-Danubia (4510 m). As from 1935 to the present the total length of the wildcat and production wells having been drilled on the territory of Hungary is about 6·1 million m.

The entire country has been measured in detailed one station sq km by gravity surveys. In this respect country-wide measurements may be considered to be terminated. Further surveys will be carried out only in case, if in the course of the prospecting activity, this would appear to be necessary.

With a view to the magnetic surveys, the country may likewise be considered to be terminated. We have a detailed map of the country's territory with one station/sq km. Additional survey of territories having undergone a detailed seismic works are still in course.

As regards electric measurement, we do not yet have a country-wide basic map. In Trans-Danubia 18 per cent of the territories having proved to be hopeful from the point of view of hydrocarbon prospecting have been surveyed by geoelectric methods; the same was done on the Great Plain to an extent of 43 per cent.

Seismic works form at present the bulk of our geophysical activity. The refraction method is mainly applied to collect information about the situation of the basement, and the reflection method is used at the detailed measurements. We have a comprehensive map of the country's territory up to a depth of 2000 m having been measured with the reflection method. At present we are working with an analogous method, but from next year we intend to switch over to the digital method.

Our well-logging service applies all methods known around the world, *i.e.* normal, micro, laterolog, proximitylog, radioactive, and sonic methods. We are often faced with rather difficult problems to be solved, since, as is known, temperature on the territory of Hungary is increasing more quickly towards the depth than in other countries. The geothermal gradient is 14–20 m/C. In the Pliocene sediments the reservoir pressure generally is hydrostatic. However, Miocene reservoirs often have an excess pressure of 60–100 per cent. Well-logging instruments suitable even in special drilling muds at temperatures ranging between 200°–260° C and at pressures of 700–1000 atm have been developed and are manufactured in the country.

DEVELOPMENT OF THE OIL AND NATURAL GAS INDUSTRY IN HUNGARY

Oil and natural gas production started in Hungary in 1937. In the period between 1937–50 production was carried on only out of the Pliocene sandstone reservoirs at Budafa, Lovászi in Trans-Danubia which are still in exploitation at present. In 1943 production amounted to maximum 800,000 tons/year. As a result of the stepped-up production during World War II, the condition of the fields deteriorated and production decreased to a considerable extent; thus in 1948 it amounted only to 483,000 tons.

In 1948 the Hungarian oil industry was nationalized and underwent a complete reorganization. In the period between 1950–55 steps were introduced in order to stop a further decrease of the yield of the old fields at Budafa, Lovászi and production started from the carbonate oil reservoirs at Nagylengyel as well. For the processing of the oil having been produced here, a refinery plant at Zalaegerszeg has been erected. Production of the Nagylengyel field amounted to: 3000 tons in 1951, 90,000 tons in 1952, 380,000 tons in 1953, and in 1955 it reached 1·2 million tons/year.

The Hungarian Oil and Gas Trust, founded in the fifties, controls 23 companies dealing with prospecting, production, transport, and refining, as well as the distribution of oil products. At present 37,000 employees are working in the oil and gas industry. Designing activity in connection with the oil and natural gas industry's demands is in the charge of the Designing Company for the Oil and Gas Industry. The activity of the Industrial Research Laboratory for Oil and Gas extends to sub-surface geology, chemistry, geochemistry, core analysis, micropalaeontology chemistry, drilling, drilling mud, cement, well logging, reservoir engineering and projecting. The above-mentioned companies are also controlled by the Hungarian Oil and Gas Trust. Reserves of the gas fields having been discovered on the Great Plain at the end of the fifties rendered a gas programme to be possible for covering almost the whole country.

Following World War II, the Hungarian oil refining industry had in fact to be built up anew, since the greater part of the modest refinery plants had been destroyed. After the beginning of the Soviet oil import and the exploration of our fields the Hungarian oil industry developed quickly. A new refinery plant was built at Zalaegerszeg (in Trans-Danubia), and production also increased in the reconstructed refineries of Szöny and Pét. In 1962 Soviet oil import started through the Friendship (Druzhba) pipeline for the processing of which a new refinery was built at Százhalombatta S from Budapest, having at present a capacity of 3 million tons/year. For the gases produced on the fields of the Great Plain gasoline plants were built up, one at Hajduszoboszló (capacity of which is 1·2 milliard cu m/year), and another at Kardoskút (750 million cu m/year). At present oil production runs up to 1·8 million tons/year and that of gas to 2·8 milliard cu m/year; production of LPG is about 130,000 tons/year. The capacity of the country's refinery plants is about 6 million tons/year at present. In the near future the capacity of the

Százhalombatta refinery will be increased to 6 million tons/year.

A pipeline is in course of building between Algyö and Százhalombatta for the transport of oil produced on the Algyö field. In the years ahead of us building of the new Friendship II pipeline will also start which will be suitable for the transport of 10 million tons of Soviet oil. The length of existing oil pipelines and those under construction amounts to 547 km; that of the mains serving the transport of oil products is 220 km. The increase of gas reserves made it possible to establish a countrywide gas network fed by four gas plants (Hajduszoboszló, Kardoskút, Szank, and Algyö). Length of the existing gas pipelines and those under construction is about 2000 km. The gas distributing network has a length of 3450 km within the towns and the settlements (Fig 7).

In the country's energy consumption the ratio of hydrocarbons amounted to 4·4 per cent in 1938; in 1950—10·9 per cent; in 1955—18·6 per cent; in 1960—21·2 per cent; in 1965—28 per cent; in 1968—39 per cent, according to our forecasts. In 1970 it will amount to 44–45 per cent; in 1975 over 50 per cent; and in 1980 63–64 per cent.

Hungary is poor in natural energy sources, but at present we are disposing such a quantity of proved hydrocarbon reserves which render it possible for a considerable part of the energy requirements to be covered by hydrocarbons. Hydrocarbon consumption of the industry shows an ever-increasing trend but the number of communal consumers is likewise growing. At the end on 1968 natural gas and LPG was used in 463,000 households (46 per cent of all the households) and gas heating was used in 83,000 dwellings. Gas consumption of the electric industry production increased fivefold in the last five years, that of the silicate industry fourfold, and that of the chemical industry was three times as high as previously.

Taking into consideration prognostic hydrocarbon reserves already known and those waiting for prospecting, as well as the present and expected hydrocarbon requirements, it can be stated that Hungary is compelled to import oil and in the future gas import will also be needed. We expect to be able to maintain our present steady oil production of 1·8–2 million tons/year on the same level for a longer period, and as to our present gas production of 2·8 milliard cu m/year, we hope to increase it to 5 milliard cu m/year.

FURTHER PROSPECTS OF HYDROCARBON RESEARCH

With respect to world classification our hydrocarbon accumulations belong into the group of small occurrences. According to our calculations up to now we know about 45–50 per cent of the country's possible hydrocarbon reserves. The thick Tertiary sediments Little Plain, SW-part of Trans-Danubia, NE and SE of the Great Plain, can be regarded hopeful from the point of view of origin, migration, and accumulation of hydrocarbons.

Investigating the oil and gas occurrences, it can be determined that the traps are the result of the Neogene sedimentation and tectonics, and the accumulation in our fields has also developed in the course of the Neogene. On the SE part of the Great Plain (at Battonya, Pusztaföldvár, Algyö, Asotthalom) the fissured Palaeozoic metamorphic basement is a subordinate hydrocarbon reservoir. In the area of Nagylengyel oil is contained in the Triassic dolomite and Upper Cretaceous fissured limestone. The reservoir rock on the N flank of the Great Plain near to Eger is Oligocene sandstone, while in the region between the rivers Danube–Tisza the Miocene limestone–sandstone conglomerate contains hydrocarbon reservoirs (at Szank). The largest former (at Budafa, Lovászi) and recent occurrences (at Hajduszoboszló, Kunmadaras, Algyö) are to be found in the Pliocene (Lower and Upper Pannonian) sandstones. In the Mesozoic (Triassic–Cretaceous: at Nagylengyel) and Palaeogene (Oligocene: at Demjén, Mezökeresztes) reservoirs accumulation occurred to a large extent in fault-controlled traps, while in the Miocene and Pliocene sandstones subordinately in sandy limestones) oil and gas reservoirs are in domes of a tectonic origin (at Budafa, Lovászi) and on the other hand in pseudodomes brought about by compaction above basement highs (at Szank, Pusztaföldvár, Mezöhegyes, Battonya, Algyö).

The age of hydrocarbons is different and it has been proved that beginning from the Mesozoic (presumably from the Palaeozoic) up to the Pleistocene there were at least four oil genetic periods. The age of their accumulation, however, may not be older than Neogene. The Pliocene (Lower Pannonian) and Oligocene clay marls of the Pannonian Basin are certainly mother rocks. In Trans-Danubia the Triassic bituminous limestone and lime-marl may also be considered to be mother rocks. Pliocene sediment of the Little Plain, in the W part of Trans-Danubia, and in the SE part of the Great Plain are of a considerable thickness; the drilling Makó-1 in a depth of 4200 m stopped in Pliocene (Lower Pannonian) formations. In the SW part of Trans-Danubia the thickness of the Miocene reaches even 3000 m. The Palaeogene is the thickest in the N part of the country. The drilling Nagylengyel-108 (4409·5 m) which run up to 2000 m in Triassic–Carbonate rocks could not perforate the Cretaceous and the Triassic formations in their full thickness.

From the beginning of hydrocarbon prospecting up to the end of 1968, 363 deep geological structures were found in Hungary, 158 of which have been explored by drillings, out of which 64 (40 per cent) proved to be successful. Supposing about the same result out of the remaining 205 structures waiting for exploration it is well possible that a further 80 fields will be discovered.

The present hydrocarbon occurrences are in connection with relative heights around the depressions

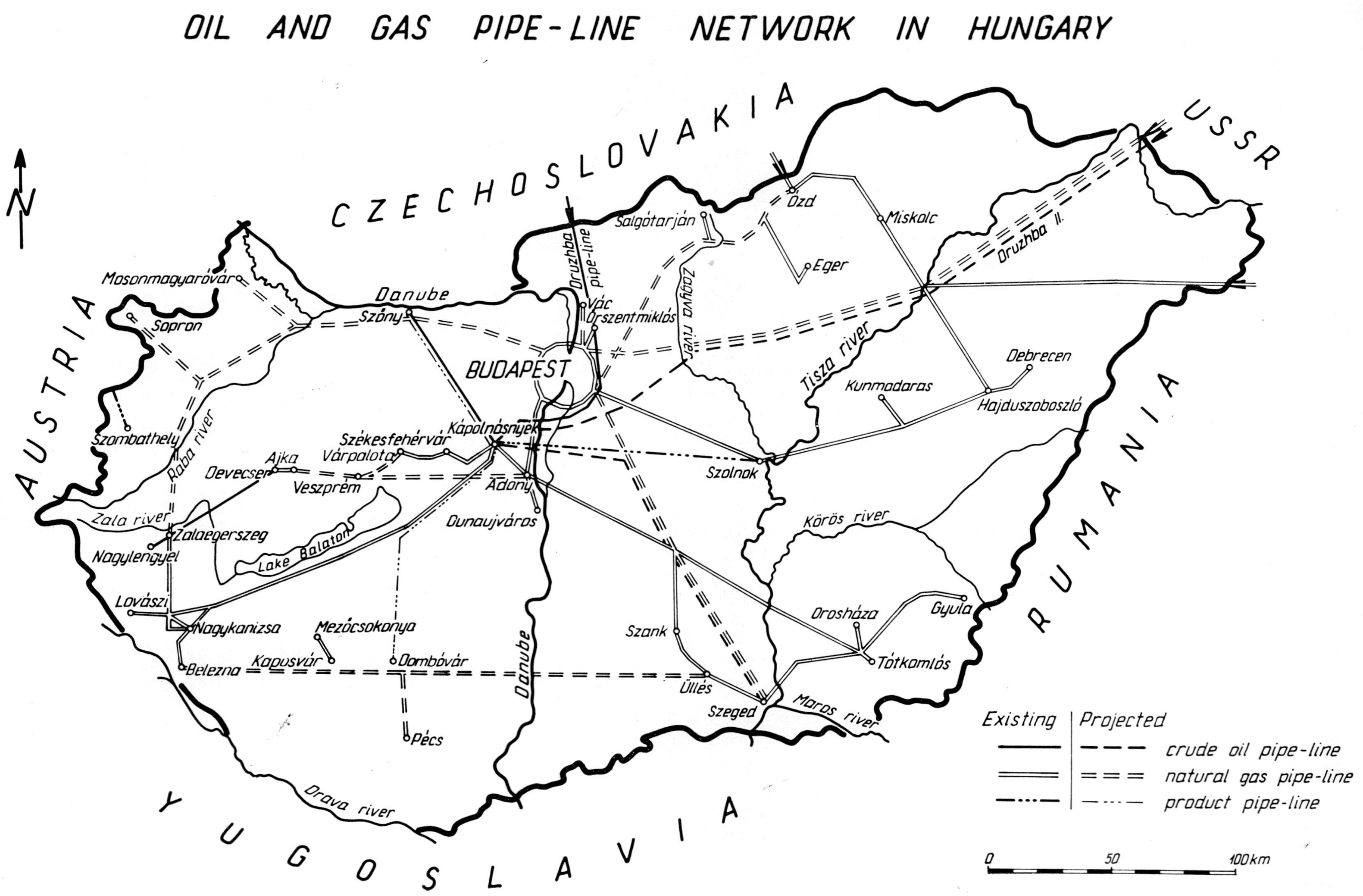
OIL AND GAS PIPE-LINE NETWORK IN HUNGARY
CZECHOSLOVAKIA
USSR
AUSTRIA
RUMANIA
YUGOSLAVIA
N
Druzhba pipe-line
Druzhba II.
Mosonmagyaróvár
Sopron
Szombathely
Rába river
Zala river
Zalaegerszeg
Nagylengyel
Lovászi
Nagykanizsa
Belezna
Kapusvár
Mezócsokonya
Dombóvár
Pécs
Drava river
Danube
Szőny
Ajka
Devecser
Veszprém
Várpalota
Székesfehérvár
Kápolnásnyék
Lake Balaton
BUDAPEST
Vác
Őrszentmiklós
Salgótarján
Ózd
Eger
Miskolc
Zagyva river
Tisza river
Adony
Dunaujváros
Szolnok
Kunmadaras
Debrecen
Hajduszoboszló
Körös river
Orosháza
Gyula
Tótkomlós
Szank
Üllés
Szeged
Maros river
Existing
Projected
crude oil pipe-line
natural gas pipe-line
product pipe-line
0
50
100km

Fig 7.

filled up with sediment masses. We are of the opinion that with a view to further explorations the continuous Tertiary sediment-forming areas supplying large sediment masses, the pinch-out zones as well as the interfaces of the tectonic units are the most hopeful.

Mesozoic and Palaeozoic formations underlying Tertiary sediments are not yet known from the point of view of hydrocarbon accumulation and their exploration is still in course.

Apart from a few above-mentioned exceptions, the Palaeozoic metamorphic basement cannot come into consideration as a reservoir rock. In principle the Palaeozoic carbonate rocks of a marine origin may be reservoirs. In the basement, however, they have a relatively small role.

Among the Mesozoic cations the bituminous Triassic and the Upper Cretaceous formations may be taken as mother rocks. As reservoir rocks first and foremost cavernous, fissured Triassic and Upper Cretaceous dolomites and limestones may come into consideration. Such rocks are known to exist in great masses in the SW part of Trans-Danubia. The Mesozoic carbonate rocks of the Great Plain, which are slightly or not at all cavernous or fissured, can be neglected from this point of view. On the Great Plain, the flysch-like Upper Cretaceous formations N from the presently known flysch-like zone covered by Miocene vulcanic tuffs are the most hopeful.

BIBLIOGRAPHY

Vadász, E. "Geology of Hungary". Budapest, Acad. Press 1953.

Kertal, Gy. *Földtani Közlöny*, 1957, (87), 383–94.

Körossy, L. *Földtani Közlöny*, 1959 (89),

Kertal, Gy. *Földtani Közlöny*, 1960, (90).

Kertal, Gy. "Typen von Erdöl- und Erdgaslagerstatten in den Ablagerungen des Mezozoikums in Ungarn". *Erdöl Zeitschrift*, 1961, (9).

Csiky, G. *Földtani Közlöny*, 1961, (91), 95–120.

Kertai, Gy. "Development of oil and gas deposits and their relation to the geological structures. The system of the Hungarian oil deposits". Thesis in the Hung. Scient. Acad. Manuscript.

Szepesházy, K. *Földtani Közlöny*, 1962, (92), 40–52.

Dank, V. "Subsurface geology of the Southern Great Hungarian Plain as shown by oil drillings". Ann. Univ. Scient. de R. Eötvös, Budapest, Sectio Geologica, Tom. VI, pp. 15–45, 1963.

Dank, V. *Földtani Közlöny*, 1963, (93), 304–24.

Körössy, L. *Földtani Közlöny*, 1963, (93), 153–70.

Oravecz, J. *Földtani Közlöny*, 1964, (94), 3–7.

Széles, M. "Paleontologische Beitrage Zur Abrenzung des Unter und Oberpannon". Magyar All. Földt. Int. Jelentései, pp. 559–68, 1964.

Kertai, Gy. *Acta Geol. Sci. Acad. Hung.*, 1967, (11), 261–78.

Pantó, G. *Acta. Geol. Acad. Sci. Hung.*, 1967, **11**, 279–81.

Szádeczky-Kardoss, E. *Acta. Geol. Acad. Sci. Hung.*, 1967, **11**, 49–58.

Juhász, A. *Földt. Közlöny*, 1968, (98), 45–53.

Kertai, Gy. "Geology of the Pannonicum". Intern. Geol. Congr. XXIIInd Session, Prague, 1968.

Srausz, L., and Széles, M. "Stratigraphy of the Pannonian", in "Geology of Pannonicum". Int. Geol. Congr. Prague, pp. 11–13, 1968.

Balázs, E., and Juhász, A. "Genetic interpretation of the Early Palaeozoic basement of the Great Plain". In the press.

Széles, M. "Summarizing stratigraphic and palaeothologic interpretation of Pliocene formations on the Great Plain". In the press.

Dank, V., Fülöp, J., *et al.* "Geological map of the Palaeozoic and Mesozoic basement of Hungary". Ed. Hungarian Geological Institute, 1967.

MS received 30 June 1969

Unfortunately, no replies to the discussion contributions have been received for this paper.

Geological Activity for Oil and Gas in the Socialist Republic of Rumania

By N. GRIGORAS
(*University of Bucharest, Rumania*)
and N. CONSTANTINESCU
(*Ministry of Petroleum, Rumania*)

SUMMARY

The geological activity developed during the last 20 years in the Socialist Republic of Rumania resulted in the discovery of oil and gas deposits in nearly all the major structural units of the country: the foredeeps of the E and S Carpathians, the Transylvanian, Pannonian, and Bîrlad Depressions, the Moesian and Moldavian Platforms.

Besides the Neogene and Palaeogene formations which have been known for a long time as oil bearing in Rumanian territory, the Mesozoic formations (Cretaceous, Jurassic, and Triassic) have now also proved to be oil or gas bearing.

Most of the oil or gas deposits are located in structural traps, but deposits located in stratigraphic or lithologic traps are also known. The reservoir rocks are predominantly of granular type.

An adequate programme of scientific research and of prospecting work by geological, geophysical, or geochemical methods, as well as of exploration activity, has resulted in the realization of the required conditions to increase the oil and gas production as follows: from 3·5 million tons of oil in 1944 to 13·2 million tons in 1967; and from 0·6 billion cu m of gas in 1944 to 16 billion cu m in 1967.

INTRODUCTION

Nineteen sixty-nine is the 112th anniversary of the first oil production in Rumania recorded in official statistics, and the sixtieth anniversary of the first discovery of methane gas fields. For the respective periods, the cumulative oil production amounts to about 370 million tons, and that of methane gas to about 160 billion cu m.

The graph of oil extraction and that of methane gas in Rumania clearly expresses its slow and wavering evolution in the period previous to nationalization, as well as its steady and ascending development after nationalization (1948).

Of the total production of about 370 million tons of oil, about 60 per cent represents the production extracted within the last 20 years, while for methane gas the figure for the same period is 90 per cent.

The explanation of these results may be found in the remarkable effort made by the Rumanian state for a methodical, scientific, and accelerated investigation of the Rumanian subsoil, an effort which is reflected in the creation of conditions for a concomitant increase of oil and gas production.

For 20 years a considerable amount of survey work, based on geological and geophysical methods, has been carried out and thus it has become possible to make forecasts for geological research drilling under reasonable conditions. Seismic surveys alone included more than 4000 party-months. Geological research drilling (prospection and exploration wells) totalled, in 20 years, more than 14 million m drilled, of which about 12 million m were for oil and more than 2 million m for methane gas. In order to appreciate rightly the effort made by the Rumanian state in connection with the geological survey for oil and gas, the above figures should be compared to the corresponding figures of 20 years ago, *i.e.* from 1928 to 1948. The number of seismic party-months for that period does not reach 100, while the amount of geological research drilling for oil and gas remains below 600,000 m.

The fact should be also mentioned that before 1948, the activity of drilling surveys was restricted only to a small area in the Mio-Pliocene zone (districts of Dîmbovița, Prahova, and Buzău) and within the Flysch zone (Moinești region, district of Bacău). Geological surveys for oil and gas carried out during the last 20 years have covered the whole country and have led to the identification of oil and gas fields in almost all major structural units of Rumania.

Finally, it should be also mentioned that before 1948 oil and gas fields were known only within the Tertiary formations. At present, besides new fields in Tertiary formations, oil and gas deposits are also known in Mesozoic formations, and hydrocarbon indications have also been recorded in some Palaeozoic members.

The contribution of the new oil-bearing objectives discovered during the last 20 years represents more than 50 per cent of the country's total oil production in 1955, more than 60 per cent in 1960, and over 80 per cent in 1968.

We particularly emphasize these results because in 1938 it was asserted everywhere that Rumania's oil reserves were on the point of being quickly exhausted.

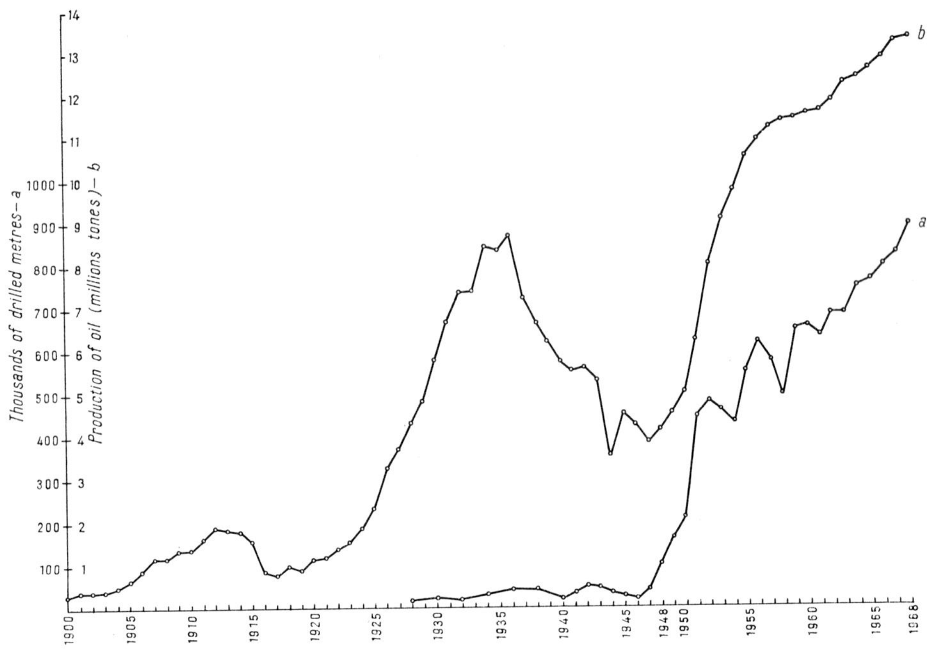

FIG 1. Geological exploration by drilling and oil production in Rumania.

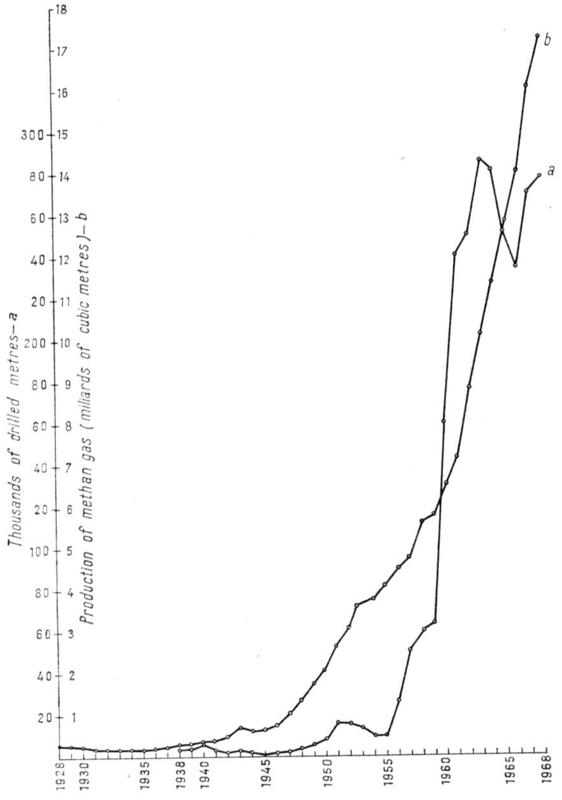

FIG 2. Geological exploration by drilling and methane gas production in Rumania.

MAIN STRUCTURAL UNITS IN RUMANIA WITH OIL AND GAS PROSPECTS

The Rumanian territory includes structural units belonging to the Carpathian Orogene and structural units belonging to the Carpathian Foreland.

The Carpathian units showing favourable conditions for the formation and preservation of oil and gas deposits correspond to sedimentation basins of the pre-mountainous type (Pre-Carpathian Depression) or of the inter-mountainous type (Transylvanian Depression and Pannonian Depression).

The Foreland units correspond either to some sedimentation basins established on platforms (the Moldavian Platform and the Moesian Platform) or to basins of the pre-mountainous type (Bîrlad Depression).

In estimating the hydrocarbon prospects of the various structural units on Rumanian territory, the following geological considerations have been taken into account:

1. The formation of oil and gas has taken place on account of the organic matter deposited simultaneously with the mineral sediment. Hydrocarbon generating series occur at various stratigraphic levels where there existed a geochemical environment which favoured the bitumination process; the identification of these levels of bituminous series in the Tertiary, Mesozoic, and Palaeozoic formations within the various sedimentation basins was a positive constituent in appreciating oil and gas prospects of the formations in question.

2. As a result of the pressure exerted by the sediments deposited in the depression zones of sedimentation basins, bitumens on the point of formation have been partly squeezed out and pushed into the peripheral zones of the basins where they are accumulating, by a process of lateral migration. Another portion of bitumens have been squeezed out

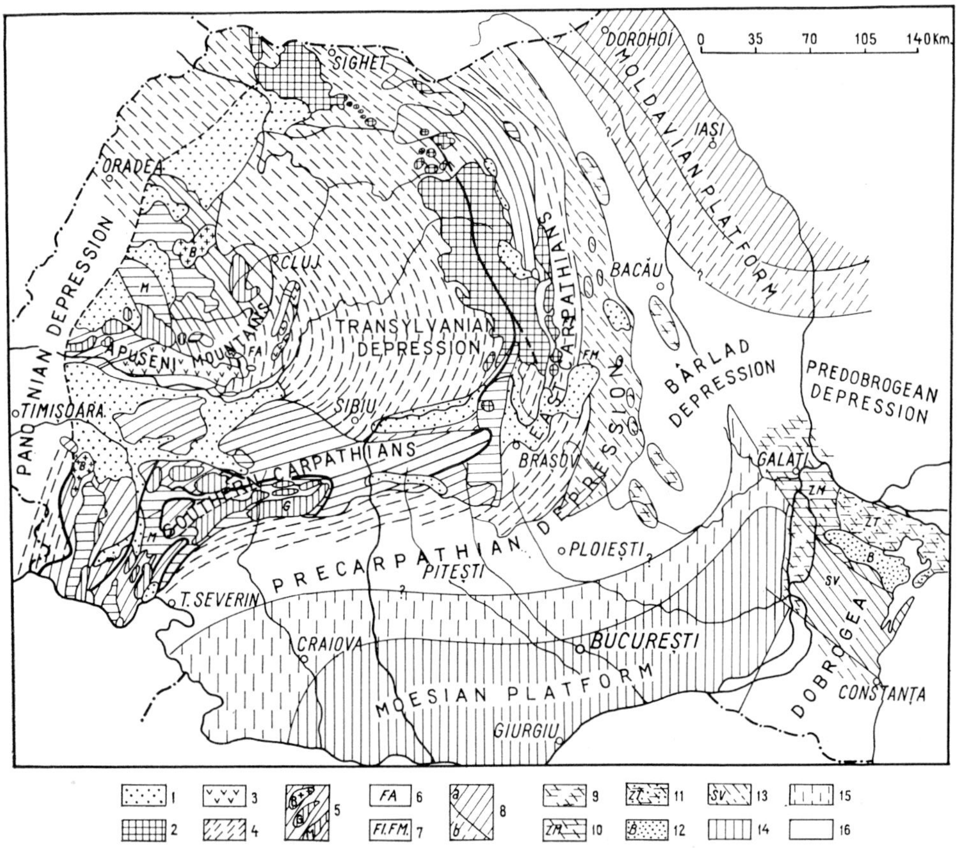

FIG 3. Tectonic sketch of Rumania.

Carpathian Zone: 1. Post-tectonic basins; 2. Eruptive Neogene; 3. Old basic eruptive (Upper Triassic-Cretaceous); 4. Subsidence zone of the Carpathian Crystalline; 5. Old Crystalline metamorphosed cores: M—Mesozoic deposits in synclinal patches, G—Intrusive Hercynic bodies, B—Banatites; 6. Zone in front of the Apuseni Mountains (Lower Cretaceous deposits in Flysch facies, and Upper Cretaceous in Gosau facies); 7. Flysch zone of the E Carpathians: FM—Marginal (external) sub-zone, FI—Internal sub-zone. *Zone in front of the Carpathian Range: Russo-Moldavian Platform:* 8(*a*). Folded Pre-Cambrian basement; 8(*b*). Subsidence zone of the Pre-Cambrian-Palaeozoic basement; 9. Dobrudjan cordilleras in the basement of the Pre-Carpathian Depression. *Dobrudja:* 10: ZM—Măcin zone (Hercynian core with green schists in basement; 11: ZT—Tulcea zone (Cimmerian foldings affect Triassic deposits overlying the Hercynian core); 12: B—Upper Cretaceous basin of Babadag; 13: SV—zone of green schists on surface. *Moesian Platform:* 14. Basement made up of Mesozoic and Palaeozoic deposits (locally); 15. Intermediate zone. 16. *Pre-Carpathian Depression.*

subsequently during foldings in the various phases of orogenesis and, under the influence of the existing pressure, it took the way either of lateral migration, thus completing and finishing the shifting started along the sedimentation process, or of cross-migration on dislocations resulting from various folding stages. In accepting this concept of the formation of oil and gas deposits, it goes without saying that the reconstitution of the geological evolution of each major structural unit on Rumanian territory, by means of lithofacial and palaeotectonic maps, constituted the obligatory way for scientifically substantiating the conclusions regarding the hydrocarbon prospects of this unit.

3. A detailed study and synthesis of all geological, geochemical, and drilling data obtained from the areas of known fields allowed the formulation of some laws concerning their geological and geographical distribution; extrapolation of such laws in new regions permitted us, on the one hand, to appreciate prospects more judiciously and, on the other, rationally to direct survey drilling programmes.

In the following is given a general, geological characterization of the major structural units where oil and gas deposits are known at present.

1. Pre-Carpathian Depression

This corresponds with the depression zone along the entire range of the Rumanian Carpathians (E and S Carpathians) which is comprised between the Crystalline cores and the opposite platforms. Delimitation of this unit is conventional, as the Crystalline cores are upthrown on the sedimentary formations of the Depression, which in their turn are overlapping the formations of the frontal platforms. It is difficult to reconstitute the original distribution area of the sedimentation basin; in the present delimitation there are included the zone of the E Carpathian Flysch (sub-zone of the Cretaceous Flysch and sub-zone of the Palaeogene Flysch), the Neogene zone of the E Carpathians (named sub-Carpathian Depression in the older geological literature), and the Getic Depression in front of the S Carpathians. By this delimitation, the notion of "Pre-Carpathian Depression" gets a double significance in space and time. The significance in space is connected with the permanently pre-mountainous position of the depression, as an absolutely necessary element, having an intermediary position between the mountainous Carpathian range of continuous evolution and the opposed platforms, the axis of the depression being continuously shifted from the interior to the exterior, towards the platforms.

The significance in time refers to the age of the depression.

In contrast to the data contained in the older geological literature, according to which the two depressions—the sub-Carpathian Depression and the Getic Depression—were considered to have been formed at the end of the Palaeogene and the end of the Cretaceous period respectively; in its new definition the Pre-Carpathian Depression is considered to have been formed during the Hercynic orogenesis, on cores which were embryonically outlined as far back as the Caledonian orogenesis.

In other words, the Pre-Carpathian Depression should not be considered as a simple Neogene foredeep achieved in front of the Carpathian range during the Alpine orogenesis, but as an old depression of a pre-mountainous character, that of a foredeep, in front of the Hercynic range, and which was preserved as such, naturally with the continuous shifting of the axis to the exterior, up to the end of the Tertiary.

Starting from this picture, appreciations concerning the oil and gas possibilities of the Pre-Carpathian Depression are of course much more favourable, as they are related to a thick sedimentary series, the composition of which includes Tertiary, Mesozoic, and Palaeozoic formations.

The idea of a more complete sedimentary sequence within the zones of maximum depth in the Pre-Carpathian Depression, expressed already in 1955 by one of the present authors (N. Grigoraş, 1957), was completed by some factors of confirmation by means of the drilling data obtained particularly in the platform areas, as well as by some surface data in the Carpathian zones. Indeed, the occurrence of a Mesozoic and Palaeozoic sequence within the Moesian Platform proved by drilling, and the occurrence of Mesozoic and Palaeozoic formations known at the surface within the S Carpathian zone, are compulsorily leading to the conclusion that in the Pre-Carpathian Depression this sequence should not be smaller but, on the contrary, that it should be more complete. In the E Carpathians, the occurrence of metamorphosed Palaeozoic deposits has recently been proved (Dessila-Codarcea, 1968), while the westward thickening and completion of the sedimentary sequence which is enfolding the basement of the Moldavian Platform have also been confirmed by drilling. In view of the fact that—according to gravity data—in the Moldavian E Carpathians the axis of the Pre-Carpathian Depression is placed right in the Flysch zone, it is to be expected that there will be found the most complete sedimentary sequence.

Tectonic movements which have taken place within the Pre-Carpathian Depressions finally also determined the facial characters of the sedimentary formation, as well as the complex structural characters presently recognized, while these characters as a whole have determined the possibility of forming and preserving oil and gas deposits.

Taking into account the history of geological development of the Pre-Carpathian Depression, we admit that the most important bituminous source series (mother-rocks) for oil and gas have occurred in the Permian, the Lower and Middle Jurassic, in the Lower Cretaceous, the Oligocene, and Tortonian. The mother-rocks forming these bituminous series are almost exclusively argillaceous and siliceous; it could be supposed that calcareous mother-rocks

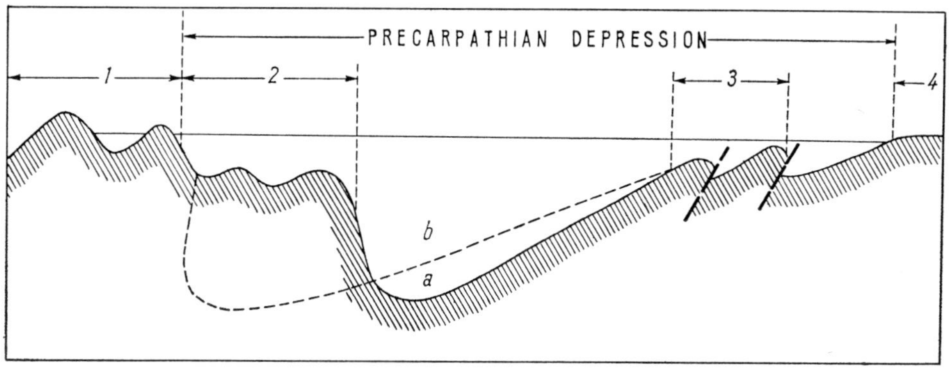

FIG 4. Pre-Carpathian Depression. (*a*) At the beginning of the Alpine orogenesis; (*b*) At the end of the Hercynic orogenesis; 1. Crystalline-Mesozoic zone; 2. Flysch zone; 3. Zone of green schist cordilleras; 4. Platform zone.

might occur locally in the Permian and perhaps the Triassic. Analyses carried out so far with regard to the content in A and C bitumens and in kerogen, to the degree of bituminization and the atomic H/C relation show that for the oil and gas deposits known in the Pre-Carpathian Depression the hydrocarbon generating series appears to be represented mainly by pelitic bituminous rocks of Oligocene age (bituminous white marls and dysodilic shales).

The reservoir rocks met with in known deposits within the Pre-Carpathian Depression are of the granular type. They are represented by calcareous and Cretaceous sandstones (sub-zone of the Cretaceous Flysch and sub-zone of the Palaeogene Flysch), calcareous sandstones of the Tarcău sandstone type in the Eocene (sub-zone of the Palaeogene Flysch), calcareous sandstones of the Fusaru sandstone type in the Oligocene (sub-zone of the Palaeogene Flysch and Getic Depression), quartzous sandstones of the Kliwa sandstone type (sub-zone of the Palaeogene Flysch and Neogene zone of the E Carpathians), calcareous sandstones, fine conglomerates, sands, and very sandy marls in the Upper Burdigalian, the Helvetian, the Tortonian, the Sarmatian, the Meotian, the Pontian, the Dacian, and even in the Levantian in the Neogene zone of the Pre-Carpathian Depression. The only reservoir of a compact character is known in the Upper Oligocene within the Palaeogene Flysch sub-zone (transition series); it is composed of menilites, silicified clays, and thin intercalations of siliceous sandstones, the complex being crossed throughout by numerous fissures.

It is to be expected that, besides reservoir rocks of the granular type, the Mesozoic and Palaeozoic formations within the Pre-Carpathian Depression may also contain reservoir rocks of the compact, carbonate type.

The cover rocks of hydrocarbon accumulations known in the Pre-Carpathian Depression are widely distributed in the geological formations of which this major structural unit is made up; a particular mention should be made of the Upper Eocene clays (Plopu Beds) overlying the Tarcău sandstones, of the dysodilic Oligocene shales overlying the Kliwa sandstones, of the Upper Helvetian marls (Cîmpeni Beds) covering the Lower Helvetian sandstones and sands, of the Pontian and Lower Levantian marls which are protecting the Meotian and Dacian accumulations, etc.

Traps are predominantly represented by the structural type, but traps of the stratigraphic and lithological type are known as well. Structural traps were for a long time—as also everywhere else—forming the decisive criterion for directing prospection and particularly exploration survey work.

In the Pre-Carpathian Depression, structures have very variable forms. For the Flysch zone the nappe structure is rather characteristic, its authochthon being frequently strongly folded and faulted as structural forms known under the name of "scales". For the Neogene zone it is at present proved that the Tertiary formations within the Depression are overthrust on newer Tertiary formations or even Mesozoic formations of the platforms, situated in front of them. The settling of the nappes in the Palaeogene Flysch zone took place in Tortonian time, while the overthrust of the depression formations on the platform formations occurred at the end of the Lower Sarmatian. By the transgressive disposition of the Middle Sarmatian or of other younger formations, the overlapping relations between formations of the depression were completely concealed and, although they were long ago intuitively believed by the Rumanian geologists Mrazec and Voitești, they could be revealed only in the course of the last 20 years by seismic surveys and drilling.

In the Neogene zone, the intensity of folding decreases from the inside to the outside. Frequently isoclinally folded and strongly faulted asymmetrical anticlines characterize the structure in Miocene formations in Moldavia. In Muntenia, the Mio-Pliocene Zone is characterized by variously shaped anticlines, symmetrical and asymmetrical ones, affected by numerous faults and of a complicated structure caused by the presence of salt bodies frequently piercing the overlying formations up to the surface.

Anticlines affected by many faults are also known

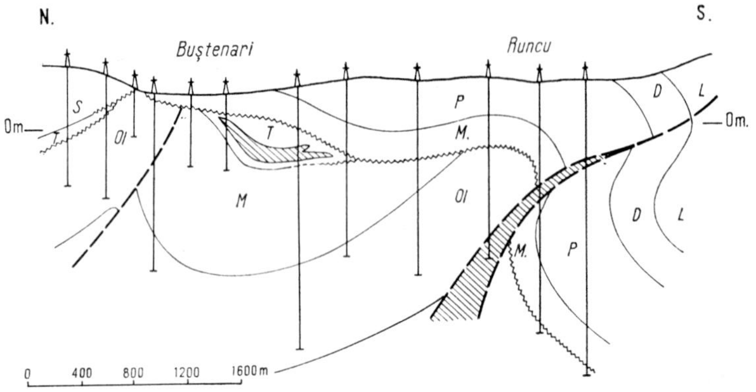

FIG 5. Geological section across the Buştenari-Runcu structures.

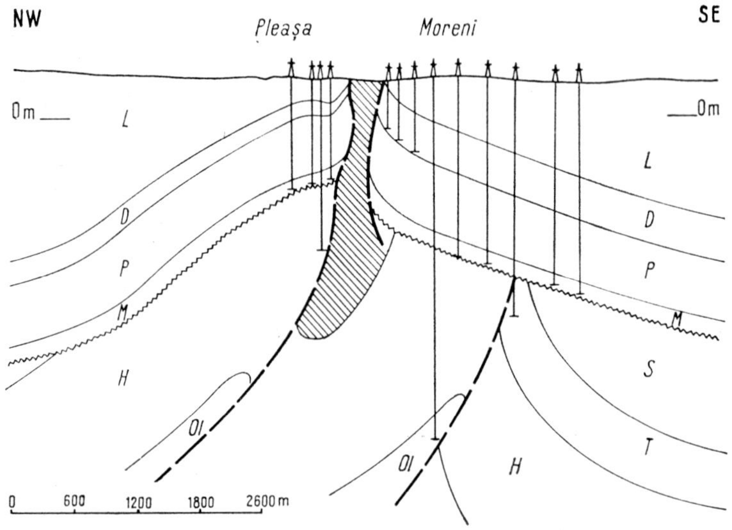

FIG 6. Geological section across the Moreni structure.

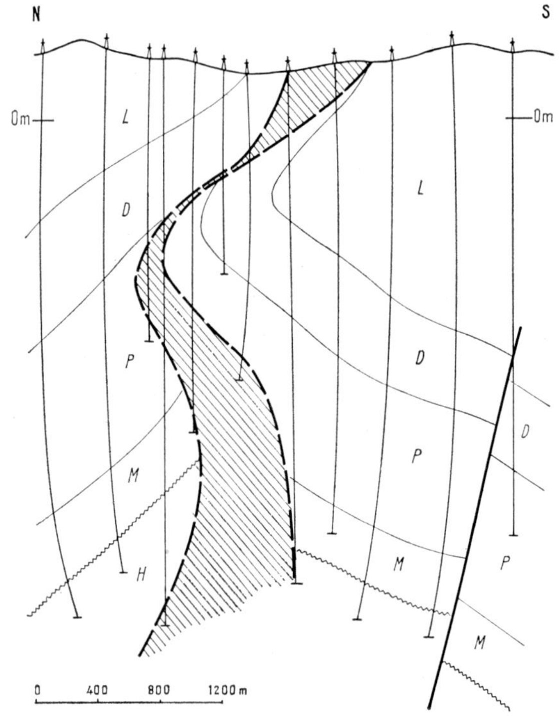

FIG 7. Geological section across the Tintea structure.

in the Neogene zone in Oltenia (Getic Depression); salt has been met with here again, but diapirism occurs very seldom, on the one hand because of the lesser thickness of salt lens and, on the other because of the less intensive folding movements in the Upper Miocene and Pliocene.

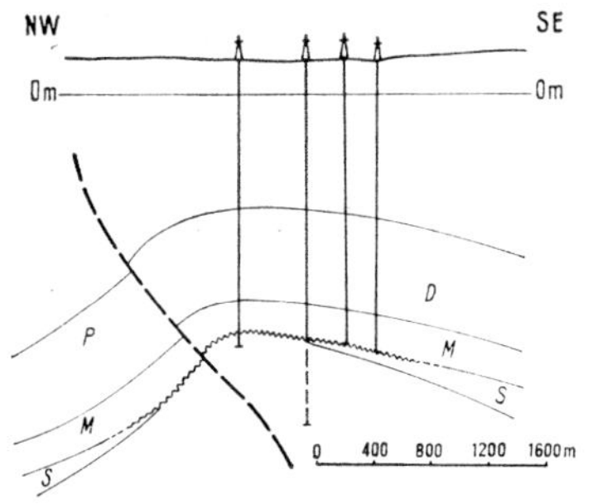

FIG 8. Geological section across the Mărgineni structure.

In the plane distribution of structural members of the Pre-Carpathian Depression there is observed a linear arrangement, in parallel with the Carpathian Range, which is also decisive for the distribution within the depression of oil and gas deposits.

Stratigraphic traps achieved by the transgressive disposition of newer formations on older ones are known in many structures within the depression. The transgressive dispositions of Pliocene deposits on Oligocene (Buştenari-Runcu, Surani-Cărbuneşti) or on Helvetian (Teiş, Moreni-Gura Ocniţei, Glîmbocel, etc.) or on Sarmatian (Bălteni-Ticleni, Mărgineni, Boldeşti, etc.) represent some examples of stratigraphic traps in which oil pools are known.

Lithologic traps are less frequent and are represented by lenticular developments within some sandstone-sand series in the Oligocene (Getic Depression) or in the Sarmatian on the outer edge of the Pre-Carpathian Depression.

Oil pool types. Most oil and gas pools within the Pre-Carpathian Depression are of the *stratiform type.* Part of these pools are vaulted and faulted. Most stratiform pools are tectonically screened (Figs 10 and 11), but there also occur stratigraphically and lithologically screened ones. In the case of overlapping pools their outlined size decreases with depth, particularly for the reservoirs which are included in the same formation (*e.g.* the Meotian M_I, M_{int}, and M_{II} pools, the Helvetian H_I, H_{II} pools, etc.). The Oligocene K_{IV} reservoirs in the Tazlău structure, the Meotian pools M_I, M_{II}, M_{int}, and M_{III} in the Mărgineni structure, the Helvetian H_I to H_{VIII} pools in the Ticleni structure, are examples of tectonically screened stratiform pools. The Sarmatian S_V reservoirs in the Ticleni structure, S_A within the Sarmatian of the Mărgineni structure, etc., are examples of stratigraphically screened reservoirs, whereas the reservoirs S_{VI} and S_{VII} in the Ticleni structure are lithologically screened. A special type of tectonic screening is observable in the Meotian and Dacian pools of the diapir structures such as the structures Moreni-Gura Ocniţei, Băicoi-Tintea, etc. (Figs 6 and 7). Diapirism has played a particularly important part in the formation of oil pools within the Meotian and especially the Dacian of the Mio-Pliocene zone in Muntenia.

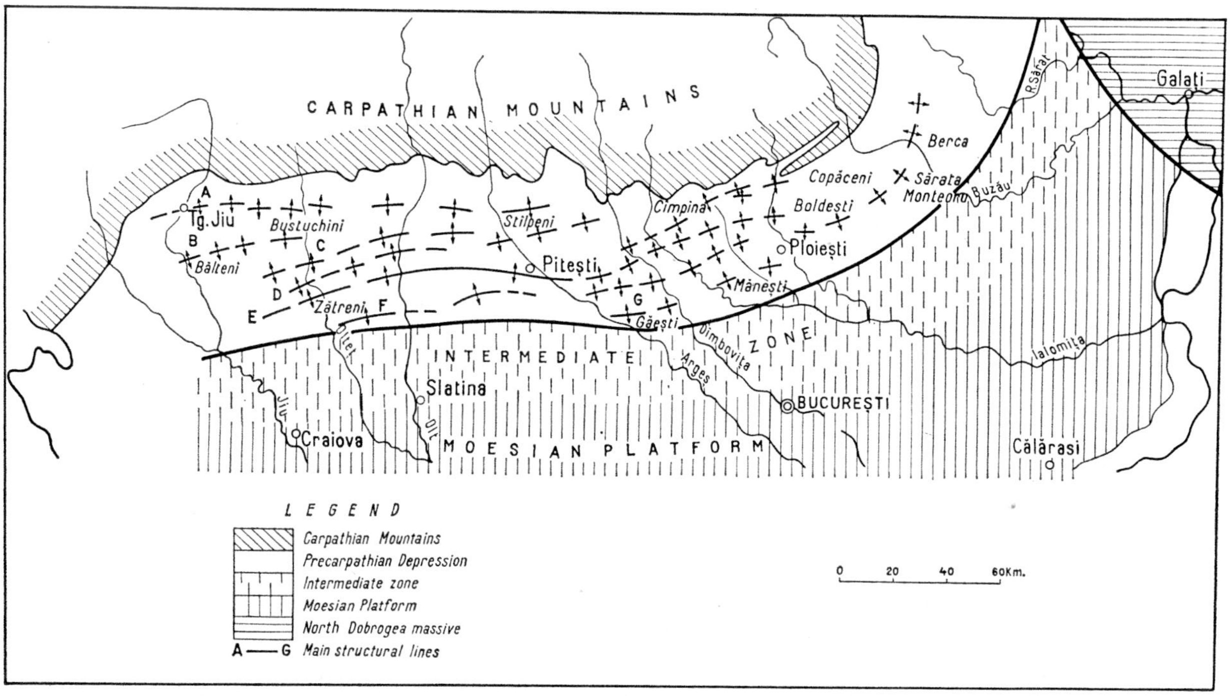

FIG 9. Main structural elements in Oltenia and N Muntenia.

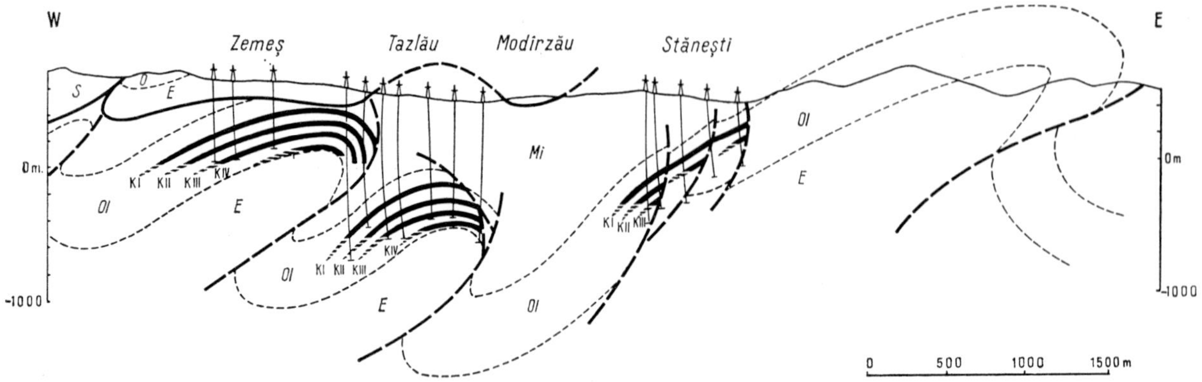

FIG 10. Geological section across the Zemes-Tazlau-Stanesti structures (after E. Hristescu, D. Culcer, and N. Grigoras). $K_I - K_{IV}$ = productive horizons.

Oil and gas pools of a massive type are exceptionally found in the Pre-Carpathian Depression. Of such a type is the oil pool accumulated in the Kliwa sandstone of Oligocene age (complexes K_I, K_{II}, and K_{III}) as well as that accumulated in the Helvetian sandstones and sands at Teiş. Although they are accumulated in granular reservoir rocks, the said pools, on account of the tabular character of the water-oil boundary, belong to the massive type.

As regards lithologically delimited pools, there can be quoted in the Pre-Carpathian Depression, as a unique example, the oil accumulation found at Gura Drăgănesii (to the W of Cîmpina) in lenticular sands associated with the Tortonian salt breccia.

Most oil pools within the Pre-Carpathian Depression are pools without any primary gas cap but rich in dissolved gas. All pools are accompanied by salt waters, mostly of the chlorocalcic type, and of a different degree of mineralization for each geological formation.

In connection with the distribution of oil and gas reservoirs known within the Pre-Carpathian Depression it is possible to formulate the following rules:

(*a*) All known oil and gas pools are without exception localized in sedimentary formations;

(*b*) The distribution of oil and gas pools is linear, along the major structural lines, the layout of which is parallel to the Carpathian Range (Fig 9);

(*c*) In the inner zones of the Pre-Carpathian Depression there are generally found oil pools which are poor in dissolved gas; towards the exterior, on the line of diapir folds, the oilfields encountered are rich in dissolved gas or have a primary gas cap; still more externally, in the Neogene formations on

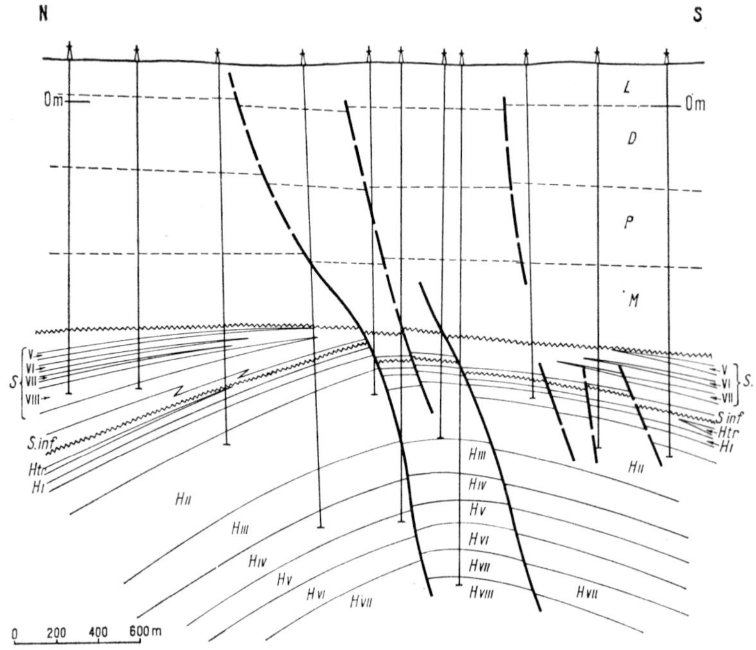

FIG 11. Synthetic geological section across the Ticleni anticline (after C. Teodorescu, L. Ponta, and N. Grigoras). $H_{VII} - H_{tr}$; $S_{inf} - S_V$ = productive horizons.

the sunken flank of the afore-mentioned platforms, there are found oil pools with a strong gas cap or exclusively gas pools.

(*d*) The stratigraphic distribution of oil and gas pools in the Pre-Carpathian Depression is not uniform: in Moldavia, Palaeogene pools are predominant; in Muntenia, the prevailing pools are of Pliocene age, while in Oltenia they are of Miocene age.

(*e*) In older geological formations or those localized at very great depth, most oils are paraffinous; the oil included in reservoirs which are localized in younger or less deep formations is semi-paraffinic or naphthenic.

2. Transylvanian Depression

This is an inter-mountainous depression which was formed at the end of the Cretaceous period by the subsidence of one sector of the Carpathian Range corresponding to the present outline of the Transylvanian Depression.

It was thought for a long time to be a single depression with the maximum depth in the central zone. The geological research activity by means of geophysical methods and drilling has, however, shown that the Crystalline basement was made up of a series of uplifts delimiting several secondary depressions; two peripheral depressions, one of which is parallel to the frame of the Apuseni Mountains, another paralleling the E Carpathians, and the third one is a central depression. Within the two peripheral depressions pre-Tertiary geological formations have been identified, *i.e.* Cretaceous formations in a flysch facies in the W depression, and epicontinental Cretaceous and Jurassic formations in the E and central ones. In the S part of the central depression even Triassic formations have been recognized.

The abovementioned activity also revealed that the Palaeogene formations, which are widely developed on the N border of the Transylvanian Depression, are poorly or not at all developed in the central zone of the depression. Here, Mesozoic or very thin Palaeogene formations, which are represented only by Eocene deposits, are transgressively overlain by the saliferous Tortonian.

In the Transylvanian Depression there are known gas pools which are accumulated in Tortonian, Sarmatian, and Pliocene reservoir rocks. It is considered that gas was generated in source rocks of Tortonian and Sarmatian age. Reservoir rocks are of the granular type, which are made up predominantly of fine sands or very sandy marls. Traps are predominantly of the structural type, and the gas pools are of the stratiform type, being vaulted and rarely affected by faults. The salt water is of the chlorocalcic type and shows a various degree of mineralization, depending on the geological formation and on the distance from the depression border. Salt water of the chlorocalcic type has been found, locally, also in some Mesozoic carbonate formations.

The gases are made up of more than 99 per cent of CH_4. There have also been determined very small amounts of higher homologues of methane gas (ethane, propane, butane). In structures with several superposed gas pools, a slight increase of these constituents with depth was observable.

As far as the distribution of gas-containing structures is concerned, here again it can be said that the arrangement is somewhat linear. This arrangement shows a NNW–SSE trend which is approximately parallel to the E Carpathian Range. Within the same gas structures, the productive areas of the various overlapping gas pools decreases with depth.

3. Pannonian Depression

This structural unit represents a large intra-mountainous depression, which was mostly formed during Tertiary time. On Rumanian territory it corresponds to the Banat Plain to the S, and to the Crişana Plain to the N. A gulf of the Pannonian sedimentation basin also extended across the N part of the E Carpathians, into the present Maramureş Basin.

Geophysical surveys and drilling carried out on the Rumanian territory of the Pannonian Depression have shown that the deep geological structure of this structural unit is very intricate. In the Crişna Plain, under the Neogene formations, Palaeogene, Cretaceous, Jurassic, and Permo-Triassic deposits have been identified. Pre-Tertiary formations have been preserved only in the depression zones of the Crystalline

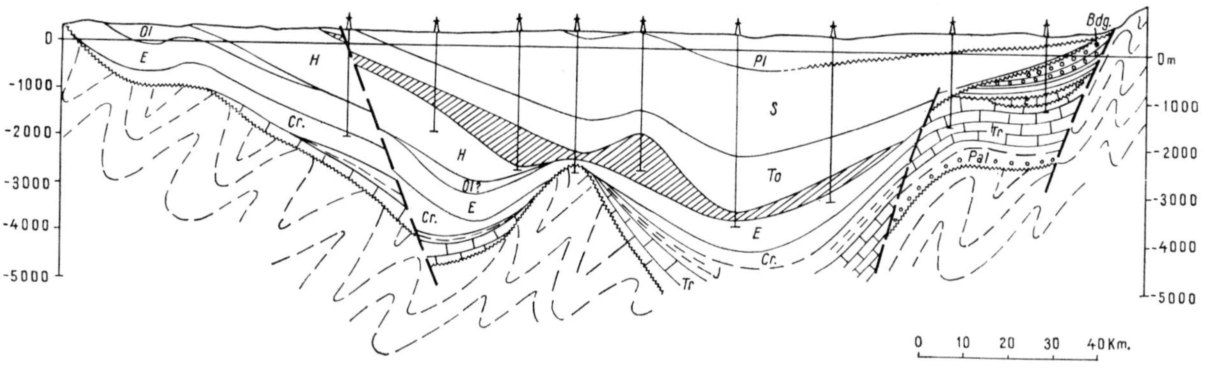

FIG 12. Geological section across the Transylvanian Depression.

basement, which is broken up by faults into many blocks.

In the Banat Plain, of the pre-Tertiary formations there were met with, locally, only Cretaceous ones. Most frequently, the Sarmatian, or even the Pannonian, is directly lying on the Crystalline basement.

Oil and gas pools have been found in Pannonian, Sarmatian, Palaeogene, and Cretaceous formations. Some small oil and gas accumulations are found quite locally also in the weathered zone of the Crystalline basement.

The series of hydrocarbon source rocks are met with in Triassic, Cretaceous, Palaeogene, Miocene, and even in Pliocene formations.

The reservoir rocks are of the granular type within the Neogene and Palaeogene, and of the mixed type—compact and granular—within the Cretaceous. Traps are of the structural type, and pools are of the stratiform or massive type.

As the exploration of oil and gas reservoirs within the Pannonian Depression is now in progress, it has not been possible to ascertain so far any rule concerning their distribution.

4. Moesian Platform

Until 1951, this structural unit, which was known under the names of "Pre-Balkan Platform" or "Pre-Balkan Land", was considered to be lacking oil and gas prospects. This opinion was based on its likeness to the geological structure of the Southern Dobrudja, where the basement of Pre-Cambrian green schists was known to be overlain by comparatively thin deposits belonging to the Upper Jurassic and Cretaceous.

Several seismic profiles of a N–S trend have revealed the occurrence within the platform zone of a sedimentary formation of about 5000 m in thickness, the latter exceeding very much that of the Upper Mesozoic in the Southern Dobrudja. The geological interpretation of seismic data led to the conclusion that the sedimentary cover of the Moesian Platform should include also other, older Mesozoic (Jurassic and Triassic) formations and Palaeozoic formations as well. More intensive seismic surveys and several wells drilled for stratigraphic information confirmed from the very beginning this geological assumption and determined the acceleration of the investigations of the Moesian Platform by means of both complex geological prospecting surveys and drilling.

The sedimentary cover of the Moesian Platform includes Palaeozoic, Mesozoic, and Tertiary formations which are separated, by stratigraphic breaks of a regional character, into the following sedimentation cycles:

The Lower Palaeozoic sedimentation cycle, with Ordovician deposits (argillites with quartzitic sandstone interlayers), Silurian deposits (argillaceous shales with graptolites), Devonian deposits (dolomitic limestones, anhydrites, argillaceous shales), and Carboniferous deposits (argillaceous shales with coal, and limestones). The Cambrian has not been recognized, but its occurrence is considered as probable in the depression zones of the platform.

The Upper Palaeozoic–Lower Jurassic cycle of sedimentation the deposits of which belong to the Upper Permian and Lower Triassic (lower red series), the Middle Triassic (dolomites with anhydrite intercalations) and to the Upper Triassic–Lower Jurassic (upper red series).

The Middle Jurassic–Cretaceous cycle of sedimentation with terrigenous deposits (argillites and sandstones) in the Dogger, and partly dolomitized limestones in the Malm and Lower Cretaceous. The Middle and Upper Cretaceous show facies changes

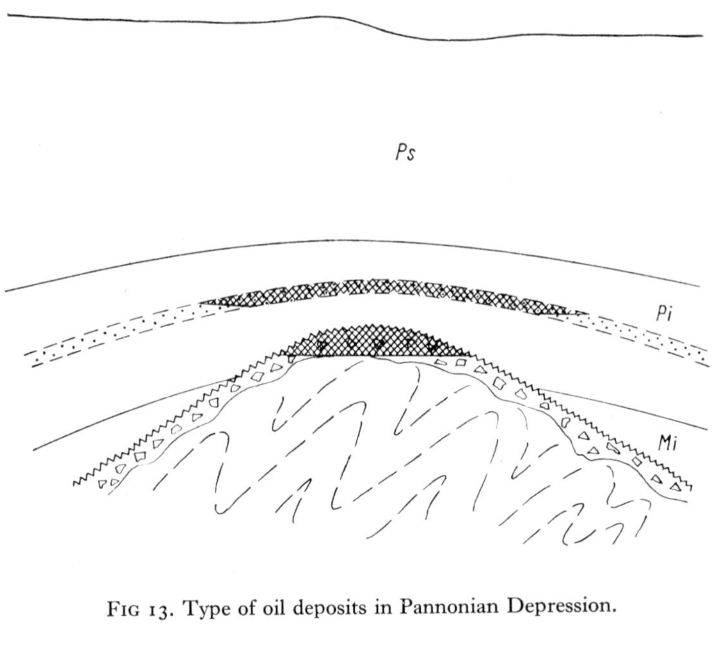

FIG 13. Type of oil deposits in Pannonian Depression.

in various zones of the Platform and are predominantly represented by marls and marly limestones, sometimes with sand and sandstone intercalations. In the W sector of the platform, this cycle also includes, quite locally, a thin layer of Eocene marly limestones. Some local stratigraphic gaps have been sometimes recorded within this cycle (*e.g.* at the level of the Jurassic–Cretaceous or of the Neocomian–Albian boundary) in the W part of the platform.

The Neogene sedimentation cycle includes the Upper Miocene (Tortonian and Sarmatian) and the Pliocene. The Neogene formations are disappearing gradually, and thus in the Danube area only the Upper Pliocene (Dacian and Levantian) deposits still occur with decreased thickness.

The Crystalline basement of the Moesian Platform is of Pre-Cambrian age.

The present tectonic picture of the Moesian Platform is a result of the cumulated effect of the orogenic movements which led to the formation of the Carpathian and Balkan Ranges.

The dip of the Silurian beds of up to 35° and the slightly unconformable disposition of the Devonian on Silurian deposits suggest the effect of some Caledonian movements which have been felt in the platform area as well. The folding movements within the Carpathian and Balkan ranges during the subsequent—Hercynian and Alpine—phases of orogenesis were also felt in the adjoining platform areas of the geosyncline zones and appeared either in the form of slight foldings or as fractures.

Going from S to N, there is observable a continuous subsidence of sedimentary formations in the platform with their simultaneous increase in thickness. The angle of the slope is small in the N part of the platform and larger in its N area of contact with the Depression (intermediary zone), while the general northward sinking is achieved by means of some major longitudinal fractures which are as a rule parallel to the Carpathian range. As regards the platform itself, the relations between the blocks separated by faults are not uniform; sometimes, the block to the N of the fault is more uplifted, in which case the sedimentary cover is faintly vaulted; at other times, the block to the N of the fault is more sunken and then the relations between compartments are those of normal fault ridges. Besides longitudinal faults there also occur cross faults, thus creating many blocks of various sizes.

On the general background of the platform's N subsidence, on the Craiova-Slatina–N București line, a major longitudinal uplift is outlined in the shape of an antheclise. Along this uplifting zone there occurred eruptions in the Middle Triassic which gave rise to lava flows.

On the platform body there also occur two depression zones, one to the W, another to the E of Bucharest.

As yet, oil and gas pools within the Moesian Platform are known in Triassic (Middle Triassic), Jurassic (Middle and Upper Jurassic), Cretaceous (Neocomian, Albian, Senonian), and in Neogene (Sarmatian, Meotian, and Pontian) formations. Oil shows in cores have been recorded from Devonian and Silurian deposits.

Oil and gas source rocks are considered to be the Silurian shales with graptolites, the Devonian argillaceous-bituminous shales and dolomitic limestones, the Middle Triassic dolomitic limestones, the argillaceous-bituminous shales of the Dogger, the Middle Cretaceous (Albian-Cenomanian) bituminous limestones, and some Sarmatian shaly clays.

Reservoir rocks of the granular type are found in Permo-Triassic, Jurassic formations, in some Albian as well as in Sarmatian and Pliocene deposits. They are of the compact carbonate type in the Middle Triassic, in the Malm-Neocomian, and locally in Senonian deposits.

Protection of the oil and gas pools is provided by the presence of some impervious rocks covering the reservoir rocks. The Dogger is productive in the W half of the platform, where Bajocian sandstones are overlain by a thick member of marly clays. In the E half of the platform, where the argillaceous-marly Bathonian is lacking, the sandstone complex is in direct contact with the carbonated Malm–Neocomian complex, both being saturated with fresh water.

Traps are of the structural or stratigraphic type; in the Neogene formations, locally, lithologic traps, are also found.

Deposits are of the stratiform type in the granular reservoir rocks, and of the massive type in the fissured, carbonate rocks.

Fig 14 represents a schematic distribution of some oil reservoirs in the platform. The vaulted compartments show a certain superposition on the vertical of reservoirs localized in the Tertiary and Mesozoic formations. Such superpositions do not occur in the compartments of a monoclinal character.

The distribution of oil pools is mostly governed by screens formed by longitudinal faults, whereby a generally linear position is observable which is approximately parallel to the Carpathian range. Some faults may also have a synsedimentary character caused during an initial phase by differential packing, which subsequently was emphasized by tectonic movements.

5. Moldavian Platform

The geological surveys carried out within the last 20 years have completed the information regarding the platform outline and its structure in depth. Thus, it was determined that the W sunken edge of the platform is accompanied by a ridge of green schists representing the NW continuation of the central Dobrudja and of the N Dobrudjan orogene. Between the S border of the platform and the N Dobrudjan orogene there is localized the Bîrlad depression. On the W margin of the platform, in the Bacău-Roman area, this depression is very narrow, but is again more developed in the N part of Moldavia, being

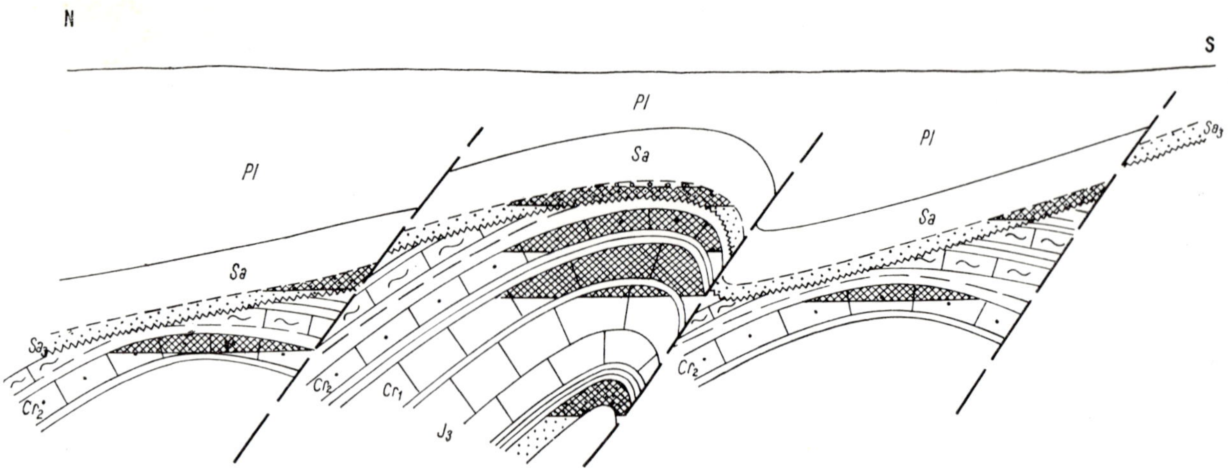

FIG 14. Type of oil deposits in Moesian Platform.

continued beyond the frontier and connected with the Lwow depression.

On the W border of the platform it was also observed that its sedimentary cover is thickening as new Devonian, Carboniferous, Jurassic, Cretaceous, Palaeogene, and Neogene formations are being added. Finally, also on the W border of the platform, its formations are overthrust by the Tertiary formations of the Pre-Carpathian Depression. The overthrusting relations were achieved at the end of the Lower Sarmatian, after which the Middle Sarmatian transgressively covered and concealed the entire deep structure.

The geological and economical results obtained so far in the Sarmatian and Tortonian formations warrant the continuation of geological surveys within this structural unit.

6. **Bîrlad Depression**

This represents a pre-mountainous depression situated in front of the N Dobrudjan Promontory. It is made up of Permo-Triassic, Jurassic, Cretaceous, and Eocene formations, and is transgressively covered by Neogene (Tortonian, Sarmatian, and Pliocene) formations.

The results obtained so far in the Lower Sarmatian are promising for the continuation of work.

The Bîrlad Depression is continued also to the E of the Prut River in the Pre-Dobrudjan Depression in which the area of the Danube Delta is also comprised.

7. **N Dobrudjan Promontory**

The basement of the N Dobrudjan Crystalline schists is sinking NW, being covered by Palaeozoic, Mesozoic, and Tertiary formations on the W flank, and only by Neogene (Sarmatian and Pliocene) deposits in the area of maximum uplift.

The results obtained in the Sarmatian and Meotian area of maximum uplift, as well as in the Sarmatian on the W flank of the Dobrudjan Promontory, are likewise promising and warrant the continuation of work.

The summary review of the geological results obtained in the last 20 years by means of the surveys carried out show that the previously made statement about a fast exhaustion of oil and gas reserves in Rumania is unfounded. In many units, survey works are still in progress, leaving open the prospect of discovering also at some future date new oil and gas deposits on Rumanian territory.

BIBLIOGRAPHY

Căpraru, C., and Alexandru, M. *Petrol şi Gaze*, 1965, (8).

Dessila-Codarcea, M. "Evolution préalpine des massifs cristallophylliens des Karpates roumaines", International Geological Congress, Prague, 1968.

Dumitrescu, I., Săndulescu, M., Lăzˇrescu, V., Mirăuţă, O., Pauliuc, S., and Georgescu, C. "Mémoire à la carte tectonique de la Roumanie". Anuarul Comitetului Geologic, Vol. XXXII, Bucureşti, 1962.

Florescu, M. *Petrol şi Gaze*, 1960, (8).

Grigoraş, N. "Géologie des gisements de pétrole et de gaz en Roumanie". Congrès Intern. de Géologie, Mexico, 1956.

Grigoraş, N. "Roci bituminoase în formaţiunile geologice din RPR". Rev. Universităţii C. I. Parhon, Seria St. Naturii No. 9, Bucureşti, 1956.

Grigoraş, N. "Rolul depresiunilor premontane în formarea rocilor bituminoase din RPR". Analele Univ. C. I. Parhon, Seria St. Naturii No. 13, Bucureşti, 1957.

Grigoraş, N. "Geologia zăcămintelor de petrol şi gaze din RPR". Editura Tehnică, 235 p, 1961.

Grigora , N., and Petrisor, I. "Consideraiţi privind legile de răspîndire ale zăcˇmintelor de petrol şi gaze din RPR". Asociaţia geologică Carpato-balcanică, Congresul al V-lea, Bucureşti, 1961.

Grigoraş, N., Pătruţ, I., and Popescu, M. "Contributions à la connaissance de l'évolution géologique de la Plate-forme moesienne du territoire de la République Populaire Roumaine". Asociaţia geologică Carpato-balcanică, Congresul al V-lea, Bucureşti, 1961.

Grigoraş, N., Petrişor, I., Hristescu, E., and Suluţiu, U. "Contributions to the knowledge about distribution laws regarding oil and gas fields in the Pre-Carpathian Depression of the Rumanian People's Republic". Oil, Gas, and Petrochemistry in the Rumanian People's Republic. Studies and articles, Bucharest, 1963.

Macovei, G. "Les gisements de pétrole". Paris, 1938.
Pătruț, I., Popescu, M., Teodorescu, C., Petrișor, I., and Anton, Gh. "Potential Mesozoic Oil Deposits in the Rumanian People's Republic". World Petroleum Congress, Frankfurt, 1963.
Pizanty, M. "Aria exploatării petrolifere în România". Tiparul "Cartea Românească", București, 91 p, 1938.
Pizanty, M. "La vitalité de l'industrie pétrolifère Roumaine". "Cartea Românească", București, 25 p, 1939.
Ravaș, Gh. "Din istoria petrolului românesc". Editura de stat pentru literatura politică, 344 p, 1955.
Ravaș, Gh. "The Centenary of Rumania's Oil Industry". Foreign Languages Publishing House, Bucharest, 124 p, 1959.
Ravaș, Gh. "The Oil Industry in Rumania". Meridiane Publishing House, Bucharest, 82 p, 1963.

MS received 2 June 1969

Unfortunately, no replies to the discussion contributions have been received for this paper.

Exploration for Oil and Gas in Austria

By R. H. JANOSCHEK

(*Exploration Manager, Rohoel-Gewinnungs AG, Schwarzenbergplatz* 16, *A*-1011 *Vienna, Austria*)

and K. G. H. GÖTZINGER

(*Exploration Manager. Österreichische Mineralölverwaltung AG, Otto Wagnerplatz* 5, *A*–1090 *Vienna, Austria*)

SUMMARY

In Austria oil and gas are found in two geological units (Molasse Zone, Inner Alpine Vienna Basin). In another three units (Flysch and limestone zones of the Alps, Styrian Basin) reservoirs may be discovered.

The Molasse Zone is an asymmetric basin. In the N, Upper Eocene to Lower Pliocene sediments are thin and overlie the Bohemian Massif Crystalline. In the S they reach thicknesses of 3000 to 4000 m and are overthrust by the Alpine Nappes.

West of Enns River, the oilfields lie in traps related to the upthrown blocks of N-dipping antithetic faults. Sandstones of Upper Eocene, Cenomanian, and Upper Jurassic age are oil-bearing and the sands of the Upper Puchkirchen Formation and the lower part of the Hall Formation are gas-bearing. In Lower Austria the Wildendürnbach gas field produces from the Middle Miocene sands.

The Inner Alpine Vienna Basin is a trough of Miocene Pliocene sediments lying on the Alpine–Carpathian Basin floor. They reach a thickness of 5000 m and are cut into blocks by numerous faults trending predominantly parallel to the basin rim.

Oil and gasfields lie in high zones along faults or are associated with buried hills. The most important oil and gas–bearing reservoir rocks are in the Alpine–Carpathian Basin floor, Upper Triassic dolomites, Upper Cretaceous sandstones and conglomerates, and the Paleogene Flysch sandstones. Sediments of the Middle Miocene to Lower Pliocene also produce within the basin. The largest oil and gasfields are Matzen, Schönkirchen Tief, and Zwerndorf. The deepest well, Schönkirchen T 32 (final depth 6009 m), discovered a gas reservoir between 4835 and 5259 m in a lower complex of the calcareous Alpine basement.

INTRODUCTION

AUSTRIA is composed of the following generalized tectonic units (Fig 1):

1. Bohemian Massif, chiefly granites and gneisses.
2. Molasse Zone, the Alpine foredeep.
3. Eastern Alps and Carpathians with their various zones and nappe systems, such as the Central Alps, Grauwacken Zone, Limestone Zone, Flysch Zone and Helveticum.
4. Neogene troughs and basins lying on Alpidic basement.

The Austrian territory west of Salzburg is not considered here because of the limited scope of the petroleum exploration carried out so far.

Oil and gas reservoirs have so far been found in the Molasse Zone, including the Mesozoic basement, in the Inner Alpine Vienna Basin, in the Flysch Zone, and in the Mesozoic Limestone Zone beneath the Tertiary Vienna Basin. No reservoirs have yet been established in the Styrian Basin or in the minor basins at the eastern rim of the Alps.

MOLASSE ZONE

The Molasse Zone of Austria lies between the Bohemian Massif on the N and the Alps and Carpathians to the S and SE. The geology and hydrocarbon accumulations of the area are known from extensive geophysical surveys and numerous deep wells drilled by Rohoel–Gewinnungs AG (RAG), a subsidiary of Mobil Oil Austria AG and Shell Petroleum Company Ltd, and by Österreichische Mineralölverwaltung AG (ÖMVAG).

Stratigraphy

Basement and pre-Tertiary. At the base of the Molasse Zone lie the crystalline rocks of the Bohemian Massif which extend far under the Alps, as shown by the deep wells Kirchham 1, Kürnberg 1, Urmannsau 1, Texing 1, Perschenegg 1, and Mauerbach 1. An abbreviated Late Palaeozoic and Mesozoic sequence (Table I) occurs as erosional remnants or in pre-Tertiary basins and is also assigned to the Molasse Zone basement. In Upper Austria and Salzburg this sequence has the outer-Alpine Germanic facies and reaches a maximum thickness of 600 to 800 m.

In the considerably narrowed part of the Molasse Zone east of the Enns River, the Mesozoic has not been encountered in the subsurface of that part of the Molasse Trough which is overthrust by the Flysch Zone; the Tertiary lies directly on the crystalline basement (Kürnberg 1, Urmannsau 1, Perschenegg 1). In the widening part of the Molasse Zone E and NE of St. Pölten numerous ÖMVAG wells have penetrated, under the Tertiary, an autochthonous Late Palaeozoic–Mesozoic sequence up to over 2500 m in thickness. This sequence, however, shows a facies different from that west of the Enns River (F. Brix and K. Götzinger, 1964; J. Kapounek *et al*, 1967).

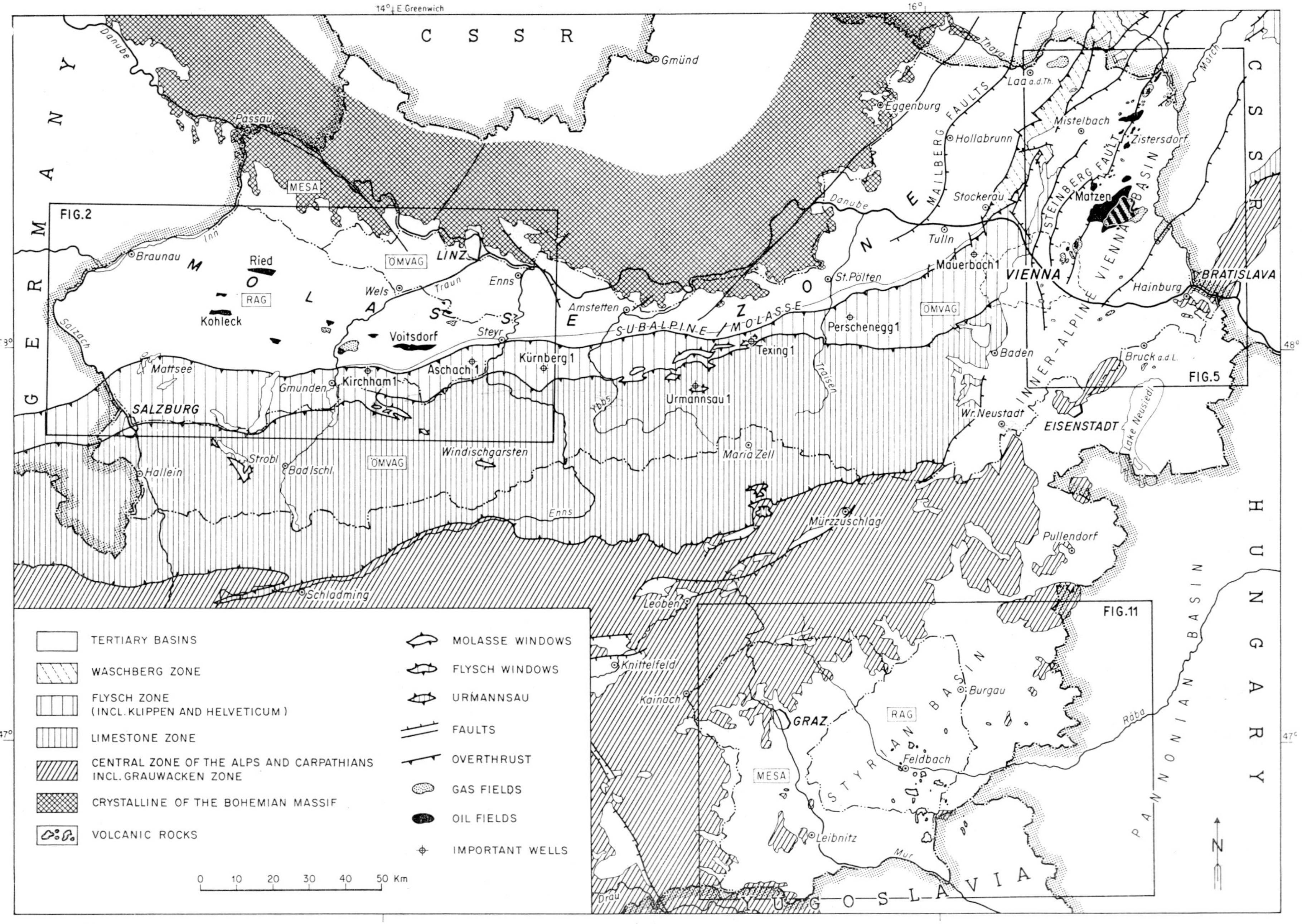

Concession Boundary — RAG Concessions of Rohoel-Gewinnungs A.G. — ÖMVAG Concessions of Österreichische Mineralölverwaltung A.G. — MESA Concessions of MESA-Petroleum A.G.

FIG. 1. Oil geological map of E Austria.

TABLE I

Stratigraphic chart of the Molasse Zone in Salzburg, Upper and Lower Austria

		Basement and Pre-Tertiary	
Mesozoic		Salzburg and Upper Austria	Lower Austria
Cretaceous	Upper	Gray shales and marls with glauconitic sandstones (Lower Campanian–Turonian) Sandstones, clay stones, thin coal seams (Cenomanian) ●	Shales and marls with glauconitic sandstones (Maastrichtian–Turonian) —
	Lower	—	Dolomites, limestones, marls, shales, clays (Hauterivian)
Jurassic	Malm	Dolomites, limestones, oolitic limestones, sandstones ●	Dolomites, limestones, chert-bearing limestones
	Dogger	—	Clay stones, sandstones, marls
	Lias	—	Clay stones and sandstones with thin coal seams and layers of volcanic rocks ("Gresten"–facies) ○
Triassic	Buntsandstein?	Vari-coloured clay stones and sandstones without fossils (Perwang 1)	—
Paleozoic			
Permian	Rotliegendes		Clay stones and sandstones with plant remains, conglomerates
Carboniferous	Upper	Shales, arkoses with plant remains conglomerates, (Westphalian)	
		Crystalline of the Bohemian Massif (granites, gneisses, etc.)	

● = Oil-bearing ○ = Gas-bearing

North of the Danube there intrudes between the autochthonous Molasse and the Flysch Zone the Waschberg Zone whose sequence reaches from Upper Jurassic into Middle Miocene. Being a continuation of the subalpine Molasse south of the Danube River, that zone—itself imbricated—thrusts onto the autochthonous Molasse and is in turn overthrust by the Flysch Zone. The end of these alpidic movements is determined by the flat transgressive marine sediments of the Baden Formation (Fig 6). (R. Grill, 1968).

Tertiary. Lying unconformably on the crystalline rocks of the Bohemian Massif or on the Palaeo–Mesozoic sediments of the basin floor is the sequence of the Molasse Basin proper. These rocks are of Upper Eocene to Lower Pliocene age and reach a thickness of several thousand metres (Table II).

In Upper Austria and Salzburg, the Molasse deposits may be divided into three cycles of sedimentation.

1. The sediments of the first cycle reach from Upper Eocene to lowermost Rupelian. At the base lie limnic beds, varicoloured clays with individual sandstone and coal layers; these are overlain by the Cerithium beds, shales and sandstones with *Cerithia, Pectinidae*, and the worm *Rotularia spirulaea* (LAM). At the top are the Nulliporae limestones, represented in the N by marine sandstones. The Upper Eocene deposits are sediments of a shallow epicontinental sea showing a preponderance of zircon in the heavy mineral spectrum, suggesting that they originate to the N, whereas most of the Molasse sediments are characterized by a preponderance of garnet, suggesting an origin in the Alpine area. The fish-bearing shales of the Lattorfian, the light marly limestone and the banded marls of the Lower Rupelian conclude the cycle.

2. The second sedimentary cycle comprises the bulk of the Oligo–Miocene sequence which consists of marine, somewhat sandy, shales (Schlier) intercalated with sands and sandstones, and also with gravels and conglomerates towards the S. The pebbles are of quartz, granitic rocks of the crystalline basement, and (crystalline) dark dolomites of the Central Alps. The beds contain no material from the Limestone and Flysch Zones. Stratigraphic units are chiefly based on foraminifera. Between the Upper Puchkirchen Formation and the Hall Formation lies a major unconformity. The brackish *Oncophora* beds conclude the cycle.

3. The third sedimentation cycle comprises limnic-fluviatile deposits lying unconformably on older Tertiary sediments. At the base are clays with some mineable coal seams overlain by gravels and conglomerates with pebbles from the Limestone and Flysch Zones. (F. Aberer, 1958; E. Braumüller, 1961; R. Janoschek, 1959, 1964; A. Papp, 1968).

In Lower Austria E of the Enns River the Molasse trough gradually becomes narrower and shallower with an accompanying reduction of the sequence. Upper Eocene and the fish-bearing shales of the Lattorfian are absent altogether. Rupelian has been encountered only in the most westerly wells Kürnberg 1 and Seitenstetten 1, and also in Texing 1 in the Texing window. The Ottnangian deposits also pinch out immediately E of the Enns River. Between Enns

TABLE II

Stratigraphic chart of the Molasse Zone in Salzburg, Upper and Lower Austria

Tertiary and Quaternary

Salinity	marine / mesohaline / oligohaline / freshwater and terrestric sediments	Salzburg and Upper Austria	Lower Austria: between the rivers Enns and Danube	Lower Austria: North of the Danube
Holocene		Gravel terraces; peat, lake clays, moraines	Gravel terraces, Loess etc.	
Pleistocene				
Pliocene	Pannonian	Coal bearing freshwater beds in the area of Trimmelkam, Hausruck and Kobernaussen Forest	-	Shales, sands, gravels
Miocene	Sarmatian		-	Shales, sands, gravels
	Badenian		Baden formation: Gravels, sands, shales	Baden formation +Nulliporae limestone
	Karpatian	?	-	Laa formation
	Ottnangian	Oncophora beds; Innviertel formation; Gray shales ("Schlier") with intercalation of sandstones and conglomerates	Oncophora beds; gray shales ("Schlier") with intercalations of sandstone	gray shales ("Schlier") ☼
?	Eggenburgian	Hall formation ☼; Gray shales ("Schlier") with intercalation of sandstones and conglomerates	gray shales ("Schlier") with intercalations of sandstone	Eggenburg formation ☼
Oligocene	Egerian	Upper / Lower; Puchkirchen formations ☼; Gray shales ("Schlier") with intercalation of sandstones and conglomerates	Melk formation: gray shales and sands	Melk formation: gray shales and sands
	Rupelian	dark gray shales; banded marls; light marly limestone	dark gray shales (in the western part only)	-
	Lattorfian	fish-bearing shale	-	-
Eocene	Upper Eocene (Ledian and Wemmelian)	Nulliporae limestone and sandstones, varicolored clays with sandstone and thin coal seams ●	-	-

● = oil-bearing ☼ = gas-bearing

and St Pölten, therefore, the Molasse trough is made up only by the Melk and Hall Formations. North of St Pölten, with the widening of the Molasse Basin, younger sediments occur again in the facies of the brackish *Oncophora* beds; N of the Danube the sequence reaches into the Lower Pliocene. In the Hollabrunn area, however, the Melk Formation pinches out and N of Hollabrunn the Eggenburg Formation lies directly on the pre-Tertiary basin floor (Fig 2). (F. Brix *et al*, 1964; R. Grill, 1968; J. Kapounek *et al*, 1965; A. Papp, 1968).

Structure

The Molasse Zone is the last remnant of the Alpine geosyncline. It is an asymmetric basin with thin Tertiary sediments overlapping the Bohemian Massif in the N but reaching several thousand metres in the S and SE, where they are overthrust by the outmost units of the Alps. In ÖMVAG's deep well Urmannsau 14 km S of the Alpine overthrust, 90 m of the Melk Formation lying between 2925 and 3015 m lie immediately on the crystalline rocks of the Bohemian Massif (A. Kröll and G. Wessely, 1967). Three tectonic units may be distinguished in the Molasse Zone.

1. The structure of the basin floor in Upper Austria and Salzburg is dominated by a NW–SE striking and SW dipping pre-Tertiary fault system which divides the Mesozoic into several basins. The NW–SE striking Landshut–Neuötting ridge also extends into Austria as a high zone.

In the E the tectonics and thickness of the Molasse deposits are chiefly influenced by the relief of the basin floor. Particularly conspicuous are the Grossgraben–Moosbierbaum–Hagenberg–Staatz ridge and the similarly NE–SW striking faults of Mailberg, Steinabrunn, and Wollmannsberg (J. Kapounek *et al*, 1967).

2. In the Upper Eocene and Oligocene, the S part of the Bohemian Massif, with its Palaeo–Mesozoic cover, subsided with the gradual northward movement of the Alps with the amount of subsidence decreasing to the N, so that in Upper Austria the shallow-sea Upper Eocene Nulliporae Limestones are about 3000 m deep (Kirchham 1 well) whereas to the N in the Ried Field they lie at about 1460 m.

During subsidence the basin floor and the Oligocene sediments of the Molasse trough were broken by numerous synsedimentary faults which strike predominantly W–E and dip partly to the N (antithetic) and partly to the S (synthetic) (Fig 2). In the Upper Eocene the vertical throws of these faults have a maximum of 300 m but this gradually decreases to the top. With the end of the Alpine movement northwards the faulting ended and it is not evident above the youngest beds of the Puchkirchen Formation.

In the S, the Paleogene sediments of the Molasse Zone are often imbricated. In the area of Perwang 1 and Voitsdorf E (Fig 3) more than 1000 m of schuppe complex has slid several kilometres northwards over the older deposits of the unfolded Molasse and is unconformably overlain by the youngest beds of the Puchkirchen Formation.

3. In the W the uppermost tectonic unit comprises the Hall and Innviertel Formations which unconformably overlie the older deposits and are not affected by the faults of the lower complexes. In the S, the beds of the Hall Formation are more steeply tilted but they are not overthrust by the Alps. Towards the E, the sediments of the Molasse trough and the age of the tectonics become gradually younger. The heavily disturbed subalpine Molasse Zone established E of St Pölten continues N of the Danube as the Waschberg Zone. Here the overthrust Laa Formation is unconformably overlain by the Baden Formation.

Oil and Gas Accumulation

The sandstones of the Upper Eocene are the main oil-bearing horizons in the Molasse Zone of Upper

TABLE III

Stratigraphic chart of the Preneogene Basement of the Northern Vienna Basin, Lower Austria

Periods	Flysch Zone and Zones of Klippen	Northern Limestone—Alps and Grauwackenzone
Eocene	Steinbergflysch and Gablitz-beds (glaucon: sandstones, clayey marls, marly shales, shales) ● Greifenstein-beds (sandstones, marls, clayey marls)	—
Palaeocene	Greifenstein-beds (friable and massive, coarse sandstones) Altlengbach-beds (friable and massive sandstones with claygalls)	Upper part of Brezova-series, Gießhübl-beds (sandstones, claystones, marlstones, conglomerates)
Upper Cretaceous	Altlengbach-beds (sandstones, clayey marls) Kahlenberg-beds (calcareous sandstones, calcareous marls, clayey marls, picrites) Kaumberg-beds (variegated shales and marls, calcareous sandstones)	Lower part of Brezova-series, Gosau-beds (variegated marlstones, sandstones, conglomerates) ○ Cenomanian-beds (marlstones and breccias)
Lower Cretaceous	Stollberg and Salmannsdorf-beds (mottled limestones, dark shales, spathic limestones)	Gault-beds (marly shales) Neocomian-beds (mottled marly limestones, arenaceous limestones)
Upper + Middle Jurassic	Aptychus-beds (radiolarites, cherty limestones) Neuhaus-beds (crinoidal limestones) Posidonia-beds (marls and marly limestones)	Malm-beds (variegated, partly nodular limestones) Jura–Hornstein-beds (chert limestones) Klaus-beds (limestones), Vils-beds (crinoidal limestones)
Lower Jurassic	Gresten-beds (sandstones, arkose, arenaceous shales)	Hierlatz–Kalke (crinoidal limestones) ○ Liassische Kieselkalke (chert and siliceous limestones)
Upper Triassic	Kössen-beds (marly limestones, arenaceous marly shales)	Kössen-beds (dark limestones and marly shales) Dachsteinkalk, -dolomit (limestones and dolomites) ○ Hauptdolomit ●○ Opponitz-beds (limestones, marly shales, anhydrites, dolomites) Lunz-beds (sandstones, shales) Aon-Schiefer (shales)
Middle Triassic	—	Partnach-beds (shales, marly limestones, limestones, chert limestones), Wetterstein-beds (limestones and dolomites) Reifling-beds (chert limestones) Gutenstein-beds (limestones, anhydrites, dolomites)
Low Triass. + Permian	—	Werfen-beds (shales, anhydrites)
Carboniferous	—	Rocks of Grauwacken-Zone (dark marly shales with thin lamellae of limestones)

Unconformities are not shown. ● = Oil bearing ○ = Gas bearing

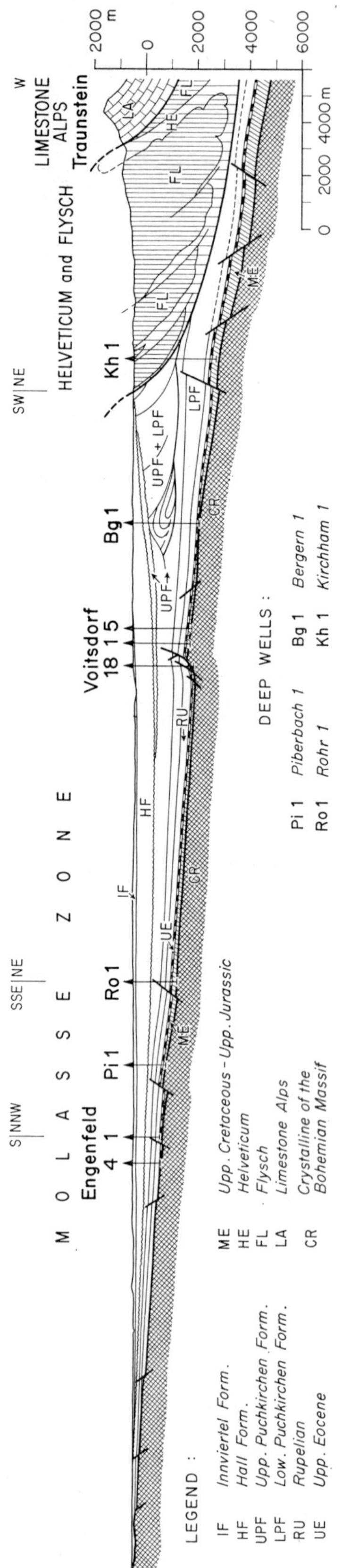

FIG 3. Cross-section through the Molasse Zone (Upper Austria).

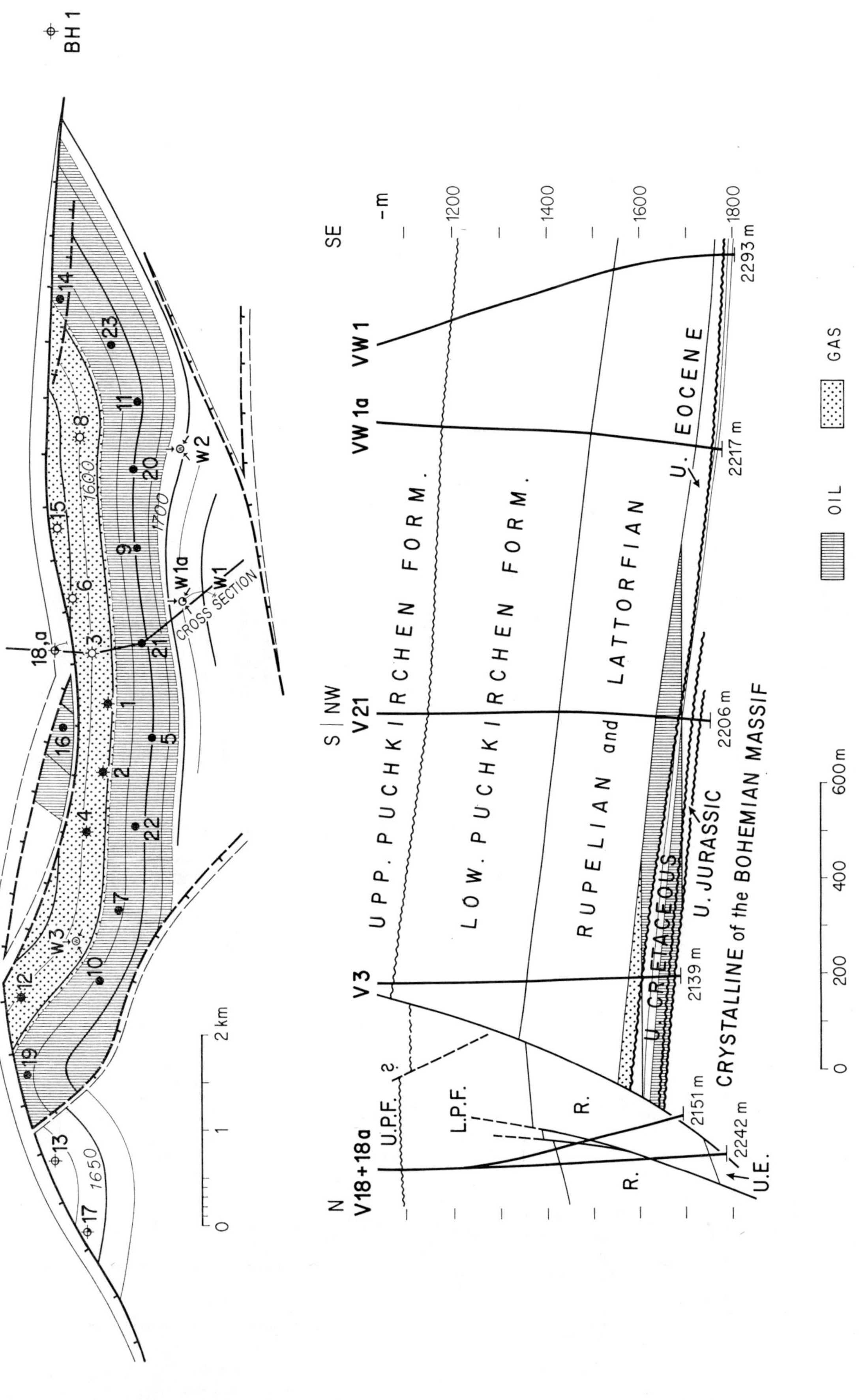

Fig 4. Voitsdorf oilfield, geological map of Upper Eocene and cross-section.

Austria. In some fields the Cenomanian is also productive and in the Voitsdorf field so is the Upper Jurassic. In Upper Austria the sands and sandstones of the Upper Puchkirchen Formation are gas-bearing, as are those in the lower part of the Hall Formation. In Lower Austria sands of the Oncophora beds yield gas and recently ÖMVAG's well Harmannsdorf 1 discovered a gas reservoir in Upper Cretaceous sands.

Oilfields. All Upper Austria's oilfields found so far lie within structures related to upthrown blocks of N-dipping antithetic faults. The largest oilfield in Upper Austria is Voitsdorf (Fig 4). From reflection seismic surveys and deep wells the fault is about 15 km long. In the crestal part of the structure the vertical of the fault is about 300 m, gradually decreasing to the W and E. The oil-bearing area is about 11 km long and one km wide.

Sandstones of Upper Jurassic, Cenomanian, and Upper Eocene are oil-bearing in this field, the porosity ranging between 12 and 18 per cent. The Lower Turonian sandstones are also oil-saturated but tight and unproductive. The trap is formed by the Rupelian shales which, at the fault, seal the oil-

TABLE IV

Stratigraphic chart of the Northern Part of Vienna Basin, Lower Austria

Quaternary and Neogene

Salinity: marine ▮, mesohaline •, oligohaline x, freshwater and terrestric ⋮			Beds		Lithology
Quaternary	Holocene	⋮	alluvial river and brook-deposits		gravels, pebbles, sands, loam, silt
Quaternary	Pleistocene	⋮	gravel terraces		loess, gravels, pebbles, sands
Pliocene	Upper Pliocene	⋮	Hungerberg-beds Rote Lehm-beds 〰〰〰		red loams, gravels
Pliocene	Pannonian	⋮	Pannonian F, G, H freshwater-beds, poorly fossiliferous-beds Pannonian D, E		freshwater-limestones, clays, sands, gravels, lignites marls and sands
		x	Pannonian A, B, C	☼ ●	marls, sands, locally conglomerates and gravels
Miocene	Sarmatian	•	Zone of faunal reduction Mactra-beds Ervilia-beds Rissoa-beds	☼ ●	clayey marls, sands, calcarenites, sandstones, coquinoid limestones; rarely gravels and conglomerates
Miocene	Badenian	▮	Zone of faunal reduction Leitha-limestones, Pötzleinsdorf-sands, Matzen-sands (Zone of arenac. foram.)	☼ ●	clayey marls, sands Lithothamnium-limestones sands, sandstones, marls
		▮	Upper Lagenidaezone		marls, clayey marls, sandstones, conglomerates, breccias
		▮	Lower Lagenidaezone 〰〰〰		marls, sandstones
Miocene	Karpatian	⋮ ▮	Aderklaa-conglomerates Aderklaa-beds in the south Laa-beds in the north 〰〰〰	☼ ●	conglomerates and breccias clayey marls, sands clays, sandstones
Miocene	Ottnangian	•	Luschitz-beds (Oncophora-beds)	☼ ●	clayey marls, sands
Miocene	Eggenburgian	▮	Schlier-beds and basal detritus (Lower Luschitz-beds)	●	arenaceous marly shales, coarse-grained sands, detritus of flysch-material
		▮	Eggenburg-beds locally in wells 〰〰〰		variegated clayey marls

〰〰〰 unconformities

● oil bearing

☼ gas bearing

bearing sandstones. The sand bodies are lens-shaped, their net thickness varying between 0·5 and 30 m. Only the Upper Eocene reservoir has a gas cap. To optimize oil recovery and production, water is injected into the Eocene reservoir through wells VW 1–3. The Voitsdorf oilfield was drilled up with 23 wells, only one of which has proved dry. Ultimate production is 3,200,000 tons.

All other Upper Austrian oilfields are similar to Voitsdorf but as can be seen from Table V they are much smaller. The oil is sulphur-free, paraffinous, and has a gravity of 0·844 to 0·872 kg/l at 15°C.

Gasfields. In the Upper Austrian and Salzburg Molasse Zone, sedimentation during the Oligocene was mainly from the S and in this direction there are intercalated into the shales of the Puchkirchen Formation rather thick sand and conglomerate beds pinching out northwards. This pinchout zone is readily recognized on the seismic cross-sections by the gradual dying out of reflections and is illustrated in Fig 2 by a dotted band. Wells Schwanenstadt 3 and 5 tapped a gas reservoir in the sands of the Upper Puchkirchen Formation, but in well Schwanenstadt 2, producing oil from the Upper Eocene, the gas-bearing sands to the S have pinched out.

Farther S in the pinchout area no gas accumulations have yet been found in the sands of the Lower Puchkirchen or Rupelian Formations. In the Lindach area a gas reservoir has been found at a depth of 1000 to 1200 m in a lens-shaped sandstone bed in the Hall Formation and uppermost part of the Puchkirchen. Other minor gas occurrences not yet examined closely or put on production have been found near Schwanenstadt, Kematen, and Harmannsdorf.

TABLE V

Oil and Gas Fields in the Molasse Zone of Austria

Name	Year of Discovery	Oil- (gas-) bearing units	Top of oil- (gas-) bearing units in metres	Number of wells			Daily production of the fields in tons	Accumulated production to 30 June 1968 in tons	Remarks
				Producing	Dry	Total			
Rohoel-Gewinnungs AG				Oilfields					
Puchkirchen	1956	Upper Eocene	2578	1	3	4	15·0	64,105	
Steindlberg	1958	Upper Eocene	1909	—	2	3	—	13,288	Exhausted
Ried	1959	Upper Eocene	1363	13	2	18	59·2	329,837	
Kohleck	1960	Upper Eocene	2240	4	3	7	50·2	144,984	
Schwanenstadt	1962	Upper Eocene	1986	1	1	2	12·8	43,680	
Voitsdorf	1963	Upper Eocene Cenomanian Upper Jurassic	2031	22	1	23	642·3	927,495	
Lindach	1964	Upper Eocene	2606	1	—	1	7·2	21,993	
Eberstallzell	1965	Upper Eocene	2042	5	—	5	49·7	23,898	In exploration
Rohr	1967	Upper Eocene	1230	—	1	2	—	124	In exploration
Kematen	1967	Upper Eocene	1150	1	1	2	9·7	531	In exploration
Engenfeld	1968	Upper Eocene	1054	1	2	3	26·3	—	In exploration
Oesterreichische Mineraloelverwaltung AG									
Piberbach	1968	Upper Eocene	1059	3	1	4	60·0		In exploration
Pram–Erdoel–Explorationsges mbH									
Leoprechting	1906	Egerian (Upper Puchkirchen Formation)	120	—	—	124	—	4,311	Abandoned
				Gas fields			Cu m	Cu m	
Rohoel–Gewinnungs AG									
Schwanenstadt	1962	Eggenburgian (Hall Formation) Egerian (Upper Puchkirchen Formation)	887	3	1	4	2690	4,407,063	In exploration
Lindach	1964	Eggenburgian (Hall Formation) Egerian (Upper Puchkirchen Formation)	863	6	5	11	59,400	14,110,979	In exploration
Oesterreichische Mineraloelverwaltung AG									
Wildendürnbach	1960	Ottnangian (Oncophora Formation)	760	10	5	17	38,550	520,337,400	
Ameis	1961	Eggenburgian (Hall Formation)	1150	—	2	4	—	1,439,100	
Hagenberg	1966	Jurassic	3040	—	1	2	—	1,608,700	
Private owners									
Wels	1892	Eggenburgian (Hall Formation)	400	4	—	100	60	62,111,600	
Bad Hall	1925	Lower Puchkirchen Formation	250	—	—	1	—	160,000	Abandoned

In Lower Austria the largest gasfield of the Molasse Zone lies near Wildendürnbach close to the Czechoslovakian border, where sands of the Upper Oncophora beds form a flat elongated anticline. Small uncommercial gas occurrences have been found in the Waschberg Zone at Ameis in Eggenburgian sands at 1310 m under an imbricated zone, and at Hagenberg in sandstones of the Gresten beds (Liassic) of autochthonous Mesozoic at about 3050 m.

Reservoir Rocks

The oil-bearing sandstones lie either on Upper Cretaceous marls or directly on the granites and gneisses of the Bohemian Massif. Three marine complexes may have been the source rocks of the hydrocarbons.

(*a*) Rupelian shales. In this case the oil would have had to migrate into older sandstones. Against this source of the oil is the fact that residual oil impregnation or other oil indicators have never been proved in Rupelian beds.

(*b*) Upper Eocene shales. Here also the oil would have had to migrate into older Cenomanian and Upper Jurassic beds. In addition, the Upper Eocene shales encountered in the structures so far investigated by deep wells are only a few tens of metres thick.

(*c*) Upper Cretaceous shales and marls. These beds are often several hundred metres thick in the pre-Tertiary basins. The contained hydrocarbons may have been mobilized by superposition of the many hundreds of metres of Tertiary rocks, and accumulated in oil reservoirs along faults. Further credence is given to this hypothesis by the fact that, from present information, there is no appreciable difference in the physical characteristics between the oils from the Eocene, Cenomanian, and Upper Jurassic.

The partly bituminous shales of the Oligo–Miocene sequence must also be considered as a potential source rock for gas. Oil impregnations have not been established either on the flanks or at the base of the gas-bearing sandstones. The associated gas of the oil reservoirs differs from that of the non-associated gas reservoirs by its comparatively high percentage of higher hydrocarbons and the two may have originated from different source rocks.

VIENNA BASIN

In July 1968 the 60 millionth ton of crude oil was produced in Austria. Of this total, about 58 million tons came from the Vienna Basin (7 million from the basin floor and 51 million from the basin sediments). Considering the volume of sediments, this abundant oil accumulation in the Vienna Basin is most remarkable (R. Janoschek 1958).

On Austrian territory the Vienna Basin is bounded on the NW by the Waschberg Zone and the northern tectonic units of the Alps, and on the SE by the central zone of the Alps. Outside of Austria, the basin is bounded on the N and E by the Carpathians. The Vienna Basin has subsided at the boundary betwee the ranges of the Alps and Carpathians, obliquely to the strike of their tectonic zones.

Stratigraphy

Basin floor. The numerous wells in the Vienna Basin have encountered basement rocks of Carboniferous to Eocene age. In the northern part of the basin the wells entered the Flysch (Cretaceous to Eocene), to the S, the Limestone Zone (Permoskyth to Paleocene) and even further to the S, the Grauwacken Zone (Carboniferous). The boundaries between Flysch, Limestone Zone and, Grauwacken Zone in the basement are shown on the tectonic map (Fig 5).

The beds of the Limestone and Grauwacken Zones encountered in the wells are listed in Table III. The members of the Flysch and Klippen Zones, however, have only been established partially by drilling. Nearly all sediments of the basin floor are of marine origin.

The most important reservoir rock in the basin floor is the Upper Triassic Hauptdolomit, overlain by limestones of Lower Jurassic age and sandstones and conglomerates of the Upper Cretaceous and Paleocene. Another important reservoir rock is the glauconite sandstone of the Eocene Flysch. The units composing the basement of the Vienna Basin have been arranged according to the current understanding of the surficial pattern of Alpine and Carpathian geology. (M. Mahel, T. Buday *et al.*, 1968; A. Tollmann, 1963).

Basin sediments. The oldest sediments of the basin deposits are separated from the basin floor by a conspicuous unconformity. The basal complex of the Eggenburg Formation has been encountered only in a few wells in the N part of the basin—a stratigraphic chart is given in Table IV. The Luschitz Formation with coarse sands and debris shows a wider distribution. These are the oldest reservoir rocks of the basin sediments and are oil-bearing in the St Ulrich–Hauskirchen oilfield. The younger members of the Luschitz Formation are represented by marine gray shales and Oncophora beds whose sandstones have often been found to contain hydrocarbons.

A second cycle of sedimentation begins in the northern part of the Vienna Basin with the marine Laa beds, but in the S there are non-marine deposits of the Aderklaa beds and Aderklaa conglomerate.

The overlying Badenian Formation commences mostly with deposits of the Lower Lagenidae Zone which, in the south, is separated from the non-marine Aderklaa beds by a distinct unconformity. In the N, where the Lower Lagenidae overlies the marine Laa beds, the unconformity is not everywhere recognizable. Succeeding the Lower Lagenidae is the Upper Lagenidae Zone which, with the still younger beds, forms the main sedimentary sequence of the Vienna Basin proper. The sands and sandstones of this sequence constitute the most important reservoir rocks of the basin.

The Badenian is overlain by gradually freshening sediments of the Sarmatian and Pannonian, consisting chiefly of sandstones, but partly also of coarse

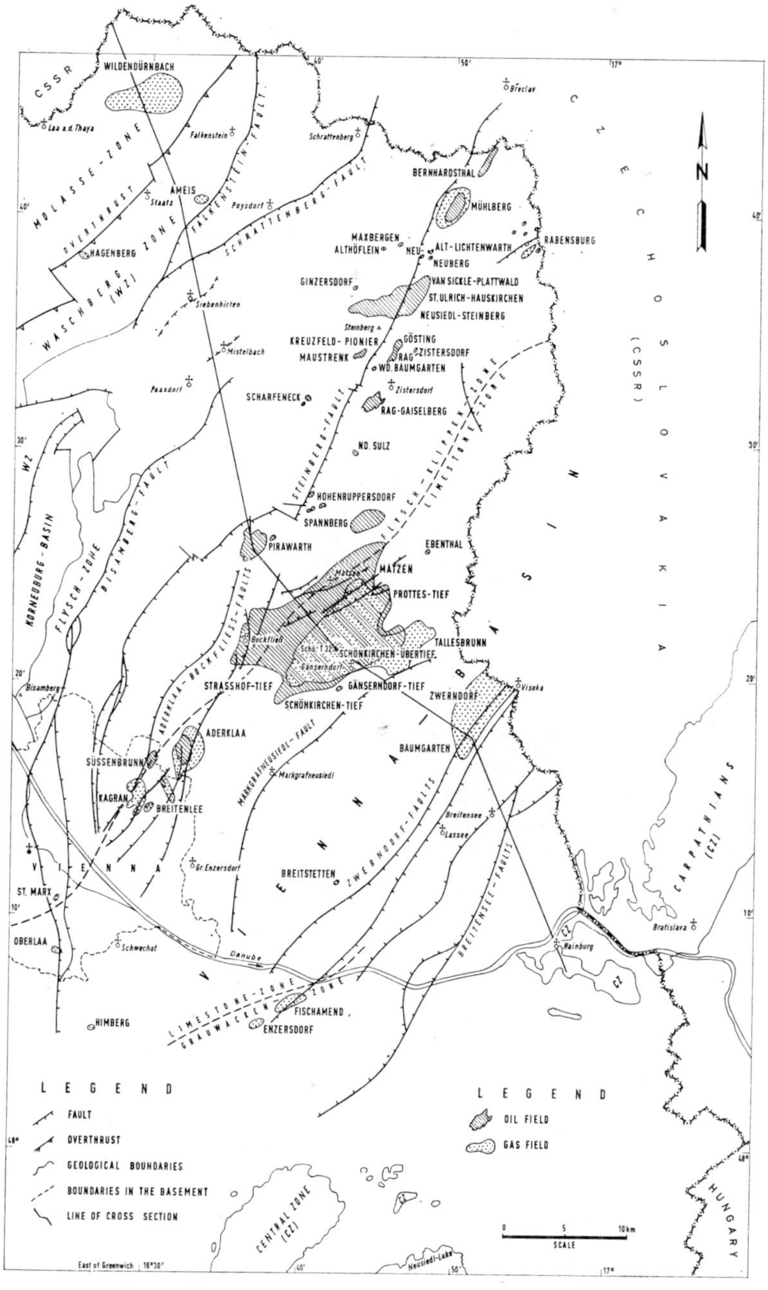

FIG 5. Tectonical map of Inner-Alpine Vienna Basin.

sands and conglomerates alternating with shales.

The thicknesses of the various Miocene formations in general increase rapidly from the rim to the centre of the basin and may vary from a few metres to over 1000 m. The subsidence of the basin ended essentially in the Pannonian, which is unconformably overlain by sands and gravels of the Upper Pliocene and Quaternary. The young sediments of the Vienna Basin are up to over 5000 m thick as shown by geophysical surveys and drilling results. (R. Grill, 1968; R. Grill, J. Kapounek, and A. Papp, 1968; J. Kapounek *et al.*, 1965; A. Papp *et al.*,1968).

Structure

The tectonics of the Vienna Basin floor is Alpidic. The Flysch Zone grows essentially wider from SW to NE, and the fold structures become flatter than at the surface. The nappes of the Limestone Alps can also be recognized in the basement (H. Küpper, 1961; J. Kapounek, L. Kölbl, and F. Weinberger, 1963).

Rocks of the Klippen Zone, of the Frankenfels, Lunz and Ötscher nappes have been proved but in the more easterly parts of the Vienna Basin they show some similarities to the Carpathians. The Alpidic system of nappes with its overthrusts, sheets, and intensive folding is dislocated by the young faults of the Vienna Basin proper. This faulting, first recognized in the basin deposits, consists of numerous fault systems, the main one of which strikes more or less from N–NE to S–SW and divides the basin into several elongated blocks. This is especially true of the basin N of the Danube, whereas S of the river the tectonics are less complicated.

The fault system most important to oil accumulation is that of the Steinberg fault to which the oldest oilfields of the basin are related. Further basinwards we find the Markgraf Neusiedl fault and the Zwerndorf fault system. At the eastern rim of the basin there is the NW dipping fault system of Breitensee NW of Hainburg (Figs 5 and 6).

The main period of activity of most of the faults was during the higher Badenian but, in the main, the marginal faults originated earlier. The dying away of the faulting, which can be proved only locally, occurred during the Quaternary. This can be demonstrated by the thick old-Quaternary and Holocene gravels in the graben between the Zwerndorf and Breitensee fault systems. Gentle folding along the faults is important to the accumulation of hydrocarbons.

The subsidence of the southernmost part of the Vienna Basin S of the Danube was essentially later, as is evident from the absence of the oldest members of the sedimentary suite. Sedimentation here begins with equivalents of the Laa Formation as proved in the area of the Enzersdorf–Fischamend gasfields. In this area also the bulk of the basin sediments consists of Badenian beds overlain by sediments of the Sarmatian and Pannonian. Basin depth decreases southwards and again gentle warping is found associated with the marginal faults.

The small basin of Korneuburg, separated from the Vienna Basin proper by a Flysch ridge, is bounded on the W by a fault. The basin sediments belong to the Laa Formation and are at most only 600 m thick (K. Friedl, 1959; R. Grill, 1968, *loc. cit.*; J. Kapounek *et al*, 1965; H. Stowasser, 1958, 1966).

Oil and Gas Accumulation

The sediments of the Vienna Basin comprise an abundant sequence of sandy reservoir rocks and sealing clay rocks. Transgression of the basin sediments and synsedimentary as well as post-sedimentary faulting caused a variety of trap types, and the alpidically folded basin floor provides buried hills as well as tectonic traps. The most important of these are described with their oil and gas reservoirs.

Tectonic traps in the basin deposits. The oldest commercial oil fields in the Austrian part of the Vienna Basin were found along the Steinberg fault. Here there is a folded zone which at the surface shows an anticlinal configuration which grades into a monoclinal structure at depth. Originally, the monoclines were thought to be drag structures and the higher anticlines were considered to have been formed by later lateral thrusting. (K. Friedl, 1936; R. Janoschek, 1958). More recently this type of structure has been explained as "roll-over" folding with attention being drawn to the remarkably regular distances between axial culminations. (H. Stowasser, 1966).

The Pirawarth field is the southwesternmost of a series of ten rollover structures which are axial culminations of a fold accompanying the Steinberg fault in the downthrown block. With distances between crests averaging 4 km, the chain of productive fields extends 35 km to the northeast up to the frontier.

The Steinberg fault system sometimes carries horsts and is often accompanied by auxiliary faults or broken into steps. The chief oilfields associated with such structures are those of Mühlberg, van Sickle, Zistersdorf, Gaiselberg, and Pirawarth. The last one of these is the most recent to have been established and will be described in greater detail as an example.

In the Pirawarth field the Steinberg fault is divided into five steps of which the deepest contains the rollover structure (Figs 7 and 8). In the Sarmatian, sixteen sands are hydrocarbon bearing, one gas, and fifteen oil, and in the Baden Formation there are eighteen hydrocarbon-bearing sands, two gas, and sixteen oil. The porosity of the Sarmatian sands is 25 to 30 per cent and the permeability ranges from several hundred millidarcies to eight darcies. The corresponding values for the Baden sands are 20 to 34 per cent and up to 9 darcies.

The reservoir rocks of the other fields along the Steinberg fault also belong to the Badenian, Sarmatian, and Pannonian. As at Pirawarth, they form multi-pay reservoirs and consist mostly of very porous and permeable, frequently unconsolidated, sands. In all the fields related to the Steinberg fault, naphthenic oils at the top are gradually replaced by paraffin-base oils at the base. The oil and gas are practically free of sulphur and hydrogen sulphide. (A. Kaufman *et al.*, 1959).

Tectonic traps in the basin floor. About 2000 m below the Schönkirchen Tief field (described later), Schönkirchen Tief well 32 (Austria's first 6000-m borehole) recently discovered a new reservoir, Schönkirchen Ultradeep, between 4835 and 5259 m. The reservoir consists mainly of Hauptdolomit with a mantle of younger debris. Production tests to date have recovered sour gas and liquid hydrocarbons. It is believed that the trap here is an alpidic structure (J. Kapounek and Sz. Horvath, 1968).

Stratigraphic traps. In the St. Ulrich area reservoirs consisting of coarse sands and conglomerates (buttressed sands) had been discovered on an upthrown block of the Steinberg fault. More recently the relationship of these onlapping beds to an underlying buried hill of Flysch sandstones of the basin floor has been recognized. Another type of primary stratigraphic trap *i.e.* exemplified in the Maustrenk field.

Combined traps. The Aderklaa structure on the E end of the Steinberg fault system, apparently related to a high on the pre-Neogene basement, is an example of a combined trap. A gas accumulation in

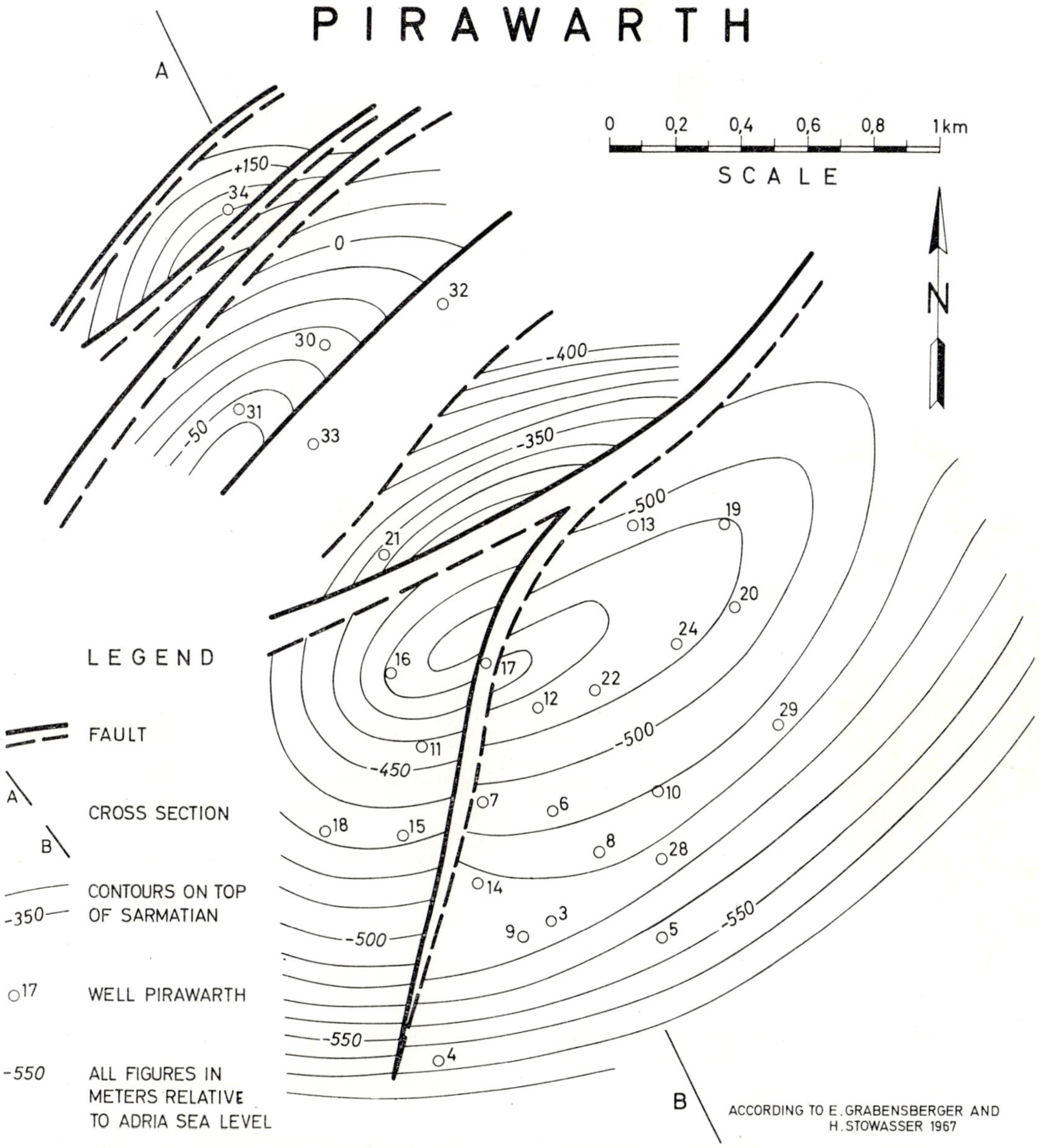

FIG 7. Structure map for the Upper Sarmatian of the Pirawarth oilfield.

the draping rocks of the lower Neogene was first discovered and then oil reservoirs in the Lower Baden Formation were found. Ultimately a large gasfield was struck, in the Hauptdolomit of the basin floor, similar to the St Ulrich–Hauskirchen field. The marginal faults of the Aderklaa structure and the associated auxiliary faults determine the limit of the reservoirs.

The Zwerndorf–Baumgarten gasfield has much in common with the Aderklaa, both as to structural configuration and exploration history. A flat anticline in the Neogene of an upthrown block of the Zwerndorf fault system was first delineated and a thick sand complex in the Lower Baden Formation was found to be gas bearing. As at Aderklaa, the basin floor was subsequently investigated and the Baumgarten gasfield as discovered on a high composed of Triassic limestones and dolomites.

Matzen, the largest oilfield in Central Europe, is similarly related to a high on the basin floor and consists of several trap types. The main oil and gas reservoirs in the basin sediments are contained in flat anticlines which are accompanied by faults on the W and N. The resulting fault blocks also contain hydrocarbons.

Sand lenses interfingering with clays frequently occur in the Laa and Oncophora Formations and form primary stratigraphic traps. As reservoirs, however, the sands are frequently displaced by the faults and the hydrocarbon accumulation is thus controlled by the structure. The basal rocks of the Oncophora Formation cover the pre-Neogene relief of the basement.

Within the 100 sq km of the Matzen field, hydrocarbons have been found in the following beds of basinal sediments:

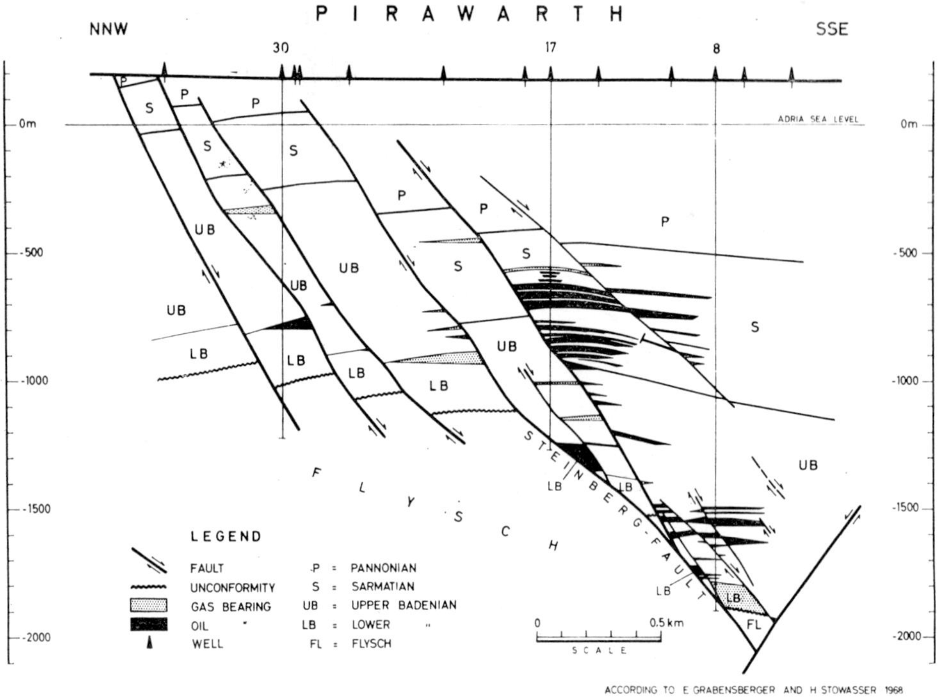

FIG 8. Cross-section through the Pirawarth oilfield.

Laa and Luschitz (Oncophora) Formations	10 oil horizons
Badenian	15 horizons
Most important the "Matzen" and Eighth Sands	
Sarmatian	7 horizons (2 oil, 5 gas)
Pannonian	4 gas horizons

The "Matzen Sand" has yielded the main production of the field and has an average net thickness of 25 m. (Fig 9). Next most important is the Eighth Sand complex, which covers the greatest area. Several horizons in the Badenian and Sarmatian extend far to the E as gas horizons and are in direct connection with the Tallesbrunn pool.

The average porosity of the Matzen Sand is 25·9 per cent and the average permeability is 1300 millidarcies, with maximum values up to 16 darcies. For the Eighth Sand the same figures are 26·7 per cent and 300 millidarcies up to 11 darcies.

The Laa and Luschitz Formations produce mixed-paraffin-base to paraffin-base oils, the Matzen and the five overlying sands yield mixed-naphthenic oils. The sands of the Upper Badenian and the reservoirs of the Sarmatian produce naphthenic oils. The gas horizons produce almost pure methane.

The basement in the northern part of the Matzen field is a ridge of dense Cretaceous Flysch and Klippen Zone rocks, but in the S there is a buried hill of Triassic Hauptdolomit with local debris containing abundant hydrocarbons. This basin floor reservoir belongs to a higher tectonic unit of the subsided and buried Limestone Alps, which in pre-Neogene time was thrust onto a deeper unit of predominantly dense rocks so that the thrust plane, together with the pre-Neogene unconformity at the surface of the buried hill, presents a fine example of a combined trap. Between the dolomite body and the pre-Neogene unconformity there lie, locally, transgressive dense Gosau beds showing an older phase of the multi-phase alpidic orogenic development. In the mostly dense overthrust unit, the small Strasshof Tief reservoir has been discovered. These southern reservoirs in the Hauptdolomit are about 1000 m lower than the barren Flysch ridge at the N end of the field due to tilting which took place before deposition of the Badenian.

Schönkirchen Tief is a good example of a basin floor field with a rich oil accumulation. The oil reservoir is in the Hauptdolomit of Upper Triassic together with local debris. A schematic relief map (Fig 10) shows the configuration of the field and its size. In the crest of the field the effective closure reaches about 220 m with a constant bottom water level at about — 2740 m.

Two small gas caps with different gas/oil interfaces (at about — 2630 m and — 2550 m) have been established. Southwest of the main structurally high area, which contains both oil and gas, the latter has been found in a flat ridge with an effective closure of about 40 m and oil/water interface at about — 2760 m. Northeast of Schönkirchen Tief is the smaller Prottes Tief field with its oil/water contact 125 m higher at about — 2615 m.

The average porosity of the Hauptdolomit and associated debris in the Schönkirchen Tief field is 9·7 per cent. The permeability to air varies considerably but appears to average around 400 to 500 mD.

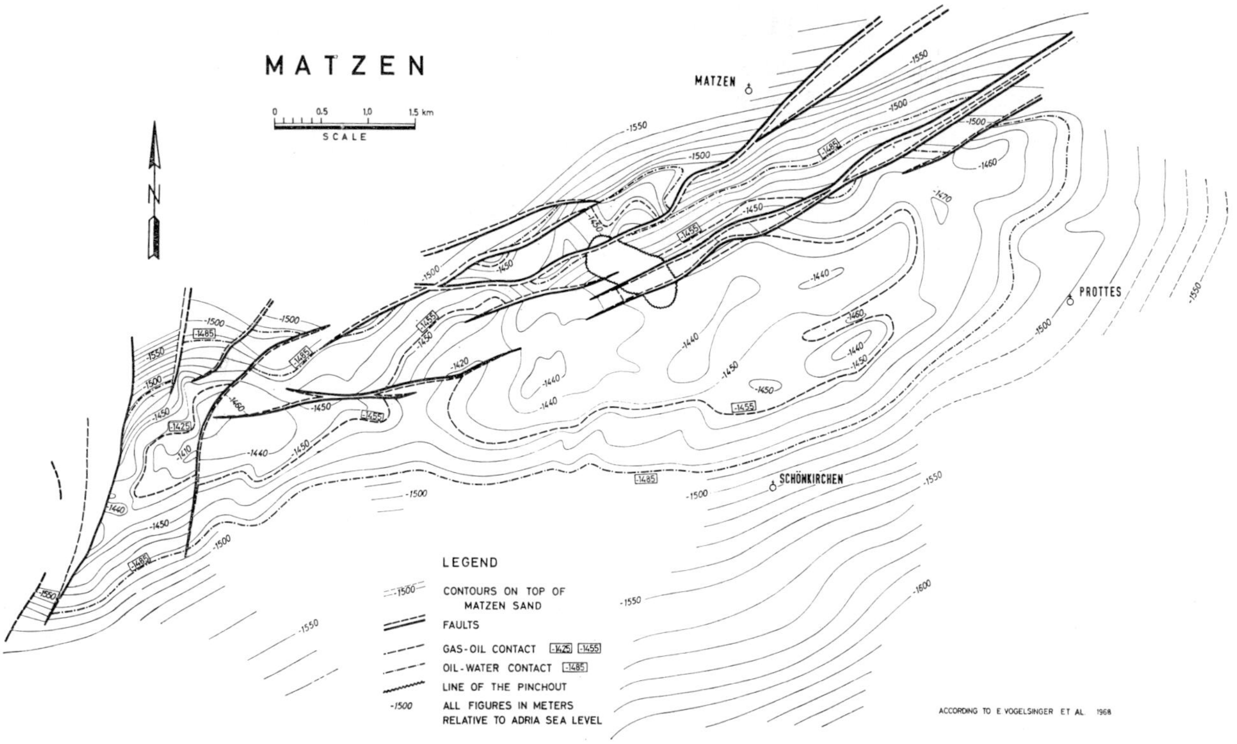

FIG 9. Structure map of the Matzen Sand.

Similar figures for other representative fields of the basin floor are as follows:

St Ulrich–Hauskirchen field—	
Flysch sandstone (Eocene)	
Average porosity	6·1 %
Permeability about	50 mD
Aderklaa gasfield—	
Basin floor Dolomite	
Average porosity	7·7%
Permeability about	100 mD
Baumgarten gasfield—	
Triassic dolomites and limestones	
Average porosity	3·1%
Permeability about 9 mD, with peaks up to 400 mD	

The Schönkirchen Tief, Prottes Tief and, St. Ulrich–Hauskirchen fields produce paraffin-base oil. The two gasfields, Aderklaa and Baumgarten, yield sour gas.

Origin of the Hydrocarbons

The general view is that the hydrocarbons of the Vienna Basin had their origin in the lutitic rocks of the basin sediments, particularly in the troughs (L. Kölbl, 1966, 1967 *loc. cit.*; H. Wieseneder, 1964). An alternate suggestion has been advanced (J. Kapounek, 1964), that the hydrocarbons may originate in the autochthonous Mesozoic of the Bohemian Massif as postulated to occur at the bottom of the Alpine–Carpathian basement of the Vienna Basin. This view gains credence from the recent discovery of the Schönkirchen Ultradeep as described above.

STYRIAN BASIN

The Styrian Basin is the largest of the Neogene basins at the eastern rim of the Alps. Extensive studies have been made of the area and some deep wells have been drilled. The basin is framed on the N and W by spurs of the eastern Alps composed of various metamorphic rocks, the Graz Paleozoic and older Mesozoic rocks in Central Alpine facies, and the Upper Cretaceous of Kainach.

The Styrian Basin is separated from the units of the western Pannonian area by the NE–SW trending South Burgenland Swell which emerges in individual Paleozoic Inselberge from under the Tertiary cover but is not conspicuous morphologically. The basement of the basin is also composed of these formations. The Sausal Swell cuts the basin in two; a smaller west Styrian and a much larger east Styrian Basin (Fig 11).

At the base of the Tertiary sequence are limnic-fluviatile deposits which probably belong to the Ottnangian. Marine deposition commences with the Karpatian shales and sands, represented in the western basin by coarse clastics and in the Eibiswald area by limnic-fluviatile beds. The boundary between Karpatian and Badenian is marked by a large unconformity representing the Styrian orogenic phase. The sediments of the Badenian, Samartian, and Lower Pannonian are similar in facies to those of the Inner Alpine Vienna Basin. Middle and Upper Pannonian sediments occur in the eastern part of the Styrian Basin but in the W only river gravels of the Middle Pannonian are found. Maximum thickness of the Pannonian in the Styrian Basin is 600 m

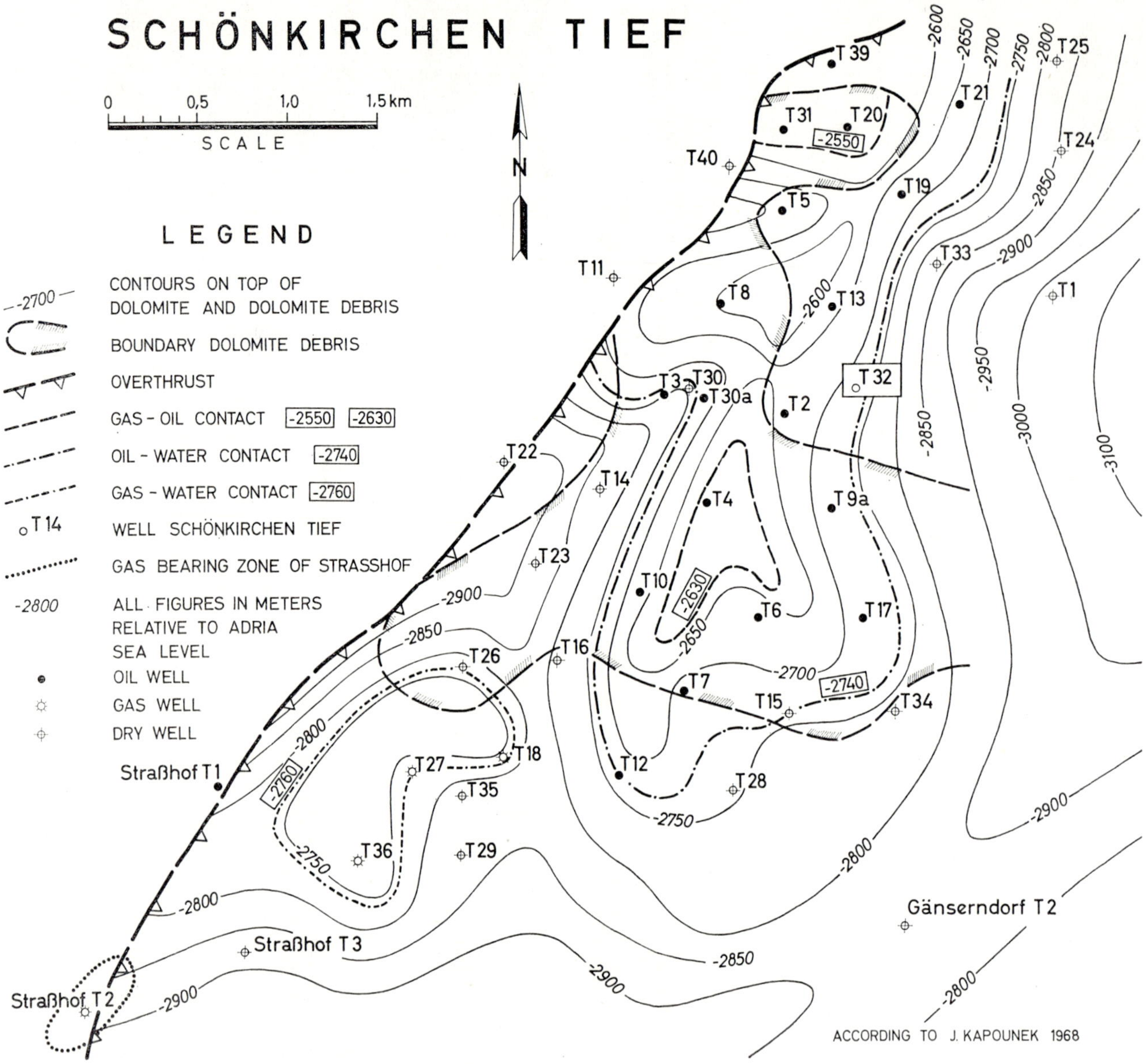

FIG 10. Structure map of the Schönkirchen Tief oilfield.

compared to more than 1000 m in the central part of the Vienna Basin.

A feature of the Styrian Basin is the Neogene volcanism which has left its mark on the landscape. Two volcanic cycles can be distinguished:

(i) Latites, trachy-andesites, dacites, quartz trachytes, etc. were discharged during an older Miocene volcanism (Karpatian to Lower Badenian). Although now mostly covered by younger Tertiary deposits, magnetometer surveys and deep wells prove that a number of large lava domes were built up by the volcanism.

(ii) A younger Pliocene volcanism is also represented on the surface by lava flows, intrusions, and tuff vents (*e.g.* Riegersburg) composed of nepheline basanites, nephelinites, etc.

Basement rises divide the eastern Styrian Basin into a number of sub-basins, Gnas, Fürstenfeld and Fehring. These features, together with the flanks of the buried lava domes, might control local accumulations of oil or gas in this basin where, contrary to conditions in the Vienna Basin, major fault systems seem to play but an inferior role.

Slight traces of oil and gas have been found in some deep wells (H. Flügel and H. Heritsch, 1968; K. Kollman, 1965; A. Papp, 1968; A. Winkler-Hermaden, 1957).

FURTHER EXPLORATION PROSPECTS

Present knowledge suggests the following possibilities for further petroleum exploration in Austria:

(*a*) The Molasse Zone, including its Mesozoic basement. Continuation of exploratory work, par-

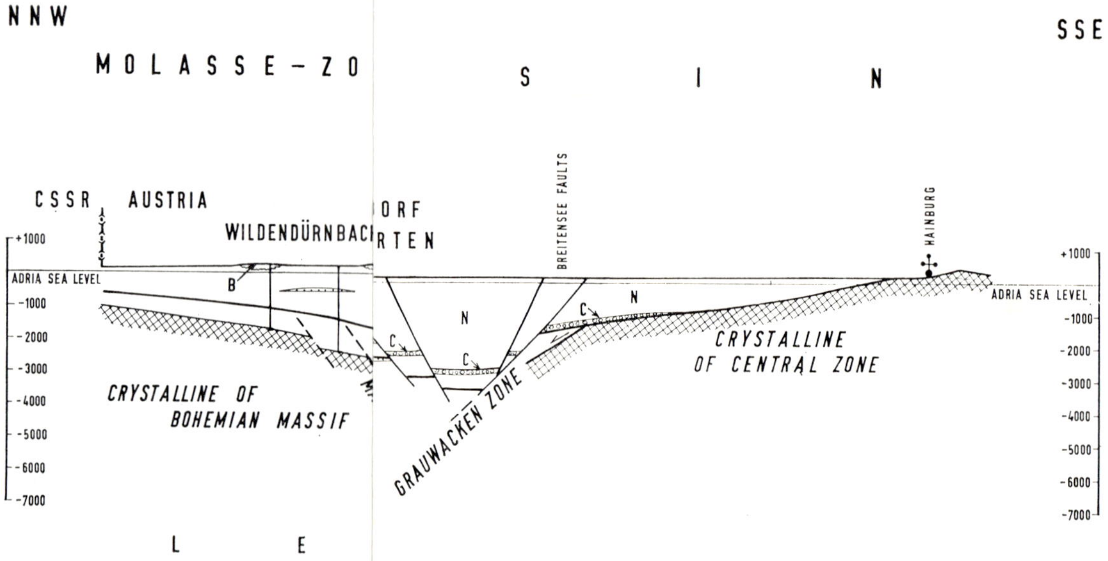

M = MOLASSE ZONE

B = BADENIAN OVERLYING MOLASSE

K = KLIPPEN

N = NEOGENE BEDS (PLIOCENE AND MIO
OF VIENNA BASIN

C = ADERKLAA CONGLOMERATE

NIWANDTER, R. GRILL, J. KAPOUNEK, A. KRÖLL, H. STOWASSER, G. WESSELY,
RIX AND K. GÖTZINGER 1969

To face page 172]

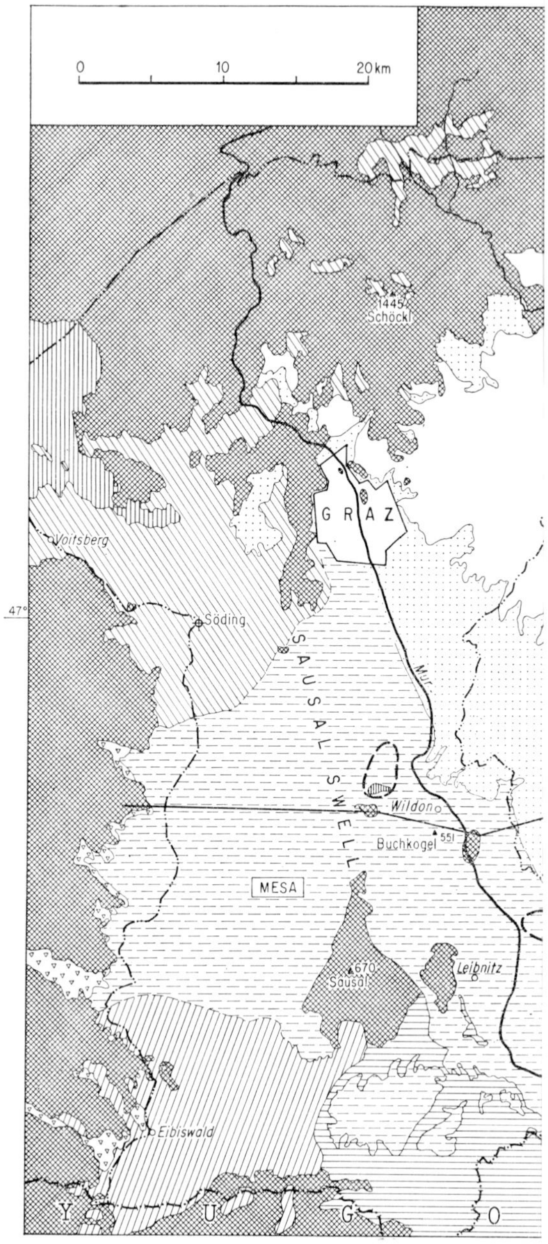

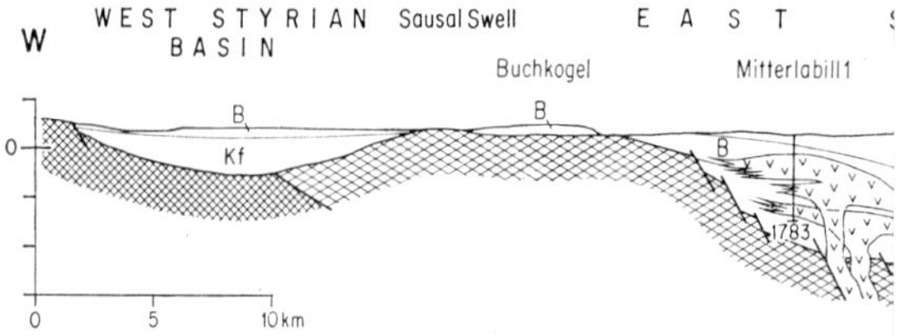

P	Pannonian	S	Sarmatian	B	Badenian, marine

To face page 176]

TABLE VI

Oilfields in the Inner-Alpine Vienna Basin

Name	Year of discovery	Oil-bearing units	Top of oil-bearing units in metres	Number of Wells: Producing	Dry	Total	Daily production of the fields in tons	Accumulated production to 30 June 1968 in tons	Remarks
Oesterreichische Mineraloelverwaltung AG									
Windisch Baumgarten	1930	Eocene	680	—	—	1	—	588·4	Exhausted
Gösting–Zistersdorf	1932	Pannonian Sarmatian Badenian Eocene	900	23	3	66	28·3	1,911,970·6	
Neusiedl-Steinberg	1936	Ottnangian Eocene	1250	5	2	8	10·1	43,234·0	
St. Ulrich–Hauskirchen	1938	Pannonian Sarmatian Badenian Ottnangian Eocene	900	127	30	376	200·2	4,789,136·3	
Maustrenk	1941	Ottnangian Eocene	900	14	7	42	24·0	336,397·6	
Hohenruppersdorf	1941	Sarmatian Badenian Eocene	700	—	16	22	—	18,030·7	Abandoned
Mühlberg	1942	Badenian	1200	74	9	137	169·2	4,832,006·7	
Scharfeneck	1944	Ottnangian	1000	—	7	10	—	3,575·6	Exhausted
Matzen	1949	Sarmatian Badenian Karpatian Ottnangian Cretaceous	1300	746	48	885	3680·4	35,959,343·1	
Neu Lichtenwarth	1949	Ottnangian	1400	1	—	1	1·2	11,380·0	
Aderklaa	1950	Badenian Karpatian Jurassic	1700	36	22	77	71·8	1,927,146·8	
Bernhardsthal	1950	Badenian	1400	2	2	5	0·2	10,288·8	
Rabensburg	1954	Badenian	2000	1	4	8	0·7	3,797·7	
Spannberg	1956	Ottnangian	2350	2	—	12	0·7	5,704·8	
Pirawarth	1957	Sarmatian Badenian	750	19	7	30	127·7	140,474·0	
Schönkirchen Tief	1962	Triassic	2705	13	4	36	1377·3	2,289,650·7	
Straßhof Tief	1964	Ottnangian Jurassic	2800	—	—	1	—	157·4	Exhausted
Prottes Tief	1966	Triassic	2740	4	—	7	155·3	85,242·0	
Ebenthal	1967	Badenian	1850	1	—	1	4·6	969·7	
Rohoel–Gewinnungs AG									
Zistersdorf	1937	Sarmatian Badenian Eocene	870	10	6	44	50·3	943,871·0	
Gaiselberg	1938	Sarmatian Badenian Eocene	1000	50	7	70	243·7	3,575,383·0	
R.K. van Sickle									
van Sickle–Plattwald	1938	Pannonian Sarmatian Badenian Ottnangian Eocene	500	39	13	108	48·3	1,419,322·1	

ticularly towards the N and NW in the pinchout area of the oil- and gas-producing formations; and towards the S and SE within the deeper Molasse which is, in part, overthrust by the Alpine nappes.

(*b*) Inner Alpine Vienna Basin. Continuation of exploration in the basin sediments and basin floor.

(*c*) Styrian Basin. Search for oil and gas prospective structures.

(*d*) Flysch Zone and Limestone Alps between Salzburg and Vienna. Exploratory work to date has not been successful but there are still possibilities of finding hydrocarbon reservoirs in these tectonic units. At present, the relative prospects seem better in the Limestone Alps than in the Flysch Zone.

(*e*) Eastern rim of the Alps. No recent exploration has been done in the small Neogene basins at the eastern rim of the Alps, although here too there are possibilities of finding oil or gas reservoirs.

TABLE VII

Gas fields in the Inner-Alpine Vienna Basin

Name	Year of discovery	Gas-bearing units	Top of gas-bearing units in metres	Number of Wells: Producing	Dry	Total	Daily production of the fields in cu m	Accumulated production to 30 June 1968 in cu m	Remarks
Oesterreichische Mineraloelverwaltung AG									
Oberlaa	1932	Karpatian	250	—	8	9	—	15,156,700	Exhausted
Gösting	1940	Pannonian	170	—	10	16	—	9,401,200	Exhausted
Alt Lichtenwarth	1940	Sarmatian Badenian	650	1	9	12	2800	21,320,000	
Hohenruppersdorf	1941	Pannonian Sarmatian	970	—	—	2	—	26,618,300	Abandoned
Aderklaa–Kagran–Breitenles	1942	Sarmatian Badenian Karpatian Cretaceous Jurassic Triassic	740	13	8	39	604,330	2,015,006,400	
Mühlberg	1942	Pannonian Sarmatian Badenian	500	15	—	15	86,940	1,195,845,700	
St. Marx	1944	Badenian	720	—	1	2	—	3,890,900	Exhausted
Niedersulz	1944	Sarmatian Badenian	1640	1	7	9	14,700	14,049,000	
Matzen–Tallesbrunn	1949	Pannonian Sarmation Badenian Ottnangian	540	28	4	70	345,110	2,163,653,600	
Fischamend–Enzarsdorf	1951	Sarmatian Badenian	700	3	11	20	16,200	165,593,500	
Zwerndorf–Baumgarten	1952	Pannonian Sarmatian Badenian Triassic	1450	19	11	43	814,610	8,937,227,000	
Rabensburg	1954	Sarmatian Badenian	1750	1	—	3	2980	35,289,600	
Alt Höflein	1959	Badenian	225	—	1	2	—	21,400	Exhausted
Himberg	1959	Sarmatian	1270	—	2	3	—	211,400	Abandoned
Ginzersdorf	1959	Ottnangian	960	1	1	2	490	3,811,900	
Maxbergen	1960	Badenian	520	1	—	1	300	1,013,900	
Schönkirchen Tief	1964	Triassic	2890	—	—	1	—	15,670,000	Abandoned
Breitstetten	1966	Sarmatian	955	1	1	2	5260	20,683,200	
Straßhof Tief	1967	Cretaceous	3340	—	—	1	—	563,200	Abandoned
Schönkirchen Ultra-deep	1968	Paleocene Cretaceous Triassic	4520	—	—	1	—	—	In exploration
Rohoel–Gewinnungs AG									
Zistersdorf	1940	Pannonian	602	—	—	3	—	9,374,000	
Gaiselberg	1940	Pannonian Sarmatian	358	2	—	4	6400	17,675,000	
R.K. van Sickle									
van Sickle–Plattwald	1944	Ottnangian	800	2	—	5	1100	19,988,500	

TABLE VIII

Austrian crude oil production according to geological units

Geological units	Accumulated production to 30 June 1968 in metrictons
Pannonian	41,394·0
Sarmatian	6,895,522·0
Badenian	42,841,868·7
Karpatian + Ottnangian	2,307,856·5
Egerian (Puchkirchen Formation)	4,311·5
Eocene	4,884,458·6
Cretaceous	475,693·3
Jurassic	106,963·6
Triassic	2,374,892·7
Total	59,932,960·9

Austrian non-associated gas production in N cu m according to geological units

Pannonian	1,658,125,000
Sarmatian	1,654,369,200
Badenian	9,427,081,500
Karpatian + Ottnangian	705,957,500
Eggenburgian (Hall Formation)	77,925,300
Egerian (Puchkirchen Formation)	4,143,500
Cretaceous	85,677,300
Jurassic	38,737,400
Triassic	1,639,996,500
Total	15,292,013,200

STATISTICS

Austria's oil and gas production has come from the Molasse Zone and the Inner Alpine Vienna Basin. Tables V, VI and VII show, by field and company, the year of discovery, producing formation and top depth, number of wells, average daily production, and cumulative production to 30 June 1968.

To indicate the relative importance of the various productive formations to Austria's overall oil and gas production, Table VIII lists cumulative production by stratigraphic age. It can be seen that the largest oil and gas volumes have so far been produced from the Baden Formation in the Vienna Basin. Detailed information on the associated gas produced from the several oilfields is not available because records of this were not kept separately before 1955. It has, however, been estimated that 7797·6 million N cu m of associated gas had been produced by the end of 1967 (F. Schippek, 1968), and cumulative production to 30 June 1968 exceeds 8 billion.

Not shown in Tables V and VI is the annual production rate, which was as follows for 1967:

Crude oil	2,684,910 tons
Associated gas	467,289,600 N cu m
Non-associated gas	1,329,932,200 N cu m
Natural gas liquids	21,785·4 tons

On 30 June 1968 Austria's reserves were estimated to be as follows:

Recoverable oil (proven and prospective)	30 million tons
Recoverable gas (gas horizons only)	10 billion N cu m

During the next five years it is estimated that associated natural gas will be produced with the oil at an annual rate of 500 million N cu m.

ACKNOWLEDGMENTS

The authors wish to express sincere thanks to their associates who have taken part in the exploration of the areas covered in the foregoing pages, and also to Österreichische Mineralölverwaltung AG, Rohoel-Gewinnungs AG, Mobil Oil Corporation, and Bataafsche Petroleum Maatschappij for permission to publish this paper.

BIBLIOGRAPHY

Aberer, F. "Die Molassezone im Westlichen Oberösterreich und Salzburg." *Mitt. geol. Ges., Wien*, 1958, **50**, 23–94.

Braumüller, E. "Die Paläogeographische Entwicklung des Molassebeckens in Oberösterreich und Salzburg. *Erdöl Z., Wien-Hamburg*, 1961, **77**, 509–20.

Brix, F., Götzinger, K., Kröll, A., and Logigan, St. "New Results of Exploration in the Molasse Zone of Lower Austria." *Proc. 6th Wld Petrol. Congr*, 1963, Frankfurt am Main, C, Sect. I, Pap. 3, pp. 247–65.

Brix, F., Diem, E., Götzinger, K., Kröll, A., Logigan, St.. Papp, A., Stradner, H., and Turnovsky, K. "Die Ergebnisse der Aufschlußarbeiten der ÖMV AG in der Molassezone Niederösterreichs in den Jahren 1957–1963." *Erdöl Z., Wien-Hamburg*, 1964, **80**, 57–256.

Friedl, K. "Der Steinberg–Dom bei Zistersdorf und sein Ölfeld." *Mitt. geol. Ges., Wien*, 1937, **29**, 21–290, 1936.

Friedl, K. "The Oil Fields of the Vienna Basin." *Proc. 5th Wld Petrol. Congr.*, New York, 1959, Sect. I.

Flügel, H. and Heritsch, H. "Das Steirische Tertiär-Becken." Geol. Führer, 2. Aufl. 1968, **47**, Verlag Gebr. Bornträger, Berlin–Stuttgart.

Grill, R., Kapounek, J., Küpper, H., Papp, A., Plöchinger, B., Prey, S., and Tollmann, A. "Neogene Basins and Sedimentary Units of the Eastern Alps near Vienna." *Proc. Int. Geol. Congr.*, 1968, Prague, Sess. XXIII.

Grill, R. "Erläuterungen zur Geologischen Karte des nordöst-lichen Weinviertels und zu Blatt–Gänserndorf. *Loc. cit. Geol. Bundesanst., Wien*, 1968, pp. 1–155.

Janoschek, R. "The Inner-Alpine Vienna Basin. An example of a small sedimentary area with rich oil accumulation." Symp. on The Habitat of Oil, Tulsa, 1958, pp. 1134–52.

Janoschek, R. "Oil Exploration in the Molasse Basin of Western Austria." *Proc. 5th Wld Petrol. Congr.*, New York, 1959, Sect. I, Pap. 47, pp. 849–64.

Janoschek, R. "Das Tertiär in Österreich." *Mitt. geol. Ges., Wien*, 1964, **56**, 319–60.

Kapounek, J., Kölbl, L., and Weinberger, F. 1963: "Results of New Exploration in the Basement of the Vienna Basin." *Proc. 6th Wld Petrol. Congr.*, Frankfurt am Main, 1963, Sect. 1.

Kapounek, J., Kaufmann, A., Kratochvil, H., and Kröll, A. "Die Erdöllagerstätte Schönkirchen Iief im Alpinkarpatischen Beckenuntergrund." *Erdöl Z., Wien-Hamburg*, 1964, **80**, 305–17.

Kapounek, J., Kröll, A., Papp, A., and Turnovsky, K. "Die Verbreitung von Oligozän, Unter- und Mittelmiozän in Niederösterreich." *Erdöl Erdgas Z., Wien-Hamburg*, 1965, **81**, 109–16.

Kapounek, J., Kröll, A., Papp, A., and Turnovsky, K. "Der Mesozoische Sedimentanteil des Festlandsockels der Böhmischen Masse." *Jb. geol. Bundesanst., Wien*, 1967, **110**, 73–91.

Kapounek, J., and Horvath, Sz. "Die Bohrung Schönkirchen Tief 32 als Beispiel für den Aufschluß einer Lagerstätte im tiefen Anteil der Kalkalpen." *Erdöl Erdgas Z., Wien-Hamburg*, 1968, **84**, 396–407.

Kaufmann, A., Kölbl, L., Kratochvil, H., and Wieseneder, H. "Reservoir Rocks, Fluids and Energy Systems of the Matzen Field in the Vienna Basin." *Proc. 5th Wld Petrol. Congr.*, New York, 1959, Sect. II.

Kollmann, K. "Jungtertiär im Steirischen Becken." *Mitt. geol. Ges., Wien*, 1965, **57**, 479–632.

Kölbl, L. "Geologische Studie über die Bildung der tortonen Zwischenhorizonte von Matzen und die Entstehung ihrer Lagerstätten." *Erdöl Erdgas Z., Wien-Hamburg*, 1966, **82**, 45–65.

Kölbl, L. "Sedimentology and the Petroleum Industry." *Loc. cit., Sedimentary Geology, Amsterdam*, 1967, **I**, 223–34.

Kröll, A., and Wessely, G. "Neue Erkenntnisse über Molasse, Flysch and Kalkalpen auf Grund der Ergebnisse der Bohrung Urmannsau 1." *Erdöl Erdgas Z., Wien-Hamburg*, 1967, **83**, 342–53.

Küpper, H. "Erläuterungen zu einer tektonischen Übersichts-skizze des weiteren Wiener Raumes." *Mitt. geol. Ges. Wien*, 1961, **53**, 1–33.

Mahel, M., Buday, T., *et al.* "Regional Geology of Czechoslovakia. Part II. The West Carpathians." Geological Survey of Czechoslovakia in Academia, Praha, 1968.

Papp, A., Grill, R., Janoschek, R., Kapounek, J., Kollmann, K., and Turnovsky, K. "Zur Nomenklatur des Neogens in Österreich (mit einer stratigraphischen

Tabelle)." *Verh. geol. Bundesanst., Wien*, 1968, pp. 9–27.

Schippek, F. "Naturgas in Österreich." Montan Rundschau, Sonderheft Energiewirtschaft f. 1968, pp. 53–56, Wien.

Stowasser, H. "Einige Bausteine zur Tektogenese des Wiener Beckens." *Erdöl Z., Wien-Hamburg*, 1958, **74**, 395–400.

Stowasser, H. "Strukturbildung am Steinbergbruch im Wiener Becken." *Erdöl Erdgas Z., Wien-Hamburg*, 1966, **82**, 188–91.

Tollmann, A. "Ostalpensynthese." Verlag Franz Deuticke, Wien, 1963."

Wieseneder, H. "Die Erdöl–Muttergesteinsfrage im Wiener Becken. *Erdöl Z., Wien-Hamburg*, 1964, **80**, 479–86.

Winkler-Hermaden, A. "Geologisches Kräftespiel und Landformung." Verlag Springer, Wien, 1957.

MS received 10 February 1969

Discussion

Marco Pieri (AGIP): Can Mr Janoschek give us some information about the methods used for the exploration of the section below the basin floor of the Vienna Basin? Were geophysical surveys carried out, and were their results clear and reliable?

R. H. Janoschek: The Alpine flysch zone and the Northern Limestone Alps continue, below the Tertiary of the Vienna Basin, in a NE direction into the Carpathians. Just as on the surface the boundary between those two nappe systems coincides, also under the Tertiary cover, with an escarpment, which means that the Limestone Alps are topographically higher than the flysch zone. The oil and gas accumulations occur in those high buried hills, in porous rocks (mainly Triassic dolomites) of the Alpine calcareous Mesozoic.

Those deep structures were first recognized by gravimeter surveys (*e.g.* Schönkirchen Tief 1). The basin floor could not be explored through conventional seismics as it does not supply any reflections. For that reason, a seismic "travel time investigation" was carried out in the deep wells. By means of that method it was established in which direction one should look for the oil-bearing dolomite. The well Schönkirchen T2 located in this manner came in productive. Telluric and electrical surveys were made only in shallow parts of the basin, for the purpose of mapping the basement.

Dr P. Bitterli (Bataafse Internationale Petroleum Maatschappij): Could you tell us something about the origin of the oil found in the Mesozoic and its relationship to the oil found in the Molasse?

R. H. Janoschek: That is a question that is very hard to answer. I shall start with the easier part of it, namely, with the relationship in the Vienna Basin. According to general opinion, all oil in the Vienna Basin originates with the Neogene shales comparatively poor in bitumen. But, with a few exceptions, all the oil and gas in the Vienna Basin is sweet. Only the oil and gas stemming from the Mesozoic basement is sour. Thus, for example, the gas from the well Schönkirchen T 32 contains 2·5 per cent H_2S and 11 & CO_2. Therefore, it is assumed that this oil and/or gas comes from the Mesozoic. J. Kapounek of Ömvag thinks that all oil in the Vienna Basin stems from the autochthonous Mesozoic overlying the crystalline of the Bohemian Massif. In such a case, one would have to assume a long-range and complicated migration through the Molasse and through numerous overthrust alpine units as, for instance, in the flysch zone.

As far as the Molasse zone is concerned, we have to consider above all the Upper Austrian area W of the Enns river. E of that river, all wells were more or less dry, with the exception of the Wildendürnbach gas field close to the Czech border. As is the case with the gas finds in the Upper Austrian Molasse zone, one may assume also that the Wildendürnbach gas stems from the Molasse sediments proper. In the Upper Austrian Molasse, the oil-bearing sections are the Upper Eocene, the Cenomanian at the Upper Cretaceous base, and sandy sections of Upper Jurassic. The Upper Eocene has a maximum thickness of 100 m, the major part thereof consisting of algal limestones. Intercalations of shales are very thin, so that it is very hard to assume that the oil stems from the Eocene itself. If one were to assume that the oil in the Eocene originated from the shales of Lattorfian and Rupelian, which do contain a certain amount of organic matter, the counter-argument is that in Austria no oil was ever found in the horizons overlying those beds, only gas. Therefore, it is my opinion that the organic matter stems from the shales of Upper Cretaceous. These are of considerable thickness. Perhaps the organic matter was remobilized partly through the tangential movements of the Alps, partly through the overburden of the Molasse sediments, and has migrated into the structures within the range of the Molasse base. An origin at even greater depth is impossible because, under the Eocene, Upper Cretaceous, and Upper Jurassic, there immediately follows—apart from thin, mostly sandy Upper Carboniferous—the crystalline of the Bohemian Massif. This is only an idea, and I am not sure whether it is right or not.

H. R. Warman (British Petroleum Co. Ltd): Can the authors give any explanation why the Graz Basin, which is similar to the other productive Neogene basins of this part of the world, has no oil or gas?

R. H. Janoschek: The Styrian Basin has more or less the same stratigraphic sequence as the Vienna Basin, but there are also some differences. In the Styrian Basin the thick cover of Upper and Middle Pannonian, *i.e.* Lower Pliocene, is missing; in the Vienna Basin, it reaches a thickness of 1500 m in the centre of the troughs. In the Styrian Basin, only the Lower Pannonian is present and consists of a sequence rich in gravels coming from the Alps. This is perhaps one reason. The second reason is the scarcity of faults, whereas the Vienna Basin is broken by many faults. I cannot say that faults are altogether missing in the Styrian Basin. For example, the W border against the crystalline of the Koralpe is formed by a fault, yet in this case there is involved but a small and shallow part-basin. Therefore one should not wonder if no oil is discovered at that fault system. The third reason could be the presence of volcanism in the sequence of the Styrian Basin. We have two cycles of volcanism, an older one in the Middle Miocene, and a younger in the Late Pliocene. The trachites and andesites of the older volcanism form the cores of buried hills. Several wells on such high zones have supplied only very poor shows.

Rustum Byramjee (CFP): Would you give some details about oil types in relation to geographical and stratigraphical location of the different pools?

R. H. Janoschek: In the Molasse zone of Upper Austria, all oils from the Eocene, Upper Cretaceous, and Upper Jurassic are paraffin-base oils with the more or less same specific gravity of about 0·850 or 35° API. In Bavaria, moreover, there is also heavier oil in certain structures.

In the Vienna Basin, there is found the whole range of most different oil types, from naphthenic heavy oils in the Upper Miocene and Lower Pliocene, with a specific gravity of 0·934 or 20° API, to paraffin-base light oils in the Lower Miocene and in the Mesozoic basement, with a specific gravity of 0·809 or 43°–44° API.

Oil and Gas Exploration and Production in the Soviet Union*

By ROBERT E. KING

(*Woods Hole Oceanographic Institution, Woods Hole, Mass., U.S.A.*)

SUMMARY

From the first oil production in Russia in 1871 until the early 1930s, efforts were concentrated on the Neogene basins of the North Caucasus. Modern exploration started in the sedimentary basin west of the Urals in 1929, and led to discovery of major oilfields, chiefly in Devonian reservoirs, in the so-called "Second Baku". Until geophysical techniques were perfected, the leading exploration method was the drilling of stratigraphic and structural holes. Since World War II exploration has greatly intensified and spread to all parts of the Soviet Union, from the Baltic to the Far East. The country has vast sedimentary basin areas, and many of these were virtually untouched until recent years. Initial exploration in some of these basins has led to spectacular restuls. Outstanding among these has been the West Siberian Basin, where large oilfields have been found along the mid-course of the Ob River, and huge gas reserves in the N near the Arctic Circle. Large oilfields have been found on the Mangyshlak Peninsula on the E coast of the Caspian Sea. Proved oil reserves in the Soviet Union may approximate 8·5 billion metric tons and gas reserves 9422 billion m^3. Much of these reserves are in remote areas, and large investments will be required to bring the oil and gas to major consuming centres. Large-diameter pipelines are under construction, and when these are completed the impact of Soviet hydrocarbon production on world markets will increase.

INTRODUCTION

THIS paper is based entirely on a review of published information as it has appeared during the last dozen years. The writer has not personally had an opportunity to visit the Soviet Union, as have so many people from Western Europe and North America in recent years.

Until the mid-1950s, reports on the Soviet oil industry were exaggerated and unreliable, with more propaganda than fact. To a large extent they were an expression of wishful thinking because real achievement lagged behind the economic plans of that period, but the reports also characterized the secretive nature of the Russians during the Stalin era.

In recent years there has been a great increase in available data from Russia, from technical papers available in translation, and from official press releases. The press releases are evidently generally factual, but they often fail to differentiate between the important and the trivial, and many of them, coming from statements by politicians rather than technicians, are exaggerated. The same problem would exist if we were to analyse the progress of the oil industry of the non-Communist world with newspapers and weekly news magazines as our only source of information.

Developments in the Soviet Union are most reliably summarized in various papers published in the proceedings of the World Petroleum Congresses held in New York in 1959, Frankfurt in 1963, and Mexico in 1967, as well as papers presented by Russian geologists at various ECAFE and other international meetings. These reports provide the background and fill-in data needed to place all the other reports in their proper perspective. Finally, Russian geological organizations have published a fine set of coloured tectonic maps that display the configuration of the sedimentary basins of the U.S.S.R.

DEVELOPMENT OF EXPLORATION TECHNIQUES SINCE 1930

Russian oil production commenced in 1871, near Baku, and development in this area was rapid because of high exploration success and flush production. From 1898 to 1902 Russia was the world's leading oil-producing nation. However, after the revolution of 1917, chaotic internal conditions prevented progress until about 1930, when geological exploration commenced outside the Caucasus region.

The leading Russian petroleum geologist of that time, I. M. Gubkin, drew attention to the large sedimentary basin W of the Urals, and proposed the drilling of a series of deep stratigraphic wells. In 1929–1930 the initial wells were drilled according to a selected grid between the Volga River and the Ural Mountains. This programme was continued and extended to various areas of European Russia until the outbreak of World War II. When normal activities were possible after the war, about 1950, programmes of stratigraphic and structural drilling following surface geological studies continued to be the leading exploration method. The information

* This paper is a revised version of "Recent Trends in Oil Exploration in the Soviet Union", published in the *Proceedings of the Institute on Exploration and Economics of the Petroleum Industry*, which was held by the Southwestern Legal Foundation Copyright © 1968, by Matthew Bender & Co. Inc., New York, N.Y., and reprinted by permission.

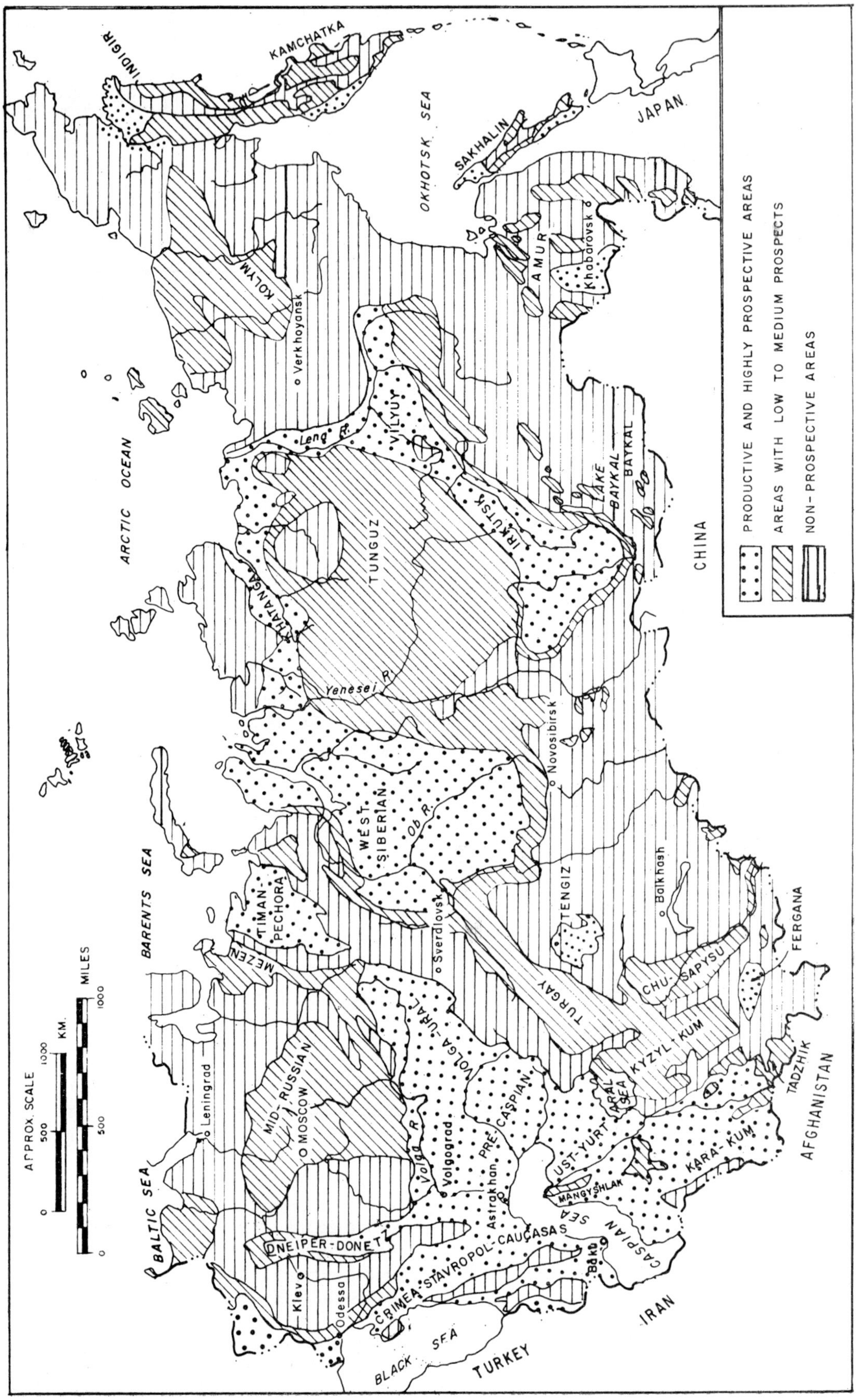

FIG. 1. Soviet Union—sedimentary basins.

wells were drilled both in grid patterns and in lines of profiles across basins. As geophysical methods were improved, the drilling was increasingly co-ordinated with geophysical surveys, and it has been the practice to drill "parameter" wells to assist interpretation of the geophysical surveys by determining rock density, magnetic susceptibility, velocity, and depth to formation contacts.

In 1959, more than 300,000 m (1 million ft) of hole was drilled in information wells. Of these, 60 per cent were more than 500 m (1650 ft) deep. The long-term trend is to drill deeper and fewer holes. Information wells are being drilled to projected depths of 7000 m (23,000 ft).

Geophysical exploration was introduced in Russia in 1925, and by 1930 there were 23 parties in the field. However, until about 1950 the role of geophysical surveys in oil exploration was unimportant. Since then, geophysical work has expanded rapidly, and by 1961 there were 850 seismic, 200 electrical, 100 gravity, and 18 magnetic parties operating. Work spread to previously unexplored basins, and co-ordinated gravity, magnetic, and seismic surveys were carried out along with the drilling of stratigraphic and parameter wells. The regional seismic surveys were generally made in profiles across the entire width of a basin.

When geophysical work was accelerated after the war, Russia was far behind the Western countries. The instruments and methods of exploration geophysics and wire-line well logging were developed rapidly, but internal research facilties were evidently inadequate to equal the rapid technological progress in the post-war period in the U.S.A. and Western European countries. Isolated from the mainstream of research in the non-Communist countries, Russia has lagged several years behind.

The vast increase in exploration has led to widespread experimentation and research along lines either not accepted or not widely used in western countries. Theories of the origin of oil from inorganic reactions in basement rocks or from organic matter in meteoric waters, long discredited in Western countries, seem to have wide acceptance in Russia. Geochemical and microbiological prospecting methods have been developed, and appear to be extensively tried. There is much greater use of electrical surveying methods than in Western countries, chiefly applied to the mapping of basement and deep sedimentary layers to determine basin configuration. On the other hand, there seems to be little, if any, use of photogeology, to judge from the lack of mention of this tool in the Soviet literature.

The importance of mineral exploration in the Soviet economy is highlighted by the fact that efforts are unified in the Ministry of Geology, established in 1965, and that 2 April 1967 was celebrated as Geologists' Day. On this day it was announced that the total number of scientific and technical employees of the geological services exceeds 500,000. In the Geological Ministry alone, there are 50 scientific institutes with more than 10,000 scientific employees. Many additional geologists are employed by the Academy of Sciences of the Soviet Union, academies of the various Soviet republics, universities, and other organizations. This army of geologists has more than 10,000 drilling rigs of all types, 15,000 geophysical units, 6000 boats, more than 60,000 powered vehicles of all types, and several hundred aeroplanes and helicopters. These geological organizations are engaged in the study of all types of mineral resources and in scientific research, but a large part of the effort is in petroleum exploration.

GEOGRAPHICAL DIVERSITY OF EFFORT

The Soviet Union has about one-seventh of the land area of the world, 23 million km^2 (8·6 million sq miles). Of this area, 14·5 million km^2 (5·5 million sq miles) is underlain by sedimentary basins, with about 10·6 million km^2 (4 million sq miles) of these prospective to varying degrees (Fig 1). This is nearly double the favourable area of the U.S.A., including Alaska and the continental shelves. The immense basinal areas of the Soviet Union are distributed throughout the country, but the largest lie on either side of the Ural Mountains and across the southern part of the country in the Ukraine, along the N border of the Caucasus, and through Central Asia to the border with Afghanistan. The geology of these basins is very diverse, with fill of sediments of all ages since the Pre-Cambrian, and with all types and sizes of structures. Salt deposits, chiefly of Devonian and Permian age, are widespread in the Ukraine, the Volga–Ural region, and the North Caspian depression. In the Lena valley of Central Siberia there are thick and extensive Cambrian salt deposits. Salt domes are abundant in several basins, and there is important oil and gas production in some places on deep-seated salt structures. The great number, size, and geological variety of the sedimentary basins provide the setting for a wide diversity of oil and gas accumulations.

The first important exploration effort outside the young Tertiary rocks of the Caucasus region started in the Volga–Ural region in 1929. In the 1930s this resulted in the discovery of some relatively small, poor quality reservoirs in the Permian and Lower Carboniferous, and more prolific oil of better quality in thick, porous, and permeable Upper Devonian sands. To dramatize the importance of this new producing region, it was called, rather inappropriately, the "Second Baku", Baku being the symbol in Russia of prolific oil production.

Exploration of the sedimentary basins of all parts of the Soviet Union began about 1950 and has continued at an increasing rate to the present. From the Baltic to the Bering Sea and from the Arctic Ocean to the Central Asian deserts, large basins (such as the West Siberian) to small basins (such as some intermontane troughs in E Siberia) have been systematically explored. Initially, the configuration of a basin is determined by surface geological, aeromagnetic, and electrical surveys, long refraction seismic lines, and a series of stratigraphic holes. To a large extent, these preliminary investigations are carried out as scientific research projects, and much of the effort is without immediate practical application. The initial basin study is followed by reflection seismic surveys, and actual drilling for oil may not start until several years after the commencement of exploration in a new area.

MAIN OIL AND GAS DISCOVERIES AND NEW PRODUCING AREAS

Exploration in the 1930s and 1940s was primarily in the Volga–Ural region. Numerous fields were found, and by 1940 12 per cent of the total U.S.S.R. production was from there. The percentage increased to 29 in 1950 and to 73 in 1960. In the 1940s a giant oilfield, Romashkino, was discovered in Tataria, between the Volga and the Urals. It was placed on production in 1953, and peripheral water injection started soon after. The field is reported to be several times larger in area than the East Texas oilfield, and to have two and a half to four times the ultimate producible oil reserves. It produces high gravity, low sulphur oil from multiple oil sands in the Upper Devonian, with additional reserves of low gravity, sulphurous oil in the Lower Carboniferous. This field is the main source of oil exported to the Central European Communist satellite countries. The oilfields of the Volga–Ural region had a great economic advantage in being located near the existing rail and water transportation network and much closer to the industrial and population centres of Russia than the older Baku oilfields.

The Dneiper–Donetz trough of the NE Ukraine was explored geophysically in the early post-war period. This led in 1950 to the discovery of the large Shebelinka gas field, with reserves of 468 billion m^3 (16·5 trillion cu ft) of gas, in porous dolomite zones of the Permian, overlying a Devonian salt swell. The productive area is 460 km^2 (174 sq miles). Put on production in 1956, Shebelinka was ideally located to supply gas to Moscow and to industrial centres of the Ukraine. Annual production from the field is about 30 billion m^3 (1 trillion cu ft). Various lesser gas, gas-condensate, and oilfields have been found in the Dneiper-Donetz trough.

Exploration in the foreland zone N of the Caucasus resulted in the discovery of major gas and oilfields in Mesozoic formations in the Krasnodar, Stavropol, and Groznyy regions, and farther W in the Tertiary of the N Crimean Peninsula. Before World War II, only the Tertiary had yielded hydrocarbon production in the Caucasus zone.

In the mid-1950s, exploration in Central Asia led to the discovery of the Gasli gas field in Uzbekistan. With reserves of 500 billion m^3 (17·6 trillion cu ft) in a Cretaceous sand reservoir, it is the largest developed gas field in the Soviet Union. Other major gas fields have been found in the same region, such as Uchkyr, Bayram-Ali, and Mayskoye.

In a more remote part of the Soviet Union, several large gas fields were found in the Vilyuy depression near Yakutsk, in the valley of the Lena River in E Central Siberia. A new gas field, Kolendo, with reserves of nearly 60 billion m^3 (2 trillion cu ft), was found on N Sakhalin Island, where there was was minor oil production before World War I.

The subsurface geology of the West Siberian Basin, with an area of 2 million km^2 (750,000 sq miles), nearly three times the size of Texas, was almost unknown as late as 1950. The basin is covered by sub-polar forests, muskeg, and swamp, entirely mantled by Quaternary deposits. The oil and gas potential of the basin was generally rated as poor. In 1953 gas was found in the Berezovo area on the NW part of the basin, leading to development of a cluster of fields with reserves of about 50 billion m^3 (1·75 trillion cu ft). In the 1950s, exploration proved that the West Siberian Basin contained a great thickness of Cretaceous to Upper Jurassic sedimentary rocks, with porous sands in the lower part, above the basement. The basin was regarded as primarily a gas area until 1960, when oil was found at Shaim, 320 km (200 miles) S of Berezovo. The real potential of the basin was demonstrated when a major oil discovery was made in 1961 at Megion on the mid-course of the Ob River, in the central part of the basin, from a Lower Cretaceous sandstone reservoir at 2170 m (7120 ft). More than 35 oil discoveries have been made in the basin since Megion. One of these, Ust'Balyk, 260 km (160 miles) W of Megion, has 16 Lower Cretaceous sandstone reservoirs with a total thickness of 100 m (330 ft) in a structure 25 km (15 miles) long and 8 km (5 miles) wide. In 1966 to 1968 a still larger oilfield was found in swamp land about 30 km (18 miles) NE of Megion. The first discovery here, Belesernoye, merged with another, Samot-Lor, into a single giant oilfield with estimated reserves of 400,000 metric tons (2·8 billion brl). The field has eight sandstone reservoirs, each tens of metres thick. Operations are supplied entirely by helicopter from river ports on the Ob.

The Berezovo gas area in the NW part of the West Siberian Basin was extended northward in 1961, with the discovery of the Punga gas field, with reserves of 90 billion m^3 (3·2 trillion cu ft). Exploration moved northward into the Arctic, and still more important finds were made near the Ob and Taz gulfs. Here the West Siberian Basin is deeper, with basement as deep as 4000 m (13,500 ft). The first large discovery in this northern region, at Tibeysale, was made in 1962. It was followed by a succession of further large gas fields. The largest of these, Urengoi, found in 1966, has estimated reserves of 2500 billion m^3 (88 trillion cu ft), and may be the largest gas field in the world. The structure is 140 km (87 miles) long from N to S and 36–40 km (22–55 miles) wide. The Zapolyarnoye field in the same area has reserves calculated at 1500 billion m^3 (53 trillion cu ft). Gas fields have been found across the entire extent of territory from the lower Yenesei valley on the E (Messo-Yakha field) to the Yamal Peninsula on the W (Novyy Port and Arkticheskoye fields). Gas reservoir sandstones in this region are chiefly in the Cenomanian (Middle Cretaceous).

A new major oil and gas province in the Soviet Union was opened in 1960 with the discovery of oil at Zhetibay on the S side of the Mangyshlak Peninsula, on the E coast of the Caspian Sea. This was followed by the discovery of a second major oilfield in the same area at Uzen in 1961. Other, apparently less important, discoveries have since been made of oil, gas, and condensate on the Mangyshlak Peninsula. The Zhetibay oilfield has proved to be second in size in the Soviet Union after Romashkino. It is 25 by 5 km (15 by 3 miles) in dimensions, with 1000 m (3300 ft) of reservoir sands in 15 zones in the Jurassic. The Uzen field, 30 by 10 km (18 by 6 miles) in size, has 22 oil and gas zones in the Lower Cretaceous between 1000 and 2600 m (3300 and 8500 ft) with a gross reservoir thickness of 935 m (3070 ft). The oil in both fields is 32°–34° API and high in paraffin.

In the Pre-Caspian depression, N of Mangyshlak, there is a multitude of salt domes. Prospecting of the crest and flanks of shallow piercement domes resulted in finding various small oilfields. In 1959 exploration emphasis shifted to drilling deep-seated salt uplifts, and this led to the discovery of the large Barankul, Prorva, and Konkiyak oilfields in W Kazakhstan.

Several oilfields have been found in Byelorussia, in Devonian sandstones, in a NW extension of the Dneiper-Donetz trough under the Prypet Marshes in W Russia. This is especially important because of proximity of the fields to the pipeline connecting the Volga–Ural region with the Central European satellite countries.

A major gas field, Vuktylsk, was found in the NE part of European Russia in 1965, in the Pechora Basin. Reserves are estimated at 500 billion m^3 (17·5 trillion cu ft). This field will contribute gas to the pipeline connecting the fields in the Ob-Taz gulf region with Western markets.

Late in 1967 a major gas discovery was made in the Province of Orenburg, in the S part of the Volga–Ural region. The field, named Krasnyy Kholm, has more than 400 m (1300 ft) of reservoir thickness in Lower Permian limestone at 1280–1800 m (3950–5900 ft), and the field is 100 by 20 km (62 by 13 miles) in size. The reserves are estimated at 650 billion m^3 (23 trillion cu ft).

Important additions to gas reserves have been made by deep drilling in the older fields of the Carpathian trough of the Western Ukraine. These are well located with respect to gas pipelines supplying Poland and Czechoslovakia.

Much publicity has been given to oil finds in the Middle Cambrian of the Baltic zone, in the Kaliningrad district, Lithuania, and Latvia, and to Lower Cambrian oil in the Lena valley of Central Siberia. These occurrences are geologically interesting, but there is no evidence as yet that they are of commercial significance.

Finally, among the discoveries since World War II, the results of drilling in the Caspian Sea must be noted. The decline of production in onshore fields of the Baku area has been offset by the development of several large fields in the shallow waters of the Caspian, where wells are drilled from fixed platforms connected with an extensive system of trestles. The most famous of these fields is Neftyanye Kamni. A newer field, Sangachily, is of nearly equal size, and several other discoveries of the last ten years are being developed. Marine seismic surveys have proved that the Baku oil producing trend extends across the Caspian Sea to the Cheleken Peninsula of W Turkmenistan. Since World War II a new major oil and gas field, Kotur Tepe, was found after a seismic survey in the Cheleken area, near some older small fields.

To date, Soviet technology has been unable to overcome problems of drilling in deep waters. The first self-elevating platform constructed in Russia was put in operation late in 1966, and a larger platform was built in the Netherlands from American designs and shipped to the Caspian via inland waterways. A second Russian-built self-elevating platform was completed in 1968. When rigs capable of drilling in deep water go into full operation in the Caspian, it may be expected that major new reserves will be found there. There have been complaints that the Caspian waters are becoming seriously polluted by industrial waste products. With increased oil operations there may be damaging impact on the caviar industry.

The success in finding large new oil and gas reserves in the Soviet Union since 1950 is primarily attributable to initiation of exploration by modern methods in large, previously unstudied, sedimentary basins. It may be compared to the success of Western oil companies in the Algerian Sahara since 1953 and in Libya since 1956, when those large basins were systematically explored for the first time. Worldwide experience shows that large structures are found in the early years of exploration in a basin, leading to early discovery of large fields. Later discoveries are on structures or in other traps beyond the resolving power of the initial geological and geophysical surveys, or in parts of the basin not accessible in the early years. Russian scientific accomplishments are of a high order, but their technology has lagged behind that of non-Communist countries. Exploration success in the Soviet Union has been achieved at great cost in manpower and drilling effort, because it is not particularly confined by economic restraints of profitability.

Eventually, as Russian drilling techniques are improved, the potential of deep reservoirs will be extensively explored. A depth of 6000 m (20,000 ft) was first reached by two wells in Russia in 1967, and a new depth record of 6700 m (21,976 ft) was reached by one of these wells, Aral Sor, in the Pre-Caspian depression in 1968.

PROVED AND POTENTIAL RESERVES

Soviet geologists and engineers estimate reserves according to a series of categories, A, B, C_1, C_2, D_1, and D_2. The A category is comparable to the reserves reported by API for the U.S.A. and Canada. The other categories have a downward range through what we would call probable, possible, and speculative, with the D reserves based on calculations of the size, configuration, stratigraphy, organic content of the rocks, and probable incidence of traps in relatively unexplored sedimentary basins. Reserve figures cited in Soviet press reports are often highly inflated, particularly if given by government officials, and are not correlative with reserve figures reported for the U.S.A. and other non-Communist countries. These reports evidently fail to discriminate between the different reserve categories.

In 1962 the proved, or A, oil reserves of the Soviet Union were about 4·57 billion metric tons (32 billion brl), and in 1961 the proved gas reserves were 2460 billion m^3 (87 trillion cu ft). This was soon after the early major oil and gas discoveries were made in the West Siberian Basin, the Mangyshlak Peninsula, and Central Asia, and these new finds were only slightly reflected in the A reserve calculations.

No recent official statement of proved oil reserves has been published, but on the basis of what seem to be reliable reserve estimates in new oil producing areas it appears that they now exceed 8·5 billion metric tons (60 billion brl). Oil reserves in the B to D categories would be considerably greater. Ultimate reserves of recoverable oil in the West Siberian Basin are estimated at 27 billion tons (190 billion brl), using a recovery factor of 55–65 per cent because of the porous and permeable nature of the producing sandstones and the probability of a water drive.

Proved reserves of non-associated gas were estimated at the beginning of 1969 at 9422 billion m^3 (332·6 trillion cu ft) and of associated gas at 1330 billion m^3 (47 trillion cu ft). Of the non-associated gas reserves, 52·6 per cent or 5150 billion m^3 (182 trillion cu ft) were calculated for the West Siberian Basin. Gas reserves of Central Asia (Uzbekistan and Turkmenistan), which supply trunk pipelines to the Ural industrial centres and Central Russia, were estimated at 1520 billion m^3 (54 trillion cu ft) and of the Ukraine at 680 billion m^3 (24 trillion cu ft). In the Soviet Union there were 548 fields of gas, gas-condensate, and gas with oil. Fourteen gas fields have reserves greater than 100 billion m^3 (3·5 trillion cu ft). The overall gas reserve figures cited above do not include major increases in reserves attributable to 1968 developments, such as the large field in Orenburg Province.

About 75 per cent of Soviet gas reserves are in Asiatic Russia, and the same proportion is probably applicable to oil reserves. Only the portions of the West Siberian Basin accessible to water transport have been well explored to date. A vast muskeg area between the mid-course of the Ob River and the waterways tributary to the Ob and Taz gulfs, hardly explored to date, evidently has a potential equal to that of the part of the basin where oil and gas production have already been proved. The large Samot–Lor oilfield is an example of the type of discoveries that may be expected when helicopter-supported exploration penetrates farther into the swamp lands inaccessible to surface transport.

Until World War II most of Russia's oil production was from the Neogene of the Baku area. After the war the Devonian rocks of the Volga–Ural region became dominant. Oilfields found in recent years in Western Siberia and the Mangyshlak Peninsula are productive from the Jurassic and Cretaceous, and these geologic systems by now have the greatest proved oil reserves of the U.S.S.R. The Cretaceous provides the great gas reserves of Western Siberia and Central Asia, and the Permian the gas reserves of the Ukraine and Orenburg.

PIPELINES COMPLETED AND UNDER CONSTRUCTION

In recent years the oil and gas pipeline network of the Soviet Union has been greatly expanded (Fig 2). The Romashkino and nearby fields of Tataria and Bashkiria are connected by the COMECON or Druzhba pipeline to Western markets. Construction of the line commenced in 1962 and it was completed in 1964. The 40-inch trunk line, 1350 km (840 miles) long, extends from Kuybyshev W to Mozyr in Byelorussia, where it divides, with 28-inch lines extending to Poland, East Germany, Hungary, and Czechoslovakia. During 1960–1964, a 28-inch oil pipeline was laid from Tuymazy in the Volga-Ural fields 3700 km (2350 miles) E to the Angarsk refinery near Irkutsk. It was the original plan to extend this line to Nakhoda on the Pacific near Vladivostok, but this extension has been deferred.

In the late 1950s and early 1960s the Shebelinka gas field of the Ukraine and the Stavropol and Krasnodar gas fields were tied by pipelines to Moscow, Kiev, Leningrad, Odessa, and other consuming centres. In 1960, following the discovery of the Gasli field in Central Asia, construction commenced on the 40-inch, 2200 km (1370 mile) Bukhara–Ural gas line to Chelyabinsk and Sverdlovsk. The line was completed in 1965, but in 1964 work was already started on looping the line. In 1965 work started on a 2750 km (1705 mile), 40-inch line from the Central Asian gas fields to Moscow, and this was completed in October 1967. Delivery of gas through the line in 1968 was about 7·6 billion m^3 (245 billion cu ft). Looping of the line with 48-inch pipe was started, as well as an extension to Leningrad. The gas fields

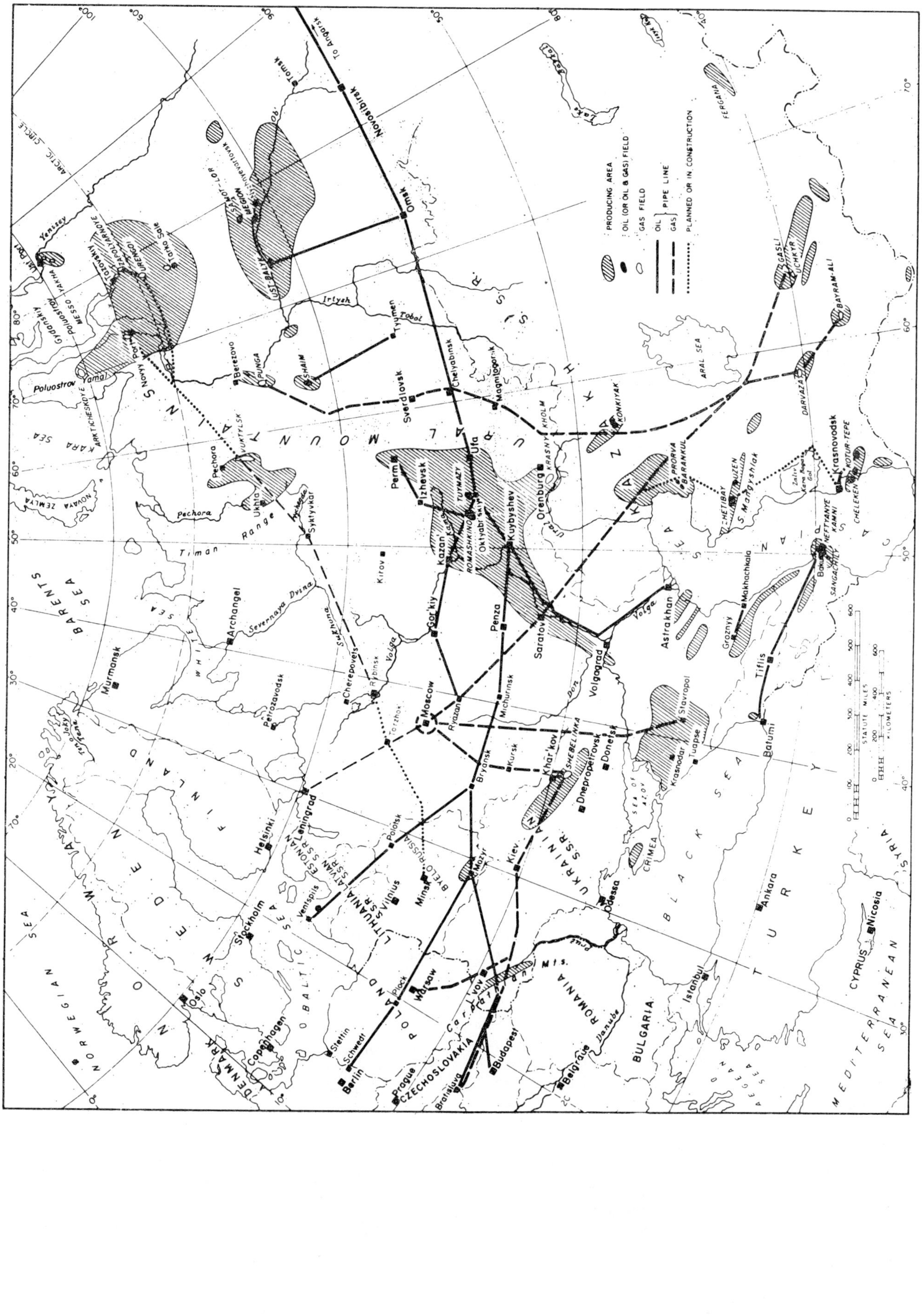

FIG. 2. Western Soviet Union—producing areas and oil and gas pipelines.

on the E side of the Caspian sea are eventually to be connected to this line.

A still more ambitious gas pipeline project, the Siyanie Severa (Northern Lights) line, has been started from the large gas fields of the northern West Siberian basin across Northern European Russia to Minsk. It will be 2750 km (1700 miles) long and of 48- to 56-inch diameter. The 1300 km (800 mile), 48-inch initial section of the line, from the Vuktylsk field in NE European Russia, to Rybinsk, 225 km (140 miles) N of Moscow, was completed in February 1969, and by the latter part of the year it was expected that another 225 km would be completed SW to Torzhok, where it will enter the main Moscow–Leningrad gas pipeline.

Restrictions on export of pipe from Western countries to the Soviet union delayed construction of oil and gas pipelines during the 1960s. The shortage was eventually partly overcome by completion of a mill for rolling large diameter pipe at Chelyabinsk in the Urals. A still larger pipe mill is under construction at Vol'sk, on the Volga near Saratov, and another near Dneipropetrovsk in the Ukraine. It is planned to manufacture 80- and 99-inch pipe for future lines, and experimental quantities have already been produced at Chelyabinsk. When the 99-inch pipe is utilized, it will transmit up to 100 billion m^3 (3·5 trillion cu ft) of gas, a ten-fold increase over the capacity of 40-inch pipe.

PRESENT AND POTENTIAL MARKETS

Oil production in the Soviet Union increased from 166 million metric tons (1162 million brl) in 1961 to 309 million tons (2163 million brl) in 1968. In 1967 Russian oil exports were 59 million tons, of which 33·5 were to Communist countries. Gas production in 1968 was 171 billion m^3 (6 trillion cu ft), and it has been increasing at the rate of 7 to 10 per cent per year.

Shortages of equipment of all kinds, particularly line pipe, and of skilled labour have retarded prompt development of some of the large new oil and gas fields. It has been difficult to provide incentives to the labour force to work in the austere environment of remote areas in Siberia and Central Asia, and there have been many complaints in the Soviet press about excessive concentration of effort in the more developed regions such as the Volga–Ural zone. However, oil and gas development is a leading Soviet economic objective, and high priority is given to exploration, drilling, and pipeline construction. Most of the major discoveries in recent years, and the great future potential, lie in Western Siberia and Central Asia, where costly construction projects are required to get the oil and gas to regions of major consumption.

The first oil pipeline from the West Siberian Basin fields was a 435-km (270-mile) line from the Shaim oilfield to the railhead at Tyumen, completed in December 1965. In October 1967, after 22 months of construction through muskeg and swamp, a 40-inch, 987-km (612-mile) pipeline was completed from the large oilfields of the middle Ob valley to the refinery at Omsk, on the Trans-Siberian pipeline. A 300-km (185-mile) extension E to Nyzhnyevartovsk was completed in 1968, tying in additional fields. The large Mangyshlak oilfields are so far connected only to a port on the Caspian coast. A pipeline will be built to Kuybyshev on the Volga, but this will require heating in order to be able to handle the high pour point crude. It will be several years before the major discoveries in new producing areas have full impact on domestic and export markets. Domestic fuel shortages in Russia are expected to persist until 1972, when large volumes of gas from Western Siberia will be available.

Plans are being made to extend the large gas pipeline from NW Siberia to Minsk onward to Czechoslovakia and Italy to supply gas in competition with Dutch, Algerian, Libyan, and perhaps Nigerian gas. A vital part of the project is for Russia to barter future gas supplies against 48- and 56-inch pipe to be delivered by Italian mills for construction of the line. Negotiations have so far bogged down in disagreement about the price and the interest rate to be paid for the pipe. The Soviets propose supplying 11·5–11·2 billion m^3 (405–425 billion cu ft) of gas per year at the Czech-Austrian border and 10 billion m^3 (353 billion cu ft) at the Italian border. As a prelude to the invasion of Western markets, on 1 September 1968, Russia started supplying gas to Austria from Soviet Ukraine fields at a daily rate of 1·3 million m^3 (45 million cu ft).

Barter negotiations have been in progress with Japan for the supply of liquefied methane from Northern Sakhalin Island to Japan, provided the liquefaction plant and related facilities are built by the Japanese. Again, the price of the gas and the interest rate have prevented completion of the deal.

With increasing internal consumption, the Soviet Union will undoubtedly absorb much of the increasing production of oil and gas during the next several years. A vast increase in fuel utilization for road transport, home, and industry can be foreseen. However, as the new oil and gas producing areas of the Soviet Union become fully developed and linked to markets by completion of the larger diameter oil and gas pipeline network, competition for markets in non-Communist countries will undoubtedly increase.

BIBLIOGRAPHY

The paper has been compiled from a number of sources. Among them are the following:

"Recent advances in the Soviet oil and gas industry." Anon. Paper 5 to the 3rd ECAFE Symposium on the Development of Petroleum Resources of Asia and the Far East, Tokio, November 1965. 12 pp.

"Methods of preparing forecast maps of oil and gas possibilities in the U.S.S.R." Paper 7, *ibid*, 14 pp.

"Oil and gas exploration in the West Siberian Lowland." Moscow, 1965. Anon. State Geological Committee of the U.S.S.R. Paper distributed at above symposium.

A valuable source of current information on developments in the Soviet Union is the newsletter *Erdöl-Dienst*, K. I. Turyn, publisher, Universitätstrasse 11, 1010 Vienna, Austria.

The following contributions from the U.S.S.R. (in English, except where noted) are published in the Proceedings of the Seventh World Petroleum Congress, Mexico, 1967:

Volume 2

Origin of Oil and Gas

"Hydrocarbons in the sedimentary mantle of the earth." N. B. Vassoevich, I. V. Visotsi, A. N. Guseva, and V. B. Olenin. Pp. 37–45.

New Oil Producing Regions

"New major oil- and gas-bearing areas of Young Platform of the U.S.S.R. (Western Siberia, Mangyshlak, etc.). M. F. Mirchink, N. A. Krylov, A. I. Letavin, L. I. Rovnin, V. P. Tokarev, M. V. Feigin, and Y. G. Ervie. Pp. 101–108.

"Gas- and oil-bearing provinces of Siberia." E. A. Bazanov, F. G. Gurari, V. D. Kozyrev, V. D. Nalivkin, Yu. A. Pritula, L. I. Rovnin, M. Ya, Rudkevich, A. A. Trofimuke, and S. G. Sarkisyan. Pp. 109–20.

"New oil- and gas-bearing region—the Prepyatsky Depression (The Bylorussian S.S.R.). V. A. Avvukumov, P. V. Ancupov, K. I. Gnedin, and G. H. Dickenstein. Pp. 121–27.

Geochemistry

"Geological-geochemical conditions of the transformation of organic matter in the sediments of the Caspian Sea." A. A. Ali-Zade and P. A. Shoikhet. Pp. 517–31.

"Oil and gas pools formation in Cenozoic sediments of Azerbaijan in the light of the latest investigations." Sh. F. Mekhtiev. Pp. 543–51.

Processing Geological Data

"Mathematical techniques in geological analysis." A. B. Vistelius. Pp. 561–66.

New Geophysical Developments and Methods

"Investigation of deep crustal structure of the oil- and gas-bearing provinces of the Ukraine by means of seismic methods." S. I. Subbotin, V. B. Sollogub, E. K. Lossovsky, A. V. Chekunov, M. V. Chirvinskaya, I. A. Garkalenko, and P. T. Turchinenko. Pp. 685–96.

"Application of electromagnetic methods in oil and gas exploration in the U.S.S.R." A. M. Alexeyev, M. N. Berdichevsky, I. A. Bezruck, V. I. Fomina, K. I. Nickitenko, M. K. Polshkov, and L. L. Vanjan. Pp. 697–708.

Exploration for Oil and Gas in the Continental Shelves

"Oil and gas bearing capacity prospect of the U.S.S.R. offshore shelves." E. N. Alikhanov and V. F. Solovyov. Pp. 863–69.

Regional Problems—Exploration and Drilling (Special Reference to Offshore Prospects)

"Principles of oil and gas territories subdivision and the classification of oil and gas." Mrs N. Yu. Uspenskaya. Pp. 00.

Volume 3

Rock Mechanics in Oilfield Geology Drilling and Production

"Deformational properties of rock as applied to petroleum geology and deep hole drilling." L. A. Shreiner, B. V. Baydyuk, and N. N. Pavlova. Pp. 27–39.

Individual Papers: Drilling

"Scientific and technical aspects of slant hole drilling." M. P. Gulizade, K. B. Shakhbazbekov, G. M. Gevinian, M. N. Makhmudov, N. K. Babaiev, and S. O. Oganov. Pp. 123–30.

"Les voix dernières du devellopment de la technique du forage à la turbine." (In French.) R. A. Ioannessian, Pp. 131–36.

Fluid Flow Phenomena in Porous Media

"Capillary and surface effects during the movement of oil, water and gas in a porous medium." M. M. Kussakov. Pp. 201–12.

Techniques for Drilling and Production in Water Covered Areas

"Exploration drilling in deep-water areas of the Caspian Sea." S. A. Orudjev and N. S. Timofeev. Pp. 259–66.

"The development of oil and gas deposits in deep waters of the Caspian Sea." A. Asan-Nuri. Pp. 285–90.

"Investigations concerning problems of offshore oil field development." I. P. Kuliev. Pp. 000-00

New Methods in Secondary Recovery

"Water injection into oil formations (flooding)." A. P. Krilov and Ju. P. Borisov. Pp. 499–506.

MS Received 30 March 1969

Discussion

Chairman: C. WILLIAMS

Dr Hans G. Kugler (Texaco Trinidad Inc.): What is the latest explanation of the ring-shaped structures of Baku which I considered to be connected with large-scale flowage of Mio-Pliocene from the E end of the Caucasus?

Robert E. King: I am afraid I am not able to answer that question, but I am sure that Dr Kugler, from his experience in Trinidad, has had more experience than I with structures of this type.

John D. Moody (Mobil): Could you comment on the reported Cambrian discoveries in the vicinity of Lake Baikal, and on the facies of the Cretaceous in the northern W Siberian Basin?

Robert E. King: I did not comment on the Cambrian oil in the Lena Valley of Central Siberia, although I made reference to the Cambrian salt that occurs in that area. This is a very interesting discovery geologically, in the region N of Lake Baikal. The discovery well there flowed a rather large quantity of oil of favourable gravity; however, it appears that follow-up drilling was relatively unsuccessful in developing or establishing large reserves. There was a great deal of exploration in this region during the period immediately before and after making that discovery, but evidently not much is being done now because there has been virtually no report in any of the press services dealing with the Soviet Union of any activity for several years. This, of course, probably does not mean the work has been suspended, but it does seem to indicate that there has been no outstanding success. The Cambrian, I believe, dolomite pay in this area is below a thick salt layer and there has been diapiric flowage of the salt above the producing zone, so this has, of course, resulted in some very severe problems in seismic interpretation.

Regarding the Cretaceous reservoirs in the Ob and Tazi Gulf region of NW Siberia, I am afraid I have no specific information on the facies of these Middle Cretaceous sandstones and associated ardulaceous deposits; they are evidently marine, however.

Dr Anthony Reso (Tenneco Oil Co.): Do you know from what part of the Cretaceous system the Gasli gas field in Uzbekistan is productive? Could it be Cenomanian, *i.e.* the same age as the reservoir sandstones in the large Zapolyarnoye and Arkticheskoye gas fields?

Robert E. King: I believe they are both from the Middle Cretaceous. However, I do not believe that there is any connection. The geosyncline in which these sediments were deposited in Central Asia was not closely related to that of the W Siberian Basin. They are far apart and in the boreal realm in W Siberia, which is probably quite different from the Tethyan facies of the Cretaceous in Central Asia.

G. Williams (Shell U.K. Exploration and Production Ltd): With the many sedimentary basins at their disposal to explore in some detail, what methods are the Russians using to evaluate an order of priority?

Robert E. King: The map (Fig 1) was adapted from a Russian map in which there is a very complex system of grading the order of merit in different basins. I simplified it a great deal. It has some 15 classifications ranging from the best to the least favourable and I suspect that this is based on a similar type of system to that employed by many oil companies in the U.S.A., in Britain, and so on, in which various parameters are evaluated and numerical ratings are given to the basins' percentage of marine sediments and so on. I have the impression that the methods that they use are probably quite similar to those we use. Of course, there is another factor too, when it comes to exploration, that there is a strategic advantage in certain areas such as the island of Sakhalin, a very important area because it is the only source of oil and gas in the Far East. They are also trying to explore in some of these small intermontane basins in the Amur River region, which, perhaps, if it were not for their strategic importance, might not be given so much attention.

D. H. Peel (British Petroleum Co. Ltd): Many of us in the West have tried to piece together the scanty information available as to the production quantities and crude oil qualities of individual fields in Russia. However, with such great distances between many of the new fields (particularly those in Siberia) and their industrial outlets in the West, necessitating construction of very long and very large capacity pipelines of the "common carrier" type, it would appear that much blending together of crudes from a host of separate fields will be inevitable. I would like to ask Dr King whether he knows of any geological or other evidence that might lead him to conclude that crudes from the same basin or area would also be of similar quality?

Robert E. King: I am afraid I cannot give a satisfactory answer to that question. Once in a while the question has arisen as to the characteristics of the crude oils in certain fields and areas in the Soviet Union, and it has taken a great deal of research to get accurate information on this. Precise information is seldom given in Russian publications. They ordinarily refer to high quality oil, and that is about it. I have the impression that the fields in the Ob Valley, now being actively developed, and from which oil is now supplied to Omsk by pipeline, have a rather uniform gravity in the neighbourhood of 35° API oil and of quality. The high paraffin content, resulting in the high pour point, of the Mangyshlak Peninsula fields is evidently such an important characteristic and is such a severe problem in utilizing these crudes and transporting them to refineries, that it has been mentioned very frequently, but with little further detail.

A Review of Petroleum Development in Scandinavia*

By THEODOR SORGENFREI

(*Institute for Applied Geology, Technical University, Copenhagen*)

SUMMARY

The exploration for hydrocarbons in Denmark, Sweden, and Norway is reviewed for the period 1935–1969. The geological results are condensed to brief outlines of the geology of the Baltic Basin, the North Sea Basin, and the Basin of the Norwegian Sea.

Non-commercial shows of hydrocarbons occurred in Upper Permian dolomites in Denmark and in Mesozoic and Palaeozoic formations in S Sweden and on the island of Gotland. Hydrocarbons were reported from five wells in the Danish North Sea shelf and from three wells on the Norwegian shelf. No pertinent geological data have been released on these wells except for limited information on the Phillips discovery on the Norwegian shelf.

Data on the onshore geology have been compiled to cross-sections, geological maps, palaeogeographic maps, etc., in order to synthesize the variety of geological features.

INTRODUCTION

THE Scandinavian sedimentary basins attracted the interest of the oil industry as early as 1935, when the first exploration well was spudded in Denmark. Some years later, in 1938, exploration work was started in S Sweden by the Geological Survey of Sweden. In hard-rock dominated Norway, however, no oil exploration activity took place until the race for the potentialities of the North Sea Basin had been initiated. The first Norwegian offshore well was drilled in 1966.

An outline of the geological results obtained during the exploration period from 1935 to 1959 in Denmark has been published by Sorgenfrei and Buch (1964). The exploration work of the Geological Survey of Sweden has not been reported comprehensively, while the stratigraphy of the tests at Höllviken has been treated in papers by Brotzen (1945, 1950), Lundblad (1949), and Ödum (1953). The preliminary report by Anderegg, Norling, and Skoglund (1968) outlines recent exploration work of the Geological Survey of Sweden.

Information about the results of the second exploration period in Denmark, which started in 1962, and the exploration on the Norwegian North Sea shelf is extremely restricted. None of the operating companies have actually released their geophysical and geological data up to now. The present review is, accordingly, not a real up-to-date account of the petroleum development in the Scandinavian area. It is exclusively based on the publicly available information and printed papers of the geology of the area.

The exploration situation as of May 1969 may be summarized as follows.

In the land area of Denmark a total of 36 exploration wells have been drilled since the start of the exploration in 1935. The location of the wells is shown on Fig 1.

Traces of oil have been encountered in dolomites of the Upper Permian Zechstein in three wells in S Jutland near the German border. Gas shows have, moreover, been reported from two wells drilled into salt domes in N Jutland. Also in these cases the gas occurred in Zechstein dolomites. The remaining wells were dry holes.

In the Danish North Sea offshore 11 wells have been drilled since 1966. Five wells encountered oil and gas, and four of these wells were tested.

In the Norwegian part of the North Sea a total of 21 wells have been completed, while three additional wells were drilling in May 1969. Three of the completed wells encountered hydrocarbons.

The location of the wells in the North Sea is shown on Fig 2. The total depths are indicated, if reported by the companies.

In Sweden, the efforts have been concentrated on Scania and on the island of Gotland in the Baltic.

A total of nine shallow core holes and seven deeper boreholes have been drilled in Scania. Weak traces of oil were encountered in the top part of the Triassic Bunter sandstone in the Höllviken 2 well and gas shows were reported from the Muschelkalk and Bunter in the Ljunghusen well.

On the island of Gotland and Gotska Sandön seven shallow core holes have been drilled. Oil traces have been reported from Ordovician limestone and Silurian sandstone.

The location of the Swedish wells in Scania and Gotland is shown on Figs 3 and 4 respectively.

GEOLOGICAL RESULTS

Regional setting

The sedimentary basins of the Scandinavian region developed around the old Pre-Cambrian Shield of Fennoscandia, which includes the Pre-Cambrian of the Scandinavian peninsula and Finland.

At the beginning of the Cambrian, the so-called

* This paper was not preprinted and was presented at short notice in place of that by Sokolowski and Tskarski.

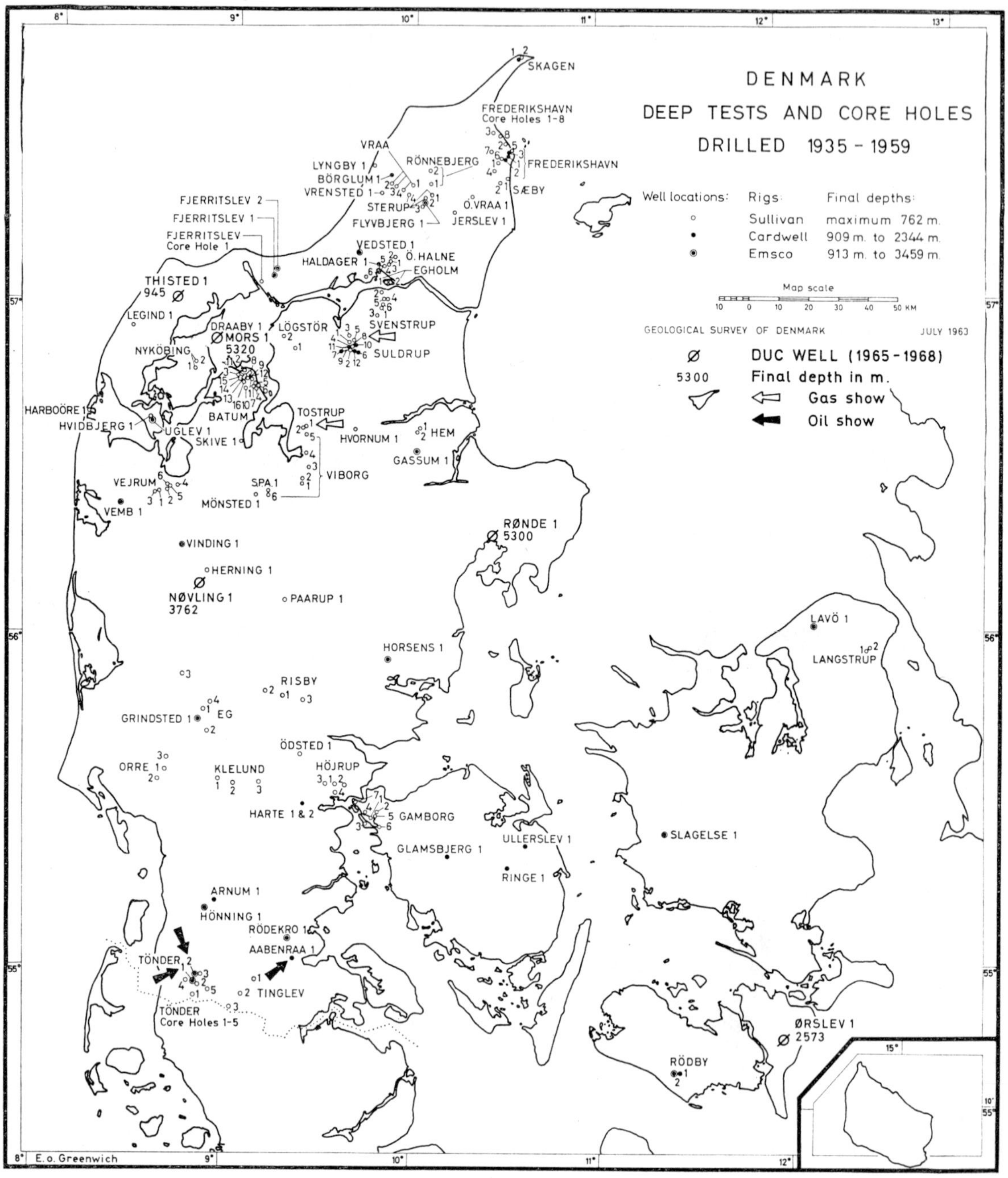

FIG 1.

Central Fennoscandian Continental Torso extended from the Oslo region NNE through the Scandinavian peninsula. The Gulf of Bothnia, S Sweden, and the Baltic were flooded by a shallow, epicontinental sea, while the Caledonian geosynclinal basin was developing NW of the Continental Torso. It is noticeable that at that time the Fennoscandian Shield was obviously subdivided into various structural units: the geosynclinal region of Norway, the Continental Torso of Sweden, the sedimentary basin of the Baltic and the Gulf of Bothnia, and the Pre-Cambrian region of Finland.

These regional units may have persisted throughout the lower Palaeozoic until the end of the Silurian, when the Caledonian folding took place, and the Caledonian Chain was built on the NW flank of the Continental Torso.

During the Devonian, a new cycle started. The Caledonian mountains were eroded, and a large continental depositional area developed away from the Fennoscandian Shield in W and E Europe while the ocean withdrew southward to central Europe.

The second important orogenetic event was the Variscan mountain-building at the end of Carboniferous time. The Armorican–Variscan Mountain Chain of W and Central Europe was formed during this period, and the Ringköbing–Fyn High emerged in Denmark. Another outcome of the Variscan orogeny was the activation of the NW–SE trending Fennoscandian Border Zone, which probably had

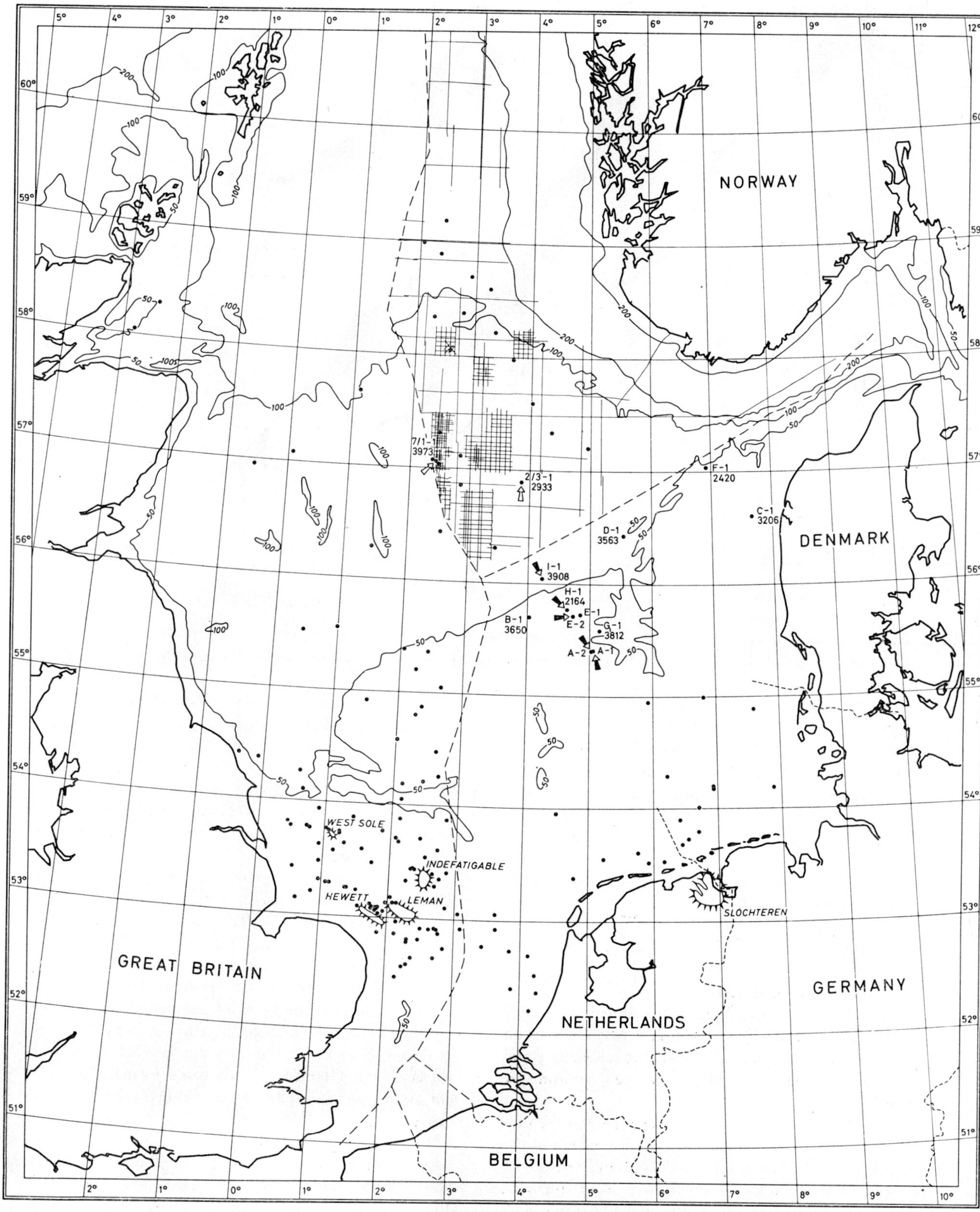

NORTH SEA AREA. OUTLINE MAP WITH MAIN GAS FIELDS AND WELLS DRILLED UNTIL DECEMBER 1968 *)

Continental shelf boundary — Water depth — Phillips Norway Seismic profiles — Well location with final depth in m. (G-1 3812) — 0 50 100 150 200 Km

Gas — Gas and oil — Gas field — *) Norwegian - Danish Shelf: until May 1969

FIG 2.

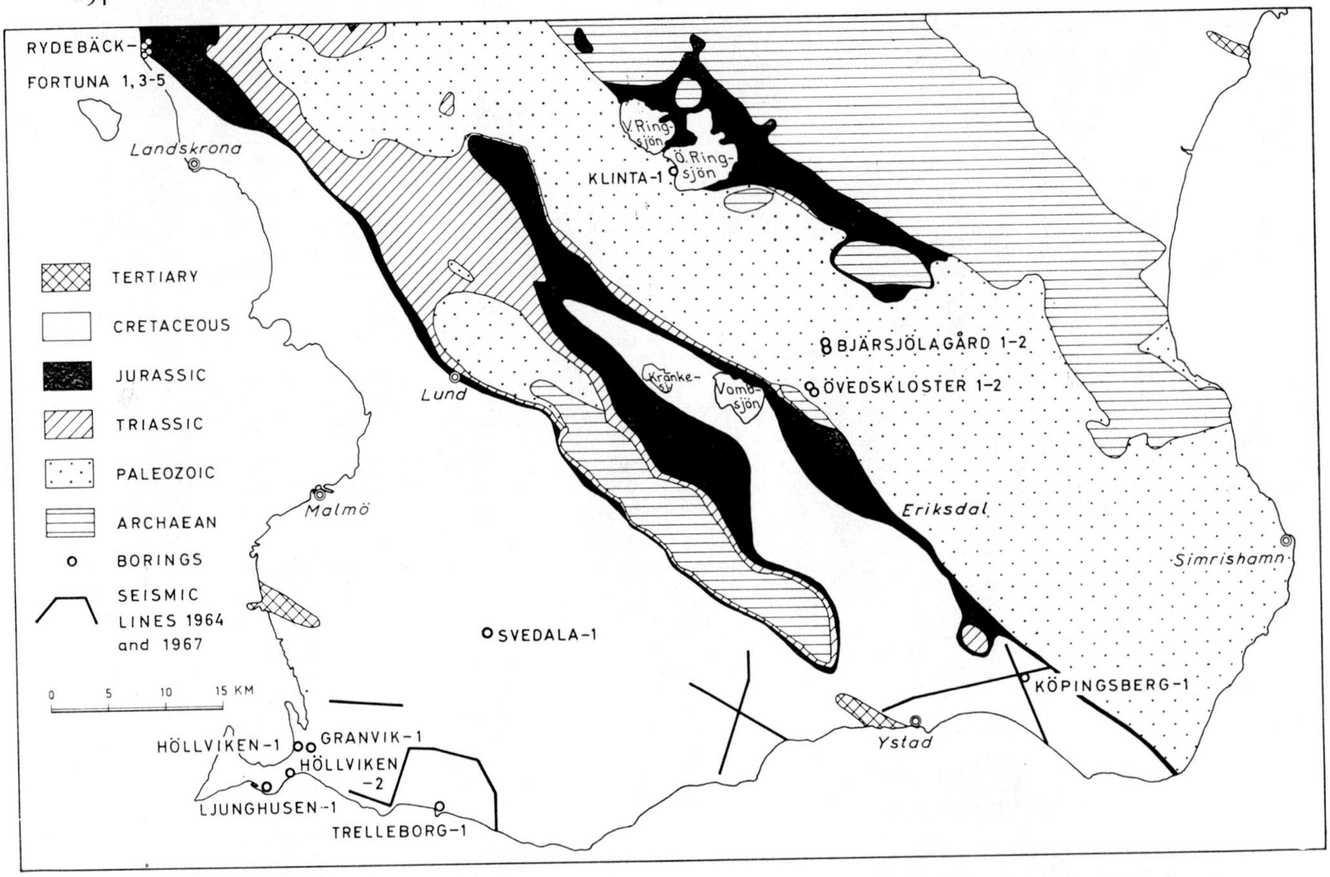

FIG 3. Geological map of Scania. Pre-Pleistocene formations and well locations. Geological Survey of Sweden.

already been laid out as a Caledonian syn-orogenetic feature.

From now on the Fennoscandian Shield was apparently a stable area of sedimentary supply. Very rarely was this area later submerged under the sea, and when it happened only small fringing bordering areas at the edge of the shield were included in the subsidence.

The regional features of the Fennoscandian Shield, its SW border zone, the Caledonian Chain, and the Ringköbing–Fyn High have to a large degree controlled the developing sedimentary basins.

The Sedimentary Basins

As was pointed out above, the sedimentary basins of the Scandinavian region initially covered part of the Fennoscandian Shield. From the Devonian and onwards, the basins chiefly developed around the shield.

The Baltic Basin is situated between the Fennoscandian Shield and the shallow Pre-Cambrian of the East European Platform, in E Poland and the Soviet Union. The Fennoscandian Border Zone forms a natural limitation of the Baltic Basin to the SW. However, the Palaeozoic formations cross this line, thereby indicating open connections between the Baltic Basin and the North Sea Basin and rather uniform sedimentary environments in the two basins.

It is convenient and logical to define the North Sea Basin as the triangular region limited by the Fennoscandian Border Zone to the NE, the Caledonian Chain to the NW, and the Variscan orogenic belt to the S. The Ringköbing–Fyn High and the Mid-North Sea High divide this basin into a number of sub-basins, namely, the Danish Embayment, the Norwegian–Danish Basin, the British Basin, and the Dutch–German Basin.

Between Norway and Scotland the North Sea Basin is connected with the Basin of the Norwegian Sea.

The Baltic Basin. The Palaeozoic Baltic Basin is a wide trough. The regional trend is NE–SW, and the axis of the basin is situated on the S shores of the Baltic in the Kaliningrad region. Here we may expect maximum thickness of sediments which may exceed 2000 m (Bölau 1959, Martinsson 1960).

To the N in Estonia and on the Swedish islands of Gotland and Öland, the Palaeozoic formations crop out at the surface. They range from the basal Cambrian to the Llandovery of the Silurian.

The core holes drilled on Gotland and Öland have proved more than 500 m of sedimentary section. Some of the bores bottomed in the Pre-Cambrian. The regional dip is SE. In the deepest part of the Basin the Silurian is succeeded by Devonian, Permian, and Mesozoic sediments. Whether these formations also reach the Swedish offshore is not known at present (Anderegg *et al.* 1968).

The Palaeozoic sequence displays several interesting features (*cf.* von Bubnoff 1959, Bölau, Martinsson).

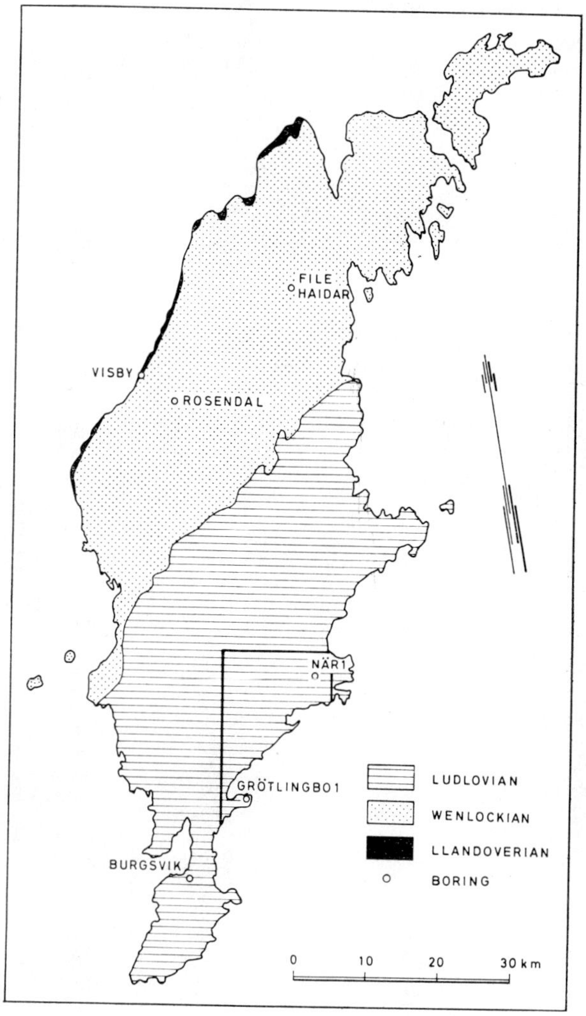

FIG 4. Geological map of Gotland. Pre-Pleistocene formations and well locations. Geological Survey of Sweden.

The Cambrian. Sandstones dominate in the Cambrian. In Estonia, the thickness is about 130 to 180 m. In the N part of the island of Gotland, the thickness amounts to 157 m and in the S part of the island to more than 160 m. The sandstones are interbedded with shaly intervals.

The Ordovician and the Silurian have developed in a different way. In the E and N parts of the basin, the sequence consists almost entirely of limestones, while both limestones and shales occur to the S and W.

In Estonia, the thickness of the Ordovician–Silurian limestones approaches 550 m. The sequence includes the bituminous kerogen marls of the Kukkersit stage (*ca* 20 m) referred to the *Chasmops* zone and correlated with the Llandeilo of the British Isles. The Kukkersit does not apparently extend into the Swedish area, where the corresponding biostratigraphic zone has developed as a limestone. However, the upper half of the 94-m thick series of the Ordovician limestones of N Gotland includes bituminous algal limestones with traces of oil. In S Gotland more than 50 per cent of the Silurian consists of bituminous shales, and in the overlying Burgsvik sandstone of Ludlowian age traces of oil were encountered in the Grötlingbo well.

The thickness of the Ordovician limestones on Gotland may be about 100 m and the maximum thickness of the Silurian shales and limestones may be on the order of 300 m in S Gotland.

A general feature of the Ordovician–Silurian of the Baltic Basin is the abundance of nonconformities, particularly in the Ordovician part of the sequence. The nonconformities correspond to rhythmic emergences of minor or larger parts of the basin. The basal beds above the nonconformities are generally glauconitic and phosphoritic. In the S part of the Baltic Basin, the sequence may be sub-divided into sedimentary cycles. A complete cycle includes three members: a basal phosphoritic and glauconitic conglomerate followed by a limestone interval and superposed by a shale interval. Either the limestone or the shale interval may be reduced or missing.

The Devonian of Latvia and Lithuania is developed in the Old Red Sandstone facies. The thickness varies between 250 and 400 m. In the northernmost parts, the typical continental ORS facies occur, while limestone and dolomite intervals, partly with gypsum, are intercalated in the sequence to the S. The sequence falls naturally into three sub-divisions: a lower and upper sandstone division and a more or less marine limestone–dolomite division in between. The lower sandstone is referred to the Middle Devonian and the two formations above are included in the Upper Devonian.

The lithological succession of the Baltic Devonian thus shows the most striking similarity with the Triassic of the North Sea Basin, which of course only suggests similarity of sedimentary environments.

While there is a faint possibility of the Devonian extending into the Swedish offshore, the late Palaeozoic and Mesozoic beds do probably not occupy Swedish areas except for the southernmost region at the Fennoscandian Border Zone. The latter divisions are therefore treated in the following chapter.

Structure. Little can be said about the local structures of the basin, except that it is reasonable to expect some block faulting.

The dominant joints of the joint system of Öland and Gotland strike SW–NE. They are crossed by NW–SE striking joints.

Local uplifts are found on Gotland but the structural type of these uplifts is not known.

The North Sea Basin. In accordance with the definition given above, the North Sea Basin includes the areas of Sweden lying SW of the Fennoscandian Border Zone, Denmark and the Danish waters, Northern Germany, the major part of the Netherlands, E England, and the North Sea.

For various reasons, data on the offshore geology in the Norwegian and Danish parts of the North Sea is much limited. Heybroek, Haanstra, and Erdman (1967) have, however, given an excellent outline of regional geological features of the North Sea Area to which reference is made. By means of the maps

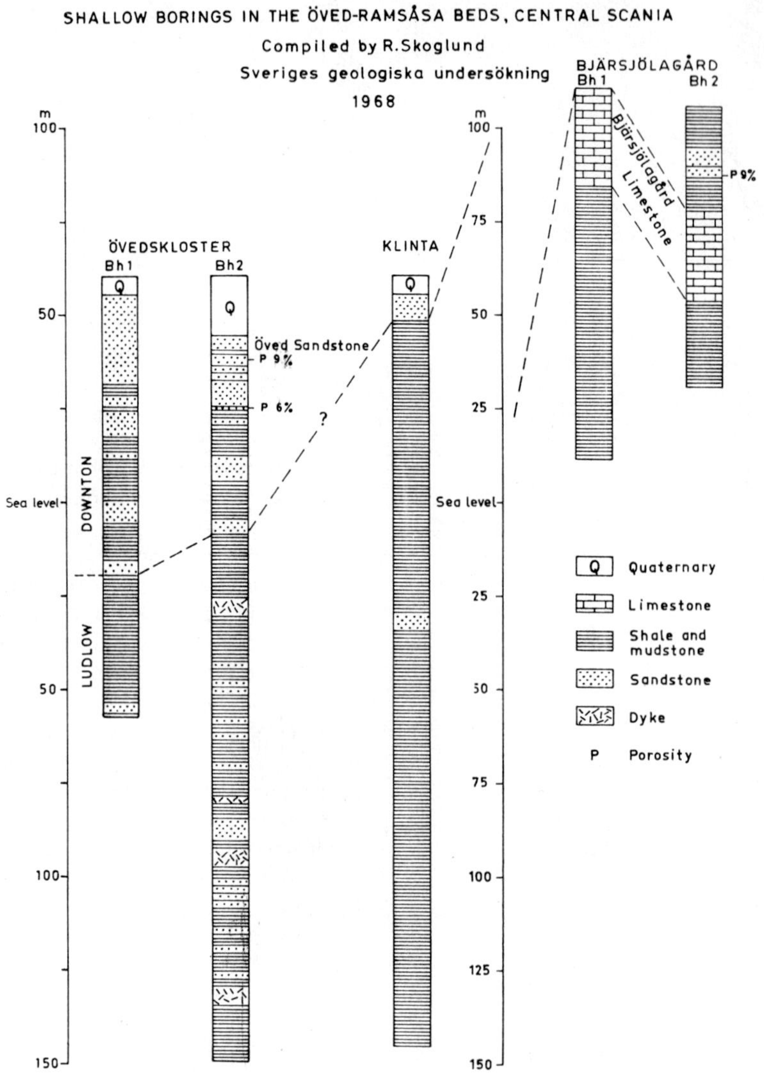

FIG 5.

accompanying the paper it is possible to draw some general conclusions regarding the potential extension of onshore sedimentary and structural features.

The following brief summary of the geological development in the land areas of Denmark and SW Sweden may illustrate some of the significant general features of the NE part of the North Sea Basin. In order to obtain a picture of the variety of geological data, the outline is relying on a number of cross-sections, generalized geological maps, etc.

The Cambrian. As it is in the Baltic Basin, the Lower Cambrian is chiefly represented by sandstones. Outcrops are found in the area of the Fennoscandian Border Zone, namely, in the Danish island of Bornholm and in Scania. To these areas may be added the Oslo region in Norway, where a reduced section is found.

In Bornholm, the basal part of the Lower Cambrian is a reddish continental arkose of 100 m thickness. Above follow about 60 m of Balka quartzite, a partly glauconitic sandstone. It is superposed by 60 m of dark greenish, marine siltstone. Six m of sandstone form the top member of the Lower Cambrian. The Middle and Lower Cambrian also parallel the conditions in the Baltic Basin. Except for two thin limestone beds at the base, the entire sequence consists of Alum shale. The thickness is 28 m.

In Scania, the Lower Cambrian sandstone is more than 125 m thick and the Middle and Upper Cambrian shales are *ca* 86 m thick. Farther W, on the island of Sjælland, about 50 m of Cambrian Alum shale have been encountered above sandstone.

The Cambrian basin undoubtedly originally covered large parts of the Danish land and sea areas. Cambrian beds are absent on the shallow basement structures of N and Central Jutland, but Cambrian sequences must be expected in the structurally low-lying portions of the subsurface in Denmark and the Danish waters.

The Ordovician and the Silurian. The Ordovician and the Silurian may cover approximately the same general areas as the Cambrian. Shales of the same type as found in the Upper Cambrian and thin limestone intervals evidence the continuation of uniform environment conditions throughout the Ordovician and the Silurian in Denmark and S Sweden.

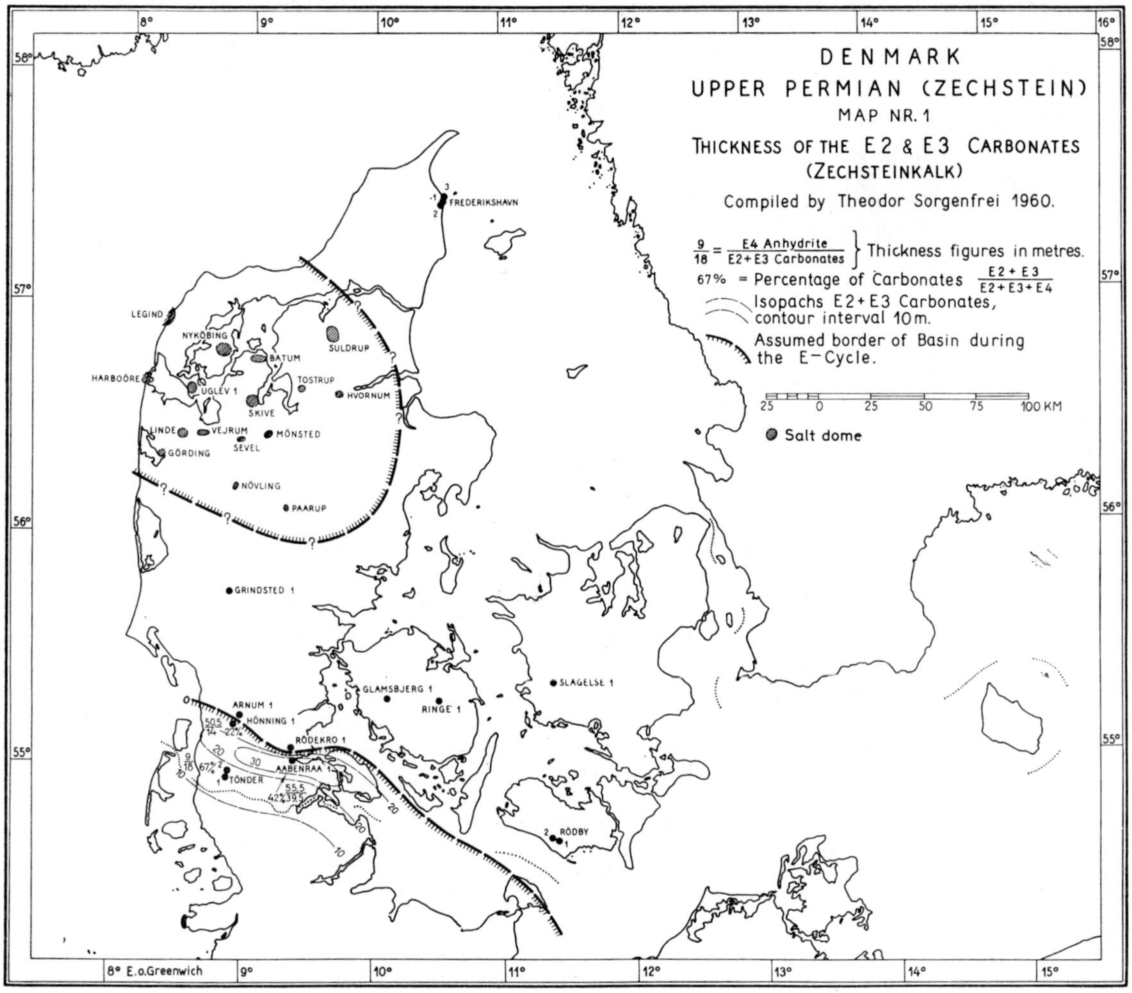

FIG 6.

A total of about 100 m of sediments is found on Bornholm. The sequence consists of shales, except for 5 m of limestone near the lower boundary. The Ordovician and the Silurian up to the top of the Cyrtograptus shales are represented. The corresponding interval in Scania is about 150 m thick. Sedimentary cycles and nonconformities as found in the Ordovician–Silurian of the Baltic Basin are observed.

A very thick sequence of about 1000 m of Colonus shales and about 300 m of marlstones, limestones, and sandstones of the Öved-Ramsåsa group follow above the Cyrtograptus shales of Scania. Fig 7 illustrates the lithology of the topmost part of the section in Central Scania. The location of the boreholes is shown in Fig 3.

The remarkable thickness of the Silurian of Scania may be due to contemporaneous faulting in the Fennoscandian Border Zone at the onset of movements in the Caledonian geosyncline to the NW.

In the island of Sjælland, the Slagelse well proved about 334 m of dipping Ordovician–Silurian shales. The upper part of the beds may correspond to part of the Colonus shale.

The Ordovician and Silurian of the Oslo region is characterized by the interbedding of shales and limestones and the greater abundance of the limestones as compared with Denmark and Scania (Henningsmoen 1960, Störmer 1967).

The Devonian and the Carboniferous. The topmost division of the Öved-Ramsåsa group mentioned above has been referred to the Downtonian which is commonly regarded as a passage group between the Silurian and the Devonian.

Thick Carboniferous sections have been encountered in the island of Rügen in East Germany adjacent to Scania and the southernmost Danish islands. It is, therefore, reasonable to expect an extension of the Carboniferous basin S of Denmark into the Scandinavian area.

The Permian. Above the base of the Permian in Denmark and above the base of the Mesozoic in Scania, the sedimentary column displays an abundance of similarities with the formations of N Germany and E England. It is, therefore, practical to some extent to rely on the well-known stratigraphy in these neighbouring regions.

The main differences are due to the controlling influence of the Ringköbing–Fyn High (RFH) on sedimentation and palaeogeographic development in the Danish area.

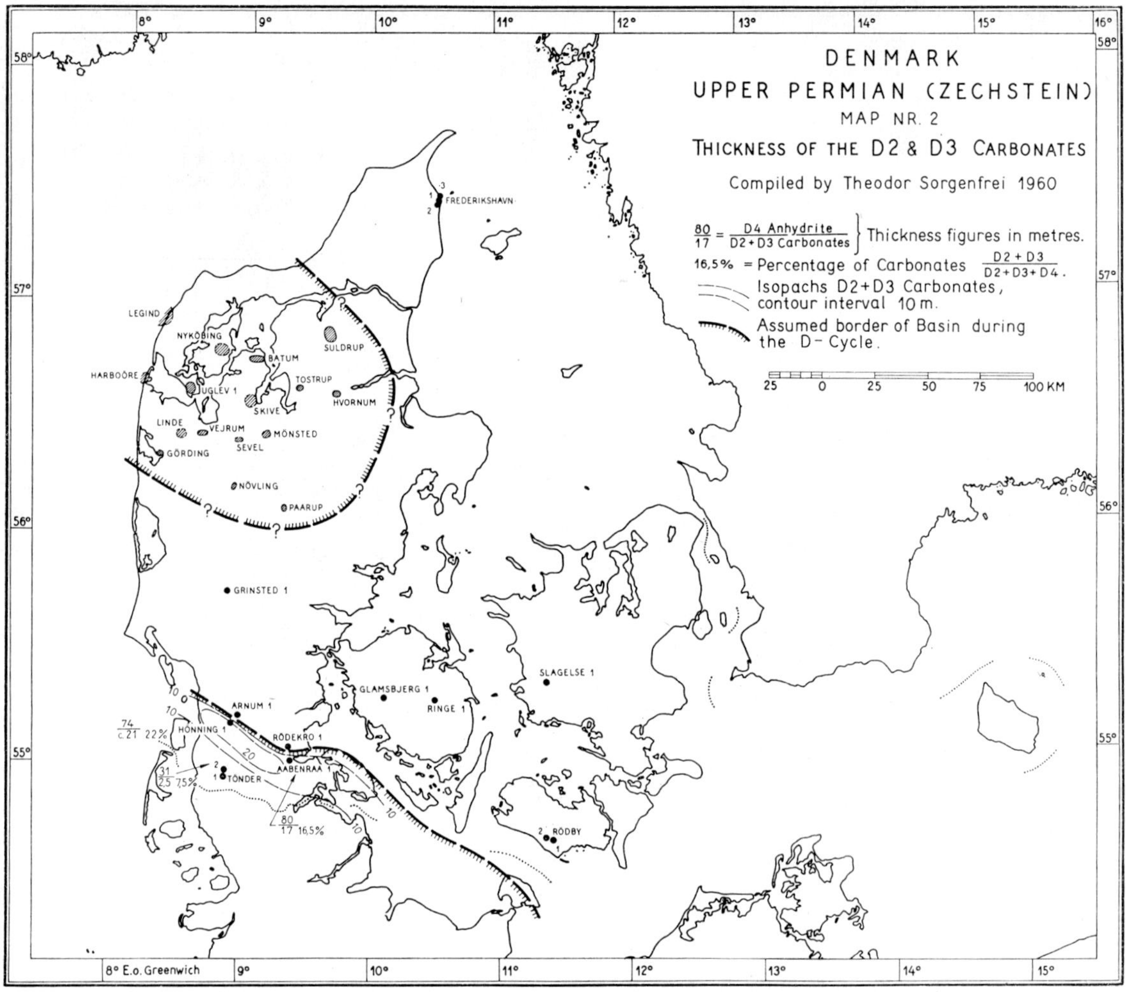

Fig 7.

The Lower Permian Rotliegendes is very similar to its counterpart in Germany. Continental redbeds, chiefly siltstones and claystones with an intercalated rock salt interval near the top, have been encountered south of the RFH. Volcanic extrusives are found below the redbeds in the Rödby No. 2 well about 120 km away from the nearest occurrence of Lower Permian volcanites in E Germany. No significant sandstone intervals occur in the drilled areas in S Denmark.

The Upper Permian Zechstein deposits surround the RFH. In S Denmark, it has been possible to define five evaporite cycles which have been lettered A to E from top to base Zechstein. The normal succession of deposits in a cycle have been numbered in ascending order as follows: (1) clastics, (2) limestone, (3) dolomite, (4) anhydrite, (5) rock salt, and (6) potassium-magnesium salts.

A normal succession may be continued by a reverse succession starting with a rock salt member (7) and ending with a clastic member (10). This happened, for instance, partly in the C-cycle.

To exemplify:

The E 3-dolomite is the dolomite member in the deepest evaporite cycle.

The E-2-limestone correlates with the Zechsteinkalk, the C 3-dolomite with the Hauptdolomite, and the B 3-dolomite with the Plattendolomite in Germany.

Oil and gas shows have been encountered in the D 2–3, C 2–3, and B 2–3 intervals.

The presence of Zechstein evaporites in the Danish Embayment N of the RFH is evidenced by the salt domes of N Jutland. However, the stratigraphy in this part of the basin is rather obscure, due to the strong contortion of the evaporites in the domes.

The isopach maps (Figs 6–11) illustrate the development of significant evaporite intervals of the Danish Zechstein where data are available. Without going too much into detail, it should be mentioned that the dolomite intervals were generally rather impermeable except for a Bryozoan limestone interval in the E 2 member of the Aabenraa No. 1 well. An interesting feature illustrated by the maps of the E, D, C, and B carbonates is the development of fringing carbonate reefs on the S flank of the RFH.

The Triassic. The three main divisions of the Triassic, the Bunter, the Muschelkalk, and the Keuper, are clearly recognized in the drilled sections in S Denmark and to some degree also on the RFH. North of the high, the Muschelkalk does not apparently occur in the typical marine limestone facies.

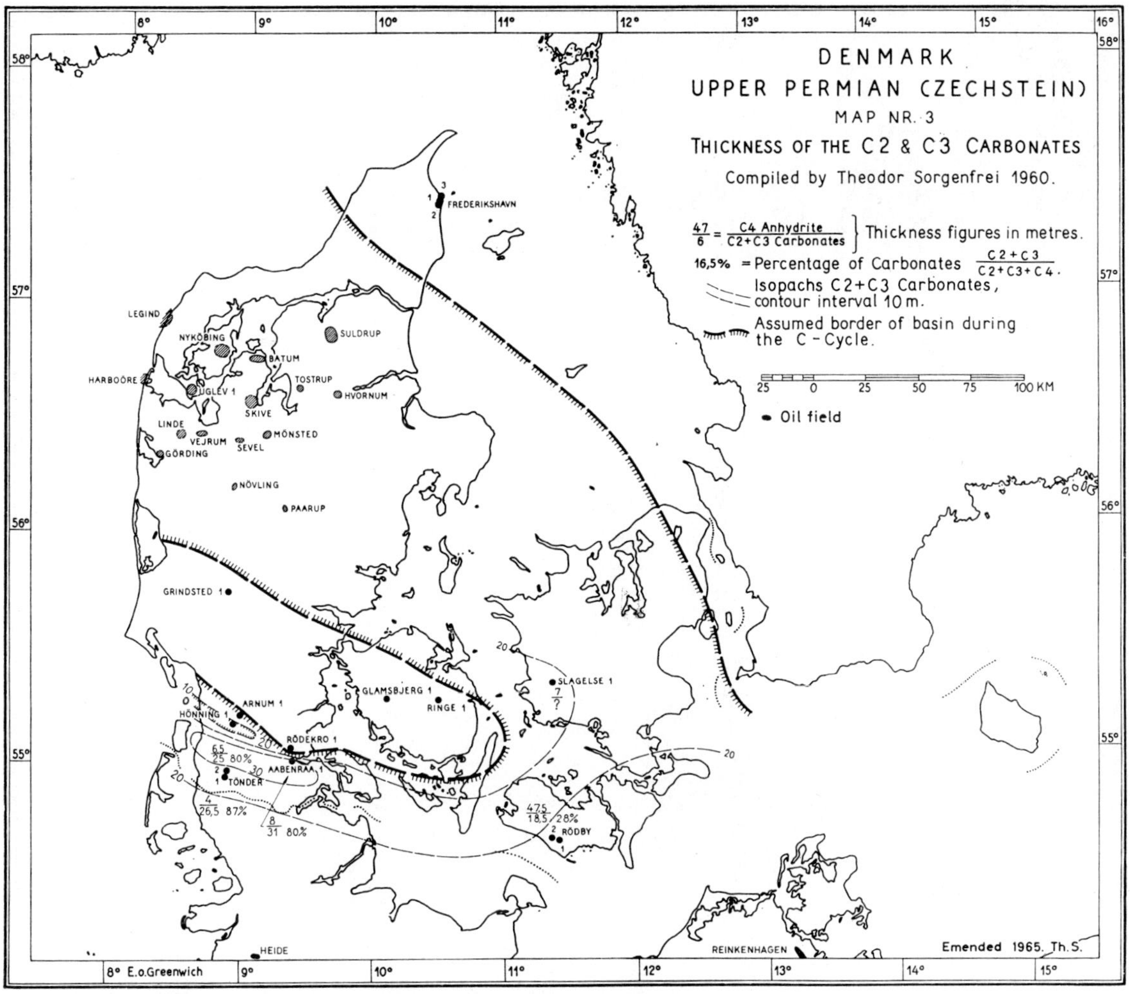

Fig 8.

Rock salt intervals in the Gassum No. 1 section suggest one or more lagunal evaporite basins in N Jutland at Muschelkalk time.

The Bunter sequence consists of continental red-beds. Pure sandstones are rare, while argillaceous sandstones, siltstones, and claystones are the rule. The topmost division of the Bunter, the Röt, represents an evaporitic stage of sedimentation in the area S of the RFH. In going from E to W, three belts dominated by dolomite, anhydrite, and rock salt, respectively, are evidenced.

The marine Muschelkalk does not exhibit particular features except for the facies differences already mentioned above. The Muschelkalk is a grey, argillaceous limestone or marlstone.

The Lower and Middle Keuper comprise reddish and grey claystones, occasionally with anhydrite, which appear to be rather uniform apart from colour differences. The lithofacies map of the interval shows that the grey claystones predominate in the areas of maximum thicknesses away from the RFH. Reddish and yellowish colours are the rule on the high. It suggests occasional sub-aerial environment conditions on the high during sedimentation or subsequent weathering during a period of emergence.

The Upper Keuper or Rhaetic actually represents the start of a new cycle of sedimentation which went on during the Jurassic and the Lower Cretaceous. The lithofacies map (Fig 12) demonstrates the distribution of the marine and deltaic facies as well as the sand/shale ratio. The Rhaetic sediments occupy the space between the RFH and the Fennoscandian Shield, and they apparently encompass the RFH.

The Jurassic. Except for the Middle Jurassic Dogger, which is brackish marine and deltaic, the entire Jurassic sequence is marine. Shales are by far the most dominant. The effect of the RFH and the Danish Embayment is obvious. Marginal areas par allel the Fennoscandian Border Zone and the RFH, and maximum thicknesses are found in the Danish Embayment (Sorgenfrei 1963).

Recent surveys in the Border Zone area have proved that the Middle and Upper Jurassic have a much wider distribution than shown on the previous maps. Reference is made to the paper by Gunnar Larsen (1966).

Of particular interest is the regional distribution and thickness of the Lower Jurassic in the entire North Sea Basin as demonstrated on Figs 13 and 15. These maps evidence the controlling effect of the

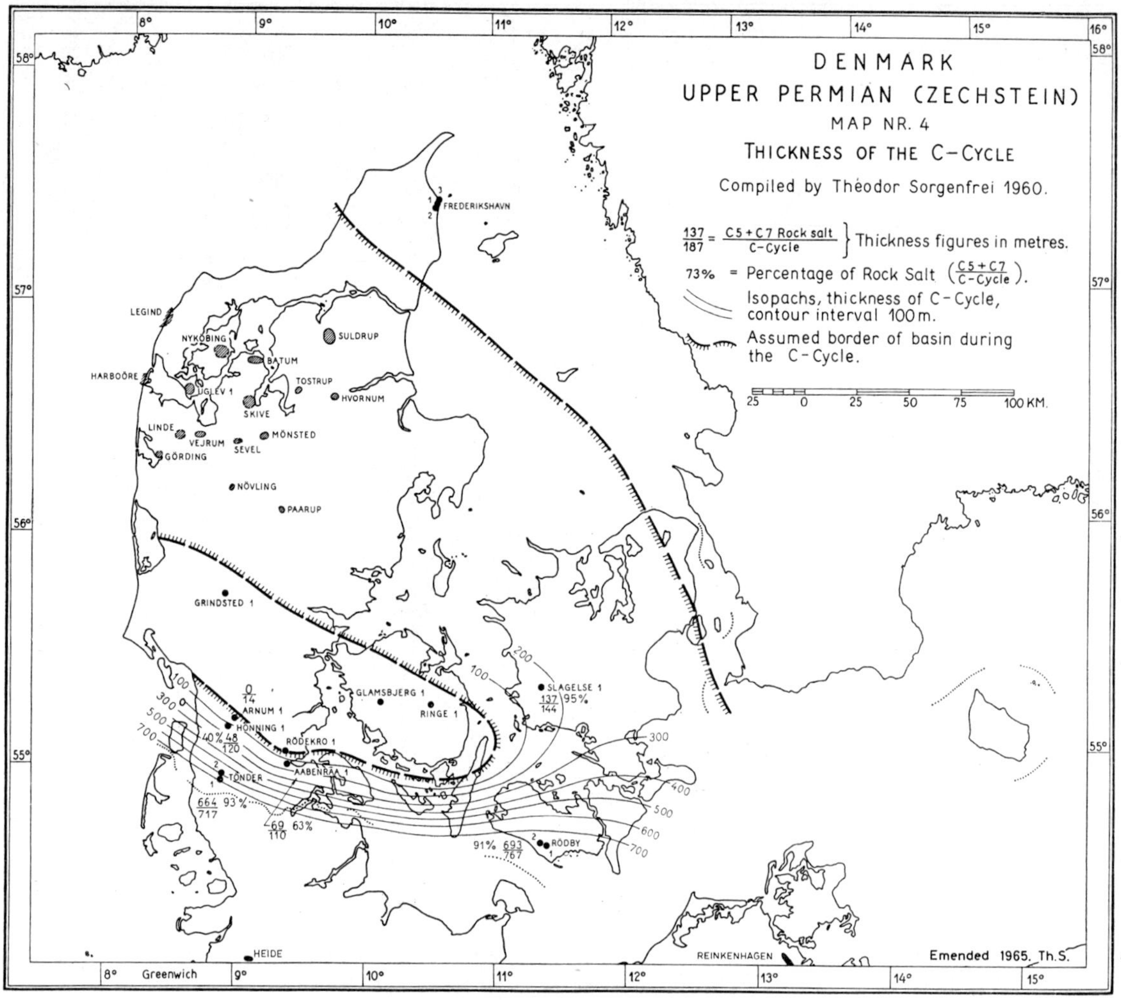

FIG 9.

structural framework on sedimentation. For further details on relations between the distribution of the Lower, Middle, and Upper Jurassic, reference is made to Sorgenfrei (1969).

The Cretaceous. The Lower Cretaceous lithofacies have been illustrated in Fig. 13. The RFH was submerged during Lower Cretaceous time. It is, however, noticeable that red colours and limestone facies are characteristic features of the reduced section on the high.

The Upper Cretaceous division is represented by a thick and uniform White Chalk and limestone blanket except for the marginal zone in Scania, where sandstone intervals occur (Fig 16). Weak gas shows have been reported from the Coniacian sands of the Köpingsberg well.

The Tertiary. No epoch-making event has been reported regarding the well-known Tertiary sequence of the Danish land area. In the case of the North Sea, the general configuration of the Tertiary basin has been outlined on a map published by Heybroek *et al.* On Fig 17, the contours of this map have been superposed on the map of the depth to the base of the Zechstein of the same authors in order to illustrate the relations between the area of maximum thickness of the Tertiary and the situation of the East Dogger Bank Graben, the RFH, and the other structural units in the North Sea area.

Structural features. The most conspicuous tectonic feature of Scania and Denmark is the NW–SE regional trend of the structural units. It is recognized in the Fennoscandian Border Zone, the direction of the Danish Embayment, the onshore portion of the RFH and the general pattern of the local structures.

Off the Danish North Sea coast, the RFH bends towards W, but the W edge of the high demonstrates the existence of the NW–SE structural pattern also in this part of the high. Little is known of the structure of the Norwegian–Danish sub-basin in the North Sea. However, NW–SE arrangement of local features is evidenced on the map of the Zechstein basin by Heybroek *et al.* (1967) in the W part of the Norwegian–Danish sub-basin (see also Sorgenfrei 1969).

The Fennoscandian Border Zone. Fig 18 clearly illustrates the features of the Border Zone in Scania. It is a strongly block faulted belt which is about 75 km wide. It consists of Pre-Cambrian basement horsts and deep grabens with thick sedimentary sequences which locally are slightly folded and faulted. The island of Bornholm SE of Scania, a

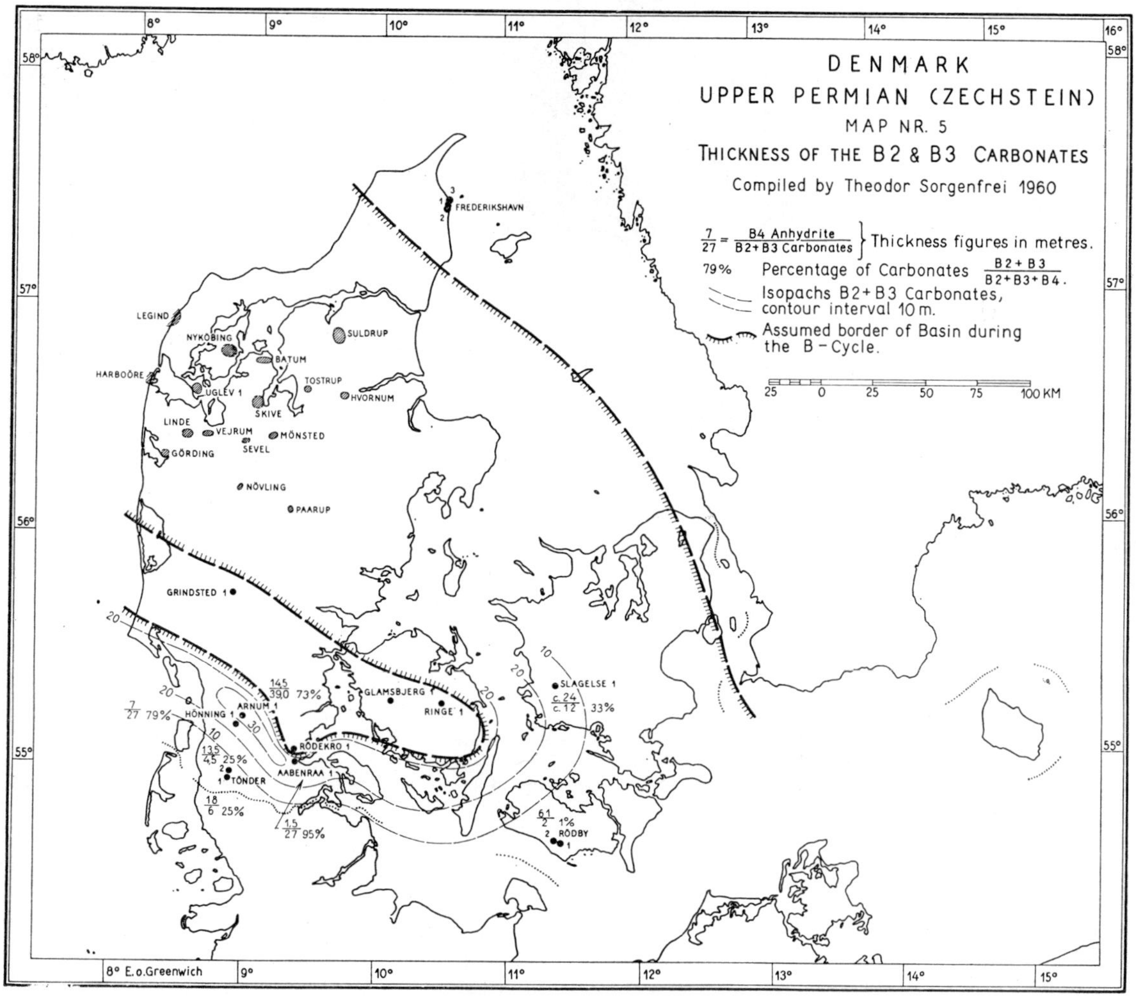

FIG 10.

basement horst surrounded by faulted and folded sediments, belongs to the same structural province. To the NW, the Vendsyssel High of shallow Pre-Cambrian covered with about 1300 m of Mesozoic sediments may represent the NW portion of the Border Zone.

The Danish Embayment follows SW of the Border Zone. The width between the Border Zone and the RFH is about 150 km. To the SE, the Danish Embayment merges into the deep part of the Polish basin, and to the NW it continues into the North Sea and the Skagerrak.

The deepest part of the basin is found in NW Jutland in the Salt Dome Province, where more than 5000 to 6000 m of sediments may be expected. The thickness may decrease somewhat toward SE.

The sediments of the SE part of the embayment are faulted and slightly folded. The same may be true for the NW part of the basin. However, here the section above the Zechstein salt is definitely influenced by salt tectonics which gave rise to the formation of piercement salt domes, salt anticlines, and interdomal features of the salt pillow type.

The Ringköbing–Fyn High is a complex structure of basement horsts and grabens. A typical feature of the high is the absence of Jurassic sediments and the strongly reduced thickness of the Lower Cretaceous. Conditions are relatively well known in the SE part of the high on the island of Fyn. NW–SE striking faults have cut the region into narrow horsts and grabens. Seismic surveys suggest that the graben fillings in this part of the RFH consist of Triassic and possibly older formations and that contemporaneous faulting occurred during the Triassic.

In Jutland, a N–S trending gravity low crosses the high. It is undoubtedly due to a downfaulted and partly folded area with relatively thick sediments.

The S flank of the high is strongly faulted in S Jutland, as shown on Fig 16.

The NE part of the German Basin. The sediments of the German Basin encroach upon the S flank of the RFH in S Jutland and on the islands south of Fyn and Sjælland. NW–SE trending salt anticlines and fault structures characterize this province, as evidenced by the Tønder and Rødby areas (see Fig 18).

Events of the tectonic evolution. The dating of the tectonic movements during which the outlined structures were built is still rather incomplete. It is, however, possible to throw some light on the tectonic history.

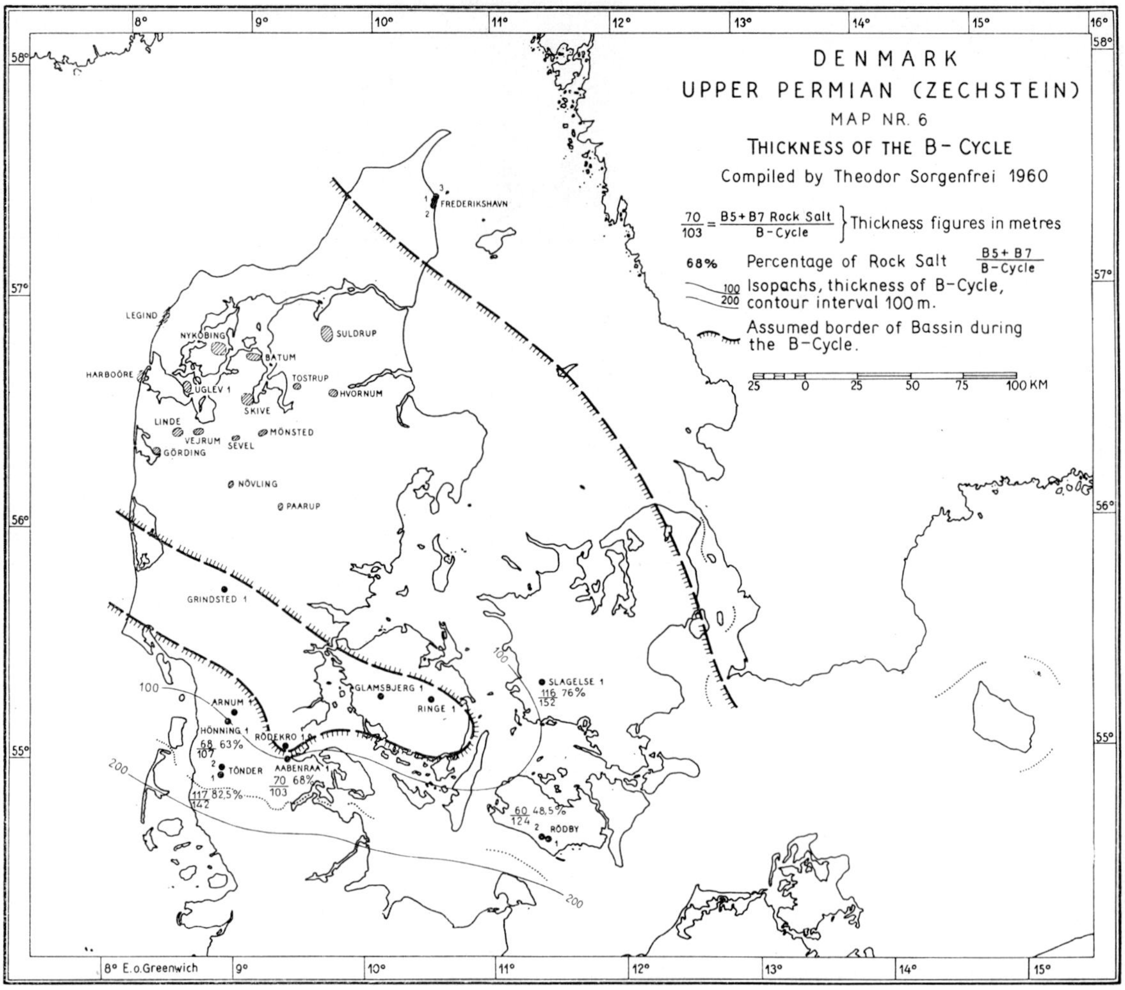

FIG 11.

Considerable interest is attached to the origin of the RFH. The angular unconformity between the lower Palaeozoic formations and the Permian of the Slagelse well clearly evidence Palaeozoic movements. However, the question whether these movements should be referred to the Caledonian or to the Variscan orogeny cannot be definitely answered at the moment. The unmetamorphosed sediments and the main direction of the RFH speak in favour of Variscan age.

It is possible that some of the movements in the Fennoscandian Border Zone are also Variscan.

As it was pointed out above, there is seismic evidence of contemporaneous faulting on the RFH during the Triassic. The situation away from the high is not known in detail. However, the thick Triassic sediments in N Jutland bear evidence of strong vertical movements during the Triassic in this general region.

Jurassic movements occurred in the Gassum area in N Jutland, as indicated by the unconformity at the base of the Upper Jurassic Oxfordian, which rests on Lower Jurassic in the Gassum well. It is, however, possible that salt flowage was partly or wholly responsible in this case.

In the Helsingør–Hälsingborg area of the Fennoscandian Border Zone, an unconformity is found between the Campanian and the Valaginian (Brotzen 1945) which suggests late Lower Cretaceous or early Upper Cretaceous movements. The extraordinary thickness of the Upper Cretaceous in the Helsingør region suggests contemporaneous faulting during Upper Cretaceous time in the same general area, which thus apparently was rather much activated during the Cretaceous.

There is ample evidence of Tertiary movements in Denmark. The new Fyn–Falster Uplift developed SE of the RFH during the Palaeocene–Eocene and early part of the Oligocene. The unconformity at the base of the Upper Oligocene marks a time limit for the formation of the uplift.

Another regional feature, the Alnarp Graben in the Fennoscandian Border Zone, probably developed during late Tertiary time.

It should finally be mentioned that several local structures on the RFH and in the basins away from the high, including basement blocks as well as salt structures, were active during the Tertiary.

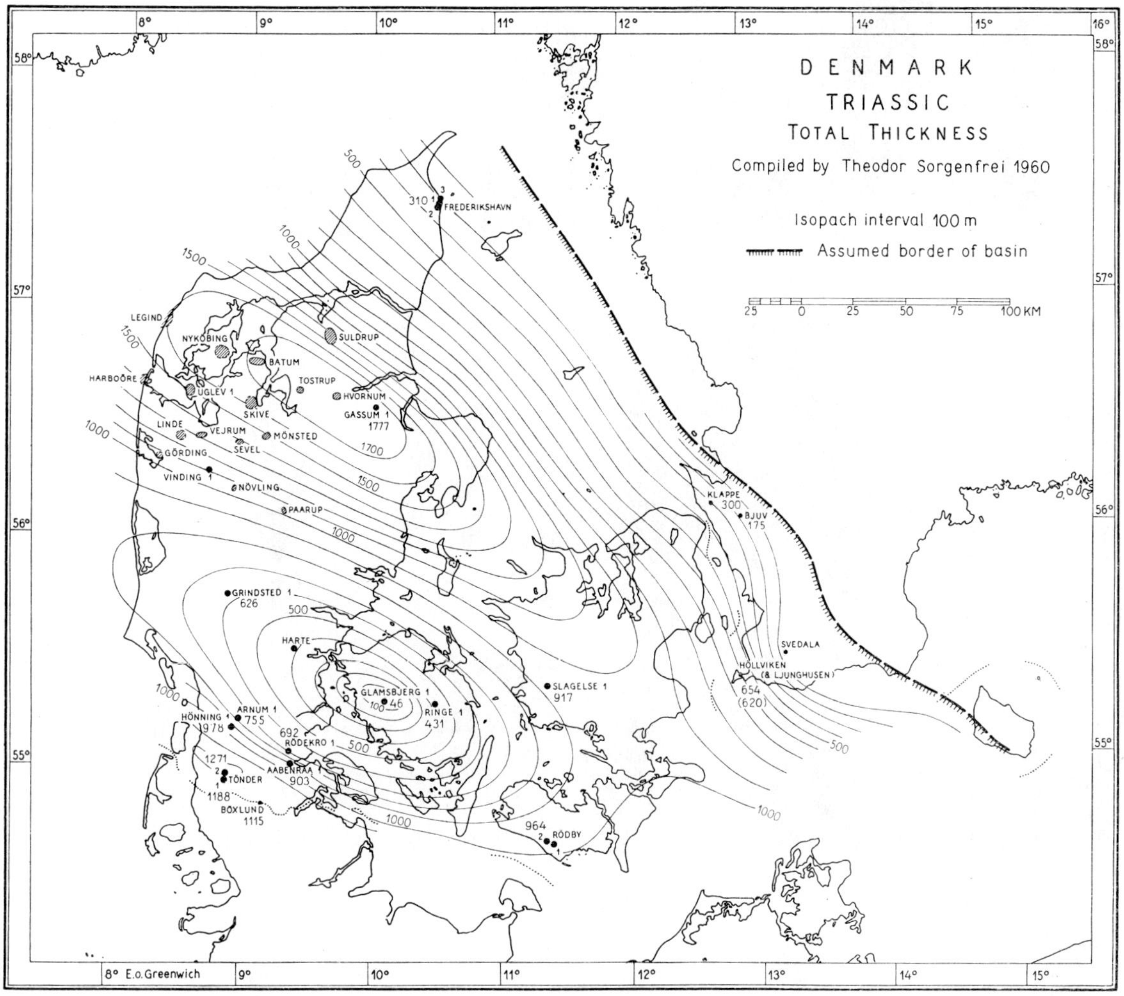

FIG 12.

The Basin of the Norwegian Sea. In the area between 61° N and 69° N, the Geological Survey of Norway has flown an aero-magnetic survey, and the University of Bergen made a shallow seismic survey with a continuous profiler. Results have not yet been published.

Little is known about the sedimentary sequence in this basin. There is in fact only one single onshore occurrence of sediments with bearing on the problem. That is the Upper Jurassic–Lower Cretaceous sequence of the Island of Andøy at about 60° 12′ N and 16° E. The formations range from the Oxfordian to the Hauterivian.

The major part of the 500-m thick section consists of sandstones. Coal seams and bituminous shale interbeds are found at the base, and 125 m of shale are found at the top of the series.

ACKNOWLEDGMENTS

The author wishes to express his gratitude to Dr K. A. Lindbergson, Director General of the Geological Survey of Sweden, for his most valuable assistance and co-operation in providing material for the present review.

He is obliged to Det Kongelige Departement for Industri og Håndverk, Oljekontoret, Norway, for kind support during the preparation of the report.

The writer is also indebted to Esso Exploration Norway, Norske Murphy Oil Company, and Phillips Petroleum Company Norway for assistance.

BIBLIOGRAPHY

Anderegg, H. J., Norling, E., and Skoglund, R. "SGU oljegeologiska arbeten 1967–1968." Sveriges Geol. Unders. Rep. Stockholm, 1968.

Brotzen, F. "De geologiska resultaten från borrningarna vid Höllviken. 1. Kritan." Sveriges Geol. Unders. Ser. C, No. 465, Stockholm, 1945; 2. "Undre kritan och trias." Sveriges Geol. Unders. Ser. C, No. 505, Stockholm, 1950.

von Bubnoff, S. "Fennosarmatia." Berlin, 1959.

Bölau, E. *Geol. Fören. Förh.*, 1959, (497), 167.

Gry, H. "Geology of Bornholm." Int. Geol. Congr. XXI Session Denmark. Guide Book IV, Copenhagen, 1960.

Hede, J. E. "Djupborrningen vid Burgsvik på Gotland 1915." Sveriges Geol. Unders., Ser. C, No. 298, Stockholm, 1919; "On the correlation of the Silurian of Gotland." Medd. Lunds Geol.-Min. Inst. No. 101, Lund, 1942; "Boring through Middle Ordovician–Upper Cambrian strata in the Fågelsång district, Scania (Sweden)." Kungl. Fys. Sällsk. Handl., N.F. Bd. 61, No. 7, Lund, 1951.

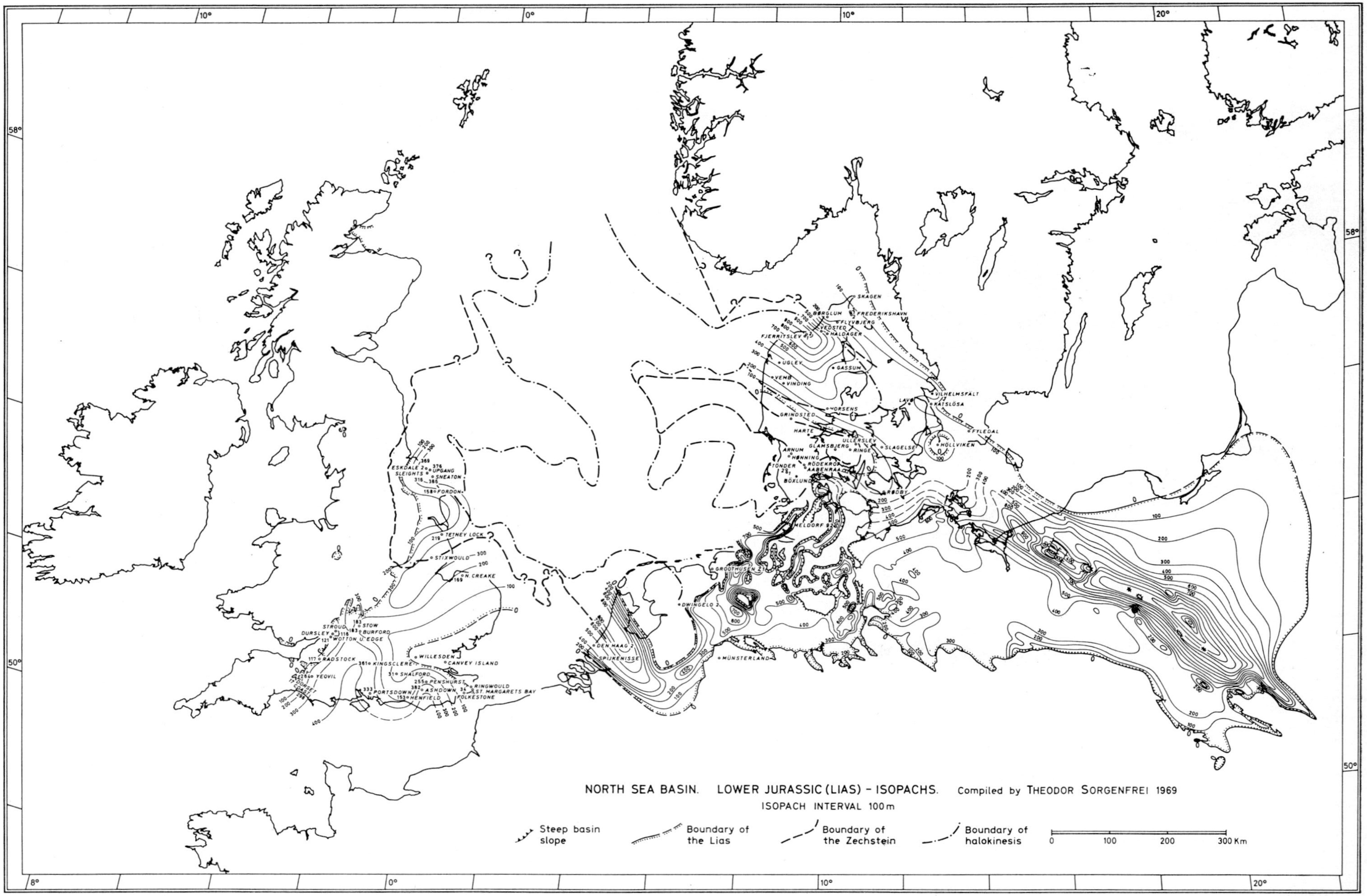

FIG 13. From Sorgenfrei 1969.

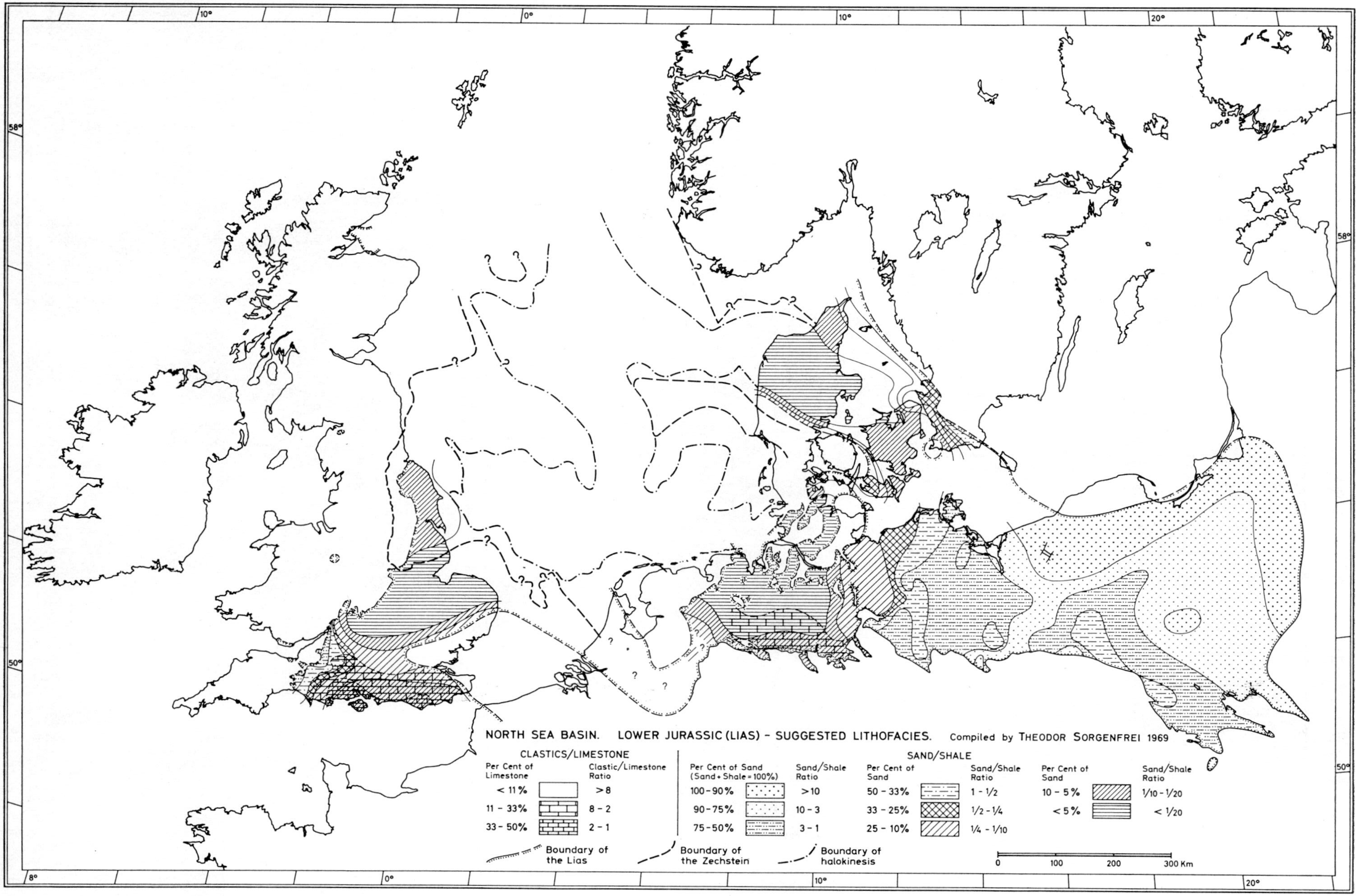

FIG 14. From Sorgenfrei 1969.

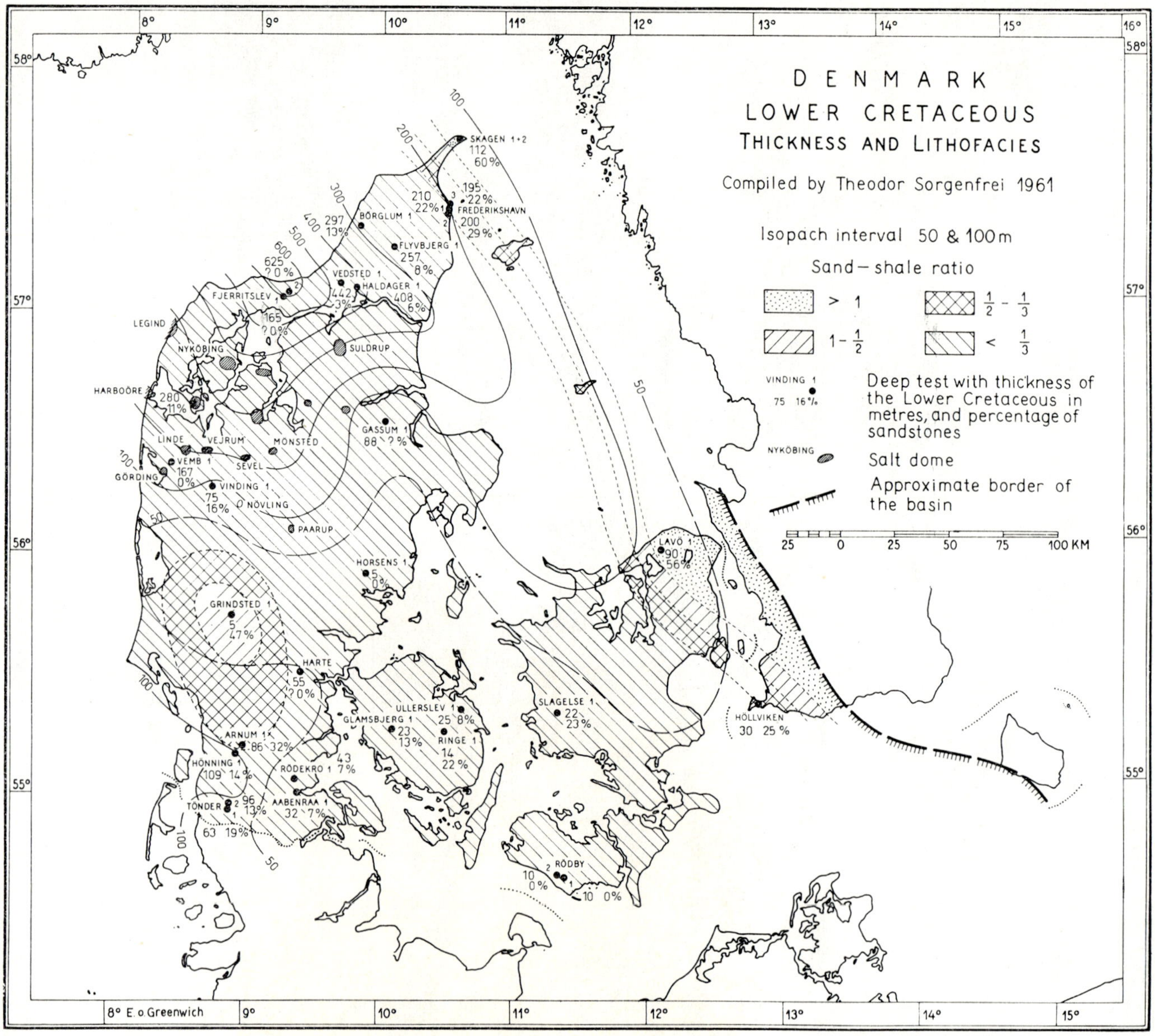

Fig 15. From Sorgenfrei 1963.

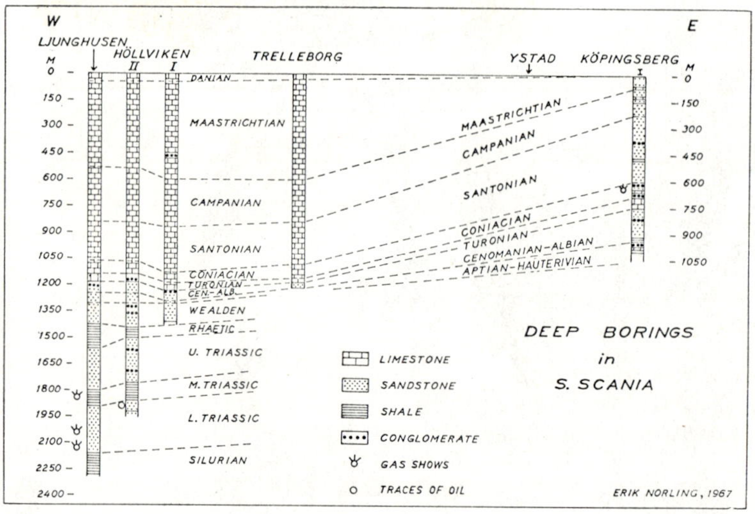

Fig 16. Cross-section Ljunghusen–Köpingsberg. From Anderegg *et al.* 1968.

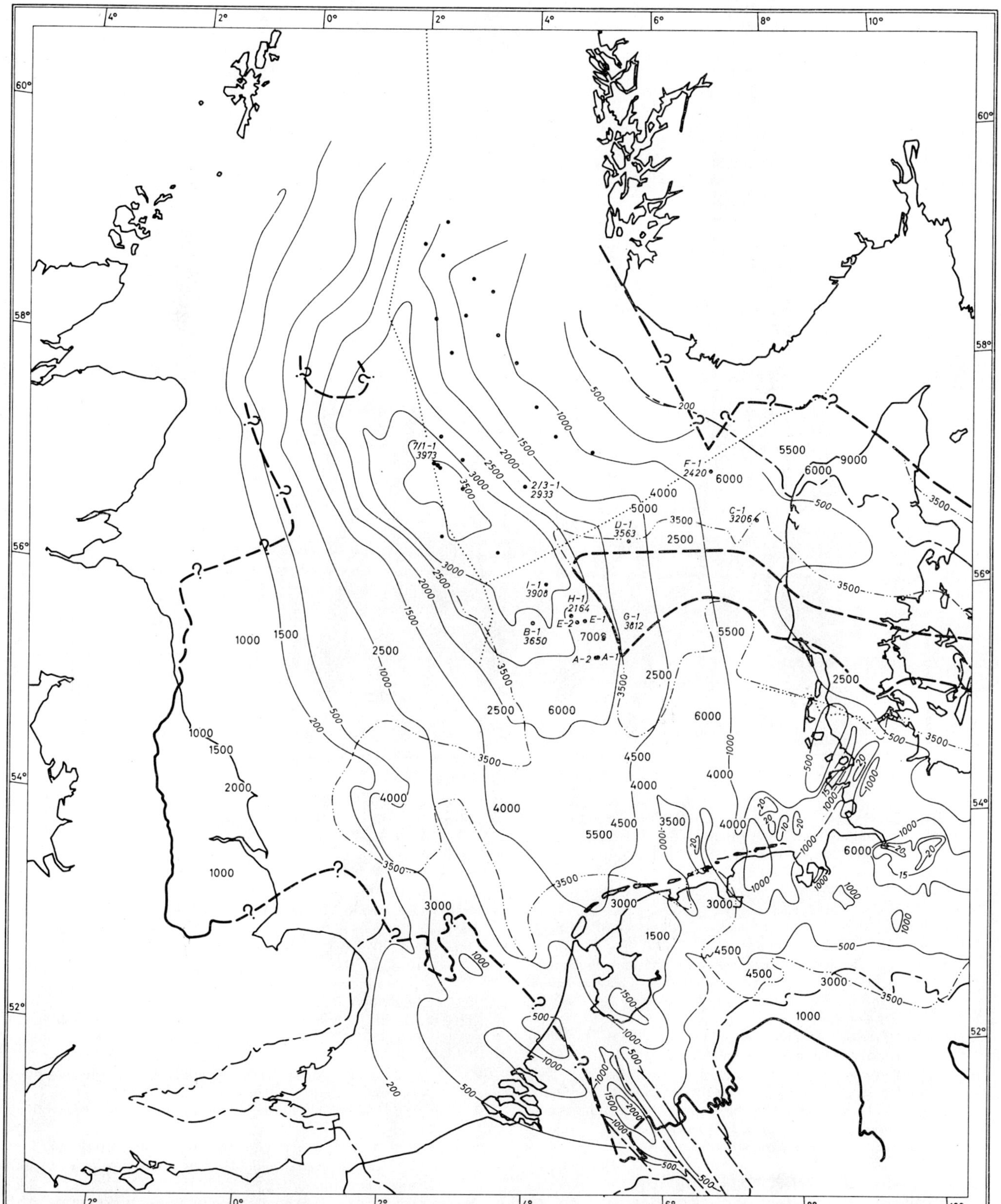

NORTH SEA AREA. CONTOUR MAP OF THE BASE OF THE TERTIARY SUPERPOSED ON OUTLINE MAP OF THE BASE OF THE UPPER PERMIAN ZECHSTEIN (Mainly according to HEYBROEK ET AL. 1967)

Contours: Metres below sea level
—500— (*italic*) Base Tertiary
·· — 3500 ·· — 3000 (roman) Base Zechstein

• G-1 3812 Well location with final depth in m.

Boundary of Tertiary

Boundary of Zechstein

.... Shelf boundary

0 50 100 150 200 Km

FIG 17.

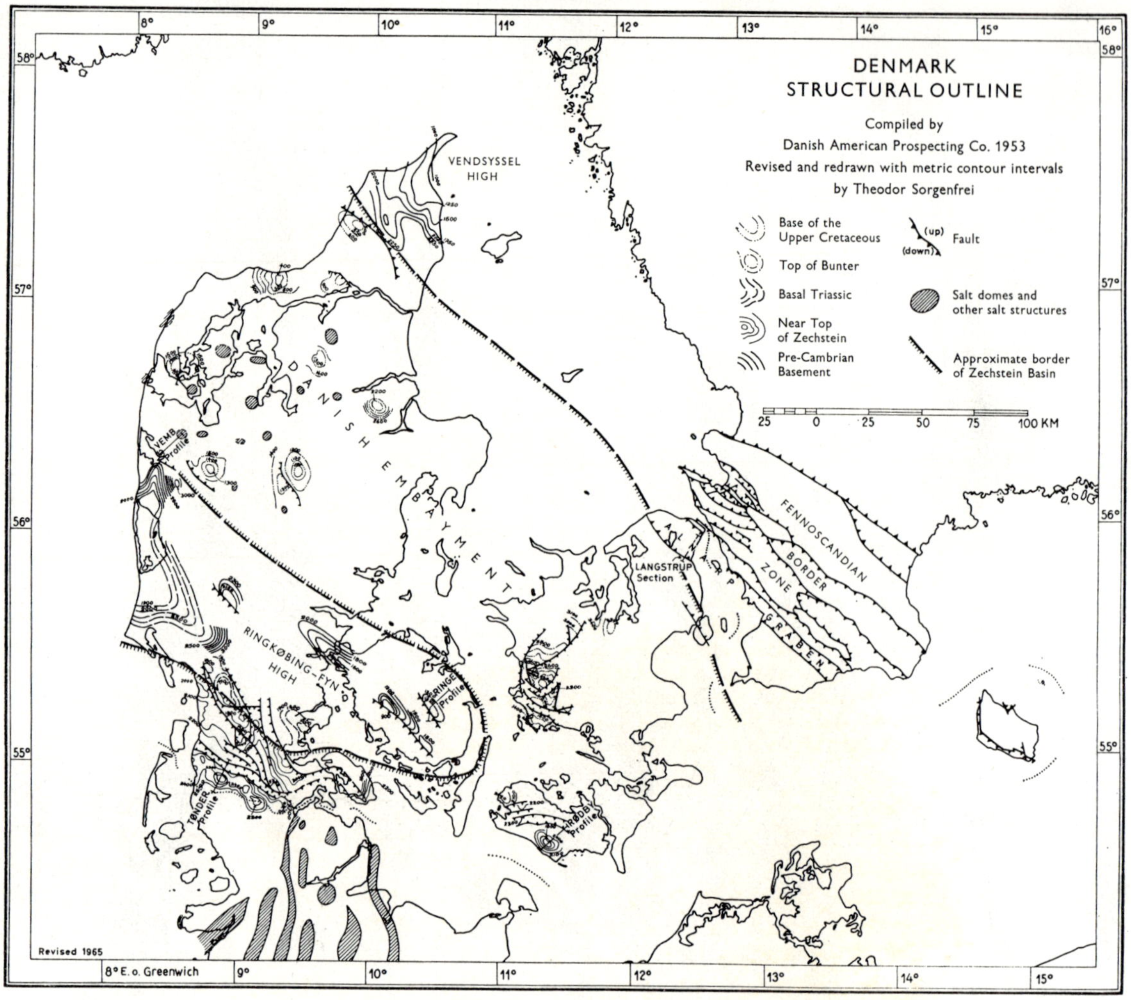

FIG 18. From Sorgenfrei 1966.

Hedström, H. "Remarks on some fossils from the diamond boring at the Visby cement factory." Sveriges Geol. Unders. Ser. C, No. 314, Stockholm, 1923.

Henningsmoen, G. See Holtedahl, 1960.

Heybroek, P., Haanstra, U., and Erdman, D. A. *Proc. 7th World Petrol. Congr.*, 1967, **2**, 905.

Holtedahl, O. "Geology of Norway." Norges Geol. Unders., No. 208, Oslo, 1960.

Larsen, G." Rhaetic–Jurassic–Lower Cretaceous Sediments in the Danish Embayment (A Heavy-Mineral Study)." Danmarks Geol. Unders., 2, ser. No. 91, Copenhagen, 1966.

Larsen, G., Bruun Christensen, O., and Buch, A. "Øresund." Danmarks Geol. Unders., Rep. No. 1, Copenhagen, 1968.

Lundblad, B. "De geologiska resultaten från borrningarna vid Höllviken. 4. On the presence of Lepidopteris in cores from *Höllviken II*." Sveriges Geol. Unders., Ser. C, No. 507, Stockholm, 1949.

Magnusson, N. H., Lundqvist, G., and Regnell, G. "Sveriges Geologi." Stockholm, 1963.

Martinsson, A. *Bull. Geol. Inst. Univ. Uppsala*, 1960, **38**, 11.

Pożaryski, W. *Kwart. Geol.*, 1957, **1** (1).

Pożaryski, W. *Kwart. Geol.*, 1964, **8**.

Regnell, G., and Hede, J. E. "The Lower Palaeozoic of Scania, the Silurian of Gotland." Int. Geol. Congr. XXI. Session. Guide to excursions, Stockholm, 1960.

Ryka, W. *Kwart. Geol.*, 1964, **8**.

Sorgenfrei, T. "Jura und Unterkreide in Dänemark". Z. deutsch. geol. Ges. Bd. 114, p. 446, Hannover, 1963.

Sorgenfrei, T., and Buch, A. "Deep Tests in Denmark." Danmarks Geol. Unders, 3, Ser. No. 36, Copenhagen, 1964.

Sorgenfrei, T. *Geologie Jhrg.*, 1966, **15** (6), 641; "Geological Perspectives in the North Sea Area." Medd. Dansk Geol. Foren. Bd. 19, Copenhagen (in press).

Störmer, L. *Quart. J. Geol. Soc.*, 1967, **V** (123), 183.

Thorslund, P., and Westergård, A. H. "Deep boring through the Cambro–Silurian at File Haidar, Gotland." Sveriges Geol. Unders., Ser. C, No. 415, Stockholm, 1938.

Thorslund, P. "Djupborrningen på Gotska Sandön." Geol. Fören. Förh. Bd. 80, p. 190, Stockholm, 1958.

Thorslund, P., and Jaanusson, V. "The Cambrian, Ordovician and Silurian in Västergötland, Dalarne and Jämtland, Central Sweden." Int. Geol. Congr. XXI. Session. Guide to excursions, Stockholm, 1960.

Westergård, A. H. "Borrningar genom Skånes alunskiffer 1941–42." Sveriges Geol. Unders., Ser. C, No. 459, Stockholm, 1944.

Westergård, A. H. "Borrningar genom alunskifferlagret på Öland och i Östergötland 1943." Sveriges Geol. Unders., Ser. C, No. 463, Stockholm, 1944.

Znosko, J. *Kwart. Geol.*, 1964, **8**.

Ödum, H. "De geologiska resultaten från borrningarna vid Höllviken. 5. The Macrofossils of the Upper Cretaceous." Sveriges Geol. Unders., Ser. C, No. 527, Stockholm, 1953.

MS received 1 July 1969

The Present State of Oil and Gas Geological Prospection in Poland*

By J. SOKOLOWSKI (*Deputy Director, Geological Institute, Warsaw, Poland*) and A. TOKARSKI (*Professor of Oil Geology, Academy of Mining and Metallurgy, Krakow, Poland*)

SUMMARY

Besides standardized Carpathian conditions, the country structure is veiled by NW oriented warping of the "Laramide incident". The post-Rotliegend content along the Polish E margin of the middle European basins is evaluated by investigating its superficial and pre-Laramide palaeotectonic features. Recent drilling (3000–5000 m) suggests very good chances of the existence of a widespread, non-denuded, relatively continuous pre-Zechstein sedimentary sequence. Here, middle Cambrian quartzites and sandstones, Devonian sandstones, and biostromes, as well as Namurian and Westphalian sandstones, appear within impermeable rock intercalations. Post-Hercynian Rotliegend sandstones have collected, displaced, and accumulated gas from their truncated basement, and the successive rebuilding of pre-existing structures, from Caledonian and Bretonian movements onwards to neotectonic ones, has reshaped traps and controlled new accumulations.

A geochemical approach supports the above geological evaluation by proving the early initiation and long-lasting mobilization of hydrocarbons in mother rocks.

A novel concept of the structural sub-division of the country is now needed for to prospect evaluation and structures sharing heterogeneous form are being purposely re-examined.

The practical effects of this are: (1) New extra-Carpathian oil discoveries have already exceeded in productivity the previous limit of 200,000 tons/year. (2) The E fore-Carpathian gas zone is not proved to extend into Poland; (3) A newly established Rotliegend gas-bearing structural polygon NE of Wrocław has been established.

DIALECTICS

HANDICAPS to geological control of oil and gas prospection in Poland are provided by the very great area to be considered at once (Figs 1 and 2) and the many problems and vectors to be evaluated and chosen between, as well as the marked inadequacy of any classical pattern. Stimulators, however, issue from the number of possible prospects, territorial overlap of different oil trap situations, and from the possibility of corrected regional exploration effort. (Tokarski, 1966).

In the multi-tiered northwards thrust flysch zone of the Polish Carpathians, the near-surface possibilities are still being discussed as well as tested currently (Fig 3), mainly by studying the implications of the varying durations of water tables (Kruczek, 1962, 1968(*a*), Tokarski, 1968).

Chances of deeper flysch horizons, now being investigated, are quite complicated (Fig 4). The more it concerns the still deeper sub-Carpathian structure of the basement of the Flysch. The latter is here no more, likewise in the U.S.S.R. Carpathians, overthrust over the inner margin of its Miocene-filled foredeep but already on its northwards ascending outer zone (Komorowska–Błaszczyńska, 1967).

Miocene deposits of the southern inner zone of the foretrough had been tectonically erased between the flysch and the basement as far as over 20 km to the S of the front of the Carpathian overthrust (Fig 5).

The whole of the remainder of the country became affected by the NW diagonal Laramide folding and warping as well as by longitudinal and cross-faulting, followed in turn by E–W post-Laramide faulting. Laramide effects, although having been rather incidental in Poland's geology, nevertheless today superficially largely mask the previous tectonics (Fig 1).

OIL AND GAS BEARING ROCK BODIES

Geophysics and a series of borings to over 3000 m enabled and is still enabling the oil and gas prospects of the Polish E end of the zone of post-Hercynian basins of Western and Central Europe to be investigated. Prospection is being carried out on both post Rotliegend superficial and palaeotectonic traps (Jawor, Kruczek, 1965, Sokołowski, 1967). In this way new petroleum indications, as well as productive zones, have already been discovered (Fig 6).

Also, following the latest set of borings, the rules of migration and trapping in pre-Zechstein rocks are being discovered. This involves mainly the permeable middle Cambrian quartzites and sandstones (Areń, 1967, Depowski, Tyski, 1968, Lędzion, 1968, Modliński, 1967, Znosko, 1965). These rocks are regionally spread under a thick impermeable and, at the same time, possibly mother rock cover, up to and including the Silurian graptolite shales (Fig 7(*a*)). Oil prospects are particularly interesting in the situations where this series has not been too deeply eroded by successive Caledonian, Hercynian, and Laramide denudations and where today it is warped up as in the Łeba uplift to the W of the Wisła mouth and beneath the adjacent part of the Baltic shelf (Dadlez, Młynarski, 1967). The same rock sequence contains Tremadocian sandstones too, above the

*This paper was preprinted but was not presented.

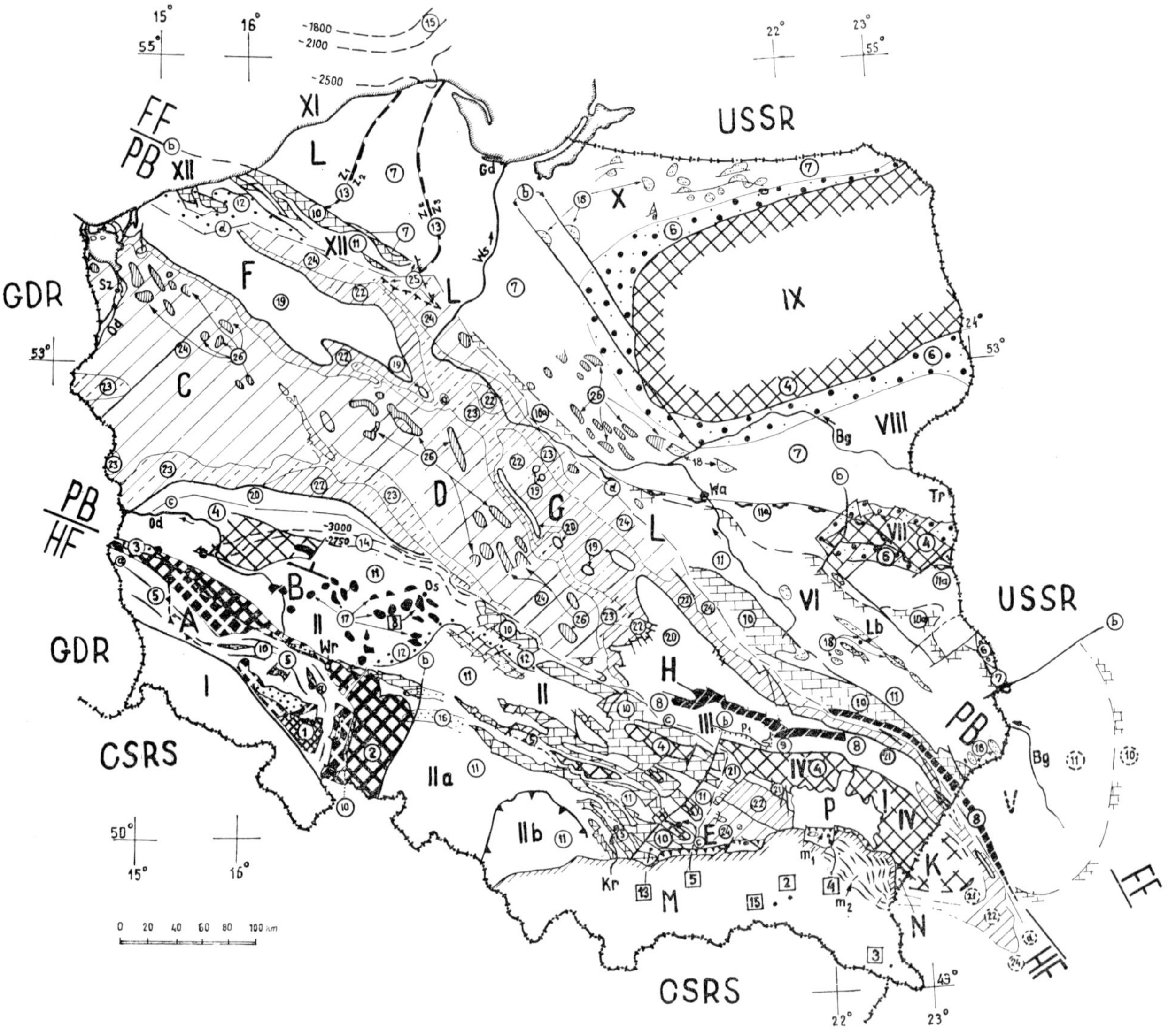

FIG 1. Synoptic map of the known subsurface conditions of Poland.

Pre-Permian pattern and its post-Hercynian revivals: HF—general Hercynian northern front, FF—general SW margin of the Fennosarmatian platform, PB—Polish Basin (Hercynian foredeep) I—Sudeten Mountains undifferentiated, II—fore-Sudetic arrangement, Cracow region included, IIa—Sileso-Moravian field of lower Carboniferous, IIb—Silesian Coal Basin, III Holy Cross Mountains, IV—fore-Carpathian sub-Miocene eastern extension of the fore-Sudetic arrangement, V—SE centroclinal closure of Polish Basin (in U.S.S.R.). VI—Lublin sub-basin (segment of Polish Basin), VII, VIII, I*X*, *X*—pre-Carboniferous major elements of the Fennosarmatian Frame (resp. Wisznice elevation, Podlasie depression, Mazury elevation, peri-Baltic depression), *X*I—Leba uplift (pre-Bunter revival of a pre-Permian dome), *X*II—Caledonian-Hercynian Chojnice-and-shore zone, 1-circled; the same applies to the following numbers—Owl-Mountain older Pre-Cambrian gneisses, 2-younger Pre-Cambrian of the peri-Owl zone (reverse black-white symbol where both outcropping and sub-Tertiary). 3-granites connected with "2", 4–Małopolska younger pre-Cambrian* and the one of the Fennosarmatian Frame, 5–lower Palaeozoic ("a"-simulated structural trends), 6-Cambrian, 7-Ordovician and Silurian, 8–northern and northeastern margin of both out—and sub-crops of lower Palaeozoic on the Hercynian side of the Lublin sub-basin, 9–undifferentiated Palaeozoic out—and subcrops southwards of "8", 10–Devonian (10a north-eastern sub-crop rand), 11-Carboniferous (11a NE sub-crop margin), 12 fragments of both southern and northern reach of Rotliegend sub-crops, $13/Z_1/Z_2$ and $2_2/Z_3$–sub-crop boundaries of Zechstein cyclothems, bottom Bunter, 14-isobaths, bottom Rotliegend, 15-isobaths, bottom Cambrian, 16 elongated palaeomorphological and/or tectonic elevations of post-Hercynian surface, 17-parallel swarm of reflected buried hills of post-Hercynian surface N or Wrocław, 18–Bretonian and younger Hercynian structures.

Laramide re-Building (sub-Tertiary or surface): A—Bottom Permian periclinal subcrops of fore-Sudetic axial elevation, B—fore-Sudetic monocline with its general NE dip indication (limb of the "A" mega-structure), C, D, E—elements of the main synclinorial zone (resp. Szczecin, Lódz, and Miechów synclinorium). F, G, H, J, K—elements of the main anticlinorial zone (resp. Pomorze and Kujawy anticlinorium, Holy Cross pericline, San anticlinorium and Stryj pericline, the latter in U.S.S.R.), L—marginal synclinorial trend (southeastwards converging with the pre-Permian form "V"), 19, 20, 21–(circled; the same applies to the following numbers)—pre-Malm series (resp. post-Permian to post-Liassie, post-Carboniferous, and post-pre-Cambrian), 22-Malm (bottom boundary emphasized), 23–Lower Cretaceous, Turonian locally differentiated (bottom boundary emphasized where encroaching), 24-Cretaceous, Danian, 25-Danian bottom pericline of Laramide elevation of Chojnice, 26 Kimmerian and Laramide structures. Post-Laramide re-building: M—Carpathians (emphasized northern edge of Flysch outcrops, m_1-northern margin of outermost Flysch element enclosed by Miocene series, m_2 structural Flysch curves of the eastern sigmoid), N—South-eastern gasiferous zone, O—fragment of northern reach of fore-Carpathian Miocene, P—intramiocene rejuvenated centre of Laramide San anticlinorium "J".

General: b (circled and the same applies to the following letters)—faults, c—N and NE edge of mental theoretically revealed unveiling the superposed Mezozoic cover, d—SW edge of mental theoretically revealed unveiling the permo-Mesozoic cover encroaching northeastwards; Numbers in squares—localization of respective figures. Cities: Gd—Gdańsk, Lb—Lublin, Os—Ostrów Wielkopolski, Sz—Szczecin, Tr—Terespol, Wa—Warszawa, Wr—Wrocław. Rivers: Bg—Bug, Od—Odra, Wi—Wiłsa.

* Differentiation as well as nomenclature kindly suggested orally by Professor J. Oberc.

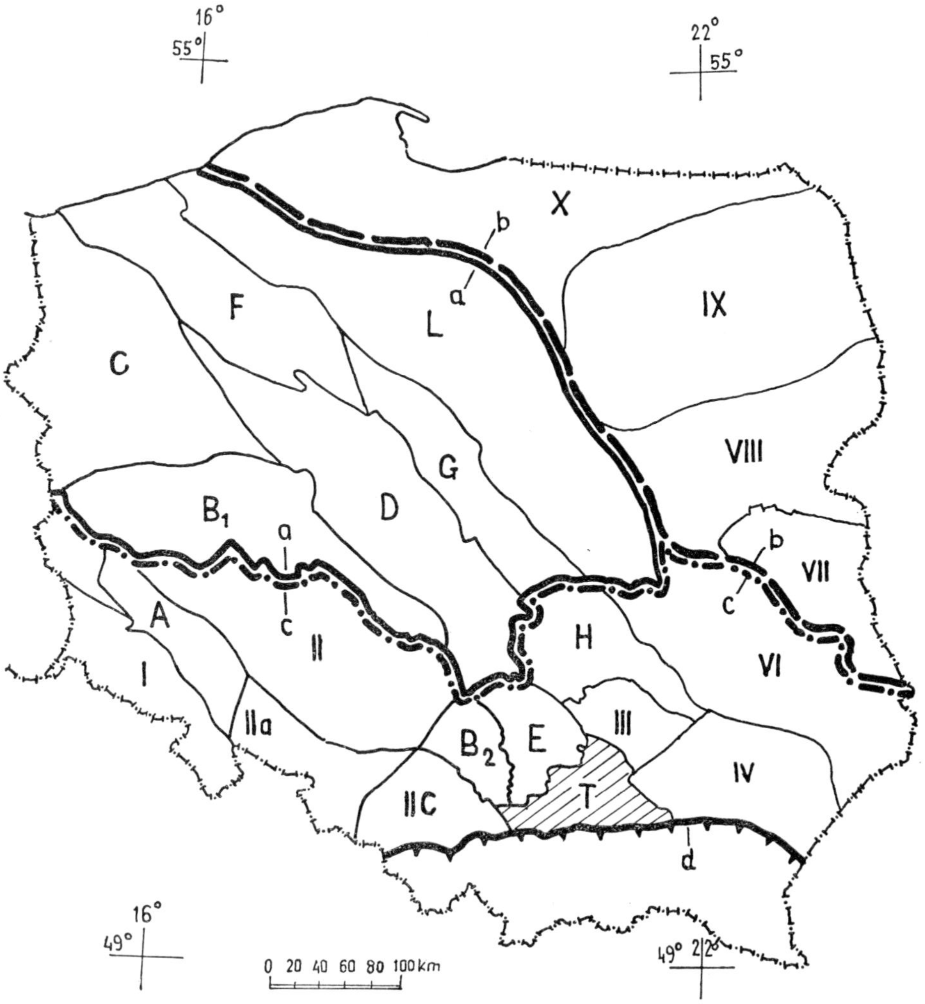

FIG 2. Prospective–structural sub-division of Poland's subsurface after J. Sokołowski, 1968, modified. First order delimitations: (*a*) outer margin of Polish Basin (ex parte the one on Fig 1), (*b*) SW margin of Fennosarmatian Frame, (*c*) northern margin of the "Fold and Block Area", (*d*) line of Carpathian northward overthrust. Second order distinctions: I–IV, VI–XI, A–H, L—megastructures corresponding, wholly or by modifications to ones on Fig 1, T—distinction not related to Fig 1.
To show location of Laramide megastructures, A, B2, E, H inside the "Fold and Block Area" with controlling predominance of pre-Permian arrangement.

post-Cambrian non-conformity, and these wedge up in the Polish near-frontier zone at Terespol from their E lower positions with oil traces in the U.S.S.R. (Antsupow, 1965, Wierzchowska-Kicułowa, Kowalska-Łącka, 1969, Znosko, 1964).

Moreover, lower and middle Devonian sandstones of a Rhine-like facies, although with a rather low permeability, constitute regional reservoirs with gas shows. The same sequence provides traps for oil in highly complex coral biostromes of middle Devonian and lower Frasnianage, under the control of their westerly primary growth direction. The position of upper Frasnian nodular limestone is regionally more uniform and seems to be a separate reservoir (Miłaczewski, Niemczycka, 1967). Bretonian denudation, to be true, in places intervenes by eroding the upper Frasnian complex but it rarely cuts down to reach the middle and lower Devonian sandstones (Fig 7(*b*)). The post-Bretonian sequence contains oil trace bearing regional sandstone reservoirs of the upper Namurian and lower Westphalian (Korejwo, Teller, 1968, Sokołoski, Żelichowski 1968).

Lastly, in an overlap position, as it were, between both post-Rotliegend and pre-Permian prospects and of overwhelming importance, is the Rotliegend cover up to over 1000 m thick, in the form of a post-Hercynian discordant Molasse (Kłapciński, 1967). It fills up basins in Poland in the E (Sokołowska, Sokołowski 1965); likewise in the North Sea area in the W (Heybroek, Haanstra, Erdman, 1967), at the termination of the middle European trend of post-Hercynian age, and serves to collect gas from the truncated basement and re-distributes it both within itself and in the overlying basal Zechstein carbonates (Fig 8). A reverse downpressing of gas into truncated basement is not to be excluded too, and this can be compared with, for instance, in Poland, such a condition of oil re-distribution by post-Kimmerian Cenomanian cover (Fig 9).

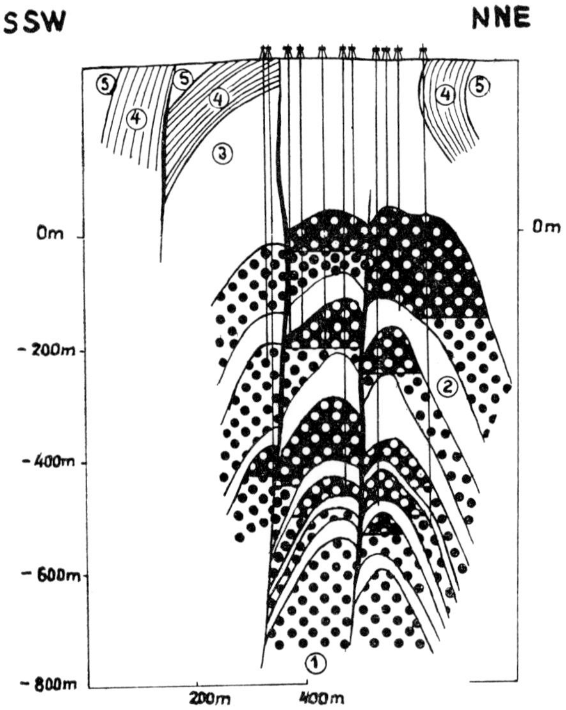

FIG 3. Cross-section of a classic oil-bearing flysch structure with water tables displaced by a post-depositional faulting (after Kruczek, 1968). Stratigraphy (upper Cretaceous to Oligocene): 1—sandstone collectors (black background where oil-bearing), 2—intervening shales (the lowermost ones dark grey, the remaining variegated), 3—top shales (variegated), 4—menilite black shales, 5—Krosno grey beds.

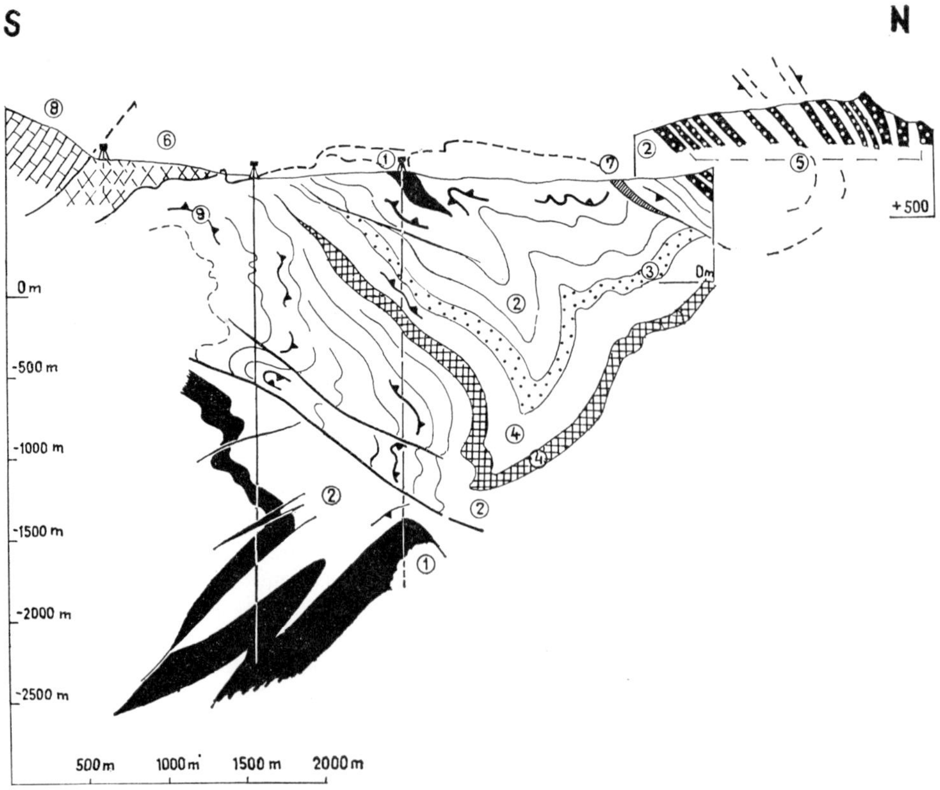

FIG 4. Complicated arrangement of prospected flysch deeper horizons (after Tokarski, 1968). Stratigraphy (Eocene to Oligocene): 1—menilite black shales prospective series lithostratigraphically below, 2—grey-black shaly passage beds, 3—key-sandstone, 4—marker black shales, 5—Krosno grey beds.

Tectonics: 6—upper digitation (summarized), 7—thrust surface breccia, 8—Cretaceous sandstones of uppermost overthrusted unit, 9—symbolic position with primary bottom indication.

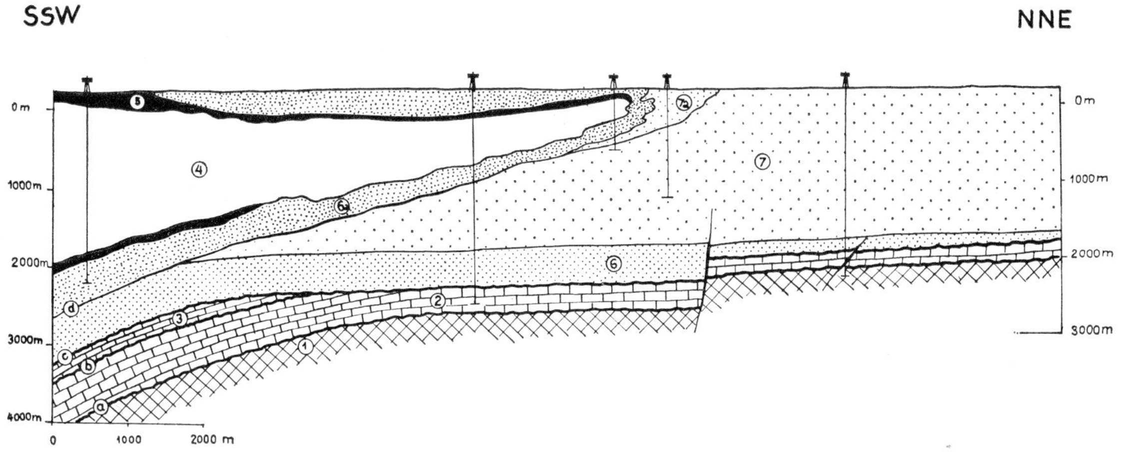

FIG 5. Cross-section of Carpathian front (after Komorowska–Błaszczyńska, 1967). 1—Pre-Cambrian, 2—Devonian, 3—Carboniferous, 4—Flysch Cretaceous–Oligocene, 5—Lower Tortonian, 6—Upper Tortonian (6a—overthrusted), 7—Sarmatian (7a—overthrusted), a, b, c—disconformities (resp Assyntian–Caledonian, Bretonian, Hercynian + younger), d—Carpathian overthrust.

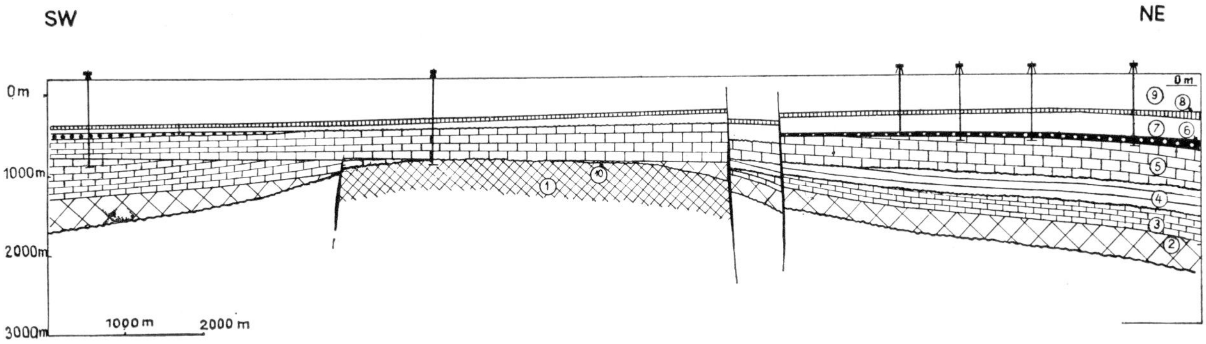

FIG 6. Cross-section of a near Cracow oil deposit (after Jawor, Kruczek, 1965). 1—Pre-Cambrian phyllites, 2—upper Ludlow graywackes and conglomerates, 3—Devonian limestones with detritic bottom series undifferentiated, 4—Carboniferous detritic with limestones, 5—Malm limestones, 6—Cenomanian sandstones and conglomerates/black back-ground–petroliferous, 7—supra-Cenomanian marly Cretaceous, 8—Lower Tortonian marls and sandstones with top anhydrite, 9—Upper Tortonian and Sarmatian clays and mudstones.

NON-CLASSIC MOTHER ROCK PROBLEM

The geochemical approach in Poland to the mother rock and primary deposit problem, mainly led by J. Calikowski (Calikowski, Gondek, 1967), contributes to an "epigenese preponderancy" pattern of an early initiated but long-lasting primary mobilization of hydrocarbons. A plot of the data concerned has been kindly communicated to the writers by the named investigator for Polish Silurian graptolite shales, *i.e.* for a widely acknowledged mother rock (Fig 10). In fact, these shales display a very wide range of hydrocarbon groups, from ones marking the long ago left local residuum, through the still migrating oil, and through already mobilized but not yet migrating oil to a raw material awaiting its future mobilization. All this means not only a possibility of "Silurian oil" to accumulate stratigraphically downwards and to be trapped today in both Ordovician and Cambrian sandstones. It also means this oil to have migrated as well as to be migrating today "primarily" from the Silurian shales into overlying covers, notwithstanding intervening periods of denudation.

The role of migration paths thus becomes emphasized, with their possibility as well as ability to gather both allochthonous hydrocabons and those of their own lithostratigraphic context.

Nevertheless, there are also Polish instances of a sharp delimitation of oil and gas of different origins between reservoirs very closely situated in a lithostratigraphic profile although not necessarily syngenetic with the rocks concerned. The wet gas of the Carpathian foreland is thus able to migrate from the sub-Miocene (its normal reservoir is in the Malm limestones (Karakowski, 1968)) and into the immediately overlying lower Tortonian consisting of Baranow sandstones and mudstones capped by a 20-m anhydrite bed. Further up (107 m) in this sequence, dry gas of Miocene origin is present in a sandy bed (Fig 11(*a*)). Elsewhere in the same Carpathian foreland the oil of sub-Miocene origin reaches still higher in the Miocene profile (Fig 11(*b*)).

In the fore-sudetic Monocline structures, conversely, the dry gas of sub-Permian origin is gathered by Rotliegend sandstones and ascends into the upper parts of massive deposits not only the lower Zechstein, Weissliegend sandstones but also as high as into the lower Zechstein carbonates, by somehow by-passing (at dislocations?) the intervening impermeable but very thin copper shale. Here wet gas as well as oil, both of them mainly of Zechstein origin, appear only higher up in the Main Dolomite above the intervening anhydrites and salt of the oldest Werra Cyclothem (Fig 8).

All this stresses the role of the study of historical tectonics both of structural zones and of single structures in its interplay with tracing and re-tracing of phenomena of introduction, migration, and trapping of oil and gas.

GAS-OIL-WATER IN MIGRATION PATHS AND TRAPS

Supplementary to the above discussed "epigenese preponderancy" pattern of Polish gas and oil reservoirs and prospects, coupled with geochemical investigations, which are the guiding features of prospection, the strong effects of renewed earth movement on older structures on gas-oil-water relationships must be emphasized. These effects started at least as old as the Caledonian and contrived to be re-formed successively until the neotectonic movements. The state and type of incorporation of every pre-existing structure has thus to be taken into account.

Also, from both spatial and historical viewpoints, the opening or braking effects of respective sets of dislocations has been studied with a positive bearing on discoveries both in the Carpathian foreland (Fig 12) and the fore-Sudetic monocline. This line of study was stimulated by the field evidence of varying positions of gas-water-oil contacts inside the same system of migration, not only as a result of post-accumulation tectonics (Fig 2) but also because of hydrodynamic and spacial conditions along the front of migration (Levorsen, 1965, Kruczek, 1968(*b*)).

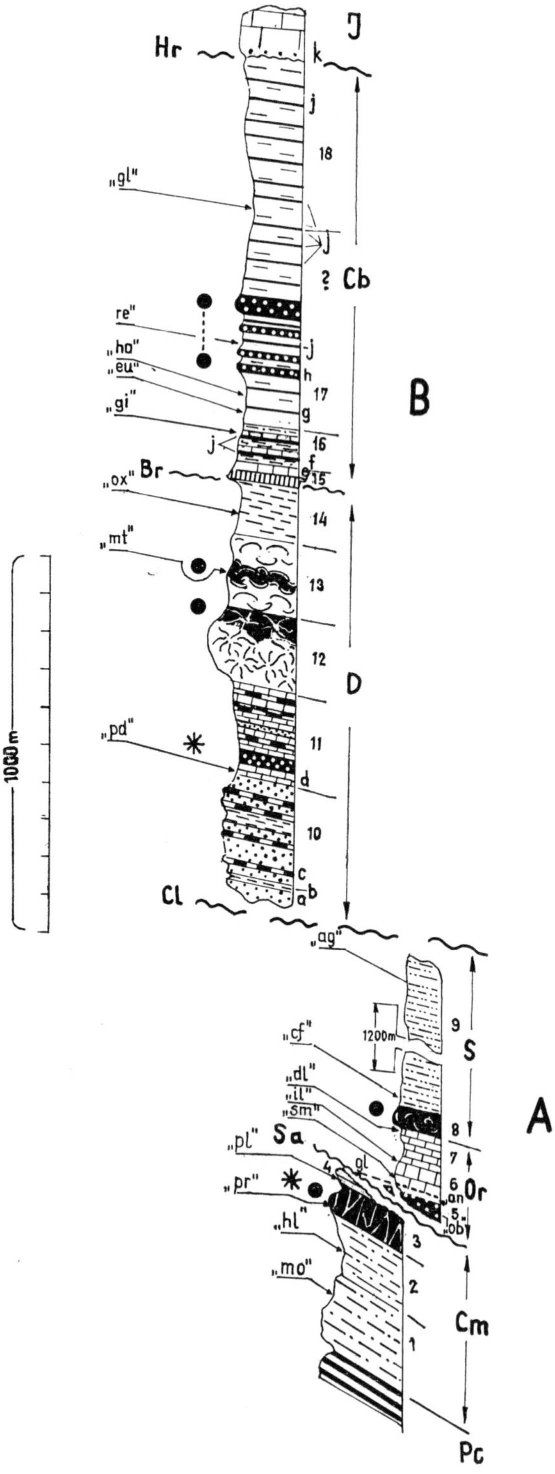

FIG 7. Synthetic borehole profile of prospective lower (A) and upper (B) Palaeozoic of Fennosarmatian Frame and Lublin sub-basin (after Areń, 1967, Korejwo, Teller, 1968, Lendzion, 1968, Miłaczewski, Niemczycka, 1967, Modliński, 1967, Sokołowski, Żelichowski, 1968, Znosko, 1964, 1965). Pc—Pre-Cambrian, laminarites arkoses, Cm—Cambrian; 1, 2–Georgian sandstone shale sub-*Holmia* beds with Mobergella-fauna "mo" and shale-sandstone *Holmia* beds with Holmia "hl", 3–Acadian sandstone beds with Peltura-fauna "pl", Or-Ordovician: 5–Tremadocian Obolus "ob" sandstone with Anisograptidae "an" shales in its top, 6–Symphysurus "sm" to Jillaenus "il" limestones, with glauconitite (gl) in their bottom part, 7–marly limestones and marls with Dalmanitina "dl" in their top portion, S—Silurian: 8–Valentian nodular limestones, 9–Graptolite shale sequence from *Pristiograptus cyphus* "cf" to *Monograptus angustidens* "ag" (real thickness over 1500 m), D—Devonian: 10–Siegenian and Emsian, 11–Eifelian and Givetian (for 10 and 11 "a"—sandstone, "b"—mudstone, "c"—anhydrite, "d"—dolomite in schematic relations, "pr"—*Productella* and *Chonetes* fauna), 12–Upper Frasnian to lower Famennian nodular limestones ("mt"—*Manticoceras* fauna), 14–Upper Famennian marly-mudstone beds ("ox"—*Oxyclymenia*), C—Carboniferous: 15–Post-Bretonian diabases, tuffs and tufoids, 16–17–18–Upper Visean, Namurian, and Westphalian (commo schematical lithology, "e"—limestone, "f"—siltstone, "g"—mudstone, "h"—sandstone, "j"—coal): faunal keys, "gi"—*Gigantoproductus*, "eu"—Eumorphoceras, "ho"—*Homoceras*, "re"—*Reticuloceras*, "ga"—*Gastrioceras*, J-Malm limestones with lowest Oxfordian transgressive breccia; disconformities and nonconformities, Sa—Sardinian, Cl—Caledonian, Br—Bretonian, Hr—Asturian to Kimmerian; gas and oil trace bearing prospective reservoirs in black; oil traces and inflows—black circles, gas traces and inflows—black stars.

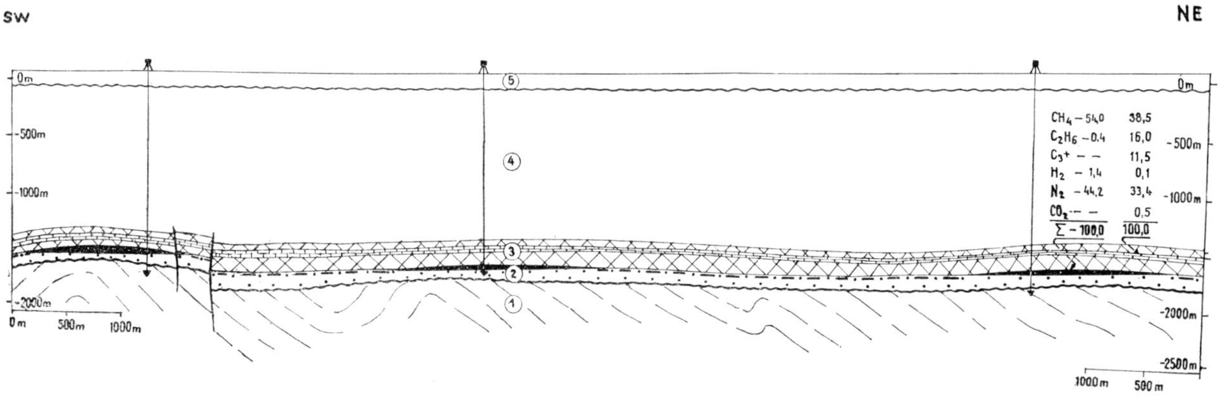

FIG 8. Cross-section of a gas deposit at Ostrów Wielkopolski (after Sokołowoska, Sokołowski, 1965, modified after unpublished reports of A. Pawłowski and J. Skibicki, 1969). 1. Carboniferous shale-sandstone series (schematically), 2. Rotliegend sandstones together with Weissliegend sandstone and copper shale above covered with Zechstein lowest carbonates (gas-filled portions black), 3. Zechstein impermeable rocks (anhydrites and salts) with gas-bearing Main Dolomite in the middle, 4. Bunter to Keuper sequence, 5. Tertiary cover.

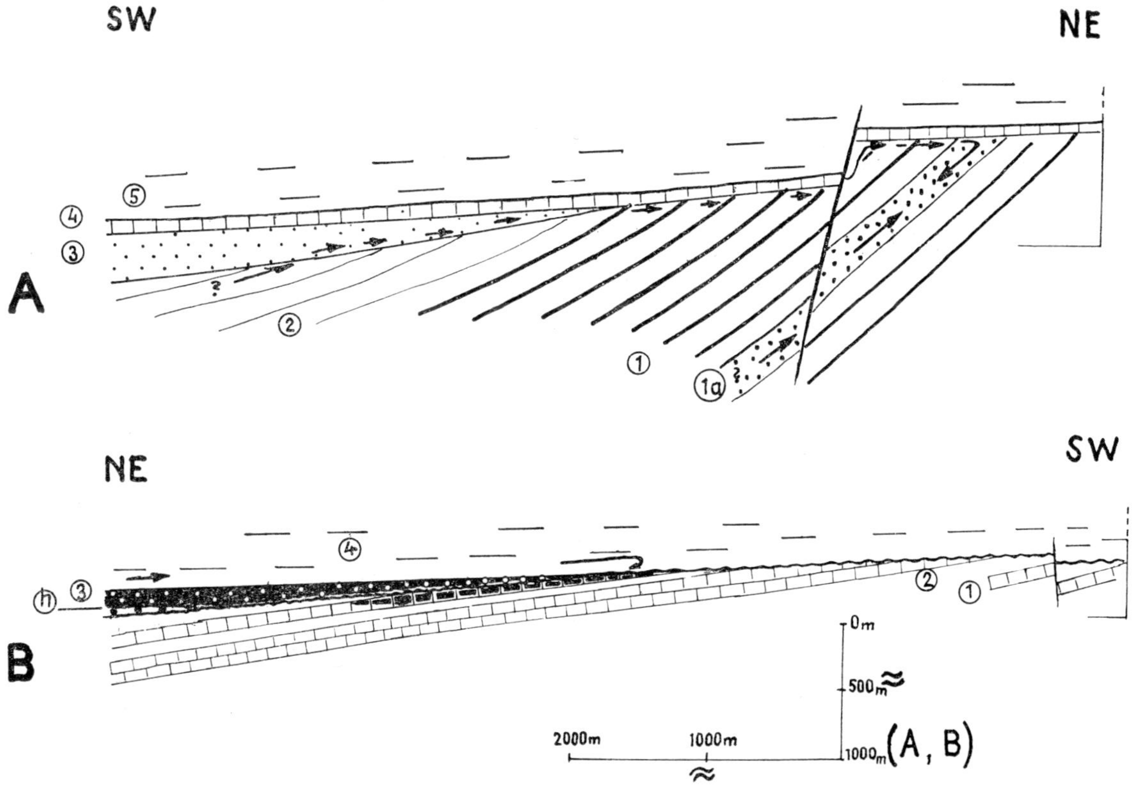

FIG 9. Schematic comparison of two accumulation possibilities in Poland. A. Chojnice-and-shore zone—dry gas found in lower Carboniferous sandstones may have migrated or still migrate either from upper Carboniferous coal-bearing series elsewhere and up-dip Rotliegend wedge as well as post-Hercynian weathered surface and then to be down-pressed or immediately up-dip of its present-day reservoir. 1. Lower Carboniferous (1a–sandstones), 2. Upper Carboniferous (presumed), 3. Rotliegend permeable wedge, 4. Zechstein unpermeable cover, 5. Supra–Zechstein impermeable sequence. B. Near-Cracow zone. (Fig 5 schematically rearranged according to oral suggestions kindly made by E. Jawor). Oil migrates up-dip of permeable detritic cenomanian wedge and therefrom down-dip into permeable complex in Malm limestones, another likewise permeable limestones remaining dry where beyond the reach of Cenomanian wedge: Malm, 1. Permeable fissured limestones, 2. Impermeable marly limestones and marls, Cretaceous, 3. Cenomanian sandstones and conglomerates, 4. Supra-Cenomanian impermeable marls, black-oil, h-water stable. Both diagrams scale approximate.

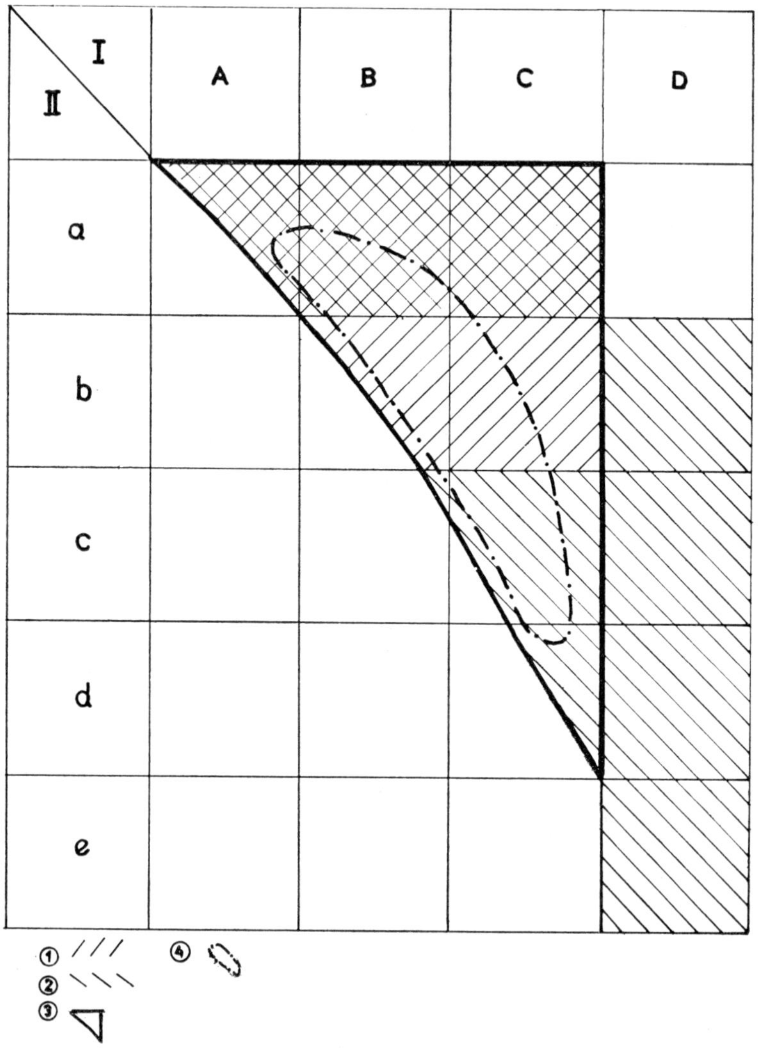

FIG 10. Diagram showing "epigenetic predominance" even for classical mother rocks (Unpublished data and views elaborated as well as kindly supplied by J. Calikowski). I—Substance classification, A—petroleums, B—Micro-petroleums, C—Bitumens highly metamorphosed, D—Bitumens middle and low metamorphosed, II—spatial relation to rocks, a—oil migrating and being trapped, b—oil passed by, c—oil not yet mobilized, d—post-mobilization residuum, e—dry rocks, 1—epigenetic bitumens, 2—syngenetic bitumens, 3—field of petroleum-like bitumens, 4—field of Polish Silurian graptolite shales.

Some regionally comparative or zonally connected models have been tested in their structural and substantial relations (Fig 13).

Also, an introduction of studies of water-filled anti-deposits in outcropping structures supplies helpful reverse models (Fig 14).

SYNOPSIS AND INVENTORY

Corresponding to prospection needs, a new natural geological sub-division of Poland has been elaborated by J. Sokołowski (1968) in which paralled to the hierarchical descent grows the exactitude of diagnostics. This descent would lead from the first order differences concerning not only the most impressing regional effects of the geological history but also the present-day hypsommetry, through the second order varieties of geological profiles and third order ones in local styles of form groups to the fourth order selections of closed contours.

The boundaries of such poly-storeyed structural areas, megastructures, structural units, and structural elements (Fig 2) may be heterogenous ones. Nevertheless, inside of the latter, structural conditions would be the same from the prospectional point of view, although not necessarily concerning lithostratigraphic as well as diastrophical homologies. Reversely, strictly enough the same would be depths to test by geophysics and boreholes the gas and oil chances in successive storeys in each sub-division.

Notwithstanding the efficiency of such an adaptation of the country to prospection management, as well as for elaborating controlled statistical indexes and for evaluation of reserves, it emphasizes some purely geological interrelations too, which would otherwise be concealed. This means some devaluation of especially the widespread apparent effects of the Laramide re-building "incident" (*cf* Figs 1 and 2).

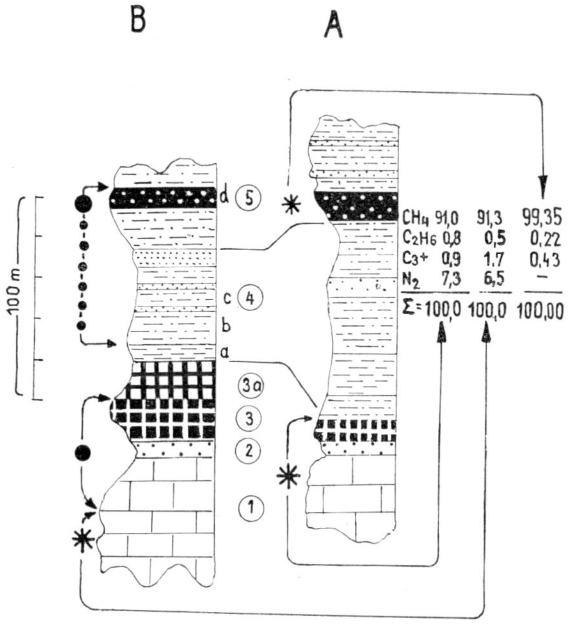

FIG 11. Synthetic profiles showing positions of dry and wet gas as well as of oil both in sub-Miocene structures and Miocene cover of Carpathian foreland (A-close to eastern frontier, B—100 km East of Cracow). 1. Upper Oxfordian pinguis limestones, 2. Lower Tortonian Baranow sandy and marly denudatum beds, 3. Lower Tortonian anhydrite (sa—dry), 4. Upper Tortonian sequence (a—Spirialis clays below, b, c—resp. mudstone and sandy Dentrophyra beds above), 5. Lower Sarmatian Anomalinoides beds (mudstone and sandy lithology as for 4), gas—black stars, oil production—large black dots, oil traces—small black dots.

EFFECTS AND CHANCES

To practical results elaborated up today by the above presented kind of geological work in Poland first belongs the start of non-Carpathian discoveries and breaching over the long time non-surpassable limit of 200,000 tons/year. Further, the guided extrapolation of oil search into the country of the E fore-Carpathian gas bearing Miocene zone must be considered. Lastly, we have the establishment of a Rotliegend gas producing structural polygon in Central Poland between Wrocław and Ostrów Wielkopolski (Fig 8).

In resumé, the Polish geological research for oil and gas has elaborated its own methods in an anti-traditional way, based upon, as well as tested by, specified geological features of the country. Moreover, this action contributes to the worldwide recognition too, by providing new non-classical deposit examples and patterns to their worldwide inventory. The latter, therefore, now tends to become a structural system handy to operate with.

ACKNOWLEDGMENTS

The authorities of both the Geological Institute and of the geological staff of the Polish petroleum industry kindly enabled the authors to dispose of the newest data as well as of newest composures, not yet published. Especially, Dr R. Dadlez, Dr F. Ekiert, E. Jawor, Dr H. Jueriewicz, J. Kucuła, and Professor J. Oberc generously supplied their unpublished

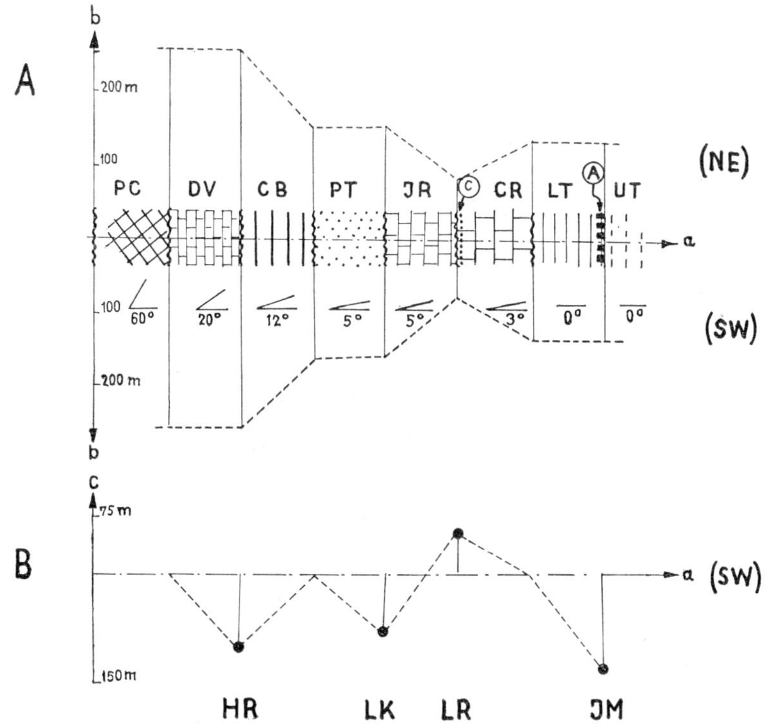

FIG 12. Scheme emphasizing the course of movements on a longitudinal dislocation in the near-Cracow zone (joint plots: A—by E. Jawor and B—by A. Tokarski, unpublished). a—time axis deformed, b—summarized displacements, c—differential displacements, PC—Pre-Cambrian (phyllites), DV—Devonian (sandy below, carbonatesa bove), CB—Carboniferous (shaly-sandy)—PT Permotriassic (detritic variegated), JR—Jurassic limestones, Cr—Cretaceous marls with detritic Cenomanian at the bottom (c), LT—lower Tortonian marly sandy complex, with anhydrite (A) at its top, UT—upper Tortonian mudstone sequence, HR, LK, LR, JM—resp. Hercynian, Late Kimmerian Laramide, and intra-Miocene movements (NE) and (SW) respective limb pictures (occasionally symmetric because of preservation of concerned series to both sides of presented dislocation).

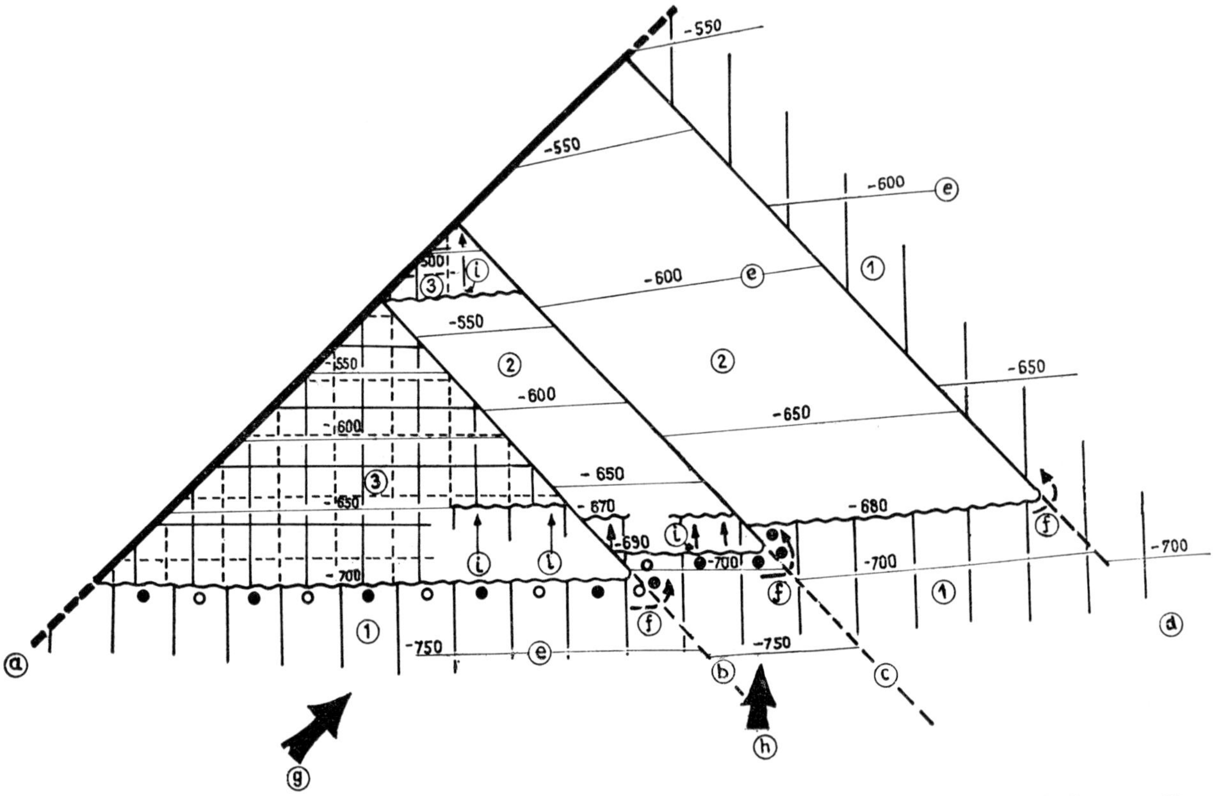

FIG 13. Scheme combining a case of primary reverse hypsommetry of boundaries of mediums (after W. C. Gussow, *Alta*, Soc. Petrol. Geol. 1953 in Levorsen 1956, p. 556) with its possible mirror-like rearrangement by new pression conditions (A. Tokarski, unpublished). 1. Water, 2. Oil, 3. Gas, a–oblique-longitudinal breaking dislocation, b, c, d–oblique-perpendicular breaking dislocations arranging shear-displaced blocks as shown by pisobaths "e" of the top of a relatively thin reservoir, with occasional spill-places "f"; from the latter each one higher situated following from left to right, g–buoyancy front at 60 atm, creating primary respective contours respectively, at −700 m (water gas), −690 m (water oil) and −680 m (water oil) from left to right as well as −530 m (oil gas) in the middle block, h—new pressure of 80 atm, rearranging "j", two left boundaries resp, to −650 m and −670 m as well as entirely cancelling gas cap in the middle block. Block dots—migrating oil, circles —migrating gas.

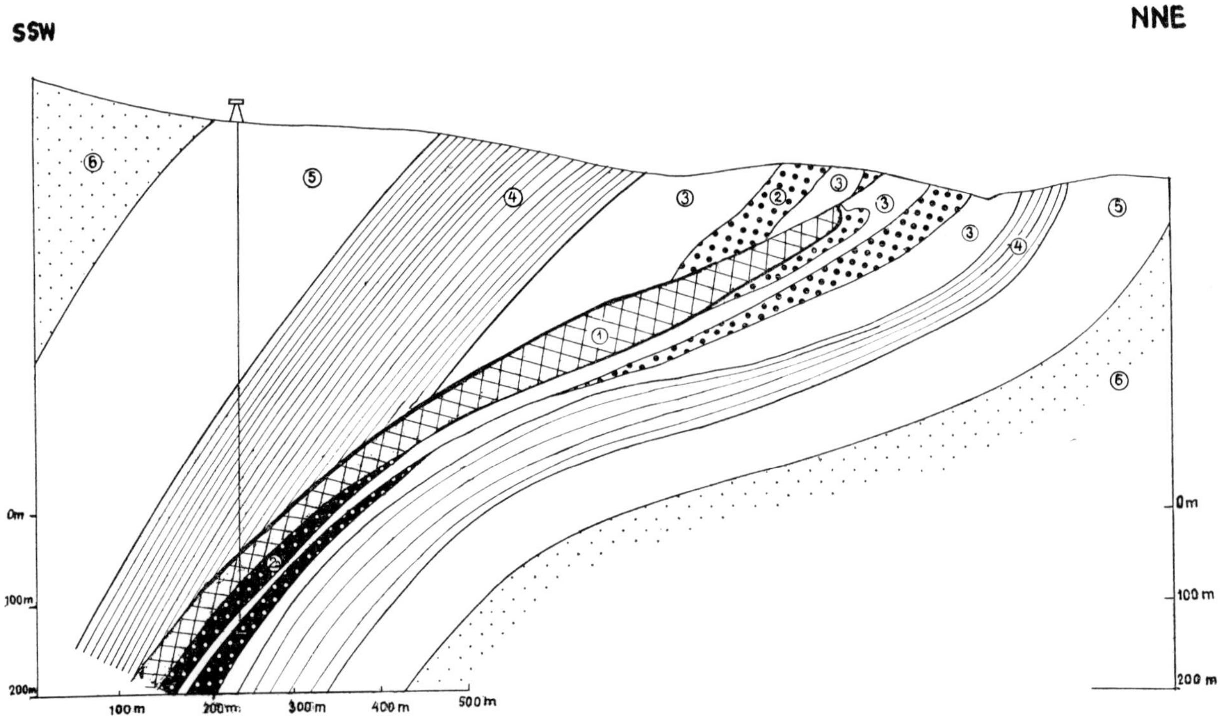

FIG 14. An instance of reversed-deposit prospection factors in Flysch (A. Tokarski, unpublished-water-bearing sandstone outcrops led to a discovery of oil deposit in the same sandstones in the subsurface). Stratigraphy (Eocene– to Oligocene): 1. Variegated shales, 2. Cergowa sandstones (black background where oil-bearing), 3. Cergowa yellowish shales, 4. Menilite black shales, 5. Passage grey-black shaly beds, 6. Krosno grey sandstone-shale beds.

maps of the post-Hercynian palaeogeology of the Baltic shore zone, W of Łeba, and the fore-Sudetic area, Cracow region included. Also Dr L. Miłaczewski and A. M. Żelichowski kindly placed their unpublished structure map of the Lublin Basin at the authors' disposal. The same has been kindly done by Dr H. Szaniawski of the Polish Academy of Sciences concerning his sub-crop map of the bottom of Bunter of the Łeba uplift.

Moreover, Dr J. Calikowski, of the Geological Institute, and Dr P. Karnkowski, of the Polish petroleum industry, discussed as well as suggested topics concerning respectively the mother rock problem and gas migration. Dr H. Kozikowski, of the Petroleum Institute, influenced the authors by his non-standard regional cartographical pictures. Dr K. Korejwo and Dr L. Teller of the Polish Academy of Sciences kindly contributed to Fig 7 by their unpublished faunistic determinations.

To all of them, as well as to many others not mentioned here, the authors express their sincere gratitude.

BIBLIOGRAPHY

Antsupov, P. V., Budnikov, N. P., and Skvartsova, O. N. *Geologya Nyephty y Gaza*, 1965, **11**, 10–16.
Areń, B. *Kwart. Geol.*, 1967, **11**, 529–46.
Calikowski, J., and Gondek, B. *Biul. Inst. Geol.*, 1967, (213), 171–84.
Dadlez, R., and Młynarski, S. *Kwart. Geol.*, 1967, **11** (3), 68–74.
Depowski, S., and Tyski, S. *Przegl. Geol.*, 1968, (7), 324–30.
Heybroek, P., Haanstra, U., and Erdman, D. A. "Observations on the Geology of the North Sea Area". *Proc. 7th World Petrol Congr*, 223–42.
Jawor, E., and Kruczek, J. *Geofizyka i Geologia Naftowa*, 1965, (4/100), 97–108.
Karnkowski, P. *Przegląd Geologiczny*, 1968, (4), 164–79.
Kłapciński, P. *Ann. Soc. Geol. Pol.*, 1967, **37**, 467–88.
Komorowska-Błaszczyкska, M. "Zaburzone utwory dolnego sarmatu w profilach nasunięcia brzeżnej strefy Karpat". Spraw. z pos. PAN, Kraków, lipiec-grudzień, p. 795–800.
Korejwo, K., and Teller, L. *Acta Geol. Pol.*, 1968, **18**, 153–77.
Kruczek, J. *Bull. Ac. Sc. Pol., Ser. Geol.*, 1962, **10** (1), 16–64.
Kruczek, J. *Biul. Inst. Geol.*, 1968, (215), 79–156.
Kruczek, J. *Nafta, Krakow*, 1968, (4), 100–09.
Lendzion, K. "The Cambrian of the East-European Platform in Poland" 23rd International Geological Congress, 1968, **9**, 109–17.
Levorsen, A. I. "Geology of Petroleum". W. H. Freeman, San Francisco, 1966, 703.
Miłaczewski, L., and Niemczycka, T. *Kwart. Geol.* 1967, **11** (3), 557–71.
Modliński, Z. *Kwart. Geol.*, 1967, **11**, (1), 68–75.
Sokołowska, J., and Sokołowski, J. *Geofizyka i Geologia Naftowa*, 1965, (10–12/106–108), 290–5.
Sokołowski, J. *Geologia Sudetica*, 1967, **3**, 267–97.
Sokołowski, J. *Surowce Mineralne*, 1968, **1**, 7–58.
Sokołowski, J., and Zelichowski, A. M. *Geofizyka i Geologia Naftowa*, 1968, (3–5/135–137), 91–92.
Tokarski, A. *Nafta, Krakow*, 1966, (9), 257–65.
Tokarski, A. *Nafta, Krakow*, 1968, (12), 345–60.
Wierzchowska-Kiculowa, K., and Kowalska-Łącka, M. *Nafta, Krakow*, 1969, (2), 33–37.
Znosko, J. *Kwart. Geol.*, 1964, **8** (1), 60–72.
Znosko, J. *Kwart. Geol.*, 1965, **9** (3), 465–88.

MS received 24 April 1969.

Recent Development in German Exploration for Oil and Gas

By ARTUR ROLL*

(*Gewerkschaft Elwerath, Hannover, Federal Republic of Germany*)

SUMMARY

Exploration is concentrated at present in the producing provinces of NW Germany and the Alpine Foreland. It is aimed predominantly at natural gas. There is no activity in the Upper Rhine graben and exploration has ceased in other prospective areas.

In NW Germany the expectations for additional Lower Cretaceous, Jurassic, and Rhaetian discoveries are related to the search for less obvious traps and palaeo-structures. Recent strikes in the surroundings of Hannover have given new impetus to the search in the Lower Lias and Rhaetian in particular.

The Triassic trough of Glückstadt, with extraordinarily thick deposits of Keuper, is being evaluated for its exploration potential.

While the Buntsandstein, Zechstein, and Carboniferous are developed within the gas-producing area and its vicinity, exploration of the Rotliegendes and Carboniferous (outside the proven area) still remains without conclusive results, except for discoveries in the extreme NW and E, respectively, in spite of extensive efforts.

The exploration of the Tertiary and the Mesozoic rocks in the Alpine Foreland continues. The Chattian has attained a leading role as a gas producer. Within the older Tertiary section high-GOR oilfields have been discovered closer to the Alps. Drilling within the folded sub-alpine belt and the northernmost Alps has been unsuccessful and is suspended at present.

INTRODUCTION

THE oil and gas fields of the Federal Republic of Germany are concentrated within three regions, NW Germany, the Alpine Foreland, and the Upper Rhine graben. The ultimate recovery from existing fields (cumulative production and present reserves—in accordance with the official statistics both proven and probable) amounts to 194 million tons of oil and 273,000 million Ncbm of natural gas at the beginning of 1968. The three regions shares are 95·4, 3·4, and 1·2 per cent, respectively. The cumulative production of oil approached 100 million tons, that of gas 15,320 million Ncbm.

The search for oil seems to have arrived at an advanced stage. The search for gas, however, is in the midst of development. In 1967 the total meterage of new and deeper pool wildcats was 95,329 m, being divided into 12 per cent for oil and 88 per cent for gas. The meterage for gas exploration first exceeded that for oil in 1961. From that year onwards the average depth of wildcats increased from 1932 m to 3219 m. During the same period the discovery rate of 1 : 8 remained almost constant. Within the past five years the growth of reserves from wildcat and outpost drilling has totalled 5,900,000 tons of oil and 220,000 million Ncbm of gas. This has been derived from actual figures by deducting the estimated volumes as due to successful secondary recovery operations in older fields.

Following the generally disappointing results in the Upper Rhine Graben, exploration has practically come to a close. Since 1968 no wildcat has been spudded. The efforts in the Carboniferous of the Saar province, in the shallow Zechstein of Central Germany, and in the Mesozoic strata of the Swabian and Franconian Alb have failed. Only NW Germany and the Alpine Foreland are therefore discussed in the following. Both belong to larger basins extending beyond the international boundaries. It has been attempted to indicate in Fig 1 the position within the regional framework.

NW GERMANY

Fig 2 represents a generalized stratigraphic cross-section through the basin. Many details of the development of thicknesses and of unconformities have had to be simplified or omitted; even the special stratigraphic features related to essential major structural elements, such as the West Schleswig Block (Hecht *et al.*, 1955) the Gifhorn Trough, or the Hunte Swell (Trusheim, 1961) could not be included. The description of structural types below the cross-section refers to their present state and does not take into account their geological evolution. As many detailed sections, palaeogeographic maps, and stratigraphic tables have already been published, showing more complicated structural features, in particular the intensive salt sectioning, it was thought advisable to concentrate on the basic facts.

Three sedimentary cycles can be distinguished. Only the unconformities separating these cycles, *i.e.* the one at the base of the Albian and the other at the base of the Permian, are indicated. It should be mentioned that these events are not precisely of the same age everywhere and may not be developed locally.

* In collaboration with Herbert Lögters, G. Deilmann AG, Bentheim, Federal Republic of Germany.

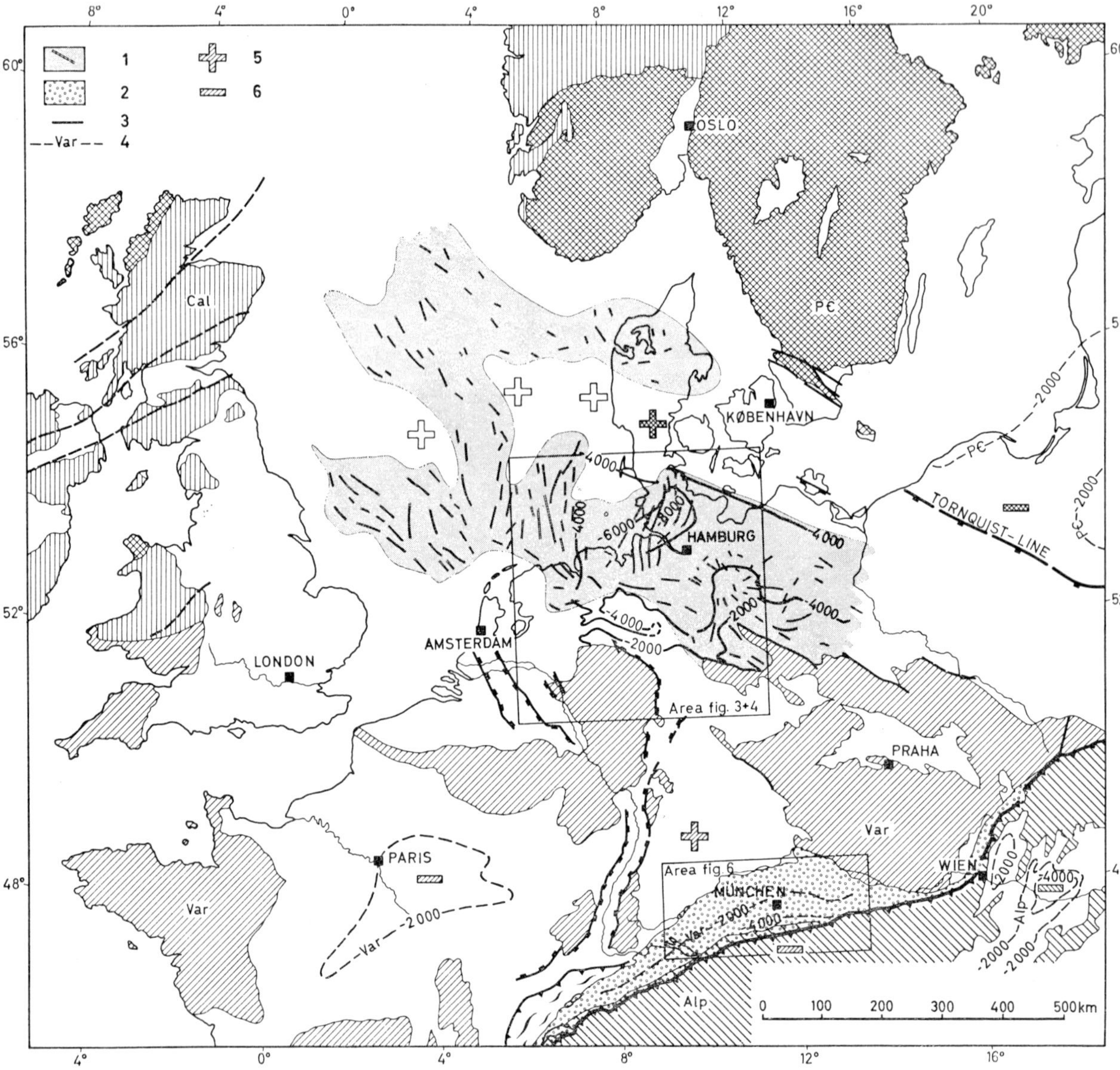

FIG 1. NW Europe. Tectonic sketch map.

Explanations

PE. Pre-Cambrian at or near surface.
CAL. Caledonian orogene at or near surface.
VAR. Variscan orogene at or near surface.
ALP. Alpine orogene at or near surface.

1. Area of mobile Zechstein salt with schematized structures.
2. Alpine Foreland—Tertiary at or near surface.
3. Contours on the base of the Zechstein in metres below sea level.
4. Contours on respective basement.
5. High block with indication of basement correlation where possible.
6. Deepest parts of basins, with indication of basement correlation.

The map was compiled using data from Albrecht (1967), Commission de la Carte Géologique du Monde (1964), Franke (1967), Heybroek *et al.* (1967), Hinz (1968), Meinhold *et al.* (1967).

C 1. Within the upper cycle only sporadic and sparse oil shows have been found so far, despite the sediments being marine. Oil in the Upper Cretaceous on top of the Reitbrook and Meckelfeld salt plugs and some Tertiary gas at Reitbrook are directly related with Dogger sediments nearby. It remains to be seen whether the North Sea exploration will eventually stimulate further prospecting.

C 2. The middle cycle is characterized by salt sequences in many members. It is present in all fields, with a few exceptions (Carboniferous) described below, and it is divided into an upper mainly oil-bearing division, a middle, so far barren, division, and a lower gas-bearing division.

C 2(a). The upper division comprises Lower Cretaceous, Jurassic, and Rhaetian. Some basic facts are

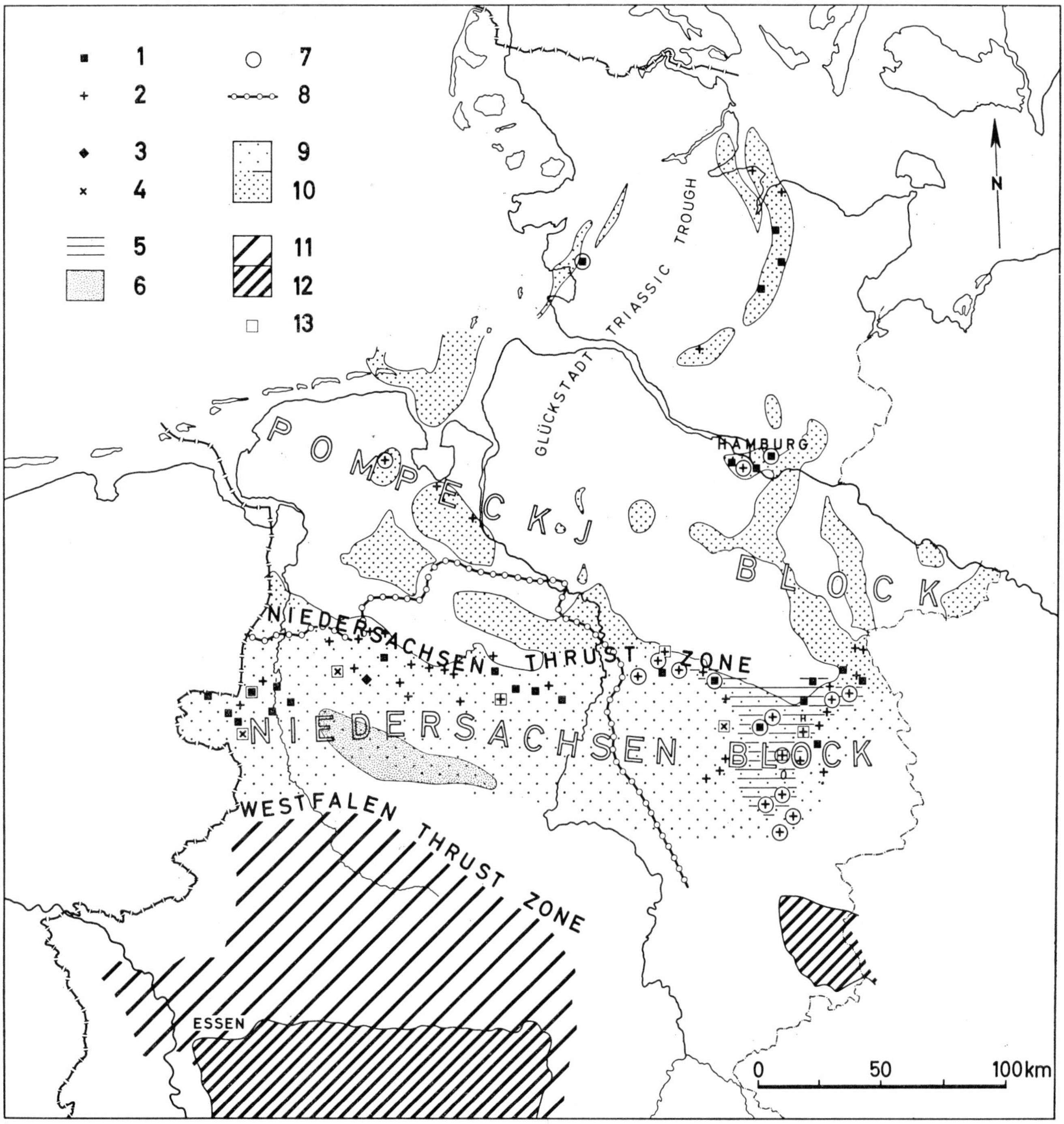

FIG 3. NW Germany. Lower Cretaceous to Rhaetian oil and gas fields. Thrust zones after Philipp (1954).

Explanations

1. Oilfield—ultimate recoverable reserves exceeding 2 million tons.
2. Oilfield—ultimate recoverable reserves between 2 million and 20,000 tons.
3. Gas field—ultimate recoverable reserves exceeding 2000 million Ncbm.
4. Gas field—ultimate recoverable reserves between 2000 million and 20 million Ncbm.
5. Hydrocarbons contained within Lower Lias and Rhaetian.
6. Bramsche massif: anthracite extension in the Wealden formation, after Teichmüller (1954).
7. Field associated with salt diapir.
8. Southern boundary of salt diapir extension, generalized.
9. Occurrence of Lower Cretaceous, Malm, and Dogger on Niedersachsen block.
10. Occurrence of Dogger and Pompeckj block.
11. Variscan orogene and foreland, subsurface.
12. Variscan orogene, surface.
13. Location of new field or new pool discoveries.

Ö. Ölheim-Süd field.
H. Hardesse field.

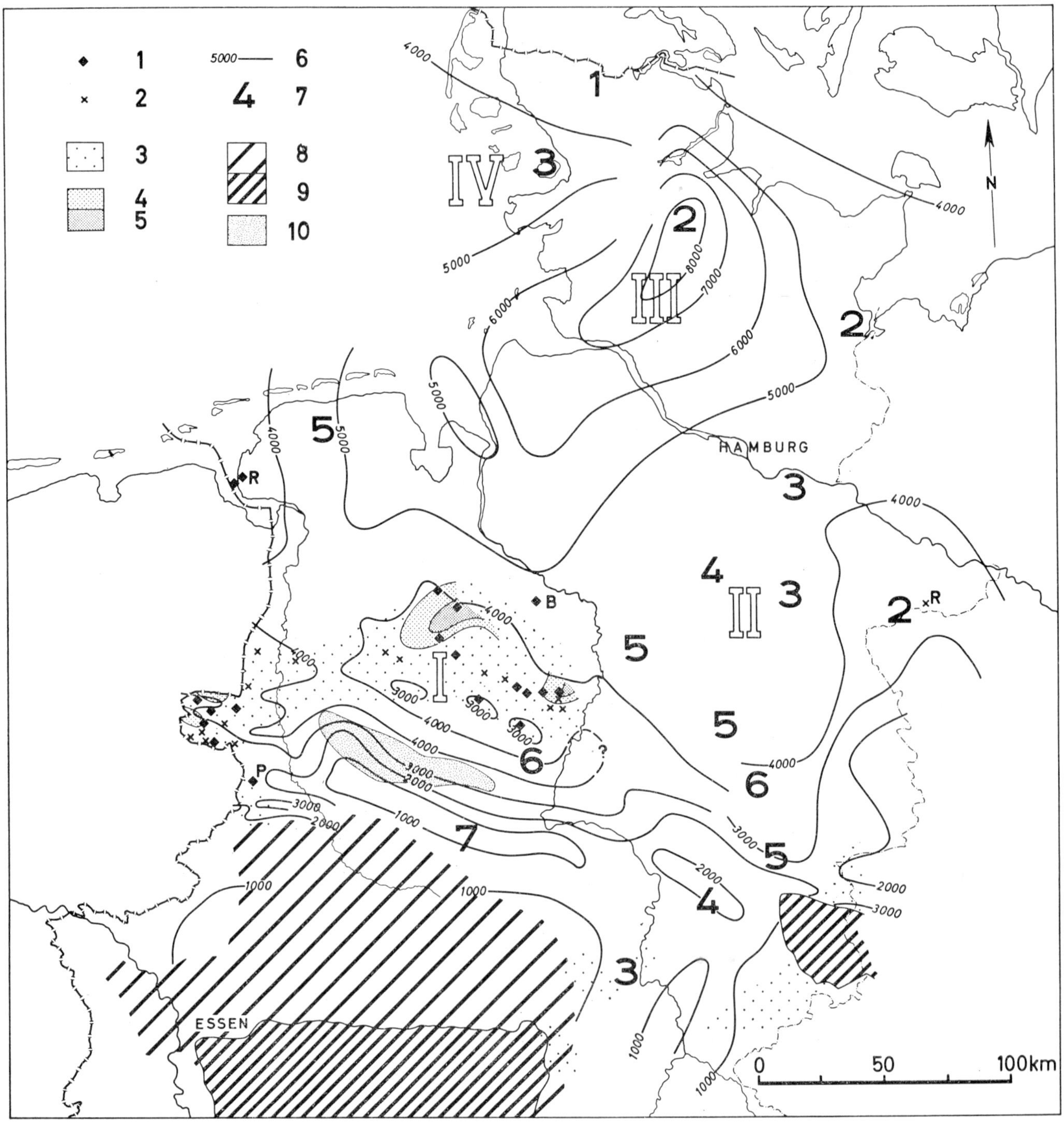

FIG 4. NW Germany. Buntsandstein and Palaeozoic gas fields.

Explanations

1. Gas field—ultimate recoverable reserves exceeding 2000 million Ncbm.
2. Gas field—ultimate recoverable reserves between 2000 million and 20 million Ncbm.
3. Algal reservoir facies within the Zechstein Hauptdolomit, as known actually (identical with the "gas province").
4. Zechstein Hauptdolomit, area of thickness proven more than 100 m.
5. Zechstein Hauptdolomit, area of thickness proven more than 200 m.
6. Contour lines on the base of Zechstein, simplified, from well data and seismic trends in metres below sea level. Faults omitted.
7. Vertical distance between the base of the Zechstein and 6000 m/sec refractor, in km.
8. Variscan orogene and foreland, subsurface.
9. Variscan orogene, surface.
10. Bramsche massif: anthracite extension in the Wealden formation, after Teichmüller (1954).

B. Isolated Buntsandstein field (Barrien).
P. Isolated Plattendolomit (Zechstein) field (Bentheim).
R. Isolated Rotliegendes field, in the W: Groothusen, in the E: Wustrow.
I. Area of Hunte swell.
II. Area of Gifhorn trough.
III. Area of Glückstadt Triassic trough.
IV. West Schleswig block.

summarized in Fig 3. With the improvement of seismic methods it has been possible since 1948 to intensively explore for prospects other than those connected with salt diapirs. The high intensity of drilling and the concentration of the prospects within restricted areas result in nearly all first-order projects having been tested. Nevertheless, several discoveries have been made in the past few years; those of oil are Barver (Malm, 1963), Rühlermoor (Malm, 1963), Ahrensheide (Valentian, 1964), Hardesse (Rhaetian, 1967, with extensive gas cap), and Ölheim-Süd (Valentian and Wealden, 1968) and, for gas, Holte (Malm, 1963), Hohenkörben (Valentian, 1963), and Thönse (Rhaetian, 1968).

Based on a conversion of one ton of oil equivalent to 1000 Ncbm of natural gas, the ultimate recoverable reserves of the upper unit are as follows: oil 92·2 per cent, natural gas, including gas caps, 3·8 per cent. The discovery rate for gas was therefore much higher than in earlier times.

Within the Niedersachsen block three groups of fields can be identified: In the west the Emsland province accounts for 41 per cent of the reserves; along the border between the Niedersachsen and Pompeckj blocks an elongated zone accounts for 20 per cent and an eastern province 17 per cent of the reserves. The fields on the Pompeckj block, widely distributed and exclusively related to Dogger preserved underneath the Lower Cretaceous unconformity, represent a share of 22 per cent. These percentages differ considerably from the oil-in-place figures, as the recovery rates are generally better on the Pompeckj block than in many fields of the other groups. Considering the intensive faulting and the complicated structural history, some blocks could exist which are now off-structure but which could have been in prominent positions at the time of migration with the respective accumulations having been sealed off. Similarly, individual faults at the flanks of salt diapirs may have trapped accumulations not yet discovered, as evidenced anew with Ölheim-Süd (Ö in Fig 3) at the diapir of Eddesse, where the flanks have been exploited since 1876. The strike of Hardesse (H in Fig 3) could be the incentive for a revival of Lower Lias and Rhaetian exploration. The area in which the Lower Lias and Rhaetian are productive, in addition to other pays, is marked in Fig 3.

There is little doubt that the hydrocarbons are derived from various Mesozoic sources.

C 2(b). The middle division is comprised of Keuper, Muschelkalk, and Upper Buntsandstein. In particular, the Keuper attains an extraordinary thickness locally, predominantly due to salt intercalations. The most typical example is the Glückstadt Triassic trough (Figs 2 and 3). It may be an important task to prove whether or not the non-evaporitic members of the section differ significantly from the beds making up the thinner standard sequences elsewhere.

C 2(c). The lower division is composed of Middle and Lower Buntsandstein, Zechstein, and Rotliegendes.

It is felt that no discussion of the Buntsandstein is necessary. It may be mentioned that some of the Buntsandstein fields are restricted to domal structures above salt pillows.

The Zechstein base is the lowermost horizon to be mapped reliably by seismic reflection throughout the basin. In general there are distinct and continuous reflections except for areas below salt diapirs. However, the conversion of travel times into true depths cannot be performed satisfactorily, as several factors combine to cause the velocity profiles of individual wells not to follow a simple systematic trend and so permit only a largely approximate reconstruction of the energy paths. The intensive salt movements effective from Triassic times onwards have led to considerable small-scale variations in thickness of many stratigraphic units superimposed upon the regional trends and not always sufficiently recognizable seismologically. This is accompanied, in places, by important changes in lithology. In addition, velocity anomalies result from the tectonic history.

It appears that north of the 4000-m contour the tectonic intensity is remarkably less than to the south. Some regional features are denoted in the generalized structural map (Fig 4).

Area I, characterized by the 4000-m depth contour protruding to the N, coincides in principle with the Hunte swell, an important element in late Palaeozoic to Buntsandstein times. There are some broad structural axes striking generally NW–SE and plunging to the NW. Faulting is responsible for the existence of individual highs on these trends. Those highs were the main objective of the exploration

The testing of isolated blocks within the Emsland province in the extreme west also led to several discoveries.

The distribution of the Hauptdolomit reservoir facies coincides roughly with the area of intensive tectonics W of the Weser river. This area lies north of the Laramie intrusion of the Bramsche massif, the heating effect of which may have influenced hydrocarbon accumulations to an unknown degree. Carbon dioxide, emitted from the intrusion, is found in porous rocks in its vicinity.

Since the latest interpretation of the exploration results was given (Füchtbauer, 1964), some new aspects have been revealed. It is generally accepted that the calcareous sediments were dolomitized at an early diagenetic stage and that the secondary porosity determining the reservoir properties is generally restricted to the algal facies of the Hauptdolomit. Locally, some porosity is developed in the very upper part of the slope deposits. Fig 4 shows the distribution of the reservoir rocks.

The Hauptdolomit was known to attain its maximum thickness (78 m) only where the underlying anhydrite member of the first cycle of the Zechstein evaporite sequence showed an increase along the boundary between the lagoonal facies and the deeper facies of the basin.

Recently, however, dolomite was found to be up

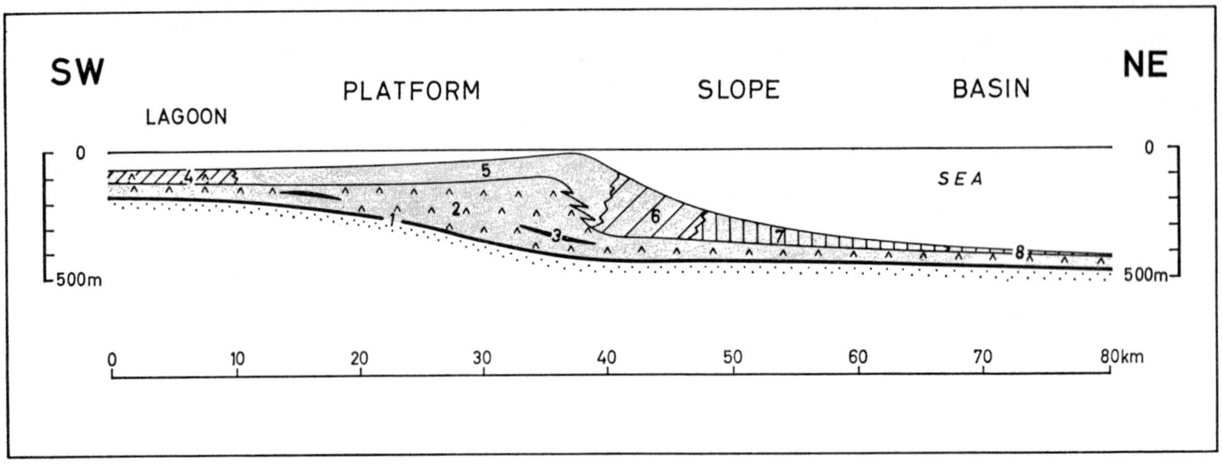

FIG 5. NW Germany. Diagrammatic cross-section through a marginal part of the Zechstein basin prior to Anhydrite 2 evaporation, referring to the situation in the SE area of increased thickness. (Basin morphology according to Richter–Bernburg, 1959.) Pre-Anhydrite 2 comple shaded.

1. Zechstein conglomerate, Kupferschiefer, and basal Zechstein limestone transgressing over Upper Carboniferous or Rotliegendes.
2. Anhydrite.
3. Rock salt.
4. Anhydritic dolomite.
5. Dolomite pay, algal facies.
6. Stinkdolomit.
7. Stink-kalk.
8. Stinkschiefer.

to 250 m thick in two areas on the basinal slope adjoining the anhydrite ridge. The observation that dolomite and anhydrite are laterally interfingering gives evidence that the carbonate deposition already started during the regressive phase of the first cycle. Since then salt intercalations were also encountered occasionally, the inter-relation between precipitation and the growth of the algal mats is probably quite complicated and needs further detailed investigations. The situation is demonstrated in Fig 5.

At several places large pay thicknesses were encountered, with rather variable reservoir properties but with large gas reserves. Since a regional belt may perhaps exist, attempts are made to locate such occurrences of increased dolomite thicknesses seismologically, although the similarity of velocities of anhydrite and dolomite causes a major difficulty.

Lithological differentiation of the upper slope and an increasing understanding of the complicated depositional pattern could encourage the search for stratigraphic traps.

Outside the extension of the actual reservoir facies, the interior of the basin seems not to be prospective for the Zechstein. This is supported not only by palaeogeographic arguments, but by the improbability that there could exist an area large enough to hold an independent platform with algal carbonate sedimentation, in view of the density of wells within drillable depths.

It may be significant that the gas fields discovered in several sandstone members of the Buntsandstein are restricted to the reservoir facies area of the Zechstein (only one field is situated slightly outside). They occur predominantly within structures productive in Zechstein.

The area productive from the Zechstein and Buntsandstein is termed the gas province in the following. In its southern part, where the Rotliegendes is generally missing, the uppermost sands of the Upper Carboniferous, directly underlying the Zechstein transgression, are productive only in blocks which also produce from Hauptdolomit.

At the beginning of 1968 the reserves of the gas province were divided between Carboniferous, Zechstein, and Buntsandstein as 8·5, 55·5 and 36 per cent, respectively.

Exploration of the Rotliegendes has probably passed the initial stage of stratigraphic fact-finding. The essential problem rests in the comprehensive understanding of the relations between volcanics, clastics, and evaporites. It appears that the general principles can be recognized. As the Rotliegendes, towards the centre of the basin, descends to considerable depth, the effects of diagenetic processes and their timing are believed to be crucial for the conservation of both sufficient porosity and permeability.

Gas fields have been discovered in the extreme west, *e.g.* the extension of Slochteren, and, of modest size, in the extreme east near Wustrow. Non-commercial gas was tested from sandstones and fractured volcanics in several wells located in deeper parts of the basin.

C 3. The lower unit has been reached outside the Zechstein–Buntsandstein province by several wells. Gas shows have been recorded in some of them in the Upper Carboniferous. In a wildcat, situated about 80 km SW of Hamburg, sandstones with low permeability and apparently containing large volumes of gas could not be stimulated sufficiently for commercial production.

The possibility of finding hydrocarbon accumulations in the foreland of the Variscan orogeny in Carboniferous and Devonian rocks encouraged the exploration in the Münsterland (Fig 4) north of the coal mining belt for many years. Although the original concept has changed fundamentally in various aspects, it may still be hoped that fields in the pre-Zechstein Palaeozoic may also be found in the German sector. These could be developed in the basin area outside the existing gas province, similar to the distribution of the Dogger fields in the upper unit on the Pompeckj block.

Below the base of the Zechstein, diagenesis has apparently led to comparatively uniform velocities. Except from marginal zones, only isolated reflections could be traced sporadically. As this represents one of the key problems of the actual exploration, much effort has been spent in applying the latest field and processing techniques for improved efficiency. Unfortunately, up till now no decisive results have been obtained.

It can be assumed that angles of unconformities, if present at all, are slight. In central parts of the basin the Zechstein can rest without break on the Rotliegend (Plumhoff, 1966) and the Rotliegendes, in turn, can overlie the Carboniferous conformably. Nevertheless, it is unsatisfactory to deduce the structural configuration of the pre-Zechstein beds only from the Zechstein base downwards, especially in view of the difficulties in depth contouring this horizon, as previously explained.

In view of this situation, it was attempted to discover by other methods the thickness of the entire sedimentary sequence and the main features of the configuration of the top of the basement. The industry jointly performed a refraction seismic survey throughout the basin. Details of the technical arrangement and related problems will be discussed by Krey and Nodop (in preparation).

Small refraction horizons could be recorded locally and, through well control, be assigned to the Devonian "Massenkalk" and the Rotliegendes volcanics, respectively. The body of the Bramsche massif intruded in the Palaeozoic appears to be depicted as well.

A refractor with a velocity of more than 6000 m/sec was followed continuously. In the extreme N it can be connected with the crystalline basement of the Fünen high. Elsewhere it is not known if and where the refractor corresponds to the top of the crystalline basement, or whether it represents an abrupt change of the elastic behaviour of sediments. At any rate, the thicknesses of the non-metamorphic fill of the basin are outlined by the order of magnitude. In spite of the wide spacing of survey lines and with the final interpretation still incomplete, representative values for the section between base Zechstein and the 6000 m/sec refractor can be given in Fig 4. The gap in the survey between the Lower Weser and the Lower Elbe rivers will be completed in the near future. The construction of a depth structure map may then be possible.

In the cross-section of Fig 2, the 6000 m/sec refractor has been delineated schematically. Its course, as well as information given in Fig 4, shows that the regional compensation of thickness in the basin, well known from younger members, seems to have been active since Palaeozoic times. It is intriguing that the base of Zechstein situated in the middle part of the sedimentary section shows the configuration of a quite regularly shaped basin (Fig 4) in which, however, certain tectonic elements are still faintly visible, the Hunte swell (I), as a positive area, and the Gifhorn trough (II), as well as the central part of the Glückstadt trough (III), as negative anomalies.

At the present state of exploration, it must be left to speculation as to what degree and in which development the several stratigraphic units attain in those areas and those depths where no well information is yet available.

As far as the Devonian is concerned, it will perhaps be one of the main tasks to look for the boundary between the Old Red Sandstone and the marine sediments of the Variscan foreland; south of this line reefs as known from the outcrops in the Variscan mountains could possibly be found. The Devonian sections in the deep wells, *e.g.* Münsterland 1 (Richwien *et al.*, 1963), situated about 60 km NNE of Essen, drilled in the northern foreland of the Rhenish Variscans are too distant to provide the relevant information.

It is an open question whether or not a Caledonian orogene is present. Considering the drilling results on Rügen (Franke, 1967) and in southern Denmark (Sorgenfrei and Buch, 1964), and in view of the more and more accepted theory that the Caledonian orogeny cannot be recognized in the Harz Mountains, unfolded Silurian, even Cambrian strata may be found.

The percentage of nitrogen in the gas is subject to considerable variations. It is not possible at present to give an explanation why no gas is found in the Buntsandstein outside the Zechstein reservoir area, in spite of the presence of both good structures and good reservoirs.

Perhaps theoretical investigations could assist in outlining the preferential projects and areas for future exploration. Reliable knowledge about source rocks and the mechanism of horizontal and vertical migration, as well as the mechanism of nitrogen input into the different reservoirs, is highly desirable.

It is valuable that initial studies to solve the abovementioned problems have been undertaken and that there is hope for good results which in their final combination may lead to a conclusive theory.

The explanation given by Patijn (1963) for the origin of the Slochteren gas may be also applied to NW Germany, in that coal seams have been located as deep and as far to the N as the present wildcat drilling progressed. Nevertheless, the geological and tectonic history has been different in NW Germany. This may be seen directly from the Zechstein contours (Fig 4). Based on investigations about the ^{13}C

content of gas both from gas fields and from the Upper Carboniferous coal bearing beds, Colombo *et al.* (1968) and Stahl (1968) are in line with Patijn and derive even Buntsandstein gas from Carboniferous sources. Müller *et al.* (1967, 1968) arrive at the conclusion that the gas within the Zechstein and Buntsandstein of Thüringen (where Upper Carboniferous coal seams are missing), as well as the gas found in the Zechstein of the NE German basin, *i.e.* the eastward continuation of the 4000 m contours of Fig 4, was generated in the Zechstein carbonates (Stinkschiefer).

THE ALPINE FORELAND

The sediments of the Alpine Foreland basin are sub-divided into two major cycles, Tertiary (the so-called Molasse) and, separated by a slight angular unconformity, Mesozoic, both dipping gently towards the Alps. A faint steepening of dip within the Mesozoic horizons along the Donau line may indicate the boundary between the slope of the SW German block and the Alpine Foreland. Numerous synthetic and antithetic faults attaining up to 100 m, and sometimes even more, of displacement extend predominantly parallel to the regional strike and thus parallel to the Alps. All fields are related to the upthrown blocks of antithetic faults.

Perhaps it is worth mentioning that towards the north the pay horizons of the Tertiary fields terminate either by onlap or shaling out; those of the Mesozoic crop out at the surface.

D 1. The Tertiary E of Munich is mostly marine, to the W it is mainly terrestrial. During the past few years some gas fields have been found in the E in sandstones of the Upper Oligocene (Upper Chattian) age, close to their eastern shaling-out line, 650 to 1150 m above the base of Tertiary.

Since 1954 the Upper Eocene "Lithothamnienkalk" and, in places, sandstones of about the same age near the base produce gas and some oil from intermediate depths. A transgressive basal sandstone of this stage is argillaceous and thin and therefore not productive. To the S, however, this sandstone increases locally in thickness and improves in reservoir properties, whereas the "Lithothamnienkalk" becomes tight in the same direction. Three oilfields (with GOR about 200 : 1) were discovered there. Their reserves remain unknown as yet but could be of a fair size. The exact situation of these Oligocene and Eocene strikes is shown in Fig 6.

The autochthonous Tertiary continues below the belt of the folded Molasse and very probably below the Alps, as it is known also from Upper Austria (well Steinfelden 1, *Erdöl-Zeitschrift*, 1968, **84**, 231, 310). The zone of maximum deposition of the lower Tertiary beds seems perhaps to be situated underneath the alpine nappes system. It is believed to be an important event that in the autochthonous Tertiary, penetrated by a wildcat situated about 50 km SSW of Munich at considerable depth below overthrust sub-alpine Molasse beds, sandstone reservoir characteristics were observed which are distinctly better than those of the laterally compressed sub-alpine Molasse.

At the moment, drilling within the sub-alpine belt and the Alps has been suspended after many years of effort.

In accordance with the facies development the W is poorly productive. Only one additional oilfield was recorded in 1964 in Upper Oligocene (Lower Chattian)—Bausteinschichten—sands resting on Malm.

Some gas strikes in higher horizons (Untere Süßwasser-molasse, Chattian) proved to be insignificant.

D 2. The poor productivity of the Tertiary in the W is compensated to a certain degree by some minor discoveries in the Mesozoic. Since 1960 several oil and gas fields with gas caps and one gas field were found in five different members of Triassic and Liassic, and some gas fields in the Dogger.

Within the drill sections oil and gas shows are widely scattered. The fields are concentrated in two small areas. In the eastern area the discoveries in Tertiary and Mesozoic fall together, although the Mesozoic and Tertiary sequences were deposited in highly different environments and in a divergent tectonic setting. Some fields are located even within the same structures.

In the E, no Mesozoic discovery has so far been made.

ACKNOWLEDGMENTS

Gratitude is expressed to all member companies of the Erdölwirtschaftsverband for the kind permission to use data not yet published.

The author is indebted to the geological staff of Gewerkschaft Elwerath, especially to E. Blissenbach J. Drong, W. Jankowsky, K. Lemcke, J. Lohr, I. Nodop, G. Stober, H. Wechsler, and J. Zimdars for contributing their ideas and the results of special studies.

BIBLIOGRAPHY

Albrecht, K. *Ber. deutsch. Ges. geol. Wiss., A, Geol. Paläont.*, 1967, **12** (1–2), 149–64.

Colombo, U., Gazzarini, F., Gonfiantini, R., Kneuper, G., Teichmüller, M., and Teichmüller, R. *Z.f. angew. Geol.*, 1968, **14**, 257–65.

Commission De La Carte Géologique Du Monde, Sous-Commission De La Carte Tectonique Du Monde. Carte Tectonique Internationale De L'Europe 1 : 2.500.000: Moscow, Congrés Géologique International, 1964.

Franke, D. *Ber. deutsch. Ges. geol. Wiss., A, Geol. Paläont.*, 1967, **12** (1–2), 83–140.

Füchtbauer, H. *Z. deutsch. geol. Ges.*, 1964, **114**, 484–531.

Hecht, F., Van Helms, H., and Kehrer, W. "Reflection-Seismic Exploration of Schleswig-Holstein, Germany, and its Geological Interpretation by Well Data". *Proc. 4th World Petrol. Congr.*, 1955, **1**, 715–30.

Heybroek, P., Haanstra, U., and Erdmann, D. A. "Observations on the Geology of the North Sea Area." *Proc. 7th World Petrol. Congr.*, 1967, **2**, 905–16.
Hinz, K. "A Contribution to the Geology of the North Sea According to Geophysical Investigations by the Geological Survey of German Federal Republic." Geology of Shelf and Seas, 55–71, 1968.
Krey, Th., and Nodop, E. "Application of Seismic Refraction to Subalt Tectonic Problems in a Deep Salt Dome Basin." In preparation.
Meinhold, R., and Reinhardt, H. G. *Ber. deutsch. Ges. geol. Wiss., A, Geol. Paläont.*, 1967, **12** (3–4), 329–53.
Müller, E. P., and Wienholz, R. *Z.f. angew. Geol.*, 1967, **13**, 456–61.
—— "Isotopen-Geochemie und Genese von Erdölen und Erdgasen im Stassfurt-Karbonat des Zechstein 2 im Nordteil der DDR (Rügen, Altmark, Südostbrandenburg) und Thüringer Becken." Ref. in *Erdöl u Kohle*, 1968, **21**, 165.
Patijn, R. J., *Geol. en Mijnbouw*, 1963, **42**, 349–58.
Philipp, W. *Neues Jahrb. Geol. Paläontol., Mh*, 1954, **10**, 440–7.
Plumhoff, F., *Erdöl u. Kohle*, 1966, **19**, 713–20.
Richter-Bernburg, G. "Zur Paläogeographie des Zechsteins in I Giacimenti Gassiferi Dell'Europa Occidentale." Roma, Atti Del Convegno Di Milano (30 Sept–5 Oct 1957), **1**, 87–99, 1959.
Richwien, J., Schuster, A., Teichmüller, R., and Wolburg, J. *Fortschr. Geol. Rheinld. u. Westf.*, 1963, **11**, 9–18.
Sorgenfrei, Th., and Buch, A. "Deep Tests in Denmark 1935–1959." Danmarks Geol. Unders., III. Series, No. 36, p. 146, 1964.
Stahl, W. J. *Erdöl u. Kohle*, 1968, **21**, 514–8.
Teichmüller, M. and R. "Die stoffliche und strukturelle Metamorphose der Kohle." Geol. Rundschau, **42**, 265–96, 1954.
Trusheim, F. *Erdöl Zeitschrift für Bohr- u. Fördertechnik*, 1961, **9**, 361–7.

MS received 12 March 1969

Discussion

Dr P. Kaye (Burmah Oil Exploration Co. Ltd): Have any hydrocarbons been encountered in the Zechstein fringe reef and bioclastic carbonates lying on the N flank of the Hunte Schwelle to the S of the well-established Zechstein 2 algal mound carbonate production?

Have any hydrocarbons been encountered in fractured basinal Zechstein dolomites?

Artur Roll: The production comes only from the algal biostromes.

Dr P. Kaye: So there would be no traces south, or have there been any traces in fractured carbonates of, say, the dolomites further out in the basin at all?

Artur Roll: No.

Abu Amr (Sonatrach): Which objectives are anticipated in the proposed deep tests in the NW German basin, and did Billerbeck 1, drilled in the Munster Basin, yield any positive results with regard to Devonian reservoirs?

Artur Roll: The objectives for the proposed deep tests in the NW German Basin are the Carboniferous and, later on, perhaps older stages of the Palaeozoic. The main objective of course now is the Rotliegendes.

I am sorry but I am not allowed to answer the second part of the question.

L. L. Nettleton (GAI/GMX): Is the 6000 m/sec seismic refraction horizon the basement? If so, has any attempt been made to determine the depth and configuration of the test surface from magnetic studies?

Artur Roll: In the N we have the connection to the basement of the Fünen high, shown by Professor Sorgenfrei this afternoon, but towards the S no definite control exists. We think that the thickness of the succession underneath the base Zechstein is so big that we should be close to basement. If the refractor is not basement, it should indicate diagenetically endured sediments, which could be called basement also. One could think on Caledonian folds, but I personally do not believe that they exist.

Such studies are being made by the Geological Survey. But, to my knowledge, up to now there is no conclusive interpretation as they have to cover a large and complicated area.

P. Heybroek (Bataafse Internationale Petroleum Maatschappij): In Dr Roll's interesting paper a connection is made between occurrences of gas in the Zechstein and in the Bunt sandstone.

Does Dr Roll think it possible that this coincidence of gas occurrences in the two reservoirs is due to the relative thin salt sequence which is present on the Hunte swell and which is probably not an effective seal for gas unifraction from the source into the Buntsandstein in this tectonically disturbed area?

Artur Roll: This could well be the case. Nevertheless, there are also Buntsandstein gas fields above Zechstein salt pillows, so the migration paths certainly were not simple ones, but of course they could exist. The interdependence could be even more complex, and I think there is no time available to discuss this question.

Ramon Querol (Kuwait Spanish Petroleum Co.): Does the porosity of the Buntsandstein change much across the NW German Basin? Do you have to define porosity trends by subsurface geological methods?

Artur Roll: The Buntsandstein sandstone members partially are good and nearly clean sands, partially dirty sands, and of course porosity trends have been established, but this does not help to detect hydrocarbons because until now these are really restricted to the "gas province", and the restriction is not a porosity problem.

Some Observations on the Initiation of Salt Structures of the Southern British North Sea

By H. E. CHRISTIAN, Jr

(Amoco U.K. Petroleum Ltd, London, England)

SUMMARY

The cause of the initiation of salt domes and other types of salt structures in slightly to moderately deformed basins has been the subject of much discussion among geologists, one of the principal questions being whether faulting or folding below the salt is necessary to trigger salt flowage. The North Sea Permian salt basin is exceptional among the salt basins of the world in that a good seismic reflection has been obtained from sub-salt beds below the domes, "walls", and pillows of the Permian Zechstein evaporites. This permits a direct investigation of the relationship between salt and sub-salt structures, and consequently an evaluation of the hypothesis that salt structures are initiated by sub-salt faults or folds. Most of the salt structures in the English Basin (southern British North Sea) are not directly underlain by faults or folds and some prominent sub-salt structures have no corresponding salt disturbance over them, demonstrating that no direct relationship exists. However, the marked NW–SE trend of the two sets of structures indicates a possible indirect connection between them, and the consistent orientation of the salt structures over much of the area and their conformity to the shape of the basin indicates that basinal influences such as regional dip may have played a role in their formation.

INTRODUCTION

In the early days (1963) of North Sea exploration, the various companies engaged in the search for lower Permian Rotliegendes gas were faced with the difficult assignment of mapping seismically the beds below the base of the upper Permian Zechstein, an evaporite-bearing formation which had been deformed by salt flowage into numerous salt structures, some of them diapiric (Fig 1). The southern British North Sea sub-basin, or the English Basin of King (1966, Fig 17), is a portion of the North Sea Basin that contains most of the problems associated with such a mapping project, and is the area considered in the present paper.

Two outstanding problems were involved in constructing an accurate sub-salt seismic map in areas of salt movement such as these. Because of the high velocity salt, most salt structures showed a positive time or velocity anomaly at the base Zechstein or top Rotliegendes level, so in order to obtain an accurate Rotliegendes structure map the reflection from that level had to be converted from time to depth. The amount of the correction was important, for under-correction would give a false anticline and over-correction might result in a false syncline. Fig 2, a large salt structure in the central part of the area, shows such a velocity anomaly. Another problem was that in some cases the Rotliegendes reflection became poor or discontinuous under salt structures, and the discontinuous nature of the event could be interpreted as fault evidence. Fig 2 is also an example of this.

Because of these problems the question of whether or not salt domes were initiated by underlying faults or folds assumed direct commercial importance, for if salt structures were induced in this way their presence would be a valuable clue to the presence of possible gas-bearing Rotliegendes structures. The advent of successful multiple coverage seismic techniques helped to resolve this question, the conclusion being that over most of the area there did not appear to be any direct relationship between the salt structures and the structure of underlying beds. The purpose of this paper is to show why this conclusion was reached and to discuss its implications in regard to the initiation of salt structures.

This paper is considered preliminary because despite the many thousands of miles of seismic line available in this large area its structure and structural history are still not completely known, well control is sparse, and work in other parts of the North Sea Basin may help to resolve some of the questions that confront us in the English Basin.

STRATIGRAPHY

Lower Permian Rotliegendes

The columnar section, Fig 3, shows the pertinent stratigraphy of the area. The lowermost formation of interest in this paper is the Rotliegendes sandstone of lower Permian age, which contains the bulk of U.K. North Sea gas reserves discovered thus far. This formation underlies the Zechstein in the English Basin and is composed of shale, sandstone, and salt, sandstone predominating on the basin flanks and shale and salt in the basin centre.

Upper Permian Zechstein

The Zechstein formation of upper Permian age is the rock unit responsible for the salt structures in the

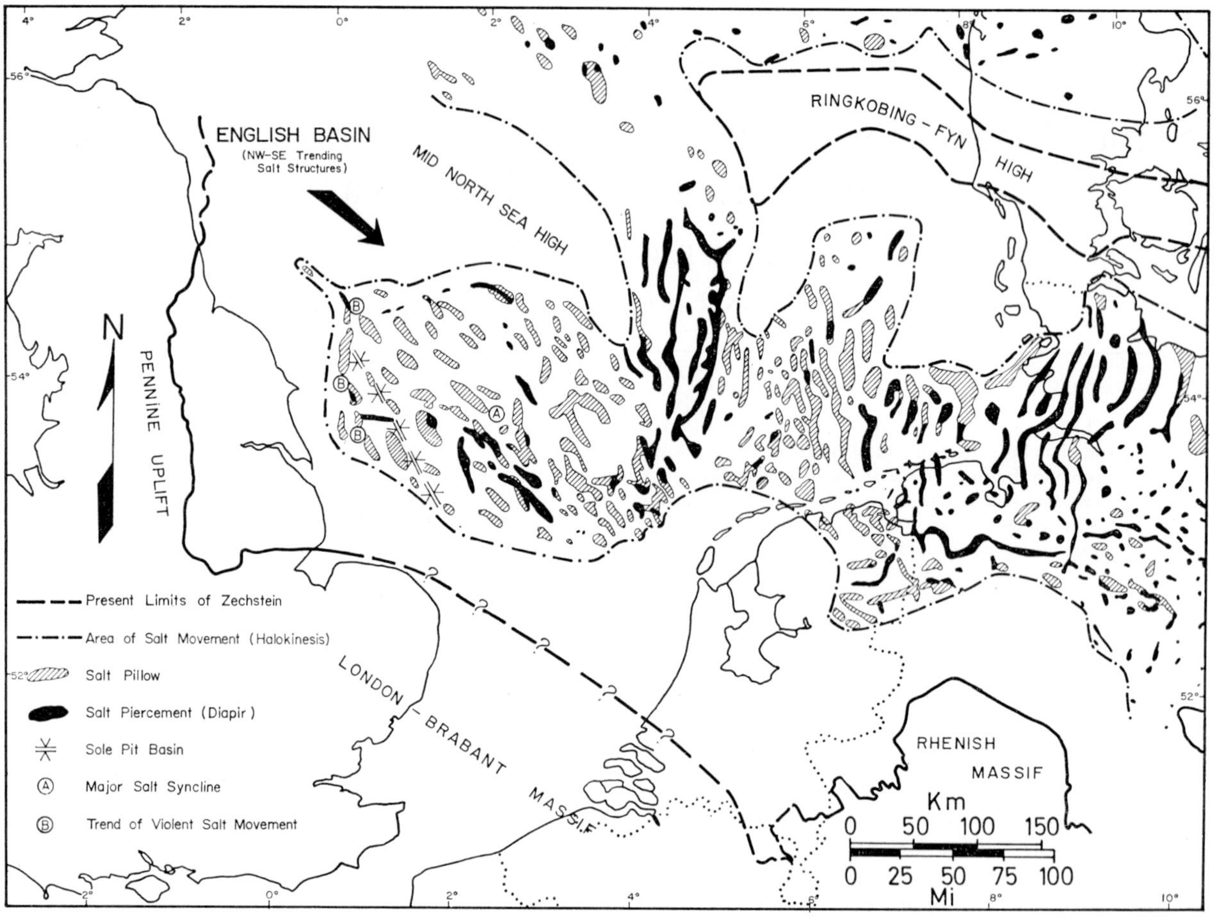

FIG 1. Salt structures of the southern North Sea (modified after Heybroek *et al.*, p. 908, Fig 3).

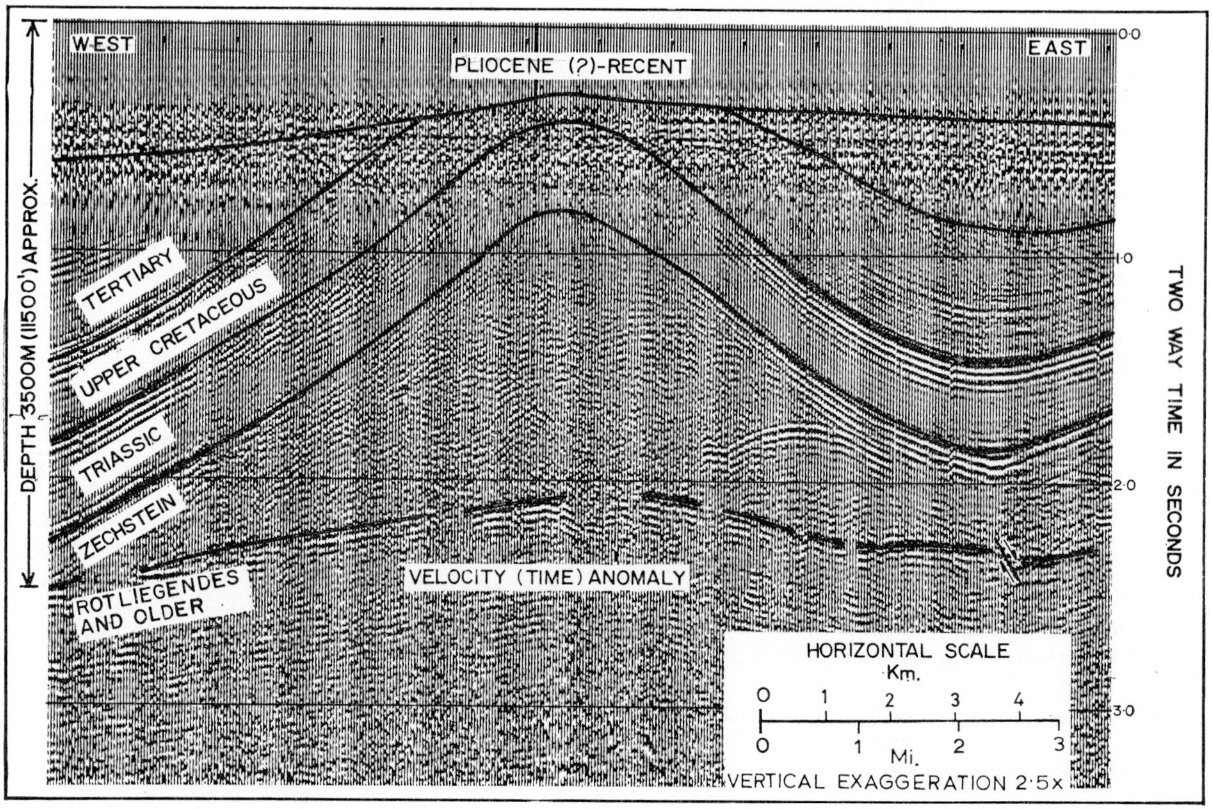

FIG 2. Large Zechstein salt pillow in central part of area. Note velocity (time) anomaly and discontinuous Rotliegendes reflection under pillow. Salt movement ceased in late Tertiary (Pliocene?) time.

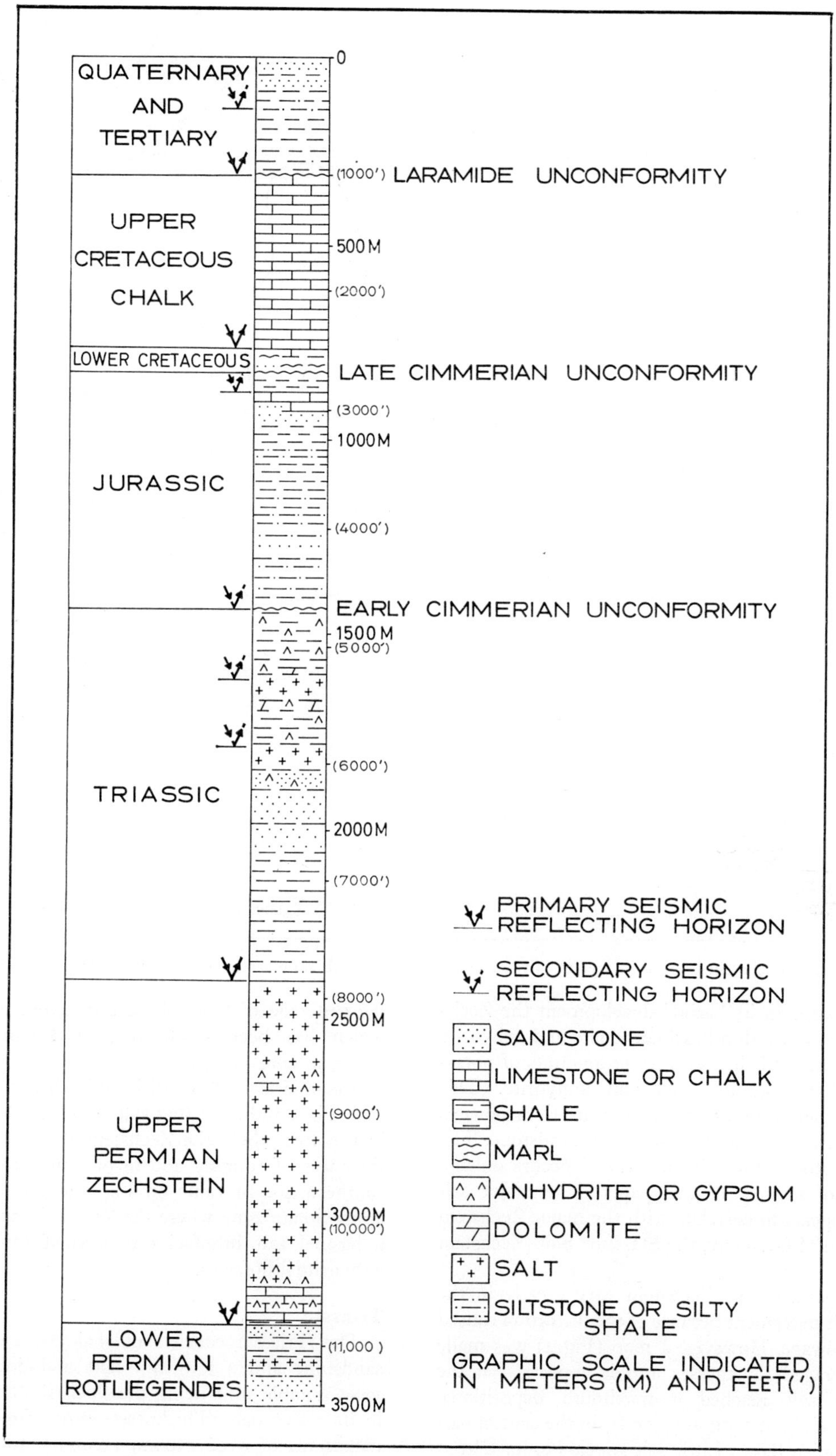

FIG 3. Generalized columnar section, southern British North Sea.

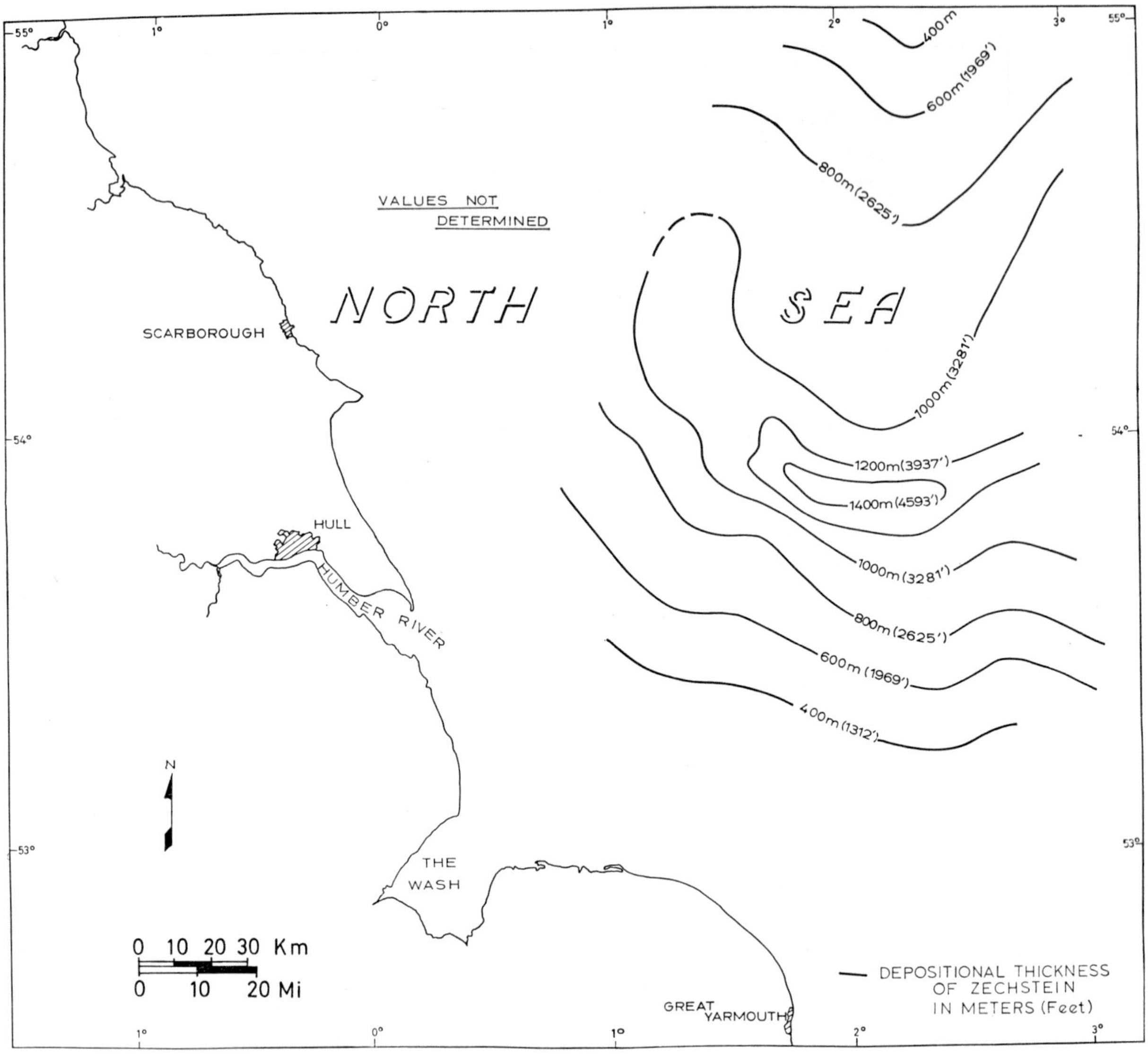

FIG 4. Zechstein isopach in area of salt movement. Averaged thicknesses were used to construct isopach.

English Basin. In its basinal development the Zechstein consists of a thin basal carbonate and anhydrite zone overlain by halite with lesser amounts of potassium and magnesium salts and anhydrite. This section changes progressively shoreward to carbonates and anhydrite and then to shale and sandstone. The shale-sandstone lithofacies in general occurs outside the area of interest of this paper. The major salt section appears to correlate with the major Zechstein salt section of Germany, the Stassfurt Salt (Zechstein II).

Thicknesses of the Zechstein vary greatly in the area of salt movement because of the distortion caused by salt flowage. However, a map (Fig 4) was made using averaged thicknesses and it appears that the Zechstein salt reached a maximum depositional thickness of about 1400 m (4500 ft) in the central part of the English Basin. Little salt appears to have been lost by erosion, so the map probably shows the original depositional thickness fairly accurately.

The limit of Zechstein salt movement is shown on the generalized Zechstein salt structure map, Fig 1, which has been modified after Heybroek *et al.* (1967, p. 908, Fig 3).

A comparison of depositional Zechstein thicknesses with Fig 1 indicates that salt structures have formed in the area where the Zechstein attained an original thickness of roughly 400 m or more (1300 ft). The southern 400-m value on Fig 4 also represents the approximate zone where the Zechstein changes from a basinal salt lithofacies to a shelf carbonate and anhydrite lithofacies.

Triassic

The Triassic consists of shale with subordinate sandstone and a few thin salts and dolomites, the most notable unit being the thick Bunter sandstone in its lower part. Thicknesses range from zero to a maximum of over 1500 m (5000 ft), the maximum values occurring in a NW–SE trending Triassic–Jurassic basin in the central part of the area that approximates the Sole Pit basin of Heybroek *et al.*

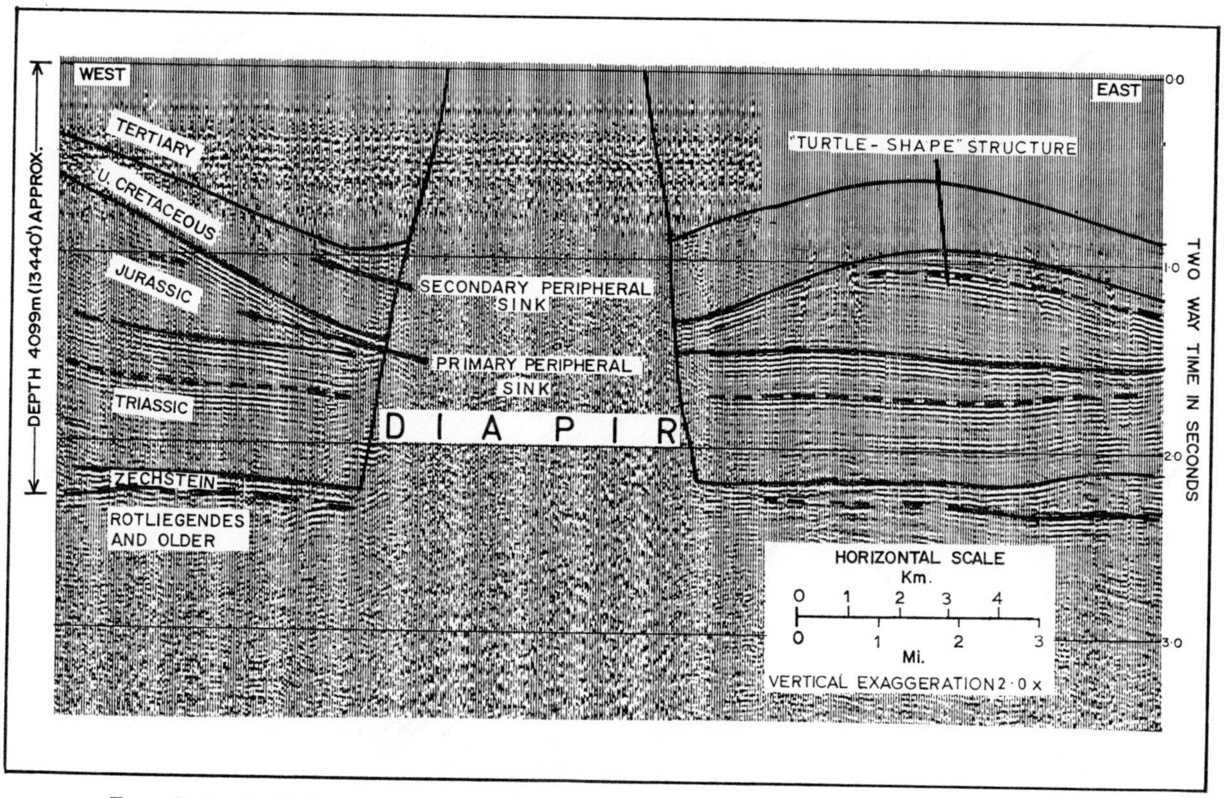

FIG 5. Zechstein diapir in central part of area. Note disappearance of Rotliegendes reflection under diapir. Minor reflections are shown by dashed lines.

(1967, p. 914, Fig 4) (Fig 1). Thin Triassic sections on the flanks of the trough are in part depositional and in part the result of an end Jurassic (late Cimmerian) unconformity, the truncation being particularly severe in the E part of the area.

Jurassic

Jurassic lithologies are also dominated by shale, but near the middle part of the section a limestone–sandstone series of Dogger age is developed. The Jurassic is deeply truncated over much of the area, however, and its present-day distribution is limited to the central and W parts of the English Basin, often only the lower Jurassic being present. This widespread truncation is the result of the end Jurassic (late Cimmerian) age unconformity mentioned in the discussion of the Triassic. Because of the many large salt structures in the area, the truncation is not uniform, and the Jurassic shows large local variations in thickness.

Cretaceous

The Cretaceous sequence consists of two distinct lithologic units, a lower unit of shales with occasional sandstones, and an upper section of the widespread Cretaceous chalk. The shale-sandstone section is lower Cretaceous in age and the Cretaceous chalk is dominantly upper Cretaceous in age. There may be an unconformity between the two units.

Tertiary–Quaternary

Tertiary–Quaternary units ranging in age from Palaeocene to Recent are present, mainly in the E part of the area. The dominant lithology is again shale and clay with occasional sandstones.

Seismic Reflections

There are four good seismic reflections in the English Basin. These primary reflections are in general continuous and can be followed over the area with confidence. Since the commencement of drilling, velocity surveys have enabled well ties to be made which have confirmed earlier geological identifications of the reflections.

The primary reflections (Fig 3) come from the top of the Rotliegendes (base Zechstein), the top of the Zechstein (top of the mobile salt), the base of the Cretaceous Chalk, and the base of the Tertiary. These are the same widespread reflections that were cited by Heybroek *et al.* (1967, p. 907) in their paper on the North Sea Basin.

Other less continuous or secondary reflections are present in the base Chalk–top Zechstein interval, which well control shows are coming from various horizons within the Triassic and Jurassic. There is also a shallow reflection that is probably coming from the base of the Pliocene. Although not as continuous and distinct as the major reflections, Triassic–Jurassic and Pliocene reflections often extend over considerable areas.

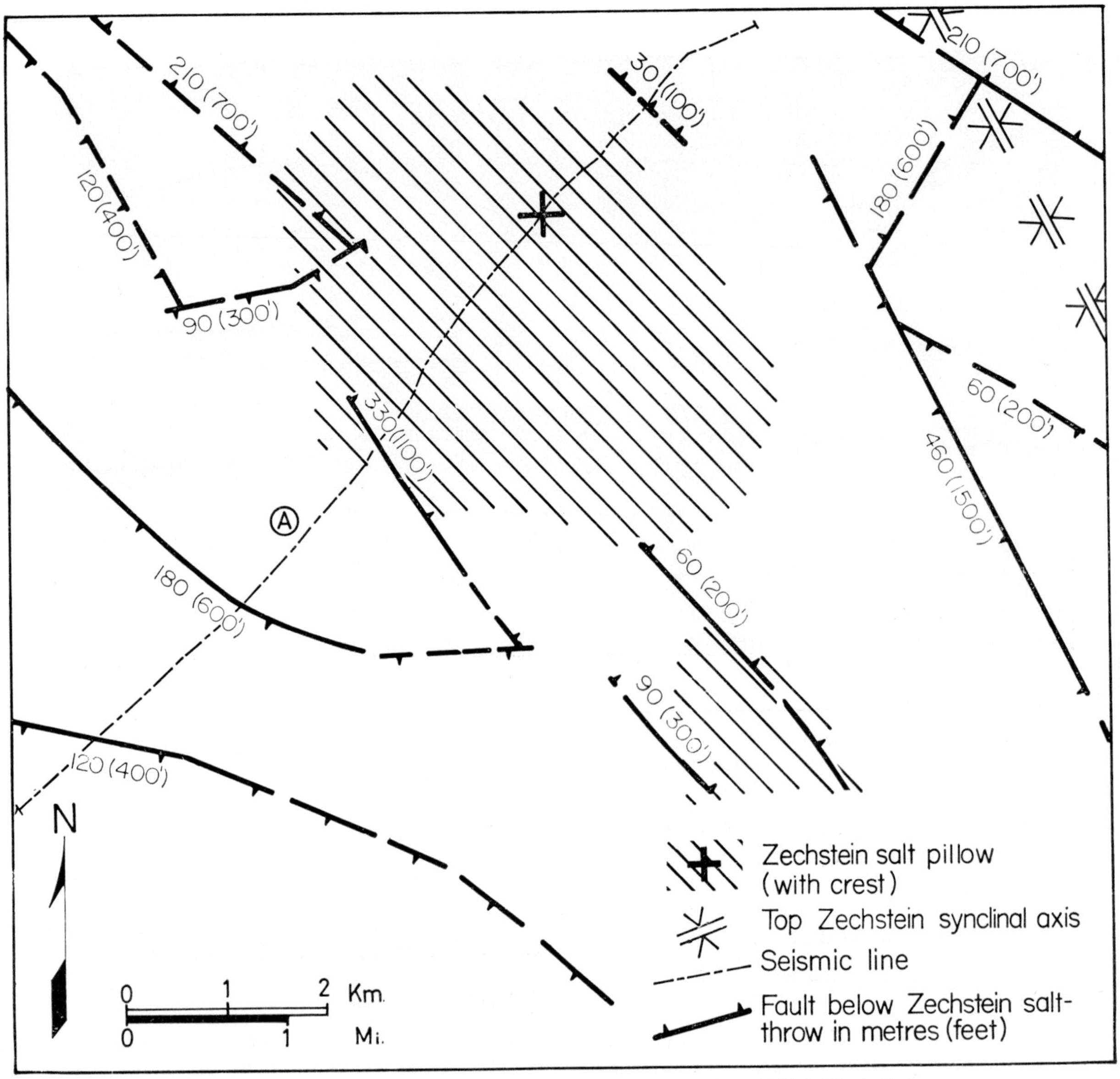

FIG 6. Simplified structural map of Area 1 (SE part of southern British North Sea).

In contrast to the strong reflection that marks the base of the Upper Cretaceous chalk, no reflection is usually detected at the base of the lower Cretaceous, although lower Cretaceous sandstones and shales are present over much of the area. In the areas discussed, the lower Cretaceous is thin, being less than 100 m (328 ft) in thickness, and is not indicated on most of the seismic sections.

Seismic isopach studies have been made in the English Basin to analyse the history of salt movement in some detail and references in this paper to age of salt movement are based on these studies.

Use of Seismic Data as Illustrations

Because of the low drilling density, seismic lines and maps derived from seismic data are the principal illustrations used in the present paper.

There are several things that should be kept in mind when viewing the seismic sections.

1. In the first place the impression should not be gained that the seismic data are invariably as good as those shown in these examples. There are areas of questionable seismic reflections, especially at the Rotliegendes level, and although the seismic data are considered adequate for the conclusions reached, there nevertheless are some doubtful areas, particularly under diapirs.

2. On many of the sections displayed, reflections disappear above about one second because of an early stacking method that did not permit accurate shallow stacking. Because of this, shallow reflections were interpreted from a single fold coverage version of the line and the data above one second sketched in on the display sections from the single fold coverage.

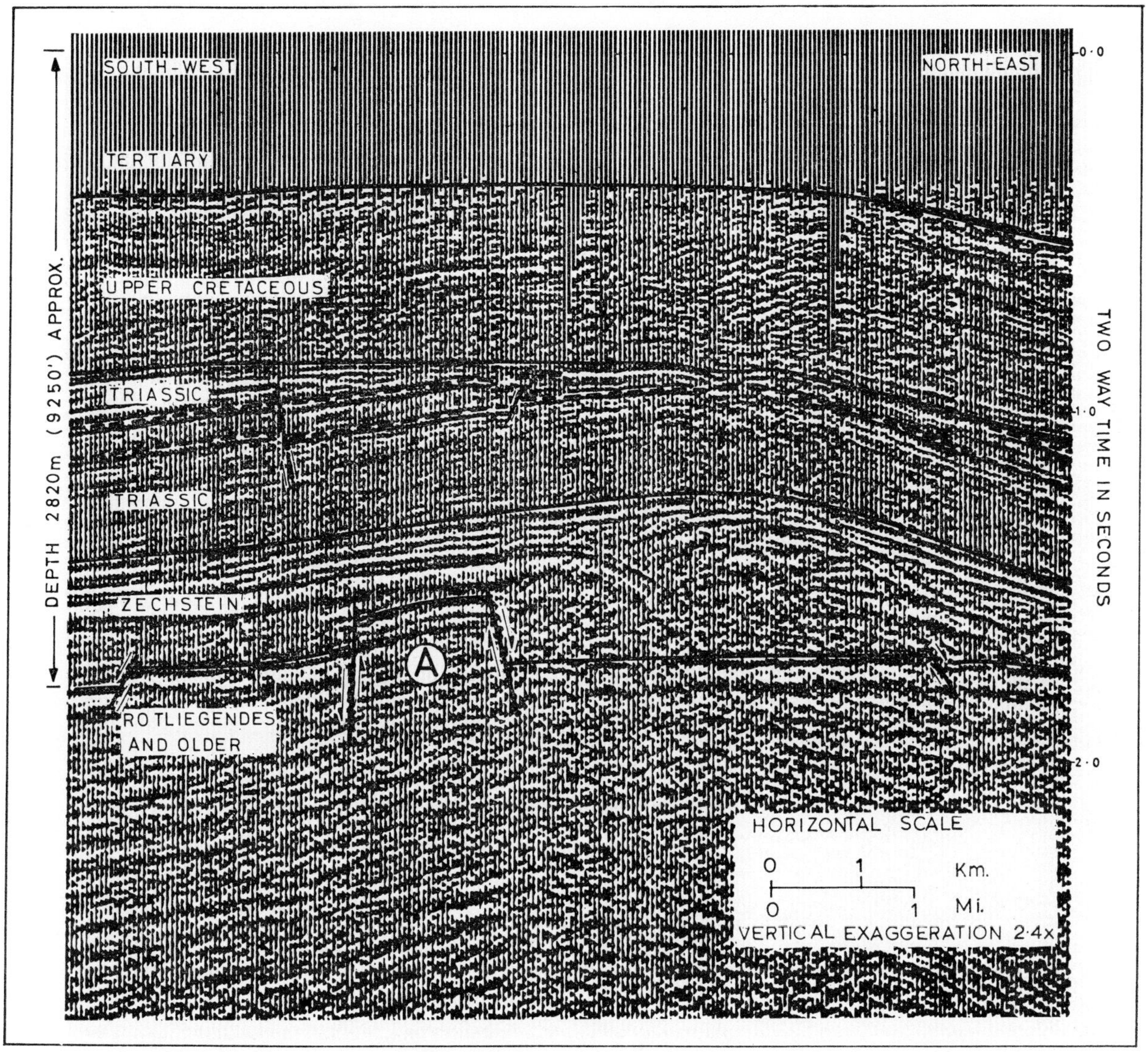

FIG 7. Seismic section in Area 1. Location of line is shown on Fig 6. No disturbance of the salt by the large Rotliegendes horst at A can be seen.

3. Another factor that may cause confusion is that the seismic sections are in time, not depth, and should not be interpreted exactly like geological cross-sections. As shown by Hornabrook (1967) this is especially true in areas like the North Sea, where there are extremely large lateral changes in average velocities to the top of the Rotliegendes because of salt movement. The presence of large accumulations of high velocity salt in a pillow or diapir causes top Rotliegendes reflections under these features to arrive sooner than peripheral reflections from the same depth that underlie thick low velocity Mesozoic and younger sediments and thin Zechstein salt, the net effect being a time anomaly under the salt structure that should be considered suspect until converted to depth. The average velocities mentioned above can vary from 3048 m/sec (10,000 ft/sec) to 4573 m/sec (15,000 ft/sec). An example of this effect is shown on Fig 2, a large salt pillow (almost a diapir) with an apparent Rotliegendes structure beneath it. When converted to depth, however, it appears that the structure is synclinal rather than anticlinal. The conversion to depth of base Zechstein time data is to some extent a matter of interpretation since variations in the velocities used can change an anticline to a syncline and vice versa. Seismic velocity studies and well velocity surveys have, however, aided greatly in the solution of this problem, and the accuracy of the depth conversions has been borne out by drilling results.

4. The seismic sections also differ from geological cross-sections in that, being in time instead of depth, the vertical scale varies with the velocities involved, and consequently so does the vertical exaggeration. An approximate idea of the vertical exaggeration of each section at the Rotliegendes level is given on

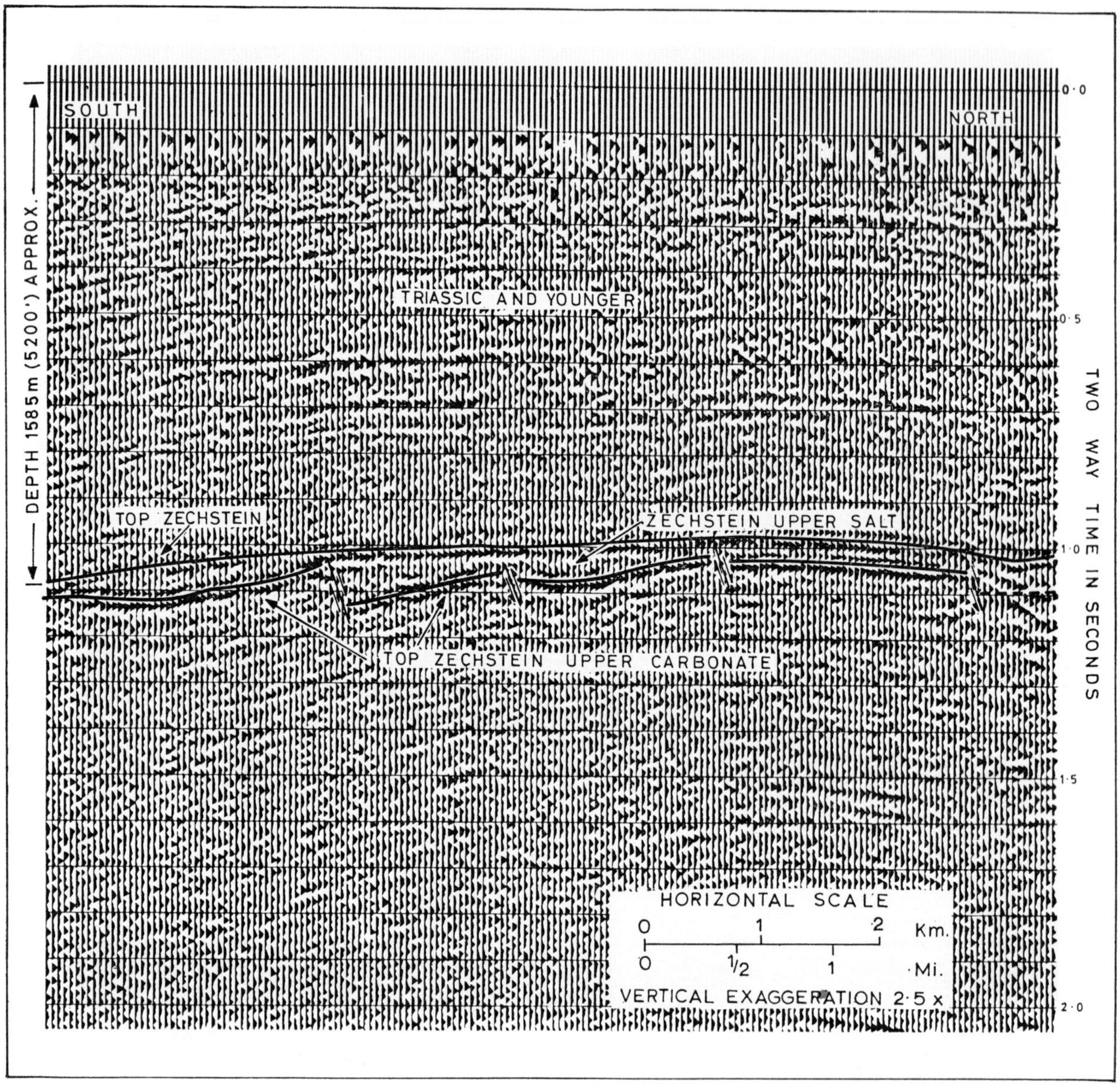

FIG 8. Seismic section SW of Area 1. Note how faults with throws up to 200 m (650 ft) are absorbed in the Zechstein Upper Salt.

each section. The velocity of the Zechstein salt, the highest velocity layer, is about 4600 m/sec (15,000 ft/sec), or about three times that of the Tertiary, the lowest velocity layer, so the vertical depth scale does change quite rapidly.

DESCRIPTION OF SALT STRUCTURES

Forms

The salt structures present in the English Basin are similar to those described by Trusheim (1960) from Germany, and his classification is used throughout this paper. The different types of salt structures present in the English Basin were described briefly by Kent (1967, p. 19, Fig 11, plate 2).

The most common type of salt feature in the area is the salt pillow, which is a swell of non-diapiric salt. Salt pillows are present on many of the seismic sections in this paper, Fig 2 being a large, almost diapiric example.

There are two types of diapiric or piercement salt features in the area. Salt stocks are round to elliptical diapirs, the equivalent of the typical U.S. Gulf Coast salt dome. Salt walls are elongate salt stocks, the only difference between the two being length, and there naturally is a complete gradation between the two types. Fig 5, an E–W cross-section in the central part of the area, is an example of an unusually well-developed piercement showing the primary and secondary peripheral sinks of Trusheim (1960, pp. 1524, 1528, and Fig 7). According to Trusheim, primary peripheral sinks are formed by stratigraphic thickening away from crests of growing salt pillows. If the pillow moves into a diapir phase, it may collapse as salt is extruded, forming a secondary peripheral sink or rim syncline next to the diapir, which shows

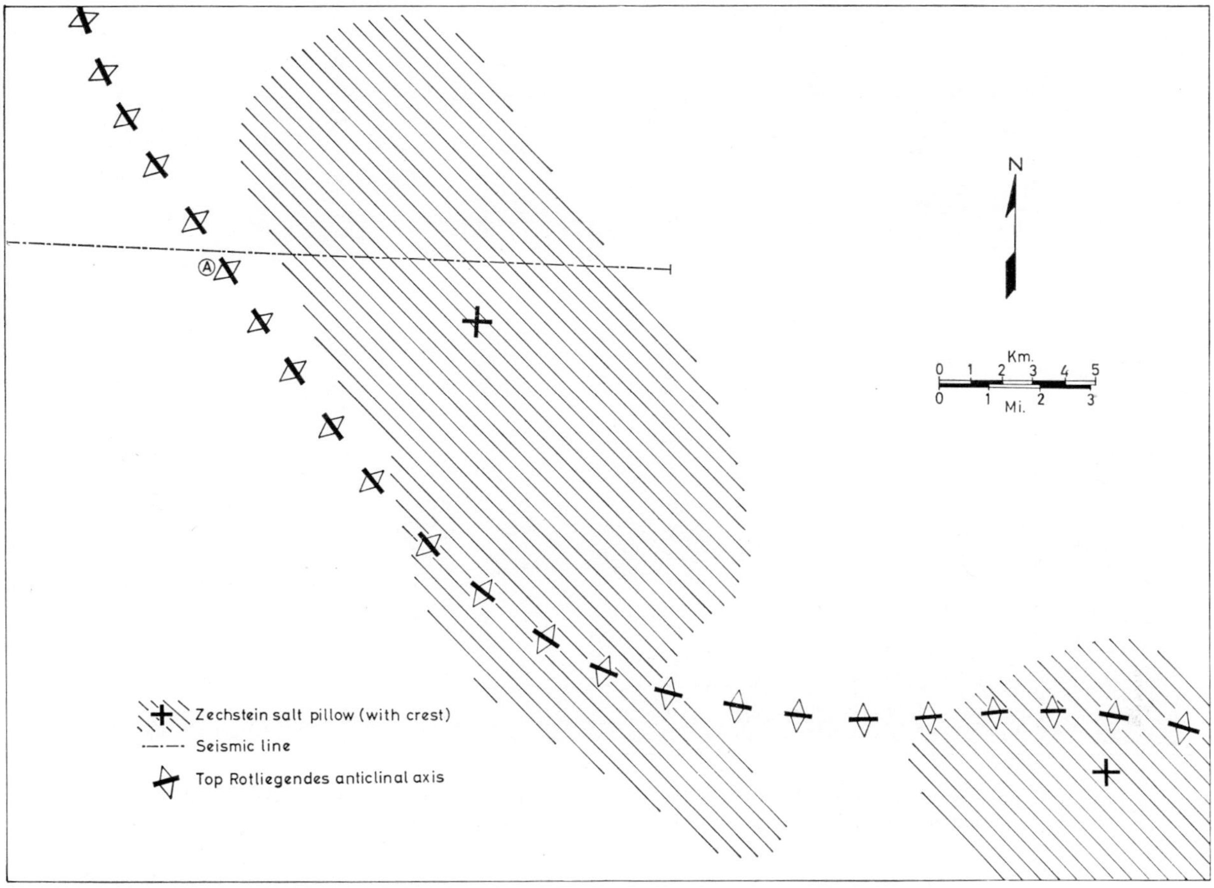

FIG 9. Simplified structural map of Area 2 (SW part of southern British North Sea).

stratigraphic thickening towards the diapir. On Fig 5 the primary peripheral sink is developed in strata of Triassic and Jurassic age and the secondary peripheral sink is developed in strata of Cretaceous and Tertiary age. Shallow seismic data indicate that this diapir comes to the sea floor, and depth calculations reveal that the base Zechstein is at about the same depth on both sides of the diapir, so probably no Rotliegendes structure is present under it.

The formation of secondary peripheral sinks on two adjacent salt structures by the collapse of the salt swells can result in what Trusheim (1960, p. 1533, Figs 18 and 19) characterized as "turtle-shape structures". These are pseudo-anticlines in that they become more gentle and even disappear with depth. An example of such a structure is shown on the E end of Fig 5.

Comparison with Other North Sea Salt Structures

The salt pillow is by far the most common type of salt structure in the English North Sea. Surprisingly, the next most common feature is the diapiric salt wall, with salt stocks comparatively rare. The most common type of salt structure in onshore Germany is the salt wall, but Fig 1 indicates that pillows predominate over large areas of the southern North Sea Basin, the only other notable area of salt wall development being that east of the Mid North Sea High.

Diapirs are most common in the S-central part of the English Basin. Their concentration in this part of the area is noteworthy, for they occupy only part of the area of thick Zechstein salt deposition shown on Fig 4. The significance of this will be discussed later.

Salt Structure Trends

As noted by Heybroek *et al.* (1967, p. 913) the majority of salt structures in the English Basin have a distinct NW–SE trend, which extends a short distance into Dutch waters. The English Basin seems to encompass this area of north-westerly trends. Exceptions to the trend occur along the W and N edges of the basin, where alignments are approximately parallel to the edges of the basin. Salt structures in other parts of the North Sea, and the onshore portion of the North Sea Basin in NW Germany, also exhibit definite trends (Fig 1), although these differ from the north-westerly trend in the English Basin. In the German and most of the Dutch offshore areas, the dominant trend varies slightly from N. In NW Germany there are two strong trends, one NNE–SSW and the other at right angles to it.

A detailed discussion of these trends is outside

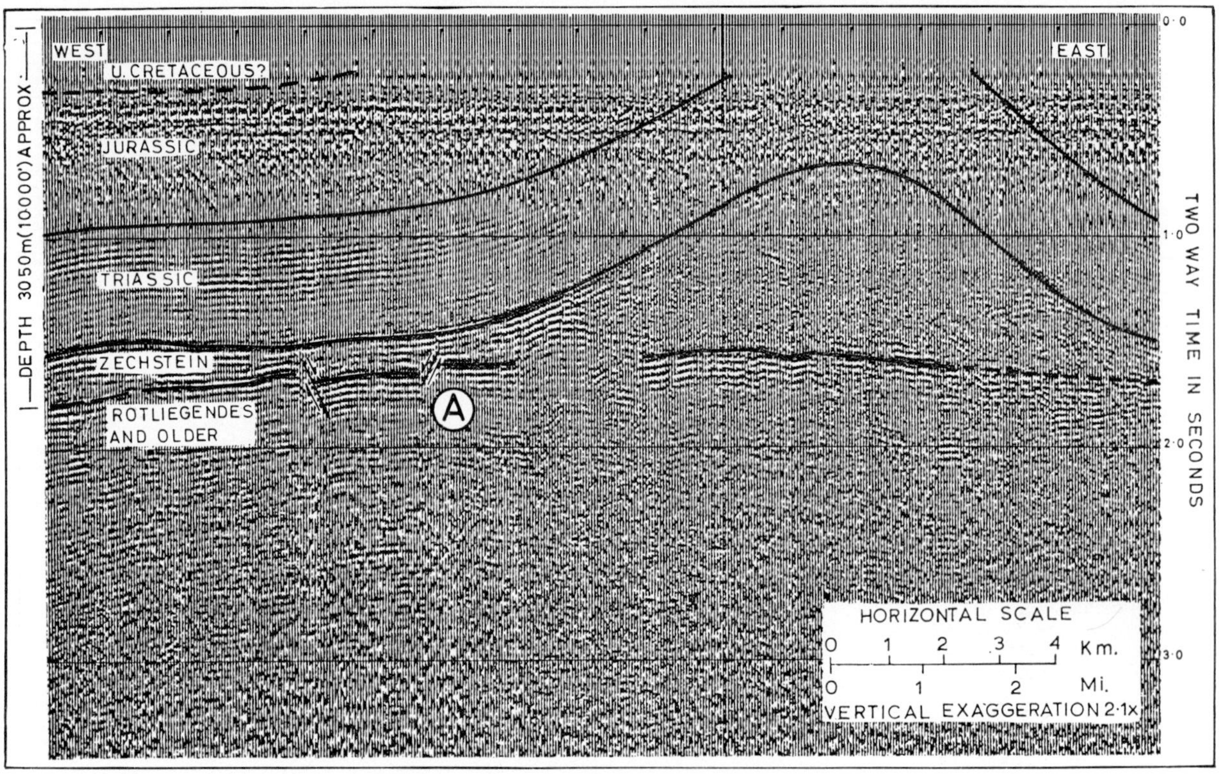

FIG 10. Seismic section in Area 2. Location of line is shown on Fig 9. Depth conversion reveals that crest of Rotliegendes anticline is at A on section, not under salt pillow, as time data on seismic section indicate.

the scope of this paper. However, the predominance of distinct trends over the southern North Sea is significant and should be explained by any hypothesis on the initiation of salt structures in this area. The search for such an hypothesis or hypotheses is not a simple one, for detailed seismic work over the area has revealed that only a few salt structures are underlain by Rotliegendes faults or folds, and conversely that some large Rotliegendes horsts underlie salt synclines. The many aspects of this problem are best illustrated by an examination of five areas scattered over the English Basin.

DISCUSSION OF AREAS

Area 1

Area 1 is located in the SE portion of the area, near the edge of the mobile Zechstein salt. The average thickness of the mobile Zechstein salt sequence in this area is roughly 600 m (2000 ft) (Fig 4).

The relationship between the Zechstein salt structure and that at the Rotliegendes (sub-salt) level is shown on the map, Fig 6, and seismic section, Fig 7. The Rotliegendes structure is characterized by numerous normal faults that form a series of fault blocks. The large horst at point A is unusual in that it appears to rise abruptly to a great height from a flat floor, but its existence is confirmed by nearby drilling. The 330-m (1100-ft) and 180-m (600-ft) faults bounding the horst are not represented at all in the overlying gentle salt structure, not even by a structural flattening, so the salt must have flowed away from the horst as it moved upward. There are numerous other large Rotliegendes faults in the area, one of which has a throw of 460 m (1500 ft), and these also do not underlie salt structures, confirming that the relationship shown on the seismic line is not exceptional for the area.

East of Area 1, however, there are indications that some Rotliegendes faults have completely penetrated about the same thickness of Zechstein, cutting the post-Zechstein beds. This appears to be the result of large fault throws and a more competent Zechstein section as a result of an increase in anhydrite compared to Area 1. On the other hand, seismic and drilling results in a highly faulted area located SW of Area 1, outside the area of salt structures, demonstrate that thin pure salt sections can have an amazing capacity for absorbing faulting. A seismic section in this area, Fig 8, shows two reflection horizons, the upper one being the top of the Zechstein and the lower one coming from a carbonate near the top of the Zechstein. Faults with up to 200 m (650 ft) of throw are present on the section, almost all of which are absorbed in a Zechstein upper salt section that has been distorted by salt flowage, but which was originally only about 100 m (330 ft) thick.

Returning to Area 1, the seismic section (Fig 7) has two reflections within the Triassic which indicate truncation of the base Cretaceous–top Zechstein interval on the crest of the salt pillow because of pre-Cretaceous salt movement. Isopach studies

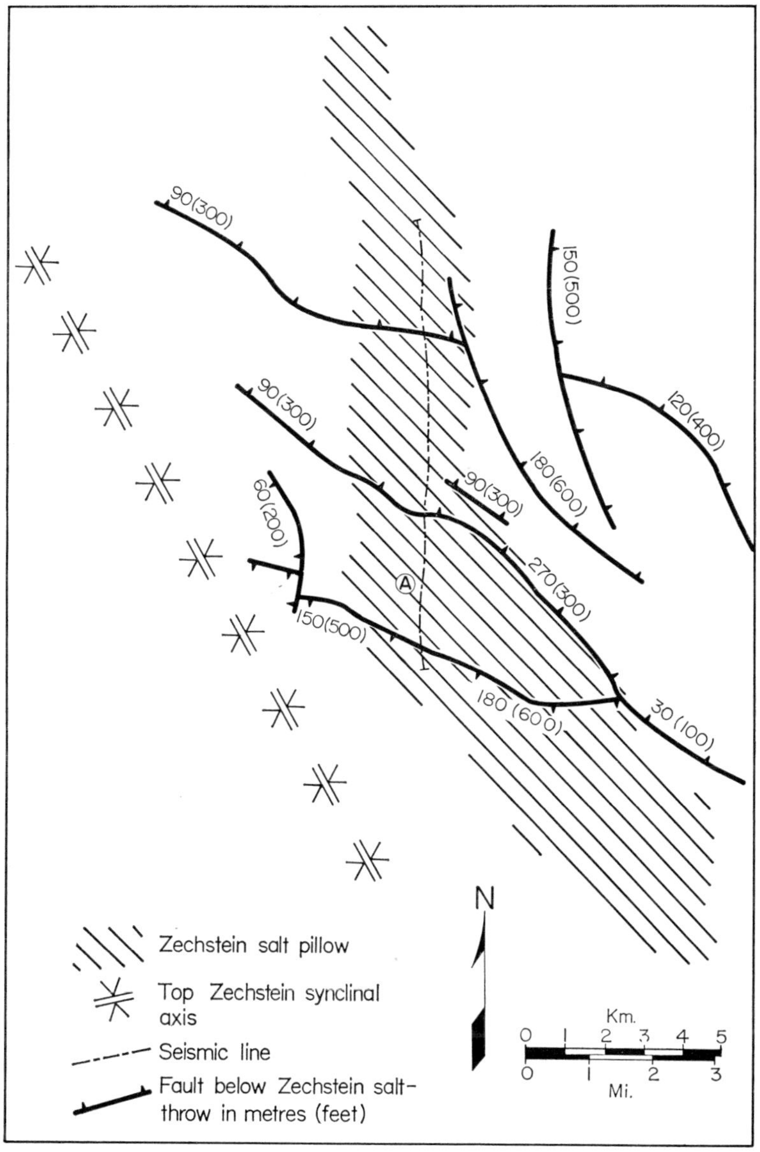

FIG 11. Simplified structural map of Area 3 (W part of southern British North Sea).

show that this salt structure was a NW plunging nose at the end of the Jurassic, remained quiescent during the Cretaceous, with the final salt movement taking place during the Tertiary.

Studies outside the area of salt movement reveal that most of the structural movement was late Cimmerian (end Jurassic) and Laramide (end Cretaceous), with some early Cimmerian (end Triassic) movement in one case. Although sub-salt faulting cannot be dated in Area 1 because the faults are absorbed in the salt, the structural trends are similar in direction to those outside the area of salt movement, and the faulting probably occurred during the same tectonic phases. If so, the pillow on Fig 7 was probably present and may have been active while intensive but disassociated Rotliegendes (sub-salt) faulting was taking place, again emphasizing the independence of the two sets of structures.

16

Area 2

Area 2 is located in the SW portion of the basin, where the Zechstein mobile salt sequence has an average thickness of 750 m (2500 ft) (Fig 4).

The structural configuration of the area is shown on the map, Fig 9, and seismic section, Fig 10. In contrast to Area 1 and the other areas that will be discussed in detail, the main Rotliegendes structure in this area is an anticline, the axis of which is indicated on Fig 9. This feature has about 600 m (2000 ft) of reversal, of which about 300 m (1000 ft) is shown on Fig 10.

Although time data on the seismic section, Fig 10, suggest that the Rotliegendes structure lies under the salt pillow, conversion of time data to depth shows that the crest of the Rotliegendes structure is actually at point A on the section. The apex of the Rotliegendes structure lies about 5·5 km (3·4 miles)

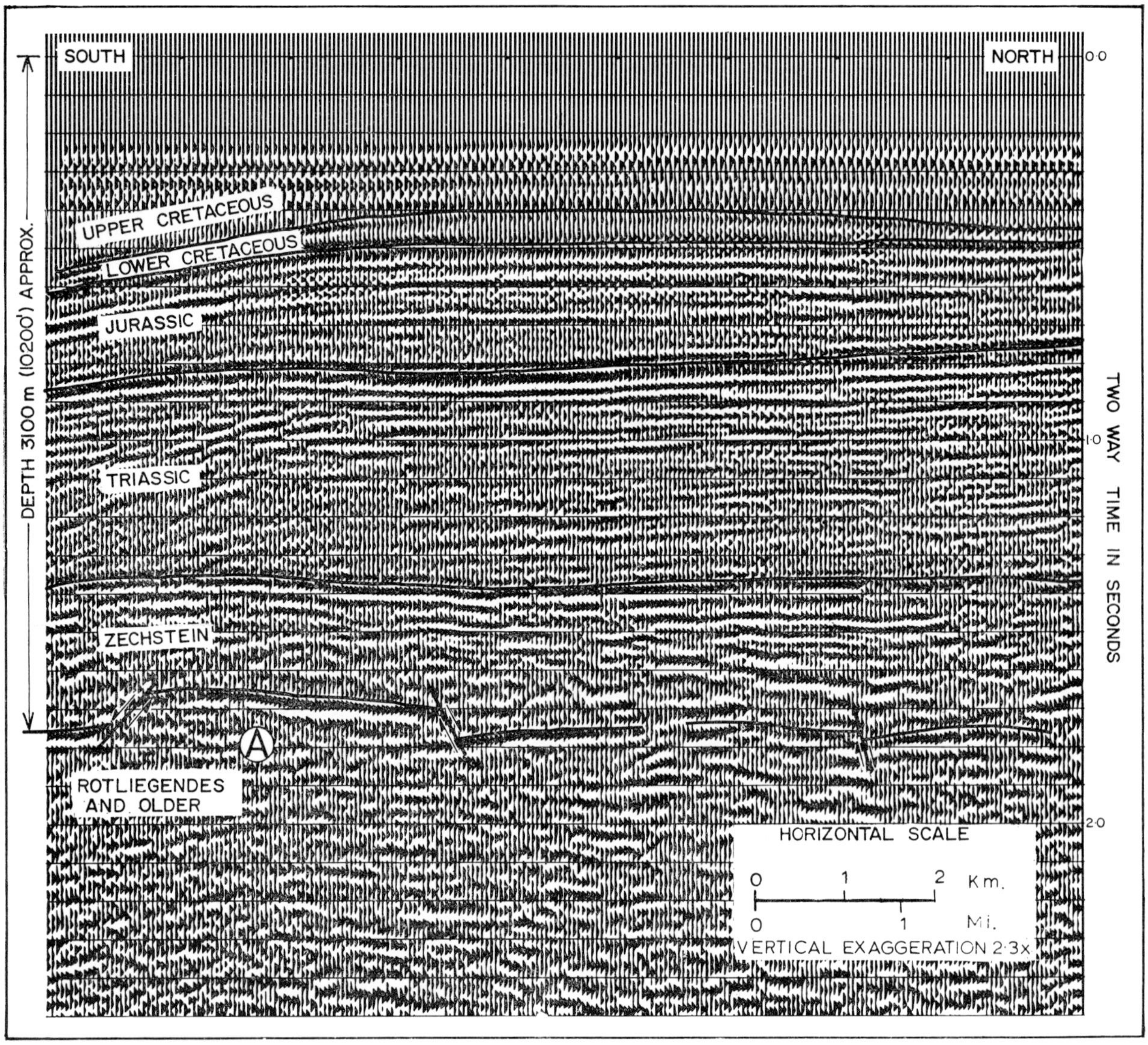

FIG 12. Seismic section in Area 3. Location of section is shown on Fig 11. No disturbance of the salt by the Rotliegendes horst at A is evident.

SW of the crest of the salt structure, so there is a considerable displacement of the crests. The salt pillow is very large, with over 900 m (3000 ft) of closure, and the reversal shown on the seismic line is about 1500 m (5000 ft).

In spite of the considerable offset of the Rotliegendes and top Zechstein axes, this is one case where a direct causal relationship appears to exist, for the continuation of the trend to the SE shows an increasingly closer association between the two sets of structural axes as the Zechstein becomes thinner. This tendency is shown by the noticeably closer association of the two structural trends in the SE corner of the map (Fig 9). It is believed that the offset of the two structures on Fig 10 is caused by differential loading resulting from a pronounced regional thickening of the Triassic and Jurassic to the NE across the area. The sedimentary load was consequently greater on the NE flank of the Rotliegendes structure, which tended to favour salt dome growth in that area.

The salt structure appears to have moved at the end of the Jurassic and in post-Cretaceous time. The Tertiary is absent in the area, so more specific dating of the structure is not possible. The end Jurassic movement coincides with one of the stronger periods of structural growth in the basin, the late Cimmerian phase.

Area 3

Area 3 is also located in the W part of the basin, but near the edge of the mobile Zechstein salt, which in this locality has an average thickness of 450 m (1500 ft).

The structural relationships in this area are shown on the map, Fig 11, and the seismic section, Fig 12. The main Rotliegendes structure, a horst, is located

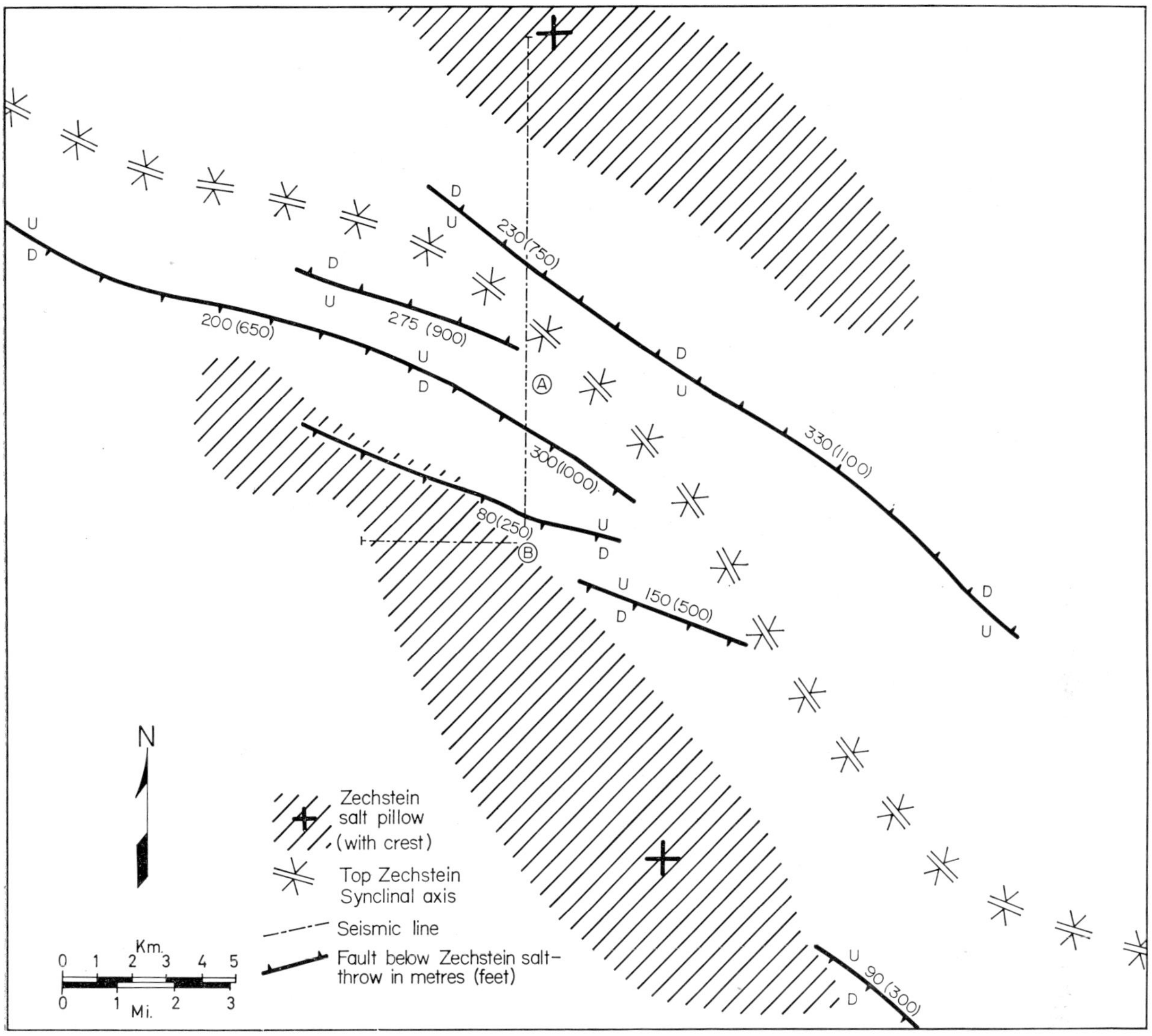

FIG 13. Simplified structural map of Area 4 (central part of southern British North Sea).

at Point A at the S end of the seismic line. As in the case of Area 1, the horst has apparently not affected the gentle top Zechstein salt structure, the NW–SE trending faults crossing the Zechstein salt structure at an angle without appreciably disturbing it. An analysis of the age relationship between the Rotliegendes (sub-salt) faults and the salt structure leads to no definite conclusions in this area.

Moving now from the edge of the Zechstein salt basin to its centre, the following two areas are conspicuous examples of large-scale Rotliegendes faulting which has been completely absorbed in the Zechstein salt.

Area 4

In Area 4 the regional Zechstein isopach indicates that the mobile Zechstein salt was originally about 900 m (3000 ft) thick.

The structural relationships in this area are shown on the map, Fig 13, and seismic section, Fig 14. At the Rotliegendes level, the most conspicuous feature is the large horst at Point A. The southern bounding fault of this horst has about 300 m (1000 ft) of throw and the northern fault 330 m (1100 ft). The seismic section shows that, like several of the other areas discussed, the Zechstein salt which lies above this horst is unaffected by the sub-salt faulting.

The thinning in the Jurassic at B on the seismic section indicates that the S salt structure was beginning to form by the end of the Jurassic. Additional relief was added during the Cretaceous, and at the end of that time there was some closure which was weak to the SE. In Tertiary time, a strong SE-plunging nose formed that complemented the earlier NW dip to give a sizeable closed structure. On the other hand, the N salt structure did not begin to grow until Cretaceous time, with additional growth during the Tertiary. As in the case of Area 1, it appears that during the most probable times of Rotliegendes fault movement salt structures were present in the area, but remained apparently unaffected by the faulting.

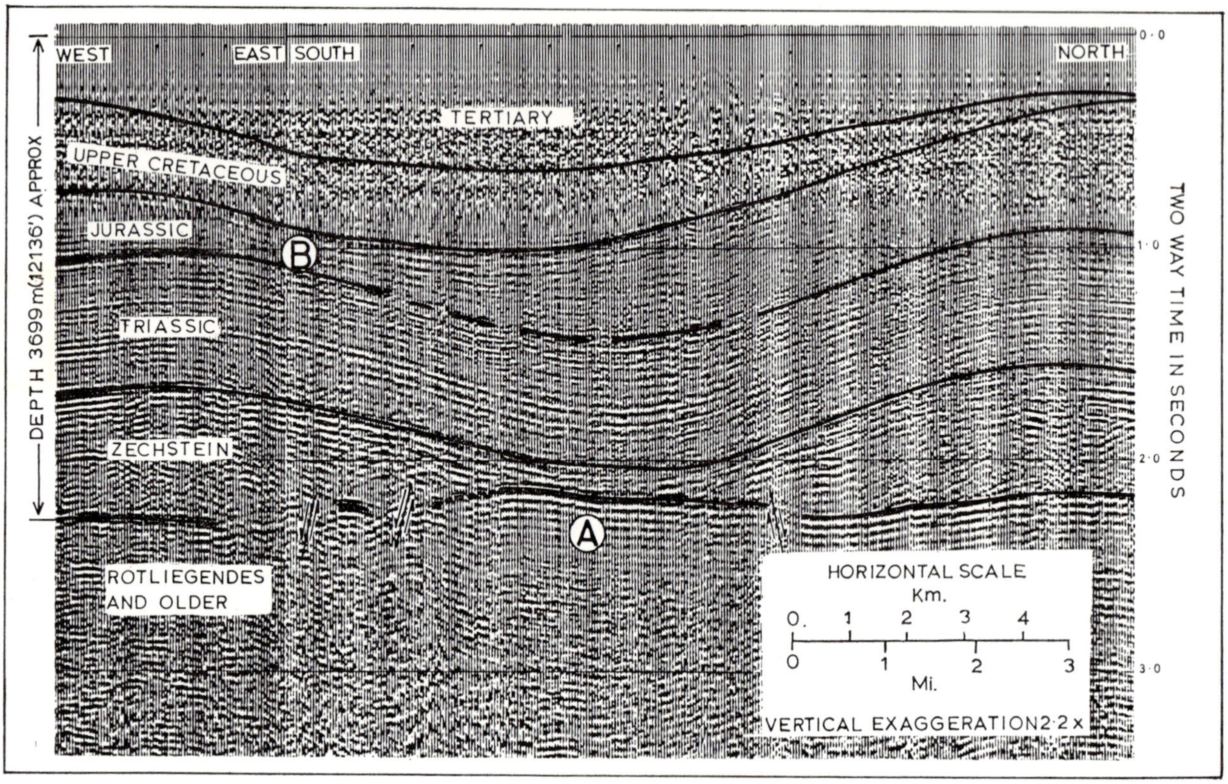

FIG 14. Seismic section in Area 4. Location of section is shown on Fig 13. No disturbance of the salt by the large Rotliegendes horst at A can be seen. The map, Fig 14, shows that the salt syncline above the Rotliegendes horst changes location with respect to the horst, so the apparent association of the horst and the salt syncline on the seismic section is fortuitous.

Area 5

This area is located in the central part of the basin near Area 4 and is similar to it in many respects. The mobile Zechstein salt is approximately 900 m (3000 ft) thick, about the same as Area 4.

The relationships of the two sets of structures are shown on the map, Fig 15, and seismic section, Fig 16. At the Rotliegendes (sub-salt) level, a horst at point A bounded by faults of up to 340 m (1100 ft) and 150 m (500 ft) of throw occurs in a deep salt syncline in a position very similar to that of the top Rotliegendes structure in Area 4, but although the Rotliegendes faults have the same trend as those of Area 4, they are not the same faults. Once again, it is apparent that a large sub-salt horst has risen into the Zechstein salt without influencing the top Zechstein salt structures.

The large Zechstein salt pillow located in the S part of the map and just touched by the seismic line has Rotliegendes faulting of lesser magnitude beneath it. The only important sub-salt fault under this dome, a fracture with 210 m (700 ft) of throw, is well down on the flank of the structure which, together with the absence of a salt structure over the main Rotliegendes horst to the N, casts doubt on the initiation of the feature by faulting. Also, prominent salt features in the W and NE parts of the map have no Rotliegendes faulting or folding under them, again supporting the idea that faults did not initiate the S salt structure.

At the end of the Jurassic a broad NW–SE trending nose was present over the S fault of the Rotliegendes horst, indicating that the sub-salt faulting may have caused it. Under this hypothesis the time of the fault would be end Jurassic (late Cimmerian), the probable age of much of the structural movement in the basin. No further detailed dating of the structure is possible, in as much as the Tertiary is absent in the area. The present-day structural configuration, however, shows that the nose disappeared in post-Jurassic time, so although the sub-salt faulting may have started some salt movement the nose did not persist, a phenomenon studies in other areas have shown is not uncommon.

The S salt structure is of post-Jurassic age, as is the NE feature, but the W salt structure was an early feature that had completed most of its movement by the end of the Jurassic. The W structure's strong early movement shows that there was sufficient overburden to permit salt movement in Jurassic time, so a lack of overburden was not the reason that the end Jurassic salt structure over the S fault of the Rotliegendes horst did not persist.

The areas discussed in detail above are extreme in that they either show a direct casual relationship between a Rotliegendes (sub-salt) structure and a salt structure, or a notable lack of any apparent relationship. It should be understood that the usual

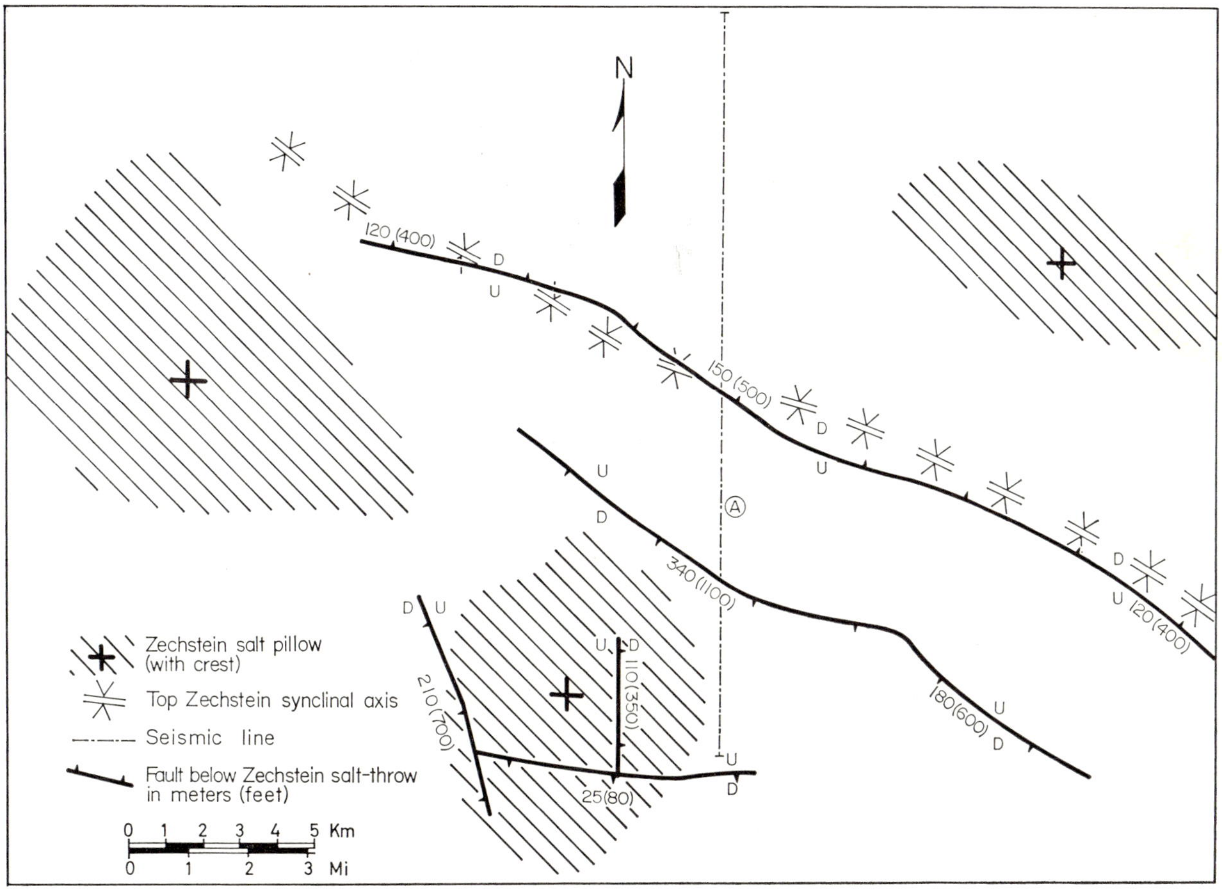

FIG 15. Simplified structural map of Area 5 (central part of southern British North Sea).

salt structure in this area is a simple salt pillow which has no Rotliegendes anticline or fault beneath or near it. These structures, together with most of those discussed in detail, have a general NW–SE trend parallel to that of the apparently unrelated sub-salt structures. The explanation for this salt structure trend is one of the main problems to be dealt with in the following section.

INITIATION OF SALT FEATURES

It is quite possible that the salt structures in this basin have been initiated in more than one way, for when salt becomes ready to move it seems likely, and model experiments have confirmed, that various processes can initiate movement. Also, if a mixture of causes was responsible for the initiation of salt movement in this basin, these causes need not be the same as those in another such basin, or at least their relative importance may change. For example, the present basin differs from the Gulf Coast Basin of the U.S.A. in that the salt structures here show a definite orientation, and hypotheses as to the initiation of salt structures of the two basins will probably vary accordingly. In order to properly evaluate the various initiation hypotheses with respect to the English Basin, it is first important to summarize the main characteristics of the salt structures.

CHARACTERISTICS OF THE SALT STRUCTURES OF THE ENGLISH BASIN

1. Both salt and sub-salt structures generally have a NW–SE trend.
2. There appears to be only an occasional association between salt pillows or diapirs and sub-salt faulting and folding. Sub-salt faulting and folding occurs anywhere from under salt anticlines to under salt synclines, but is more often found underlying structural flanks. In some cases sub-salt faulting or folding trends at an angle to the axis of the salt structure (Area 3).
3. A number of salt structures appear to have no sub-salt faulting and folding anywhere near them.
4. Diapires tend to concentrate in the S central part of the basin.

GENERAL HYPOTHESES OF INITIATION

The following hypotheses are the principal ones that have been advanced to explain the initiation of salt structures:

1. Irregularities in the upper surface of the salt.
2. Variations in the density of the overburden.

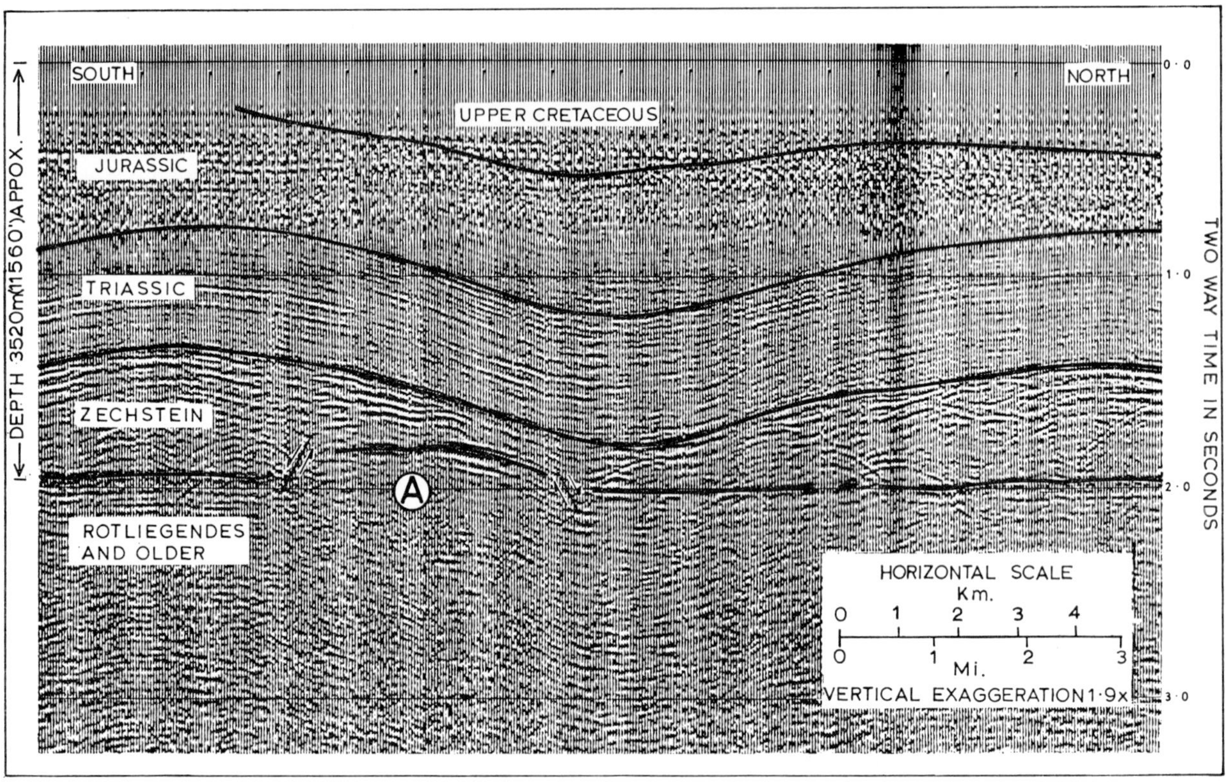

FIG 16. Seismic section in Area 5. Location of section is shown on Fig 15. No disturbance of the salt by the Rotliegendes horst at A is evident.

3. Progressive plasticization of salt by heat.
4. Initiation of one or more salt features by another (salt structure "families").
5. Tectonic (faulting and folding).
6. Regional dip.

APPLICATION OF HYPOTHESES TO THE ENGLISH BASIN

Irregularities in the Upper Surface of the Salt Variations in the Density of the Overburden

Hypotheses 1 and 2, irregularities in the upper surface of the salt and variations in the density of the overburden, are not believed applicable as general explanations in this area, for it seems unlikely that they would vary so often and in such a consistent way as to cause the notable NW–SE trend of the salt structures.

Progressive Plasticization of Salt by Heat

Gussow (1968, p. 42–44, Figs 23, 25), has advanced the idea that salt does not diapir until it has reached depths of about 7600 m (25,000 ft), where it reaches a temperature of about 300° C and becomes plastic. This change in state is thought to occur abruptly and in a subsiding basin could be responsible for successive sets of diapires towards the basin margins as subsidence progressively raises the temperature of the salt above the critical point of 300° C. This hypothesis does not seem applicable in the English Basin because it is doubtful if the salt was ever buried below 4600 m (15,000 ft), and most of the structures are not diapirs. The geothermal gradient of the basin is similar to that assumed by Gussow. Conversely, from the relatively shallow depths in this basin it would appear that diapirs do occur above the depths postulated by Gussow.

Salt Structure "Families"

Experiments by Parker and McDowell (1955, p. 2420) demonstrated that model diapirs often had a ring of secondary diapirs form around them as the result of differential loading, indicating that one salt diapir could initiate a whole series of others. If true, such a process has the advantage of avoiding having to ascribe a specific geologic event or condition to initiate each salt structure. The results of Parker and McDowell's experiments were applied by them to the Gulf Coast Basin of the U.S.A., and later Sanneman (1968) applied the idea to the onshore German portion of the North Sea Basin. In Germany the salt structures consist of long ridges instead of circular salt stocks, but the basic idea was the same.

It is doubtful that this idea has much application in the English Basin. The hypothesis appears to demand diapirs with secondary rim synclines to set up differential loading stresses sufficient to initiate other salt features, but there are only a relatively few diapirs in the basin and they tend to be concentrated in the S central part. However, the two salt walls SW of A on Fig 1 may have been initiated in this manner.

The western of the two diapirs has the only really well-developed secondary rim syncline in the basin, and isopach studies show that this syncline was forming at about the same time as the E diapir was rising. It is possible that detailed well information might show that the two events are closely related in time, which would support the idea that the E diapir was initiated by the W one.

Tectonic Hypotheses

The initiation of salt structures by tectonic means probably has occurred in at least one instance (Area 2), where a series of salt structures was initiated by an anticlinal trend. For purposes of discussion the types of tectonic activity can be divided into three categories: sub-salt faulting, overburden faulting (faulting affecting only the post-salt beds), and folding.

The distinction between sub-salt faulting and overburden faulting is an important one for, as has been shown in the discussions of Areas 1, 3, 4, and 5, large sub-salt faults are often absorbed in the salt. Parker and McDowell (1955, p. 2410, Fig 8) showed hypothetically how a sub-salt fault that penetrated through the salt and affected the younger beds could initiate a salt structure because of a differential loading effect, but the difficulty in applying this idea in the English Basin is that there such faults do not often penetrate the salt. The reason for this appears to be that in the deeper portions of the basin the salt has behaved essentially like a fluid in its response to large scale sub-salt fault movements. In the shallow basin area E of Area 1 large-scale faulting appears to have penetrated about the same amount of Zechstein as in Area 1 and initiated salt movement. Conditions are different in this area though, because the Zechstein is changing to a shelf facies of anhydrite and carbonate and is less capable of absorbing faulting.

On the other hand, it is possible that sub-salt faulting may have had an indirect causative effect. Earthquakes emanating from the sub-salt faults as they moved may have been sufficient in some cases to set off salt movement in nearby areas where the salt was more unstable, or caused salt structures already present to start growing again. Although no mechanism has been suggested that would result in salt structures with a definite orientation, some support for this idea is given by the concentration of diapirs and sub-salt faulting in the S central part of the basin.

Experiments by Parker and McDowell (1955, p. 2411, Fig 7) and Tanner and Williams (1968) showed that overburden faulting could also initiate salt movement. This type of faulting may have initiated salt movement in two localities: (1) a N–S trend of violent salt movement along the W edge of the basin (B on Fig 1); and (2) two salt walls in the S part of the English Basin, SW of A on Fig 1. These localities were in local zones of steep dip on the flanks of the Triassic–Jurassic Sole Pit Basin (Fig 1). It is postulated that the steep dips may have caused faulting in the overburden, which in turn initiated the salt movement. A possible difficulty with this idea is that the salt walls cited under (2) occupy only the N part of the zone of steep dip, but this is thought to be because of the regional thinning of the Zechstein salt S of the diapirs.

The salt walls listed under (2) were also discussed in the section on salt structure "families", where the possibility was considered that the W salt wall initiated the E one, so the E salt wall could be the result of either of the two processes.

Large-scale folding has been shown (Area 2) to have been a probable cause of salt movement, but its influence outside of the area cited is believed negligible.

Regional Dip

It seems reasonable to expect that a consistent regional trend such as that exhibited by the salt structures should have some kind of regional cause. Aside from the regional trends of the sub-salt structures, which have the same NW–SE orientation as the salt structures, there is one other regional factor, regional dip. This has been mentioned by several authors, including Trusheim (1960, pp. 1523–24) and Halbouty (1967, p. 35), but no one appears to have actually applied this idea in detail to a salt basin nor is it clear just how regional dip would initiate individual salt structures.

It appears, however, that there could be something to the idea. For one thing, the regional grain over most of the area does run NW–SE, the strike of the salt structures. Also, the salt structures along the rim of the basin tend to conform to the change in strike of the basin, although this might be explained by other causes such as local steep dip causing overburden faulting (locality 1 in faulting discussion). Another factor to consider is that at least one other basin, the East Texas salt basin (Halbouty, 1967, p. 72, Fig 5–5) shows a change of salt structure trends with regional strike, so the phenomenon is not restricted to the North Sea. This would appear to be a fruitful area for detailed investigation, perhaps including model studies.

Regional studies show that the area has had two main stages of development in post-Zechstein time.

1. The Sole Pit Basin (Fig 1) existed in the area from Triassic to end Cretaceous time.
2. In Tertiary time the Sole Pit Basin was tilted strongly to the E, depressing and destroying its E rim.

The regional tilting to the E in Tertiary time may have been a factor in the development of a major salt syncline shown at A on the salt structure map (Fig 1) and the large salt structures around it, almost all of which are of Tertiary age. This was the only area of thick, undeformed Zechstein salt remaining in post-Cretaceous to pre-Tertiary time, and it is thought that the strong eastward tilting, combined with the additional load of Tertiary sediments, has resulted in the intense deformation seen today. The reason for the strong NE–SW trend of the major syncline, which is at right angles to the dominant NW–SE trend of the other salt structures, is not known.

The formation of the Sole Pit Basin resulted in local zones of steep dip along its flanks. These were discussed under faulting and, as concluded in that section, the steep dips are believed to have initiated strong salt movement through faulting of the overburden. It is possible that the development of the Sole Pit Basin and the Tertiary tilting may have caused additional overburden faulting and initiated more salt structures than those specifically attributed to it. Overburden faults are not characteristic of many salt features, but evidence of them could have been destroyed by later salt uplift and erosion.

Conclusion

Despite the exceptional advantage of good sub-salt seismic data, in only a few cases has it been possible to explain satisfactorily the initiation of specific salt structures in the English Basin. However, the main conclusion of this investigation, that despite similar trends there is no direct connection between the salt and sub-salt structures, is significant in that it demonstrates that the problem of initiation is more complex and difficult than was supposed. In any attempted solution of the problem, the salient characteristics that should be explained are the consistent NW–SE trends of the salt features and their general conformity to the shape of the basin, including the ancient Sole Pit Basin, and a detailed investigation of a possible indirect connection between the salt structures and the sub-salt structures would also be of value.

ACKNOWLEDGMENTS

The author is indebted to Ian Macartney, R. L. Blanton, and Dr W. E. Humphrey for their constructive criticism of this paper, and to J. C. Van Wagner for his many suggestions for improvements in the test. The geophysical section of Amoco U.K. Petroleum Ltd, headed by H. G. McCleary, provided much of the data used in the paper. Mr Van Wagner and Dr M. H. Mason helped in preparing the illustrations.

Permission of American International Oil Co., Amoco U.K. Petroleum Ltd, The Gas Council, Amerada Petroleum Corpn, and North Sea, Inc., to publish this paper is gratefully acknowledged.

BIBLIOGRAPHY

Cook, E. E. *Geophysics*, 1965, **30** (4), 495–510.

Gussow, W. C. "Salt Diapirism: Importance of Temperature, and Energy Source of Emplacement", *in* "Diapirism and Diapirs, A Symposium" Tulsa, Okla., Amer. Assoc. Petrol. Geol., Mem. 8, pp. 16–52, 1968.

Halbouty, M. T. "Salt Domes: Gulf Region, United States and Mexico." Houston, Gulf Publishing Co., 425 p, 1967.

Heybroek, P., Haanstra, U., and Erdman, D. A. *Proc. 7th World Petrol. Congr.*, 1967, **2**, 905–16.

Hornabrook, J. T. *Proc. 7th World Petrol. Congr.*, 1967, **2**, 837–56.

Kent, P. E. *Proc. Yorks. Geol. Soc.*, 1967, **36** (1), 1–22.

King, P. E. *Bull. Amer. Ass. Petrol. Geol.*, 1966, **50** (8), 1625–80.

Murray, G. E. "Salt Structures of Gulf of Mexico Basin—a Review," *in* "Diapirism and Diapirs, A Symposium." Tulsa, Okla., Amer. Ass. Petrol. Geol., Mem. 8, pp. 99–121, 1968.

Parker, T. J., and McDowell, A. N. *Bull. Amer. Ass. Petrol. Geol.*, 1955, **39** (9), 2384–470.

Sannemann, D. "Salt-Stock Families in Northwestern Germany," *in* "Diapirism and Diapirs, A Symposium." Tulsa, Okla., Amer. Ass. Petrol. Geol., Mem. 8, pp. 261–70, 1968.

Tanner, W. F., and Williams, G. K. "Model Diapirs, Plasticity and Tension," *in* "Diapirism and Diapirs, A Symposium." Tulsa, Okla., Amer Ass. Petrol. Geol., Mem. 8, pp. 10–15, 1968.

Trusheim, F. *Bull. Amer. Ass. Petrol. Geol.*, 1960, **44**, (9), 1519–40.

MS received 16 April 1969

Discussion

Chairman: MICHEL T. HALBOUTY

Dr P. Kaye (Burmah Oil Exploration Co. Ltd): What was the age of the faulting now seen in the Rotliegendes and how was this age determined? Did the tread of the faulting follow lines of earlier movement and structural weakness? Could any facies changes be detected between the crestal areas of the Rotliegendes horst blocks and the basinal areas?

H. E. Christian, Jr: From isopach studies and structural relationships outside the area of salt movement, we believe that the principal ages of Rotliegendes faulting were Late Cimmerian (end Jurassic) and Laramide (end Cretaceous). We have not been able to determine if these follow lines of earlier movement. As to lithofacies changes in the Rotliegendes from horsts to grabens, we have not observed any.

P. J. Walmsley (BP Petroleum Development Ltd): Mr Christian is to be congratulated on his lucid description of some of the U.K. North Sea salt structures and the problems associated with their growth. When preparing his paper he was unaware that another paper by R. G. W. Brunstrom and myself on Permian evaporites in the North Sea Basin was being published in the April 1969 *Bulletin* of the AAPG. In part our paper covers similar ground to that of Mr Christian, so that it is no doubt a relief both to him and to us that we find ourselves in close agreement with each other.

A discussion on the initiation of salt structures can but be hypothetical. No doubt a number of factors combine to cause salt movement. In our paper Brunstrom and I suggested faulting as the primary cause but we were thinking in terms of salt movement as a whole rather than faults being associated with each individual salt plug. We believe it of significance in the British basin that the oldest salt structures appear to lie close to the margin of the basin, and the younger nearer the centre.

The margin of the basin, *e.g.* to the W of West Sole and in the vicinity of the Hewett field, is controlled by deep-seated faulting. It is here that the facies of the Zechstein changes from predominantly anhydrite/carbonate to predominantly salt (thus providing a barrier to salt movement). It is here that the thickness of the overlying Triassic and Jurassic begin to thicken rapidly north-eastwards. It is here that faulting occurs below and above the salt and that the oldest salt movement can be recognized. Once the salt has moved the inbalance in the overburden it becomes more pronounced, leading to secondary salt uplifts on the basinward side in the manner of Sannemans salt families. This would account for their conformity to the shape of the basin, regional faulting no doubt controlling the trend in the more basinward area.

We ourselves have not had the time to study each individual salt structure in detail in order to prepare maps showing the ages of the various peripheral sinks. I am sure such a study would be most rewarding. I would be interested to know if Mr Christian has had such an opportunity and, if so, whether or not the overall conclusion that the age of the salt structures decreases basinwards can be supported in detail.

H. E. Christian, Jr: We have determined the ages of all the salt structures in the U.K. portion of the English basin and are unable to see a progressive decrease in age basinwards, except that there is considerable Tertiary (young) movement in the present central part of the basin near the U.K.–Dutch median line. However, this area lay on the flanks of the Triassic–Jurassic Sole Pit trough, only receiving a thick layer of sediments in the Tertiary, and the predominance of Tertiary movement could well be related to that. The Tertiary is only present in the central and E parts of the basin, so the post-Jurassic history of the salt structures in the W part of the basin cannot be determined in detail. It can be said, however, that the W salt structures do show a fair amount of post-Jurassic movement, some of which could be Tertiary. There is considerable Tertiary salt movement right up to the point where the Tertiary becomes too shallow to map seismically. As Mr Walmsley states, the margin of the basin in the West Sole–Hewett field area was a depositional hinge line in Triassic–Jurassic time (although probably not in Zechstein time). In accordance with Mr Walmsley and as suggested in the present paper, either sub-salt faulting or overburden faulting, more likely the latter, probably initiated salt movement in this area. However, most of the salt structures in the English basin are salt pillows and lack the pronounced peripheral sinks thought necessary to initiate other salt structures. In addition, almost all of them have had several periods of growth, and their early relief and associated peripheral sinks were generally considerably less than their present relief. The ages of movement of adjacent salt structures and the occasional disappearance of salt pillows with time (Area 5 of the present paper) also do not support the hypothesis. I should mention that an important difference between the Walmsley–Brunstrom paper and mine is that they believe that there is an indirect relationship between the sub-salt faulting and the salt structures. For a time this appeared to me to be an attractive idea, and it still is a fair possibility, but the problem is that in the N part of the basin the salt trends are still present but there is little sub-salt faulting, which is difficult to explain. This is the reason why the regional dip hypothesis was suggested, which could fit fairly well in that area. It appears that these salt structures have been initiated in a number of different ways, and there is therefore no single explanation that will fit the whole pattern.

L. L. Nettleton: What is your conception of the mechanism by which Rotliegendes faulting is absorbed by the salt? Does the salt simply flow away from the uplifted area?

H. E. Christian, Jr: Yes, I think the salt must have simply flowed away from the uplifted area, but it is remarkable that this could occur. It would not be so unusual if these were all small faults, but some of them are quite large, 300 m (1000 ft) or more, and this is one of the greatest surprises that we had in the North Sea.

Theodor Sorgenfrei (Technical University of Denmark, Copenhagen): I agree with the main viewpoint that the salt structures are related to the geological structure of the sub-salt formations. Broadly speaking, the salt features of the British Basin parallel the Charnian trend of Britain. In going E, other regional trends of the salt structures occur. In the area of the East Dogger Bank graben N–S directions prevail. Several other directions appear further E in the off- and onshore portions of the German Basin. The origin of the onshore trends in Germany has repeatedly been discussed by German geologists (Richter-Bernburg, Schot, Trusheim *et al.*). In my own attempt to contribute to the solution of the problem, I have suggested shearing stresses in the crust and horizontal movements of the blocks below salt, resulting in both components of stress and tension. The salt structures may well be the combined result of salt buoyancy, thickness of the salt, depth of the basin, vertical movements of the sub-salt structures, and horizontal movements in the crust. The apparent *én echélon* arrangement of the salt structures in relation to the deeper structure may be evidence of horizontal movements.

H. E. Christian, Jr: I have never considered a possible relationship between salt movement and shearing stresses. Not being a structural expert, it is hard to comment, but I would think the salt would act like a liquid and simply absorb the shearing stresses.

Victor Church: Is there any direct relationship between the salt structures and the gas accumulation in the Rotliegendes or other older or younger formations?

You indicated that some of the faulting was of Cretaceous age; however, none of the sections showed faulting penetrating through the Zechstein. What is the evidence for dating these faults of later age?

You mentioned "steep dips" in regard to the flanks of the Sole Pit Basin. What is the order of magnitude of these dips?

H. E. Christian, Jr: There is no direct relationship between the Rotliegendes gas accumulations and the salt structures, but several Bunter sandstone gas accumulations occur on salt structures.

In regard to the age of the faulting, you generally cannot determine this in the area of salt movement because the faults are absorbed in the salt. The probable ages of fault movement were determined from studies outside the area of salt movement.

As to the magnitude of dips on the flanks of the Sole Pit Basin, the steepest dip found was on the E side of the Sole Pit trough. This dip was about 6000 ft in 20 miles, or about 300 ft per mile.

D. W. MacLean (Mobil North Sea Ltd): It was of interest to note the absorption of faults in the Zechstein salt section, even in the area of thin salt on the edge of the basin, as in Fig 8. Has Mr Christian seen this effect elsewhere, and does he have any idea of the age of the faulting?

H. E. Christian, Jr: In this particular area, the age of faulting probably coincided with the age of uplift of the structure, which was at the end of the Jurassic.

D. W. MacLean: I was wondering if you had noticed this relationship elsewhere along the fringes of the Zechstein salt basin, where the salt might be thin?

H. E. Christian, Jr: Yes, we have. One such area is E of Fig 8. Then, on the other hand, there are other areas with thicker salt that faulting penetrates. I think it is mainly a function of the thickness of the salt, plus the throw of the fault. When you have the right combination the salt will be penetrated, but it is extraordinary the size of the faults that can be absorbed in the salt.

P. Heybroek (Batanafse Internationale Petroleum Maatschappij): We should thank Mr Christian in that he shows us convincingly that lateral salt movements are very important in the Southern North Sea.

It is therefore not surprising that pre-salt highs show reduced salt, as in the first instance the salt will tend to flow laterally away from the high in the lower and synclinal areas of the pre-salt structure. This relationship is clearly demonstrated by some of the examples given by Mr Christian, and would tend to indicate a direct connection between pre-salt structure and halokinesis.

It is clear, however, that salt flowage could be a prolonged process influenced by a series of other factors. A more regional approach is needed, and I hope that the oil industry will find ways to make such a regional approach possible.

H. E. Christian, Jr: I do not believe that pre-salt highs consistently show reduced salt, as Mr Heybroek states. One of the principal points of the paper was to demonstrate that each salt structure is not invariably underlain by a sub-salt structure or fault, so the examples selected tended toward the other extreme, but in reality the spatial relationship of sub-salt structures to salt structures with few exceptions appears to be random. Their trends are, however, similar.

Michel T. Halbouty: Dr Trusheim has done much work in the areas of salt domes and general salt studies and among his many contributions are the terms halotectonism—bedded salt deformed by compressive salt tectonic forces—and halokinesis—salt structures which are the result of autonomous, isostatic salt movement. These terms are widely accepted and used by those of us who have made considerable studies on salt structures and salt masses. In the North Sea, do you have halotectonism or halokinesis, or both?

H. E. Christian, Jr: In the U.K. North Sea I believe the salt movement has been halokinetic, just the upward movement of the salt caused by buoyancy and geostatic loading, with no compressive forces evident.

Michel T. Halbouty: I also noticed in one of your illustrations that you had the isopach thickness of salt from 400 m to 1300 or 1400 m. The illustration, as presented, would give the impression that the actual thickness of the salt is uniform throughout the basin as represented by the isopach thickness. However, is it not true that you do not have a graduated uniform thickness throughout this basin? In other words, in some areas there may be thicknesses of 1000 ft or 2000 ft and maybe several hundred feet away the salt will diminish rapidly and have considerably less thickness. Are these assumptions correct?

H. E. Christian, Jr: Yes, in this figure we tried to restore the Zechstein to its original thickness, because at present it varies greatly in thickness as the result of salt flowage.

Dudley J. Hughes (Hughes & Hughes): Mr Christian's paper is undoubtly a major contribution to the science of geology. In the Gulf of Mexico Basin of the U.S.A. we have been concerned almost exclusively with salt structures during the long exploration history of this basin. The oil and gas production in the Gulf of Mexico Basin is from beds above the salt and, due to the depth and thickness of the salt, we have very little information as to what lies below the salt.

There has been much speculation as to the triggering mechanism of salt structures, and one widely accepted theory is that faulting in underlying beds is a major cause. Mr Christian's study discredits the hypotheses.

In my own study of faulting in beds overlying salt structures, I have found no cases where faults cut into the salt. The faulting is always confined to the overlying beds and is the result of stresses caused by the underlying salt movement.

You stated in your paper that "there are indications that some Rotliegendes faults have completely penetrated about the same thickness of Zechstein, cutting the post-Zechstein Beds". Do you have sufficient control to verify this or could the faulting in post-Zechstein beds be only the results of salt movement and not an upward extension of Rotliegendes faulting?

H. E. Christian, Jr: Our seismic data are such that we cannot be entirely sure, but it appears likely that in the exceptional area referred to in the paper Rotliegendes faulting completely penetrated the Zechstein. As the paper stated, however, there is more anhydrite in the Zechstein in that area, rendering it less capable of abosrbing faulting, and fault throws are large.

Michel T. Halbouty: Has salt movement in the North Sea in any way affected or controlled accumulation of hydrocarbons?

H. E. Christian, Jr: There is one recent discovery and one non-commercial accumulation in the Bunter sandstone on salt domes, so salt movement does control the accumulation of hydrocarbons to a minor degree.

Oil Shale Occurrences in Western Europe

By L. E. SCHLATTER

(Bataafse Internationale Petroleum Maatschappij, N.V., The Hague, Netherlands)

SUMMARY

It seems appropriate to add to a symposium on the oil possibilities of Europe and North Africa a short review of non-conventional crude possibilities. These include oil shale, which is considered a subject for long-range planning by governments and oil companies alike. The purpose of this paper is not to add new details to a long list of existing publications, but briefly to compile the scattered information and to point to some geological and economic factors which will influence the planning of future oil shale utilization.

No oil shale exploitation of any importance is taking place in Western Europe at present. However, in 12 countries more than 50 oil shale deposits of various sizes are known in formations ranging in age from Cambrian to Tertiary. They will be summarily described and shown on a map.

No reserve estimates are given because it is realized that most oil shale deposits are insufficiently known. For this reason, the need for their proper exploration is emphasized and prospecting methods are discussed. Only careful geological evaluation with modern methods will eventually provide more reliable figures for an economic assessment of Europe's numerous oil shale resources.

INTRODUCTION

At a meeting devoted to exploration for petroleum in Europe and North Africa one might as well briefly review some future possibilities at present known under the name of "unconventional fuels". Among these, shale oil is the most important and is regarded by many governments and oil companies as a worthwhile subject for long-range planning.

We are all well aware that shale oil has not yet found a place in a free economic climate because the costs of extraction are still higher than those of ordinary crude production. There are, however, two factors favouring oil shale. The first is the increase in cost of producing petroleum arising from costly offshore and future deep-sea drilling and production. The second is the reasonable possibility that with increased technological research the costs of shale oil extraction, both from mined shale and from *in-situ* processes, can be considerably reduced. Thus it is only a matter of speculation whether it will be in five, ten, or 20 years from now that the gap between these two trends will be closed. In other words, at some time in the future the curve of increasing petroleum production costs will meet the curve of decreasing costs of shale oil extraction.

Attempts to produce shale oil date back about 200 years and in the past 100 years we have seen medium-sized plants for shale oil extraction set up in various countries such as France, Scotland, and Spain. Under restricted and wartime conditions other countries, such as Sweden and Germany, have temporarily produced shale oil, but the fact remains that these operations have all been uneconomic and have closed down completely.

It is not the purpose of this paper to go into detailed descriptions nor to add new data to a long list of existing publications on oil shale in European countries. Instead, I intend only to compile the more relevant information and to summarize the oil shale deposits of Western Europe.

However, before going on to enumerate the individual deposits, I have to explain why this paper does not contain any reserve estimates. The reason is that in most cases reliable figures are simply not available, in spite of a great many attempts at evaluation. This is because, with very few exceptions, no serious exploration with modern means has been carried out which would enable estimates of reasonable reliability to be made.

Nevertheless, figures have been compiled in many publications, but a careful literature study has made it abundantly clear that most of these data are mere guesses and that the majority have hardly any substantial basis. Major parameters, such as the thickness of the deposit, its extent, and assay often relate to a few spot observations, while significant variations of these factors are hardly ever considered. In some of the rare, well-studied cases, figures usually refer to a particular property or restricted mining area and often fail to include the entire deposit or the exploration of the whole sedimentary basin. Many estimates are pervaded with inadequate assumptions, resulting in different cases in both gross over-estimation and under-estimation of oil shale reserves.

Moreover, many readers may raise the point that they have read of a great number of oil shale localities not listed in my compilation. This is another disturbing problem if one bases one's knowledge mainly on a study of literature, which refers to a profusion of local place names. In many cases a large number of local oil shale outcrops are cited, all, however, belonging to the same principal deposit. This may create the impression of a large number of individual deposits, but screening of these localities usually results in a greatly reduced number. Among the various examples I may mention only the literature of France, in which more than 100

Oil Shales of Western Europe in TIME + ENVIRONMENT

		marine	paralic	limnic
CEN.	Tertiary			❶ ❶ A+D ② E ④ ❶ F+I
MESOZ.	Cretaceous	❷ I	❶ F	❶ E
	Jurassic	❶ ❶ F+GB Lias	❶ GB	
	Triassic	❶ ❶ CH D ❶ I	❶ A ❶ N	
PALAEOZOIC	Permian			❷ F
	Carboniferous		❸ F ❺ GB	❶ E
	Devonian	❶ GB		
	Cambr. Silur.	❺ S		

FIG 1.

names are found in various publications. These, however, refer to hardly more than the 14 occurrences shown in this paper (see Fig. 1).

DEFINITION AND GENESIS OF OIL SHALE

To start with, we should perhaps first answer the simple question: what is an oil shale? This seems the more justified, because the term "oil shale" is a typical misnomer which led to the famous definition that "an oil shale is neither a shale nor does it contain oil". Nevertheless, the term is well entrenched in the English language, it is brief and convenient, and I see no need to replace it.

Several efforts have been made at a more positive definition but in the past there has been simply too little knowledge of the real nature and the generation of oil shale. Recent progress in geological and geochemical research gives us better insight and perhaps the best we can formulate today is that under oil shale we understand "a sapropelite containing a variable amount of kerogen which, on destructive distillation, yields shale oil".

Oil shales are also often referred to as "bituminous shales" or "bituminous rocks". I consider that an unfortunate expression and should rather like to have it abandoned, because the term "bituminous" has such a wide range of meaning as to be far from being specific. (It is generally applied to both naturally occurring and artificially derived products and is equally used for both: organic matter of sapropelic as well as of humic origin.) Therefore, my suggestion is to use in future only the more specific term of "kerogenous rock".

As to the origin, no more fundamental difference between an oil shale and a petroleum source rock is postulated. Both are recognized as kerogenous rocks of the same type and we should also drop the still widely held idea that there is a difference in the aquatic domain of their origin: kerogenous rocks can be formed as well in marine, brackish, and in fresh water.

The two most important prerequisites for the formation of kerogenous rocks are (1) an abundance of suitable organic matter, and (2) anoxic conditions for preservation of the dead organisms and for their transformation into kerogen.

The organic matter favourable for oil shale contains mostly compounds of the lipid group like fats, fatty acids, steroids, waxes, etc. Although these are also found in higher plants and animals, it is the plankton, and especially phytoplankton such as algae, which is particularly prolific under certain conditions and can produce enormous amounts of organic matter. In fact, the total volume of plankton is considered to be several times greater than that of all other aquatic life combined and its growth rate produces a quick turnover, especially in warm climates with a high level of sunlight and a considerable supply of nutrients.

The second prerequisite for the formation of oil shale is the preservation of a part of the lipid-rich organic matter. Immediately after the death of the plankton, it is—like all organic matter—attacked and

FIG 2.—Situation and Age of oil shale deposits in Western Europe

devoured by bacteria, starting a biochemical process which at first is mostly oxidation, leading to the decay of the organic remnants. Therefore the preservation of organic matter in the sediment depends mainly on a low content of oxygen, both in the bottom waters and in the sediment.

Biochemical processes are principally responsible for the existence of oxygen minima in the water body and a lack of circulation is responsible for its maintenance and position. Under such anoxic conditions, further biochemical alteration of the remaining organic matter takes place through the action of anaerobic bacteria, the end product being the insoluble, solid kerogen.

The most favourable environments for such reducing conditions are of course completely stagnant basins such as parts of the Black Sea, which has given its name to the "euxynic" facies, or some silled fiords and many lakes. There the bottom waters are completely devoid of oxygen but contain considerable amounts of sulphuric acid. The sapropelic mud of these waters is usually very dark coloured or black.

However, there are environments where the bottom water is not completely stagnant but with layers having a sufficient oxygen deficiency for the preservation of organic material. Such muds can often have less pronounced dark colours with shades of greenish brown.

Another restrictive factor in the formation of oil shale is the amount of mineral matter which is intimately mixed with the preserved organic debris. The rate of admixture of inorganic constituents affects the quantity of kerogen in a rock, and since we consider an oil shale to be a particularly rich kerogenous rock, dilution by minerals must be restrained.

Bacterial action decreases gradually and becomes very weak when the sediment reaches a certain depth. This boundary is not exactly known and must be transitional, but bacteria generally do not survive temperatures of 80 °C. It is believed that at this comparatively early stage of organic metamorphism, where the biochemical gives way to the geochemical processes the transformation of the organic matter into kerogen is already accomplished.

EXPLORATION AND ASSESSMENT

If we ever wish to assess oil shale deposits as important energy resources of the future, it will be indispensable to apply accurate geological prospecting techniques. This exploration will, however, be somewhat different from the search for oil, because oil shale is an autochthonous, solid material and the geologist will have to combine his petroleum experience with some mineral prospecting methods.

Proper exploration should, of course, start with a regional geological investigation of the sedimentary basin and the reconstruction of its tectonic and depositional history. This should lead to the interpretation of the most favourable environment for oil shale formation and to the outlining of those parts of the basin where undisturbed layers will allow exploitation over a wide area.

Apart from mining considerations, which belong in the realm of production, specific exploration methods include the recognition of oil shale, detailed stratigraphic examination, proper sampling, and correct assay of the potential oil yield within each section. This is followed by the evaluation of its lateral extent and the detection of variations in facies, thickness, and yield. A thorough knowledge of the geological and physical properties of oil shale, both in fresh and in weathered conditions, is the first prerequisite in the field, at the well-site, and in the laboratory.

One of the most significant properties of oil shale is the common interbedding of rich, lean, and barren intervals on a small, medium, and larger scale. Thin slides reveal microlamination of organic-rich and mineral-rich parts, a macroscopic study often shows varved texture, while on a larger scale several richer seams of oil shale often alternate with lean and barren beds. Hence the study of this relationship is of great importance for the economic evaluation of a deposit.

It is quite obvious that only a minute examination of each section can provide the basic information needed for assessment. This is often very difficult to accomplish by surface geological methods, and more often it is not possible at all. In any event, the best, or even the only reliable way to accomplish this task, will be continuous coring with a light slimhole drilling outfit. This method also produces fresh, unweathered samples which are often hard to get at the surface, but which are indispensable for proper assaying.

Considerable attention should be given to the selection of samples for assaying the yield, because each sample must be representative of the entire interval chosen for analysis. The best solution again comes from using cores, of which for every foot (or similar convenient small interval) a segment is cut away perpendicular to the bedding plane. This slab is crushed and thoroughly mixed in order to avoid the predominance of any unrepresentative intervals within it.

The determination of the yield of shale oil is usually done by destructive distillation of the crushed rock in the so-called modified Fischer retort. This is the most reliable and convenient test and quite efficient, when automated batteries of retorts are used.

Since coring is an expensive method, drill cuttings from slimholes have also been used and results are satisfactory, provided even more precautions are taken to select a truly representative mixture of samples,and that the assays are calibrated and corrected using assays from continuous core holes.

Several attempts have been made to replace the Fischer assay method by more rapid and cheaper processes for approximate yield estimations, such as pyrochromatographic tests and volumetric and photometric analyses. These all have in common a saving in time and in cost and also the merit of needing only minute amounts of sample.

However, the feature which at first seems advantageous proves to be the danger of such processes. This is the fact that such small quantities almost entirely preclude the sample from being representative of the whole interval. This hazard must always be borne in mind when applying rapid analytical methods.

Another helpful means for the evaluation of oil yield, not so much a substitute for as a supplement to the processes referred to and to the Fischer Assay, is the application of some specific log analyses, especially of sonic and gamma-gamma ray logs. But do not overlook the fact that, favourable as such auxiliary methods may be, they should never be used alone. Only careful calibration and frequent checks with Fischer assay values can guarantee adequate results.

A somewhat awkward situation is found in the expression of assay values. According to the American Standard, assay is reported in U.S. gallons of shale oil per short ton of oil shale, usually just given as gallons per ton. It must be realized, however, that in literature many analytical results are measured not in U.S.

gallons but in British Imperial gallons and, to make matters worse, one finds short tons as well as long tons!

This situation clearly calls for a change to the decimal system and indeed many assay results are now given in percentage by weight, which is a more acceptable denomination. The oil industry, however, is bound to adopt the expression in per cent, or rather in per mille by volume of rock in place. This is for a purely practical reason, because it would be the logical base for a specific yield of any oil shale deposit; a unit which states the amount of oil in place per decimal square unit per cent vol would, for instance, mean litres per square metre, or cubic metres per hectare or 1000 cu m/sq km. In such terms the simplest mental arithmetic will assess any deposit in an instant.

Contrary to the practice in petroleum geology, calculated or estimated reserves of oil shale are usually expressed in terms of "oil in place". How much oil will eventually be recovered depends on improvements in technology and will change with time.

Proper exploration will thus provide the basic data for any future decision on the value of oil shale resources. However, one has to bear in mind that final evaluation of a prospect depends on a variety of non-geological factors. I shall not, however, expand on such technological and particularly economic aspects.

THE LOCATION AND THE AGE OF EUROPEAN OIL SHALE DEPOSITS

This compilation has to be restricted to Europe west of the Iron Curtain because of the lack of sufficient information from Eastern European countries. Among the latter, oil shale is reported from Czechoslovakia, Rumania, and especially from Bulgaria. Better known, but so far of more local importance, are some deposits in Yugoslavia. As to North Africa, the picture is more straightforward: no oil shale deposits of any importance have up to now been reported between Morocco and Egypt.

In Western Europe over 50 oil shale deposits of various sizes are known in 12 countries and from formations ranging in age from Cambrian to Tertiary. They are all shown in Fig 1, together with an indication of tion of their geological age, using the usual abbreviations. There is only one oil shale horizon of almost continental extent, namely, the black shales of the Lias, which is indicated by a special type of shading. The other occurrences are of more local extent and are shown in circles. Please note that the countries are marked with their almost official initials as seen on their respective car licence plates.

Austria (A) has three different oil shale deposits all situated in the tectonically much disturbed Tyrol alpine region, which would render large-scale extraction very difficult.

The oldest and richest occurrence is of upper Triassic (Norian) age and is best known from Seefeld, where it is distilled for pharmaceutical use. Up to six seams, totalling between 40–120 m, are intercalated in a 700- m thick marly formation of shallow marine origin. Assay values are reported between 15–30 per cent vol. A non-carbonate organic carbon content is described which varies from 3–26 per cent wt (of dry sample).

Another oil shale formation occurs in the lower Jurassic in marine Toarcian beds, best known from Baechental. The formation is up to 35 m thick, of variable richness assayed to a maximum of 15 per cent vol, with up to 10 per cent wt organic carbon.

An Eocene (priabonian) deposit of limnic origin is found overlying coal in a syncline at Haering (near Kufstein). It is only 6 m thick, assays up to 15 per cent vol, and contains 1–5 (exceptionally 26) per cent wt organic carbon.

Belgium (B). A small portion of the Lias of the Paris Basin (see France) reaches into Belgium's southernmost province of Luxembourg.

Denmark (DK) has no oil shale.

Germany (D). In the Federal Republic, the oldest oil shales occur in the Permian of the Saar-Nahe Basin, but are practically unexplored and appear to be very thin layers only.

Triassic oil shale of Austrian Seefeld type, actually belonging to the same deposit, is also described from the highly folded Karwendel mountains in the Bavarian Alps.

By far the most important oil shale is the Lias, with extensive outcrop and subcrop areas in SW and NW Germany. The marine Toarcian is in a black shale facies 5–8 m thick, having a variable, comparatively low yield from 12–16 per cent vol.

The richest oil shale of Tertiary (Eocene) age used to be quarried at Messel, near Darmstadt, and is now practically exhausted. It was a local limnic deposit up to 300 m thick with a very high yield.

Finland (Suomi) (FS) has no oil shale.

France (F). The oldest occurrences are of Carboniferous age: the one at Esterel (Var Department) was insignificant and is now exhausted, but some similar pockets may exist elsewhere in the tectonically disturbed area. A thin Carboniferous occurrence is described at Faymorreau (Vendée), reportedly yielding about 30 per cent vol. The third Carboniferous deposit is in the subsurface at Littry (Manche), an oil shale or sapropelic coal associated with humic coal. Its extent has not been explored and no assay data are known to the author.

Of more significance are the extensive Permian (Autunian) sapropelites from Autun (Allier *et al.*). Although only 1–4 m thick, their yield is 19–25 per cent vol; their subsurface extent, however, has barely been explored. A similar deposit, also in a continental intermontane basin, is found further west at Aumance (Allier). It is only 1–2 m thick and its yield is equal to that of the Autun deposit. Both were exploited to some extent in the past. The Lias (as elsewhere, the Toarcian) provides the greatest areal extent of kerogenous rock in France. This unit transgressed over the old Hercynian massifs and is now found outcropping between them and the younger sedimentary basins (*e.g.* Paris Basin, Aquitaine Basin, *et al.*). The thickness is 10–15 m, the assay yield rather low (10–14 percent vol), and although the deposits have been sporadically exploited, they are not sufficiently explored.

Another marine Jurassic occurrence at St Pompon (Dordogne) is local and considered insignificant.

The only Cretaceous oil shale is reported from Vagnas (Ardèche), probably in paralic facies and thin, but said to assay between 25–30 per cent vol.

In the older Tertiary, marine Oligocene fish shales at Froidefontaine (Ht. Rhin) contain 6–30 m of oil shale of unknown yield and the whole deposit is practically unexplored. A second Tertiary (Oligo-Miocene) sapropelic deposit is known from a small basin near Manosque (B. Alpes), 3–5 m thick, yield 20–30 per cent vol, little explored.

Finally, there are several limnic, purely local—though at times thick and rich—deposits of Miocene age in southern France, mainly, La Boutaresse (P. du Dome), Lavayssiere (Dordogne), Malvezy (Aude), and Menat (P. du Dome). The latter is of the Messel type of Germany and is 70–100 m thick.

Great Britain (GB). The oldest oil shale in the United Kingdom is an unexplored, probably small, marine Devonian occurrence, reported to be 2 m thick in the Scottish Highlands.

The Carboniferous occurrences are more important; these are in a transitional facies, where rich oil shales and thin seams of sapropelic coal are associated with humic coal which is extensively mined. The oil shales of the Lothians, near Edinburgh, are well known for their long history of exploitation! They are only 1–4 m thick and have yielded an average of 23 per cent vol. Their extent is restricted, but thin layers (up to 1 m) of probably similar rock are found in the Scottish oilfields and sapropelic beds are reported from Flintshire, North Staffordshire, and from Antrim in Northern Ireland.

In southern England, marine Lias is found outcropping in a large belt, but contrary to the case in continental Europe its organic content is lower, the kerogenous beds being only 1–6 m thick and yielding about 10 per cent vol. Of better quality but hardly exceeding 1 m in thickness is a sapropelic seam of considerable extent in the marine Kimmeridge yielding up to 25 per cent vol. Another upper Jurassic deposit, finally, is of purely local value, a sapropelic bed associated with humic coal at Brora in NE Scotland.

Greece (Gr). No oil shales of any significance are known in this country.

Eire (IRL) is also without oil shale deposits.

Italy (I) the oil shale occurrences are not well documented, but admittedly the author's knowledge of the relevant literature is not too profound. It appears that none of the mainland deposits are of great significance, while only Sicily has more prominent resources.

On the Italian mainland, a purely local, insignificant marine Triassic deposit is exploited for pharmaceutical oils at Besaro (Lombardia province). It is about 1 m. thick and said to yield 24 per cent vol or more Several other small and tectonically equally disturbed Triassic occurrences may be found in the alpine provinces.

Thin marine upper Cretaceous oil shale has been found at Mollaro (Trentino) and exploited since the 17th century on a local scale. Another is found in the Metauro Basin (Marche). Only 1–2 m thick, it yields from 18–20 per cent vol.

Unexplored Tertiary oil shales are reported from Chieto (Abruzzi), where up to 20 m of low yield (13 per cent vol) rock occurs in the Eocene, and from Taranto (Basilicata), on which no particulars are available.

The best known Italian oil shales are those on the east side of the island of Sicily. The shales are called "tripolite" and are also of Tertiary age, occurring as seams some 5 m thick in a transitional facies capped by evaporites. Descriptions exist of a great many local occurrences spread over a large area, which has led to estimates of several billion barrels of oil in place. This possibility cannot be ignored, but the area—which is tectonically complicated—needs a serious exploration effort before any assessment can be made.

Luxembourg (L). Like the southern tip of Belgium, the Grand Duchy of Luxembourg embraces a small part of the Lias, but so far the area has not been favourably appraised for exploitation.

The Netherlands (NL) has no oil shale deposits on or near the surface.

Norway (N). From the mainland only minute and doubtful oil shale indications are reported but on the Arctic island of Svalbard (better known as Spitzbergen) some 25 m of kerogenous marine Triassi care known to outcrop, but no assay is reported and the occurrence is completely unexplored.

Portugal (P). There are outcrops of marine Lias centred around Coimbra and the neighbouring provinces. These have never been studied for oil shale properties and the occurrence appears to be thin and of low yield.

Spain (E). The Carboniferous oil shale deposit of Puertollana in Ciudad Real province was exploited from 1922 to 1966. Three thin seams of oil shale occur together with coal in a continental-intermontane series of upper Carboniferous age. Only one seam reaches a thickness of 2–3 m and this is assayed at up to 25 per cent vol; this is the only one which was worked. Reserves were estimated at about 60 million brl of oil in place.

The Lias, still untested and unexplored but partly

in a kerogenous facies, reaches 30–50 m in thickness in Santander province in the Cantabrian region.

At Ronda in the province of Malaga there is a small Jurassic oil shale locality which is mentioned in literature, but without any specific data.

A local, rather insignificant oil shale occurs in continental lower Cretaceous beds at Oviedo in the province of the same name.

Two Tertiary oil shales are known in continental limnic facies. One is of Oligocene age, thin and largely unexplored, but assayed at 26 per cent vol in the surroundings of Ribesalbes (Castellon); the other, still less known, is reported in the Miocene beds of Rubielo de Moro (Teruel).

Switzerland (CH). An oil shale deposit of very restricted size is known at Merida (Ticino). It is similar to the one in neighbouring Besaro (Italy), 1 m thick, average 27 per cent vol, of marine Triassic age in the highly folded alpine region.

There are outcrops in NW Switzerland of a segment of marine Lias which makes the connection between France and Germany. The structure is partly folded and partly monoclinal, and the shale is some 15 m thick but of low assay (±12 per cent vol).

Sweden (S) has done some pioneering wartime exploitation of its considerable resources of marine upper Cambrian oil shale which often also have a rather high uranium content. They are found in several layers in the following regions: in Naerke (Orebro province): 10–15 m of 15–19 per cent vol reported yield; in Vestergotland (Skaraborg), 11 m thick, 15 per cent vol, and in Omberg (Ostergotland), 18 m thick, 12 per cent vol. On the island of Oeland the oil shale layer is thinner and appears to be stratigraphically somewhat higher, occurring at the Cambro-Silurian boundary, while for the island of Gotland no data are available.

Local, tectonically disturbed and dissected remnants of the same Cambro-Silurian oil shale have also been mapped in Scania (Kristianstad province). From the same area there are also reports of some sapropelic shales of Triassic age found in boreholes at Sandakra, but these were never further explored.

BIBLIOGRAPHY

The preparation of a useful bibliography on this topic is a task of great difficulty. This is a result of the unfortunate fact that although hundreds of articles on European oil shales have been published, very few of these are recent, and still fewer deal with a whole country or Europe in general. The following list is therefore only a tentative selection of references, many of which themselves contain bibliographic lists.

Bitterli, P. *Geologie en Mijnbouw*, 1963, **42,** 183.

Bitterli, P. *Proc. 6th World Petrol. Congr.*, 1963, I/30.

Breuer, P. K. *Brennstoff Chemie*, 1921, **2** (2), 106–7.

Brockamp, B. "Zur Palaeogeographie und Bitumenführung des Posidonienschiefers im deutschen Lias." Reichsamt f. Bodenf.-Lagerstättenarchiv Heft 77, 59 pp., 1944.

Charrin, V. "Les roches bitumineuses métropolitaines." Le génie civil, 1954, **CXXI** (23), 449–53.

Cortese, E. *Rassegna Mineraria e Metallurgica Italiana*, 1930, LXX (12), 133–35.

Duncan, D. C., and Swanson, V. E. "Organic-rich Shale of the U.S. and World Land Areas." U.S. Geol. Survey Cir. No. 523, 1965.

Duncan, D. C. "Geologic setting of oil shale deposits and world prospects." 7th World Petrol. Congr. Mexico, Panel Disc. 14, 1967.

Guillaume, C. H. *Ann. Soc. Géol. Belgique*, 1936, **59**, 5–6.

Guthrie, B. "The Oil Shale Industries of Europe." U.S. Bur. Mines Rep. Invest. No. 4776, 1951.

Holzer, H. F. "Oil Shale Deposits of Austria." U.N. Symp. Dev. and Util. Oil Shale Resources, Tallinn, section I, 1968.

Jaffé, F. C. *Bull. Colorado Sch. Mines*, 1962, **5** (2), 1–11.

Matthess, G. "Zur Geologie des Oelschiefervor-kommens von Messel bei Darmstadt." Abh. Hess. Landesamt f. Bodenforschung. H. 51, 1966.

Ozarov, G. "Progress and Prospects in the utilization of Oil Shale." U.N. Dept. Econ. and Soc. Affairs, New York, 1965.

Rickenbach, E. "Vorkommen von bituminösem Schiefer." Beitr. Geol. Schweiz.—Erdölgeologische Untersuchungen, Teil 3, Abschn. 1, 1947.

Schlatter, L. E. "Oil Shale Deposits of Western Europe." U.N. Symp. Dev. and Util. Oil Shale Resources, Tallinn, Section I, 1968.

Schlatter, L. E. '"Exploration Methods and Geological Assessment of Oil Shale." U.N. Symp. Dev. and Util. Oil Shale Resources, Tallinn, section I, 1968.

Schlatter, L. E. "Definition, Formation and Classification of Oil Shale." U.N. Symp. Dev. and Util. Oil Shale, Tallinn, section II, 1968.

Sundius, N. "An account of the oil shales and the oil shale industry of Sweden." Ymer (Svenska Sällskapet Antrop. och Geol.) 1943, **63** (1), 1–16.

Thorne, H. M., and Kraemer, A. J. "Oil Shale in Spain." U.S. Bureau Mines Rep. Invest. No. 4736, 1950.

"Oil Shale and Cannel Coal, I. Institute of Petroleum, London, 1938.

"Oil Shale and Cannel Coal, II. Institute of Petroleum, London, 1951.

Vié, G. "L'industrie des schistes bitumineux en France." Le génie civil, 1946, **CXIII** (46), 277–79.

MS received 17 February, 1969

Discussion

Dr C. F. Knutson (CER, Geonuclear): Yield is an important factor in oil shale evaluation. Fisher assay is sometimes considered quite biased since it ignores the gaseous by-products. Could Dr Schlatter suggest an alternate technique that may be less biased?

Complete coring and analysis is the best technique for oil shale evaluation. However, it is quite expensive and supplementary techniques are called for. What would Dr Schlatter suggest as the most efficient supplementary evaluation techniques?

Dr L. E. Schlatter: I realise that the Fisher assay is not necessarily the best way to evaluate, but it has the tremendous advantage that it is used almost worldwide. Therefore it provides the best figures for comparison. We have heard in other lectures that once distillation is carried out on a larger scale it is not unusual that yields are reported as, say, 110 or 120, or perhaps only 50 per cent of Fisher assay. This reflects the efficiency of the processes used for extraction. I do not think that the Fisher assay, for the time being, should be replaced, if only because it gives us a worldwide means of communication.

As to your second question, the substitutes of the Fisher assay, I have mentioned a few in my paper and am almost sure that each company dealing with oil shale has developed its own quick assay methods. However, one has to realize that by using such substitutes, or by using cuttings instead of cores, one will only get reliable figures if the results are calibrated with Fisher assays of cored sections. These are the only ones giving a truly representative mixture, especially if very small intervals are assayed. In a typical oil shale I would not take more than one ft per sample.

Dr P. Bitterli (Bataafse Internationale Petroleum Maatschapij): If I remember rightly, some large-scale assessment of the economic possibilities of the Liassic Posidonia shale by a core hole drilling campaign was undertaken some years ago in NW Germany. I wonder if Dr Schlatter or somebody in the audience could tell us something about the outcome of this study?

Dr L. E. Schlatter: I am sorry to disappoint you. I am not familiar with such a study in Germany.

Lewis F. Wells (Consultant, Salt Lake City): Can you tell us the potential yield of the various oil shale deposits as referred to in your paper, particularly the Sicilian oil shales, and compared to other areas of Europe?

Dr L. E. Schlatter: No, I mentioned Sicily mainly as a large potential for its areal extent. I may have mentioned some figures in my paper, I am afraid I do not know them all by memory, but the richness or the yield of the Sicilian oil shales are by no means very high. It is their regional extent which makes them a potentially good resource.

H. C. Potter (King's College, London): Has Dr Schlatter's study covered any secondary deposits of solid hydrocarbons in Europe, such as gilsonite and manjak?

Dr L. E. Schlatter: I do not consider gilsonite or manjak as oil shale. In my opinion they belong to the group of asphaltites and, as in the case of the Utah gilsonites (which are the most extensive ever found), they most probably have to be regarded as a product of early hydrocarbon expulsion products from the underlying Green River oil shale. I do not have any notion of extensive gilsonite in Western Europe, although one sometimes finds minor amounts of asphaltites or asphaltic materials within some rocks.

Rustum Byramjee (CFP, France): In petroleum source rock studies we evaluate the oil potential by determining through physical and chemical analysis the degree of evolution of the kerogen. Do you use similar techniques for oil shale evaluation? Do you have new methods for studying kerogen?

Dr L. E. Schlatter: I could not go into all these details in my short paper but of course it is natural that everybody looks at the maturity of the organic matter. As long as kerogen is immature for the natural generation of oil, it is richest in organic matter still concentrated in the rock. As soon as organic metamorphism has advanced to a stage where oil is generated and expelled, it leads of course to a depletion of the oil shale and therefore the yield of the shale becomes increasingly smaller.

As to the question concerning methods of analysing kerogen, our knowledge of kerogen has advanced only in the past four or five years and we do not yet know sufficiently of the structure of this complex organic matter. I think, therefore, that it is premature to propose any standard method at this time.

M. A. Reynolds (Broken Hill Proprietary Co. Ltd, Melbourne, Australia): Your criteria for the formation of kerogenous rocks are very similar to those which are commonly quoted for source rocks for hydrocarbons. Is the containing mechanism for the kerogenous material within the rocks due to lack of permeable outlet for fluids or are other factors involved?

Dr L. E. Schlatter: I think that in the case of the oil shale it does not matter at all, because we consider that an oil shale, being mostly in an immature state, has no liquid hydrocarbons. Your problem would only come into the sphere of our interest if we talk of oil source rocks and the expulsion of the oil from the source rock, but I am afraid that this is beyond the scope of my paper.

Michel T. Halbouty: Are there any retorting or refining operations going on now in Europe?

Dr L. E. Schlatter: Not as far as I know. The last, which has been closed down, was at Puertollano in Spain. There have been a few operating in this century, such as those in Britain, Sweden, and Spain, and also, during wartime, in Germany, but they mainly did it under special economic pressure or only on an experimental basis.

Michel T. Halbouty: If you are familiar with oil shale in the U.S.A., how do you compare it to reserves of those of the whole of Europe?

Dr L. E. Schlatter: That seems a difficult question as we do not have reliable reserve figures for Europe. But I can answer your question by saying it is only a very small part of it. The potentially recoverable oil from oil shale in the U.S.A. is in the range of one trillion (10^{12}) brl of oil in place, while I think in Europe we would be happy to have a few billions (10^9). Now there is one point which has never been sufficiently studied here. All the oil shales reported in literature are usually from near-surface or from surface outcrops. Only if we start to investigate the subsurface extension of these oil shales in view of applying subsurface *in-situ* exploitation methods could we arrive at a better evaluation of our resources in Europe.

A History of the Exploration for Hydrocarbons within the Territorial Boundaries of the Netherlands

By H. A. STHEEMAN and A. A. THIADENS*
(*Netherlands Geological Survey, Haarlem, Netherlands*)

INTRODUCTION

THE authors considered the need for another review of exploration in the Netherlands because, thanks to the liberal policy of NAM, the oldest exploratory company in the country, a great number of publications covering geological and production results have been printed, and most of these also contain the historical setting. It seems hardly worthwhile to repeat these publications. However, they describe, perforce, a specialized subject over a limited period from a special angle. For geologists tend to appear clever, rational, and logical in their publications, whereas sometimes important discoveries have been due to chance or even sheer fluke. We do not know, for instance, whether exploration would have been continued in Slochteren after finishing the Ten Boer well had the wrong casing-scheme not been used in the latter, thereby swabbing-in the well when pulling drill pipe from an unpromising formation, which definitely did not look testworthy.

Indeed, such things occur, but they can hardly be recorded in company-sponsored publications, which are not meant to voice self-criticism publicly. In that respect they are hardly "history".

Moreover, the important decision to search for oil and gas in a country covered for 95 per cent of its area by thick and sterile Quaternary, Tertiary, and Upper Cretaceous formations (Fig 1), devoid of any seepages, tar sands, or tar pits, cannot have been based on geological prospects alone. The supply of the necessary capital must have been the result of a variety of considerations, among which are geological ones which, however necessary, are not more than background music.

Those moments of decision are not only important for Dutchmen from a historical point of view, but also of general interest, because they offer examples of the approach to the unknown by the human mind, an approach always needed every time a new enigma has to be tackled.

We might compare our historical problem with the geological mystery presented by the Hallstatt Man, who, a millennium before Christ, dug shafts in the summit of an Austrian Alp to reach solid salt, a commodity he had never seen *in situ*. The parallel with our case seems to lie in the creation of great wealth by producing in the midst of an inland market an otherwise seabound product, imported from the far shores of the continent.

The parallel, however, partly fails because in the 1930s the supply of oil was ample, the freight cheap, and the home consumer market small. A set of different reasons must, therefore, have lured the shrewd oil investor to launch a costly exploratory campaign for bitumen in the thirties, a time of profound recession of the economy.

In order to understand the right dimensions, it should be kept in mind that the inland consumption of the Netherlands at that time was some 15 gal/head of the population per annum.

At this very same time future giants of the oil industry, however, moved into prospective areas instead of into proven areas, which was the policy round about 1913. Technological developments made this possible, even for a country like Holland. Our famous gravity expert, Professor Vening Meinesz (who never thought of oil exploration), finished his gravity-survey over the Netherlands in 1923, which he had started in 1913 for geodetic not for prospecting reasons, proving thereby by the well-established anticlinal theory could be applied in conjunction with geophysics in the search for oil in Holland.

Investment in risky exploration is, however, to a great extent ruled by the laws in force, as well as by the attitude of the government.

The geologist must only submit a geological picture, which limits the search to a well-defined subarea. It is for these reasons that our history must go back to at least 1913, the year our famous Dr Van Waterschoot van der Gracht published his "Draft of a Tectonic Sketch Map of the Dutch, Belgian, and Westphalian Coalfields and the Bordering Area to the North", a fundamental tectonic outline of the subsoil of the Netherlands, in essence still valid today, which exactly meets that requirement.

If, however, we wish to understand the influence of the legal aspects regarding exploration and the attitude of the government, we must go back to the very beginning of the century. It is an indispensible phase of our history, because it will partly explain the late arrival of international competition (1963). We say "partly", because the oilmindedness of the industry as a whole laid emphasis on liquid oil, and gas was

* As neither of the authors was able to attend the conference, the paper was presented on their behalf by Professor J. J. Dozy of the Delft University of Technology, who also answered questions in the discussion.

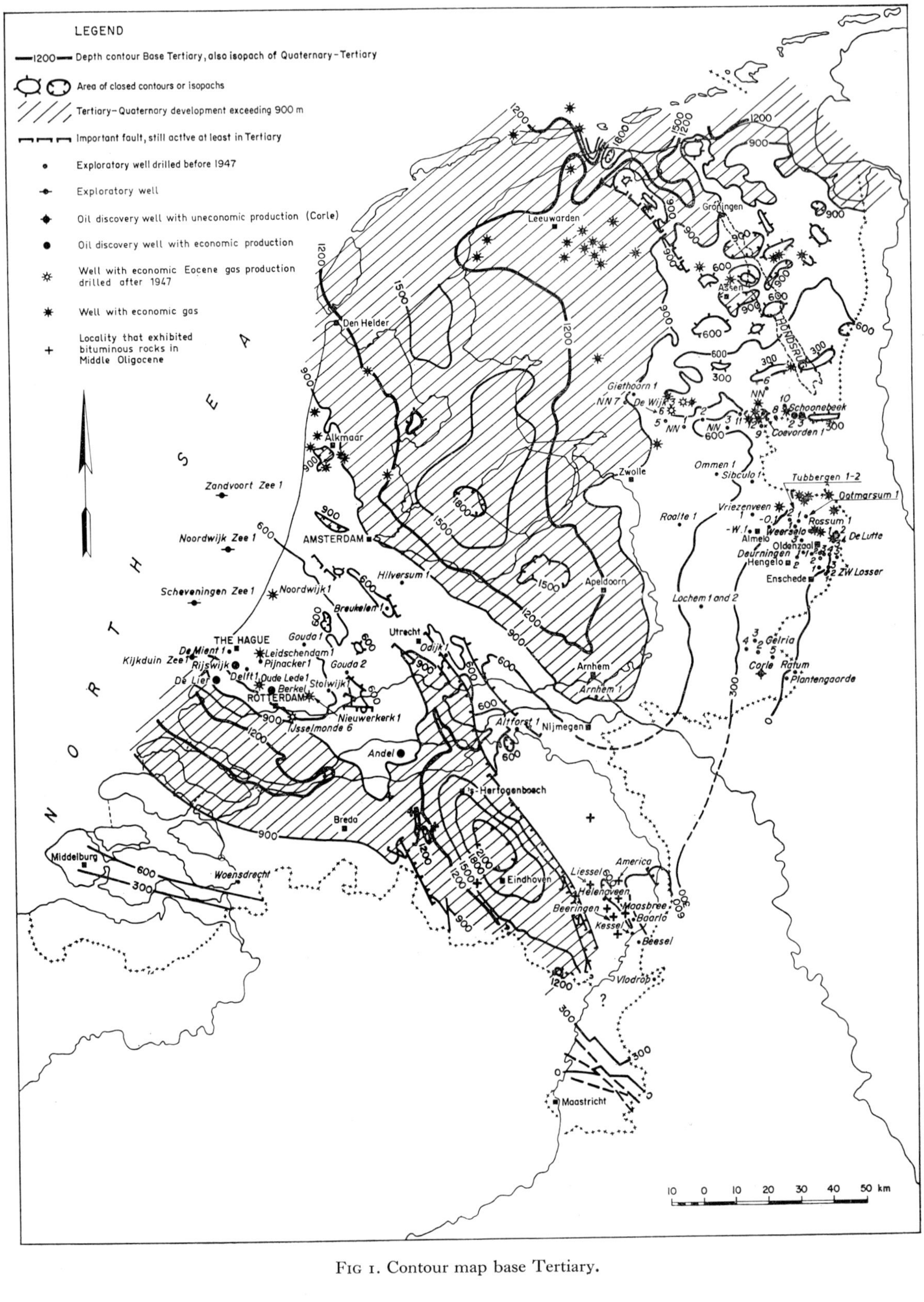

Fig 1. Contour map base Tertiary.

for a long time seen only as competition to, rather than as a complement and substitute of oil, until the Suez crisis and the up-swing of central heating in Western Europe. Industry again carries part of the debt, as the Dutch Government, which mitigated its restraint of free sales of gas with a simultaneous return to rational prices only in the early sixties, closed in the 1950s a gas contract fit only for very moderate quantities, imposing prices which were only acceptable for gas as a by-product.

PROTECTING THE COUNTRY AGAINST FOREIGN PENETRATION (1902–1923)

In 1902 the Government decided to supply the working capital for a state-owned coal industry to exploit new concessions in the S of the country, the Dutch investor abstaining from investment in mining in the home country and thus creating a vacuum for the benefit of foreign investors. In 1903 the Government took the next logical step by starting a state entity for the exploration of minerals, simultaneously passing a law which would abrogate the freedom of exploration, guaranteed by the old mining law of 1810, an inheritance of the Emperor Napoleon himself. The establishment of this state service, a predecessor of the present Geological Survey, was instigated by two Dutchmen, the mining engineer Beyerink (who became its first chief) and Dr Verbeek, the grand old man of the Bureau of Mines in the former Dutch East Indies, with the responsibility of the search for new coal reserves and later salt reserves, if possible of the potassium type.

By law a free hand was given for six years, excluding all private exploration in the four eastern provinces concerning the search for coal and salt only, but thereafter extended in 1908 to the whole country and to all minerals for a further period of 15 years. It is an unsolved question whether or not the tacit inclusion of oil was done on purpose; some bituminous rocks were believed to have been found in the Carboniferous in boreholes in 1905 and 1906 and the inclusion of oil may have been influenced by Van Waterschoot van der Gracht, who took over from Beyerink in 1906, and whose constant preoccupation with Carboniferous possibilities of gas and oil was to last until at least 1936.

Beyerink resigned, utterly depressed, after locating two boreholes in a deep Quaternary–Tertiary graben.

His successor, however, was in many respects a remarkable man. He applied, for example good and modern oil exploration practice. He considered the above reverse not a negative but a positive fact, namely, that the Carboniferous surface might not just be un undulating series of valleys and bulges, but was on the contrary moulded by step-faults and modelled by fault tectonics into a graben and horst system. He studied this system in neighbouring Germany and postulated in line far-away in Dutch territory faults which, as they might have been mobile for a long period, might still be visible at the surface. Quite rightly, he had recognized the coincidence of the delimiting fault of the German Erkelenz–Elmpt horst with remarkable steep zones in the landscape, separating Tertiary on the horst from younger Quaternary on the graben side.

He decided to base his exploration philosophy on that fact, and studied the very good surface countour map belonging to the report of the State Committee for irrigation of 1896. On these data he assumed his "Peel" high, a continuation of the productive high of Erkelenz, and started surface work in Quaternary to young Tertiary formations, though his objective was the Carboniferous. (We mention this especially, because it will be repeated once again and lead to Groningen).

Thereafter he did something which every geologist should do from time to time, namely, impress the investor. In his case, it was Parliament that needed persuasion, as it was restive after two failures.

He located a wildcat far away from the nearest German test holes—and drilled in 1906 No. 5a Helenaveen, discovering at 940 m the top of the Carboniferous, and proving in two months six workable coal seams.

He had now found and could distinguish between three structural elements (Fig 2), namely, the N flank of the Barbant Massif, the Central Graben (which extends from Roermond to the NW and is supposed by him to reach the North Sea coast), and his "Peelhorst".

The Central Graben is a very disturbing feature in Dutch tectonics because, having once been a high with mainly Dogger and older preserved, it has sunk irregularly since then, offering hardly any Cretaceous prospects. Only in 1938 did it accidentally become evident that its supposed Western continuation in the province of S Holland must be of quite a different nature, a fact which could not have been deduced from Vening Meinesz' gravity-survey without drilling.

The third element mentioned is the high block, a continuation of a German horst, comprising, among others Van Waterschoot's Peelhorst, and found to continue as far as the NW of Amsterdam.

In search of coal and salt, he then stepped out to the E of the province of Overijssel and discovered the so-called "Trias Swell", in reality the SE extremity of the now well-known "Netherlands Ridge" which runs straight to about the island of Texel. This high was, as a matter of fact, found in the previous century by the Baron of Twickel who, when drilling in 1886 for water, ran into Triassic salt.

After engaging Dr Tesch, who performed a great deal of surface work and succeeded him in 1915, he then located the borehole Plantengaarde, finding, besides Zechstein salt of the first evaporation cycle, coal-bearing Carboniferous.

It was through his genius, however, that he furthermore initiated a thorough investigation of the salinity of surface waters in the Northern provinces, long

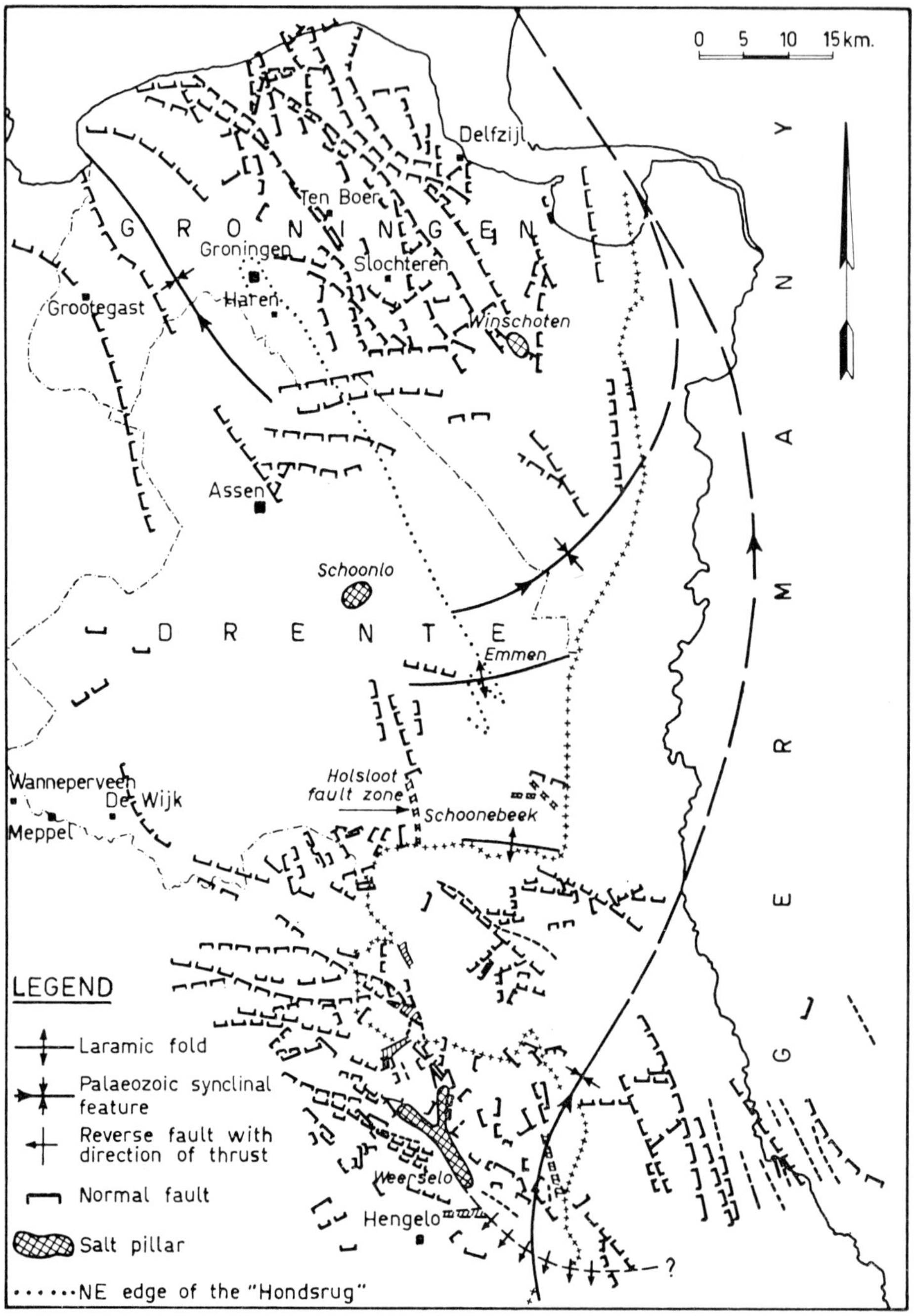

FIG. 2 Fault systems in NE Netherlands.

before Vening Meinesz' gravity survey had shown the possibility of salt domes in the Northern Groningen area. He did this just because two salt domes had been located in NW Germany. From the results he quite rightly postulated an area of halokinesis, welded on to the Netherlands Ridge by a zone of folding of the "Saxonic fold-and-fault system", which was described in 1913 by Professor Stille after the latter's investigation of a thrust feature in Germany, which would later appear to run on into the Netherlands. This latter zone contained the anticlines, which had just been universally proclaimed the favourable traps for oil, confirmed by optimal results in the same year in California.

THE APPEARANCE OF OIL IN THE DUTCH PICTURE

1913 is quite different from 1923. Two future giants of the oil industry not only moved into each other's markets, but also tried to obtain holdings in the then-proven fields, which are limited to exposed anti-clines in countries with surface oil shows.

Standard Oil New Jersey moved into Sumatra and Sir Henry Deterding, the head of Royal Dutch Shell group of companies, made a counter move into the U.S.A., sending over a Dutch engineer, B. H. van der Linden (who later, in 1928, joined the Hague office and built Shell's exploitation engineering department) to Coalinga and engaging Van Waterschoot van der Gracht in 1915 as a director for a subsidiary company (Roxana) to work W of the Appalachians in the Mid-Continent.

Before leaving, the latter mentioned the word "oil" in connection with Holland for the first time in his final report about his work for the Government. At that time it was a far cry, because the only oil production in Western Europe was derived from three fields near Hanover, essentially traps against salt pillars, first found in 1859 and producing 100,000 tons/year in 1908, dropping off to some 40,000 tons by 1920. The oil is shallow and not thought to exist below 300 m.

The first well to reach 500 m and to achieve Lower Cretaceous production was drilled in 1923 in a new and vehement effort to increase production with the help of the anticlinal theory, against the background of Professor Stille's Saxonic folding.

We have now come to the year 1923, in which the exploration period in the Netherlands expired and Dr Tesch was allowed just one last borehole, which he located in another Saxonic thrust more or less ending the Netherlands Ridge, along which Zechstein, with its many oil traces in dolomites and anhydrites, comes into contact with Carboniferous sands. He did not find the potassium salts he was looking for but, on the other hand, oil traces, finally recovering two barrels after pulling the casing. This caused quite an exaggerated stir, but after all, it was the first oil produced in the country.

A CRUCIAL PERIOD

At this juncture we must describe a legal phase, which may have discouraged would-be explorers for oil, notwithstanding the restored liberty of exploration.

There was another Dutchman named Koster, actively engaged in the oil industry, who had quickly availed himself of the opportunities offered by the expiration of the law of 1903–1908 and, together with an Amsterdam bank, had drilled four test holes near Dr Tesch's stamping ground. He applied quickly for a concession not only for coal and salt, but also for oil and gas! By mentioning gas he showed himself not only free of the usual inhibitions of the oil industry, but in fact years ahead of it.

He had actually tested two of his test holes for oil, but one collapsed and the other showed a badly leaking water shut-off.

This had the unexpected effect that the Government drew the attention of Parliament to the possible damage to oil reserves should exploration not be immediately followed by exploitation but, so the Minister said, that could be overcome because in any case an explorer should have a reasonable chance of obtaining a concession when he had found oil deposits which, in accordance with the law, could be usefully exploited.

The Minister said so in connection with a law in which oil was not mentioned at all; it was a law resulting from stocktaking of legal rights acquired through Government exploration. They had to claim the finders' rights stipulated by law, *i.e.* in case a concessionaire should acquire territories in which the Government, by means of its State service, had proven the presence of exploitable coal and salt, the concessionaire had to pay costs and premium to the State.

Parliament, however, made the Minister swallow this whole digression on oil, perhaps not so much to safeguard the Crown against any claims, but more so because Parliament was divided on the subject of concessions granted to private concerns, and the alternative possibilities of State exploitation or exploration contracts to third parties. (In the first case the mineral resources become the absolute and permanent property of the concessionaire, but not in the last. The Government had already applied this latter solution, when Royal Dutch Salt asked for a salt concession on the Trias Swell, requiring, moreover, State participation in the capital.)

The result was that the Government not only took back its words, but thought it wise to dedicate one separate article of the law to eventual oil in the territories concerned, which restricted the possibility of exploration for oil to holders of special permits, promising at the same time to consult Parliament when an application for a concession would be filed. (Almost in a nutshell the present law concerning the North Sea Shelf.)

This law, which was passed in 1924, concerned only four designated territories and exploration remained free in the rest of the country.

We think this story may have had some adverse influence on foreign investors unfamiliar with Dutch politics and the moral standards of our Government.

However, it made a thorough questioning of the Government's intentions advisable. We enter modern times and the modern idea, that investments made to the benefit of the country ought to be rewarded (if possible), was gradually gaining the field.

THE PRELUDE TO ACTUAL OIL EXPLORATION

Oil production was improving in Germany, production nearing the 200,000 tons/year level, which would be passsed in 1930.

We are certain that in 1928 the possibilities of sharing risks of exploration, among other places in Germany, but also in Holland, were discussed between Royal Dutch Shell and Standard New Jersey.

We shall find these groups as partners in the originally Dutch-owned German company of Brigitta and, delayed by World War II, in 1947 in NAM, the joint exploration and production outfit in the Netherlands.

In view of the foregoing, it is only natural that talks were opened by BPM, the Dutch operative company in the Royal Dutch Shell Group, with the Ministry concerned. We do not know exactly when, but may surmise in the year 1929 or 1930, because early in 1932 a law was submitted to Parliament to restrict again freedom of exploration in five eastern and northern provinces, this time concerning exploitation by drilling only, to the benefit of BPM. Such laws mostly require a few years to be prepared.

We can hardly believe that detailed geological facts have had nny decisive effect. We discard entirely the influence of the finding of some oil in far-off Volkenroda in Thuringen.

The ultimate proof for the validity of our supposition is that the report covering oil possibilities in the Netherlands was drawn up only a year and a half after the law of 1933 was presented to Parliament. That suggests some hurry, even though nobody as yet sincerely believed in favourable prospects, not even the able writer of this accurate report, Dr Klein (a disciple of Van Waterschoot van der Gracht).

It is guesswork but there is a possibility that the presence of Mr Koster, with his drilling and contracting firm, who had not only intimate knowledge from his contract work for the Government, but who had also acquired holdings in, for example, Rumania and Argentina, and who not always sold his properties to Deterding, had possibly a speeding effect.

It made BPM, moreover, very aware of the special danger caused by the necessary preliminary stage of geophysical work. A wildcat located on geophysical considerations was thereafter no more a pure wildcat but, indeed, an invitation to competition. This reason, submitted to the Government, brought about the law of 1933 which was approved by Parliament without even resorting to voting.

Once launched, BPM started with gravity work in 1934, followed by shallow drilling at the end of 1937 so shallow that it was almost useless. Of the 12 wells drilled before World War II, two remained in the Tertiary, nine in the sterile Upper Cretaceous, and only one, which should have carried the number Schoonebeek I, revealed 50 m of prospective Lower Cretaceous. The cover was thicker than expected and, to be frank, the gravity methods applied all over the country were only good enough for broad culminations. However, a purely accidental discovery led to the inclusion of three more western provinces under the law of 1933 just before World War II in 1939. BPM decided to put up a demonstration drilling rig in 1938 at "De Mient" near The Hague, which revealed an unexpectedly shallow bottom of the Tertiary, with oil traces, in the midst of the supposed extension of the Central Graben. This suggested the possibility of a Cretaceous basin and enhanced prospects in the W. In this case action was quick, the motive clear, the origin a lucky coincidence, albeit that the unknown promoter may have been very clever indeed. The Mient boring was described by Tesch in 1939 and van Voorthuysen in 1950.

WORLD WAR II

BPM had transferred its legal headquarters out of enemy reach, to Curaçao, and the occupying authority had nominated a Caretaker of the "Reich". Naturally, the story is confused, but by that time the mapping of the bordering area in Germany by Dr Bentz and Dr Bartling was bearing fruit, and Elwerath, a German company, was pushing on to the extension of the Schoonebeek anticline in Germany, guided by the forceful and able Mr Brockhaus. Frontiers did not exist anymore, certainly not for lengthy, still experimental seismic refraction work.

Moreover, there was definitely pressure, countered with ambiguous methods. On the one hand, there was the wish to keep Dutch drillers employed in Holland, and also to claim stakes, if stakes had to be claimed (as in the case of Schoonebeek). On the other hand, there was the wish to delay and to avoid proving oil.

The result was that in 1942 drilling was resumed and the Schoonebeek field proved in 1943 by a small show in Dogger, right on the denuded culmination. The first to strike Cretaceous oil was Wintershall, which had concessions on the S flank, followed by BPM in 1944. A concession was applied for by BPM, with "seat in The Hague" in 1944 (but exchanged for a new one with "seat in Curaçao" after liberation in 1945).

The remaining work done in those war years spread out over the whole of the country, although it was restricted mainly to the belt of Saxono-type anticlines in the E, never finding anything in not less than 41 holes.

The only reconnaissance of value was made in three holes in the Western Cretaceous basin, where indeed the lower part of the Lower Cretaceous was found, covered by Oligocene on a gravity high, whilst two outsteps into the basin showed a fairly continuous series in the top of the Lower Cretaceous, in all cases with oil traces.

DEVELOPMENT OF THE WESTERN BASIN

About this we wish to be very short, limiting ourselves to some highlights, because its story has been amply described. One of the important prerequisites was the improvement of seismic reflection work introduced in 1947. Through variable area recording, multiple- and pattern-shooting, etc., such improvements could be achieved that complicated structures of even small sizes could be located in the W, whilst from 1955 more or less reliable reflections could be recorded below the Zechstein salt in the E.

Quantitative well-logging calculations reduced the overlooking of testworthy horizons, whilst better drilling equipment and techniques improved speed and accuracy, so necessary when drilling strings are continuously reduced in number.

Even though the freedom of exploration was restored to nationals and foreigners in the country by the definite abrogation of the law of 1933 in 1947, the operating company felt itself fairly safe against any outside competition, not only on the basis of its great lead in geophysical work, but also on the basis of services rendered to the general economy of the country, which proved in later years a rather presumptuous and hasty conclusion.

In the beginning success came very slowly. This was partly due to the fact that exploration still stuck to the old gravimetric highs and, therefore, to the edges of the basin. Lateral migration was insufficiently recognized as a major factor controlling accumulations, as also was the possibility that the edges of the basin might never have been invaded by migrating crude oil, receiving only gas, or small and heavy oil accumulations in the deepest (Dogger) layers.

Notwithstanding these drawbacks, an effort to explore the basin into the underlying Liassic source beds was already made in 1952 (Berkel). This test well, which was only later made productive, showed clearly that the big unconformity, noticed in the Lower Cretaceous, became less serious basinwards, whereby the best productive Valanginian–Hauterivian sands were proved productive at Rijswijk in a small narrow, and not too deep anticline.

Nevertheless, costly time was again dedicated to the edge at Andel, whereby some heavily paraffinic oil in sands of Dogger age and also asphaltic, but uneconomic oil was found next to gas.

These rather poor, scattered discoveries were thought to be deceptive. The presence of the Liassic epsilon and alpha source formations had no great appeal. Unmathematical, semi-quantitative appreciations played a constant depreciating role Dr Klein already commented on the thinness of source beds, keeping the whole source problem rather vague, a definite error in exploration. However, we still meet today.

In 1954 Stheeman, the first author, was transferred to NAM. Originally a specialized ore geologist, he received his basic practical training in group companies in Rumania and Sumatra and arrived in Holland imbued with the belief that the locating of source beds is a primary necessity, as also with the innate conviction that there must be a genetical meaning behind the sequence "gas-asphaltic oil followed by heavier and thereafter lighter paraffinic oil", which is often encountered when vertically penetrating basins, but also when stepping out into basins from the edge. Naturally he was also entirely sold on the preponderance of lateral migration over vertical, except for gaseous hydrocarbons.

He took a more optimistic view of results in the western basin, pushing steps-out in the basin, whereby quick results were obtained, not least through the perseverance and obstinacy with which his production manager, Van der Lely, tested the smallest indications.

Besides a number of smaller accumulations, two sizeable fields were found in 1956.

Nevertheless, the very fact that the oilman explores downwards against the current of migrating oil means that the whole secret is often revealed only at the very end. It was only after drilling the Spijkenisse reconnaissance hole in 1960 that the jigsaw-puzzle was solved, although in not less than 26 boreholes the Liassic had been previously investigated. This very necessary investigation could be cheaply effected, every time geologists lured the exploration on to the wholly uneconomic Cornbrash sands in the Dogger, which had already been pointed out by Mr Klein as a most likely reservoir, as he was deeply impressed by the fact that these sands were just discovered in Germany in the Hanoverian Basin when he wrote his report.

The total benefit of 39 trials resulted in two small producers. This imaginary possibility was another reason for reverting once again to the edge of the basin and also to launch a special drilling campaign in the Central Graben (1958, 1959), where erosion had spared these rocks, but no production was found.

Stheeman's views on the oil occurrence in the Western Cretaceous basin were fully published on the occasion of the Jubilee Convention in 1962. At present the authors are quite open to the acceptance of additional migrational factors, namely the French suggestion of bacterial influence (Gabon) and the easier migration of asphaltic components by better solubility in formation water.

THE STORY IN THE EAST

In the East exploration was incited by two discoveries, namely, the Schoonebeek one and the gas found by Deihlman as early as 1938 in a Zechstein dolomite, only 300 m deeper than the level reached in 1905. This last important find, which supplied Germany with the raw material for her synthetic rubber tyres during World War II, also became one of our important objectives after the war.

It meant quite a change of Dr Klein's prognostication, which had been made entirely on the basis of oil, in which respect he quite rightly considered the Zechstein unfavourable, thereby degrading even more possibilities in Groningen, where these dolomites turn into stinky shales (*cf* also Visser and Sung).

After the formation of NAM in 1947, the new management went all out for this gas, discovering almost immediately in 1948 the important dolomite bound gas field of Coevorden, which was followed by a number of similar finds between 1951–57. The favourable facies of primary as well as secondary porosity was found to exist in a narrow belt only. This discovery would have a considerable influence on exploration. Whereas the first manager of NAM, Bolderdijk, dreamt of a profitable trade in natural gas, the oil-mindedness of the industry as a whole obliged the Government to devise a curious supply contract, with downwards sliding prices, which in the end had no connection at all with either fuel value or exploration costs. Prohibiting free sales, it meant, on the other hand, that the Government had to buy, and to pay up for any new reserves in 20 years. It was hailed, however, by municipal and other gas boards, who were labouring under an almost destitute state of uneconomic public gas distribution systems.

Stimulating the State's eagerness for profits, this contract proves simultaneously that gas still had a low rating as a source of profitability, which may only hold good for some shallow, low-pressure gas reserves.

Not only these dolomites made NAM explore a number of gravity highs, now additionally detailed by seismic, but also the possibility of Cretaceous oil, thereby forgetting Klein's and Van Waterschoot's warning that Liassic source beds could hardly be present except near Schoonebeek. This exploration revealed no oil but much more gas, this time not only in Zechstein but also in Carboniferous, the Bunter, Lower Cretaceous, and even in shallow Tertiary.

The first of these was still discovered under Bolderdijk's management in De Wijk. They were quite rightly explained by Visser and Sung at the end of 1954 in the AAPE publication "Habitat of Oil" as a result of vertical migration along the typical Hercynian fault system, which prevails in the Netherlands with a W–E to a NW–SE direction and runs almost parallel with the N flank of the Brabant Massif. They are frequent in this very zone of gas fields (*cf* Fig 3), which is more or less the N boundary of true Laramic folding. Visser gave the gas a Zechstein origin, although its sweetness and nitrogen content was in contradiction to the normal composition of sulphurous Zechstein gas proper. Moreover, one truly Carboniferous source was discovered which, in accordance with the investigations made in 20 Carboniferous deep tests, could better be explained as a result of downward migration, *i.e.* unless one accepts the verdict of Van Waterschoot van der Gracht, published in the *Bulletin* of the AAPG (1936, **20** (11)), which said "The writer believes that Palaeozoic accumulations of petroleum may be expected in Europe". This belief was shared by the first author, without denying additional sources of gas as later described by Dr R. J. H. Patijn.

THE DISCOVERY OF GRONINGEN

This state of affairs made a profound impression on the first author, who considered that these secondary gas accumulations might represent only a small part of the amount that had possibly escaped in the air, realizing that the secondary gas reservoirs were filled to spill point and capacity and possibly collected only a fraction of the stream that had passed upwards. He shared this belief with Van Waterschoot van der Gracht, also expressed in the same paper of 1936, in the sense that upward migration of volatile hydrocarbons might be impaired by supposedly impervious strata but not prevented. He was strengthened in that belief by the discovery of a fringe of small gas fields in the Western basin (Noordwijk–Lower Valanginian, 1955), Leidschendam (Hauterivian, 1957), Nieuwerkerk (Barremian, 1958), Oude Leede (Barremian, 1959), apart from Albian gas in De Lier and Barremian in Rijswijk, but most of all by the discovery of a Tertiary Eocene accumulation straddled above the N boundary fault of IJsselmonde.

He shared the very cautiously expressed opinion of Dr Klein, who had already hinted at the possibility as early as 1933, that the marsh gas in Western Holland could be of petroleum origin, temporarily held up in its upward migration by almost impervious lenses of Quaternary peat.

This vision might indeed have prevented the first author from proceeding with the Groningen exploration because one could visualize the Rotliegendes as a carrier rather than a trap for gas. Stheeman, on the other hand, had from previous field work a fair knowledge of the Palaeozoic in Europe ending with the Rotliegendes. He was consequently immediately struck by the Variscan direction of a gentle Quaternary feature, the "Hondsrug" along the E boundary (Fig 2), which quite "unnaturally" runs straight as an arrow in a definite NNW–SSE direction, *i.e.* more or less parallel to the edge of the Russian Platform. This appeared an inviting possibility, but as an objective only the unpromising stinky shales of the Zechstein, reputedly showing oil traces and yet insufficiently tested in Groningen, could be chosen. As a first step a seismic cross-section was shot perpendicular to the supposed Variscean Carboniferous fault blocks, the movement of which might have initiated halokinesis, the tectonic origin of which he had, experienced in Rumania. From that experience (*inter alia*, in Tintea) he hardly believed in the activity of salt as a geological process, considering salt to be utterly passive, at least in the initial stage.

A test well was to be located on the first pre-Zechstein bulge, keeping also in mind the Rotliegendes, which fulfilled all the requirements set out by Van Waterschoot van der Gracht in 1936, who said "Lateral migration, especially in widely spread, continuous blanket sands of high porosity, must be admitted"; it certainly should be below now really impervious salt. The latter objective was, however, not yet convincing and feasible only from a purely theoretical point of view.

Seismologists at that time still laboured under the difficulties of the conversion of time in depth, so ably described by NAM in its recent publication about the Groningen gas field. The two phases of halokinesis, whereby *inter alia*, a lot of Bunter sandstone is erratically cut out, makes the location of true culminations of Lower Permian and Carboniferous very difficult indeed.

The first testhole near Ten Boer was consequently erroneously located on the edge of the Groningen gas field, revealing the unprospective stinky Zechstein shales exactly as predicted. As already mentioned, gas was found by accident in that formation, evidently having penetrated therein along a cracked fault zone, along which Rotliegendes could come in contact with Zechstein. The gas appeared to be free of sulphur and relatively rich in nitrogen, and evidently of extraneous, deeper origin. Contrary to several publications, the well was brought into production during a number of years, delivering daily small quantities of gas to municipal gasworks, thereby proving the definite absence of sulphurous components.

Van der Lely had wanted to immediately deepen the well, though it was not in a very good mechanical condition, to the Rotliegendes, but it was considered too risky to hang the whole future of the north of the country, and possibly part of the North Sea Shelf, on a well which was evidently not located on a true culmination.

This risk was enhanced by the probability that we were dealing with a small reservoir and even by the possibility that gas of Lower Permian origin had already been dissipated in the possibly many small secondary fields of the De Wijk type, for nothing was as yet known about the influence of tilting movements. Finally, the whole proposition was rather contrary to international geological beliefs, strengthened by the investigation in many Carboniferous deep tests.

A reverse in Ten Boer would have meant the end of any further exploration of the N and the adjoining shelf. Continuation, on the other hand, held the promise of a new possibility, which might even assume fantastic proportions on this Carboniferous high and offer in that case an additional attraction, namely, that it could serve to show the Government that the prevailing gas contract was born from low expectations and running against private as well as State interests. This could be done by proving in one stroke such a great quantity that the Government would be compelled to abolish a contract which restrained sales. It would at least be a possibility that could serve as an alternative for the grinding work of proving such a quantity by discovering many secondary accumulations, which up till then, and after 21 years, had brought the total gas reserves of the Country only up to some 30,000 million cu m.

But how to proceed? Gas having been proven, be it ever so little, in the stinky Dolomite, another trial could be promised to the Government representative on the board of NAM as an offer to help with the greatly needed industrialization of Groningen.

The next location was plotted near the shore at Delfzijl, but soon Dr Brueren, the chief geologist, could assure that the highest culmination was probably situated near Slochteren, which was drilled in 1959 with the well-known result. Shortly thereafter in 1963, the deepening of Ten Boer, after having been put in the right shape, confirmed Van der Lely's optimism, but it served at the same time as a warning of the hazards of exploration, unprepared for the high pressures encountered.

THE AFTERMATH OF SLOCHTEREN

A kind of pandemonium broke out. In the first place, uncertainty and almost international disbelief was felt, mixed with disturbing publications in the press. In the second place, the field, if really as big as was thought possible, was bordering on Germany in an area where no agreement had ever been reached on the exact frontier line as from the birth of the kingdom of the Netherlands in 1813! This had to be fixed first.

The next point was the change of the gas contract, which had to be adapted to the possibly enormous size of the new discovery, which required a definite energy policy before the matter could be put on a sound footing, plus also consideration as to retail prices. Before all this was really arranged, international competition became aware of possibilities, which by all means must be considered later.

The very freedom of exploration caused chaos. The connection with North Sea possibilities was evident, and NAM had already shown its hand after drilling Slochteren by stepping out into the North Sea as a token move of great importance (Kijkduin 1961, followed by a few more in 1962).

What happened afterwards is hardly history. The Government felt compelled to introduce again a system of exploration permits "by drilling", as a preliminary to granting concessions on land, which move, however, was coupled to the drafting of suitable legal rules on the North Sea shelf.

From the rather scanty data obtained, amongst which only those received from Petroland and NAM were of real help, the authors can guess that the "Golden Lane" of secondary gas accumulations next to primary Permian accumulations, which was once thought to start somewhere near Bentheim, running in a westerly direction at a certain distance from the line where halokinesis shows up on Variscan type fault blocks, have been followed up and investigated by all competing companies very neatly to the province of N Holland. The only real original discovery appears to be that of a Cretaceous basin by Petroland in Friesland, where the Lower Cretaceous, lying out of reach of Liassic source beds, had, instead of oil, collected waste gas of deeper origin.

THE UPPER CARBONIFEROUS AND SOME FINAL OBSERVATIONS

In this story we have truthfully described the approaches to the unknown, and also the fact that other than purely geological arguments play a decisive role at the supreme moment of decision. We have perhaps yet insufficiently set out the enormous influence of even obsolete slogans. The Rotliegendes and the Bunter sandstone, condemned, not only by Klein, because of their continental nature and supposed "dryness", were given insufficient attention for a long time.

The question may be asked whether we disregarded the possibilities of the Upper Carboniferous, the Stephanian, which was only encountered in three cases in Hollend, but which should have possibilities along the E border on the flanks of a Triassic syncline postulated by Dr Haanstra. The Germans were first to prove it productive. In its upper reaches NAM also proved it prospective near Tubbergen in 1967.

In mentioning Dr Haanstra we should point out that he had indeed the chance, as well as the capacity, to arrive at a coherent geological picture of the Netherlands, first presented on the occasion of the Jubilee Convention in 1962. From his publication it is rendered manifest that constructive geological thought starts from below, as opposed to the direction of the drilling bit and from which we "normally" collect our information. Moreover, he unravelled the differences in the Lower Cretaceous unconformities, which seem to cut off oil migration in the E but, being of an earlier data, promoted accumulation in the W. Both Stheeman and Haanstra set out that migration must have happened before the deposition of the Upper Cretaceous, and that migration is almost entirely governed and set in motion by the Lower Cretaceous overburden, and hardly so by any later developing overburden.

All this serves to prove that sweeping, conclusive statements are only achieved in the end, statements which are almost unknown when a crucial decision concerning a new effort must be taken. It is because of this aspect that the history of exploration in any part of the world turns into a mystery almost as soon as the verbal word is forgotten.

SOME AFTER-THOUGHTS ABOUT THINGS UNSAID

Every oilman has experienced the difficulty that most important but speculative considerations must be kept out of the conversation about new exploratory investments in order to avoid a disturbance of the trusting realtionship between investor and professional. Those things therefore play no direct part in a historic review but to a certain extent they may have been indirectly important.

Among these subjects we mention the importance of the early pre-Zechstein but late Carboniferous orogenic phase, the possibility of very early migration preventing diagenesis, and the palaeogeographic distribution of reservoir properties in connection with tectonics.

Another fundamental consideration concerned the nitrogen content of gas in connection with the presence of salt bodies in the subsoil. This connection is undeniable in Western-Europe. The effect of the Lower Permian turning into an evaporite in the N led to an early prediction of increasing N_2-content of Lower Permian gas. (Jubilee Convention, 1962).

Just as the problem of source formations, such considerations often enter the open conversation as "mustard after dinner", as the Dutch saying goes. The selection of convincing arguments is often a difficult problem for the professional geologist when advising management.

BIBLIOGRAPHY

Brouwer, G. C., and Coenen, M. J., "Nederland–Aardgasland". Amersfoort, A. Roelofs, van Goor, 301 p, 1968.

Haanstra, U. "A Review of Mesozoic Geological History in the Netherlands." Verh. KNGMG, G.S. 21–1, 1963.

Patijn, R. J. H. *Erdölu Kohle*, 1964, **17**, 2–9.

Schierbeek, P. "Oil and Gas in the Netherlands and Under the North Sea." Amsterdam–Rotterdam Bank N.V., 38 p, 1965.

Stheeman, H. R. "Petroleum Development in the Netherlands, with Special Reference to the Origin, Subsurface Migration, and Geological History of the Country's Oil and Gas Resources." Verh. KNGMG, G.S. 21–1, 1963.

Stille, H. "Die taxonische 'Faltung'." Zeitschr. Deutsch. Geol. Ges., Band 65, Jg. 1913, Monatsbericht Nr. 11.

Symposium on the Groningen Gas Field. Verh. KNGMG G.S. 25, 1968.

Tesch, P. "Verslag over de onderzoekingen der Rijksopsporing van Delfstoffen gedurende het tijdvak 1920–1923." Supplement op het eindverslag der R.O.D. in Nederland 1903–1916. Den Haag, Alg. Landsdrukkerij, 1924.

Tesch, P. *Geol. en Mijnb.*, 1939, (1), 14–15.

Vening Meinesz, F. A. "Observations de pendule dans les Pays-Bas 1913–1921." Publ. Comm. Géodés, Néerl. J. Waltman, Jr, Delft, 146 p, 1923.

Visser, W. A., and Sung, G. C. L. "Oil and Natural Gas in the North-Eastern Netherlands." Habitat of Oil," Amer. Ass. Petrol. Geol., 1958, 1067–90.

Voorthuysen, J. H. van "The Quantitative Distribution of the Plio-Pleistocene Foraminifera of a Boring at The Hague (Netherlands)." Med. Gel. St. N.S., No. 4, 1950.

Waterschoot van der Gracht, W. A. J. M. van. Jaarverslag der R.O.D. over 1913. Electr. drukkerij "'t Kasteel van Aemstel", Amsterdam, pp. 36–98, 1913.

Waterschoot van der Gracht, W. A. J. M. van. Eindverslag over de onderzoekingen en uitkomsten van den dienst der Rijksopsporing van Delfstoffen in Nederland 1903–1916. Electr, drukkerij "'t Kasteel van Aemstel, Amsterdam, 1918.

Waterschoot van der Gracht, W. A. J. M. van. *Bull. Amer. Ass. Petrol. Geol.*, 1935, **19**, 1824–25.

Waterschoot van der Gracht, W. A. J. M. van. *Bull. Amer. Ass. Petrol. Geol.*, 1936 **20** (11) 1476–93.

MS received 24 April 1969

Discussion

Hugo R. Kamb: Has a geological article been published on the Groningen gas field?

Professor J. J. Dozy: Yes, there was published, about a year ago, a symposium on the Groningen gas field. It appeared in the Transactions of the Royal Netherlands Geological and Mining Society (Verhandelingen Kon. Ned. Geol. Mijnb. Gen., Geol. Serie, 25 (1968)). It is available in most libraries and you can order it from the Treasurer of that Society at Delft, Mijnbouwstraat 20.

Michel T. Halbouty: Has there ever been a case where an exploratory well was drilled in the Netherlands which resulted in a discovery of hydrocarbons of substantial quantities and the discoverer did not end up with the concession?

Professor J. J. Dozy: I do not think it ever happened, but for a long time (up to 1962) there was only one company interested in the search for hydrocarbons (NAM). I believe, however there was a case where an operator was not awarded a concession asked for on the basis of previous drilling results. (This refers to the "Gelria" concession near Winterswijk. The applicant had encountered shows of hydrocarbons in some of his exploratory wells. The main objective was the possible exploitation of rock salt. The application covered coal, rock salt, and hydrocarbons. The Government granted a concession (1930) for the two first-mentioned commodities and refused to include hydrocarbons.)

Theodor Sorgenfrei: What was the objective of the first well in the Groningen area? Was the gas discovery here as accidental as the oil discovery in the Hague?

Professor J. J. Dozy: No. The Groningen gas discovery was the logical consequence of a policy to explore the whole sedimentary sequence of a basin, in this case down to the top of the carboniferous.

Exploration in Algeria: Past, Present, and Future

By A. MEGATELI, A. SAID, and D. G. SARBER

(*Direction Recherche et Production, Sonatrach, Algeria*)

SUMMARY

Oil seeps have been exploited in Algeria since the time of the Phoenicians, over 3000 years ago. Modern exploration started in the Chelif Basin of NW Algeria in the late 1800s and early 1900s. Non-commercial oil shows, World War II, and logistical problems discouraged intense exploration activity in Algeria until the late 1940s, when Oued Gueterini was discovered in N Central Algeria.

Prospecting spread to the storied Sahara and in 1956 the boom was on, after the two important strikes of Edjeleh and Hassi-Messaoud focused the oil exploration attention on this country.

From 1960 until mid-1966 the exploration activity in Algeria indicated a very sharp and critical decline due to unsettled conditions and withdrawal of companies from Algeria. However, after independence in 1962, with the creation of Sonatrach, the national oil company, in December 1963, and following the signature of the Franco—Algerian Agreement of July 1965, the exploration was once again on the move but below the level of 1958–1959 on an area of 180,000 km^2.

Since then exploration effort has been increased by Sonatrach acting alone and also in association with Getty Petroleum Company over an acreage of 50,000 km.

Exploration effort of one type or another (*i.e.* drilling photogeology, surface geology, geochemistry, subsurface geological studies, and geophysics) is now being conducted in practically all corners of Algeria. An estimated \$100,000,000 will be spent on exploration alone in Algeria for the year 1969.

Two important discoveries have been recorded in 1969 at the time of writing this abstract.

It should be pointed out that this paper will present very briefly the general exploration picture, past, and present, without attempting to go into a great amount of detail.

Oil exploration and exploitation, if we use the definition "seeking and utilizing", can be traced back as far as the Phoenicians, who arrived on the beaches of Algeria around 1200 B.C. They were in turn followed by the Romans in 200 B.C., the Byzantines in the 5th century, the Arabs in the 7th century, and the Turks in the 11th century.

Historically, invaders use those natural resources at hand. Asphaltic residues near the numerous oil seeps in N Algeria have been utilized to caulk their boats since the times of the Phoenicians. Can we say that the oil industry in Algeria is over 3000 years old?

Getting back to exploration in Algeria, Fig 1 shows us what hydrocarbon exploration has, to date, created in Algeria.

As evidenced by Fig 1, it is a rather large country (approx 1 million sq miles), and about 80 per cent of it could be considered as possible oil prospective.

We have superimposed a map of England on the same scale, not to boast, but to honour our host country and to further confuse those proponents of continental drift.

As shown, a fairly well-developed evacuation network has been established, with additional capacity already in construction or on the drawing board.

Existing refineries are at Algiers and Hassi-Messaoud, with a liquification and ammonia plant at Arzew, and another liquification plant under construction at Skikda.

Modern subsurface exploration in Algeria began in the late 1890s in the Chelif Basin of NW Algeria, not far from the city of Oran. Several shallow holes were drilled, some reporting oil shows, but the play was eventually given up as non-commercial.

The next splurge of activity was started after World War II, still in N Algeria, which resulted in the first exploitable oil accumulation of Oued Gueterini, discovered in 1948 in Eocene sediments, at a depth of 100 to 400 m.

Exploration then spread to the Sahara, and in August 1956 the Edjeleh field, producing from the Carboniferous and Devonian, was discovered near the Libyan border. This was quickly followed by the Cambrian giant of Hassi-Messaoud in late 1956 and the Triassic gas condensate giant at Hassi R'Mel.

As seen in Fig 2, several known sedimentary basins do exist. To the left of this figure is the Tindouf Basin, which is about the size of the state of Oklahoma, with 7000–8000 m of sedimentary rocks, primarily Palaeozoics. To date there have been eight wells drilled in this large basin, with some slight scattered hydrocarbon shows.

Sonatrach is currently drilling a deep wildcat on a very large structure (400 km^2) on the N flank of this basin.

To the E are the Reggane, Hassi Homer, and Ohanet Basins, each with approximately 5000–7000 m of sediments, and each with less than ten deep exploratory wells (1500 m†) having been drilled there. Non-commercial shows of oil and gas have been encountered in several of those wells.

To the N is the Bechar Basin, with 15 deep dry holes, most of which have reported oil and or gas shows; however, to date there is no commercial production in this basin.

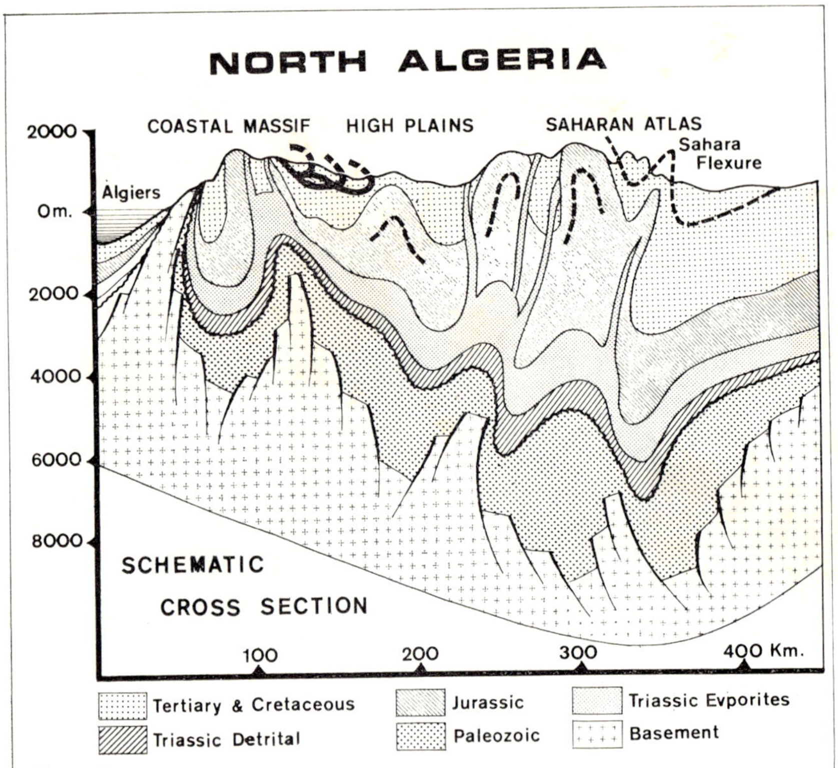

Figure 5

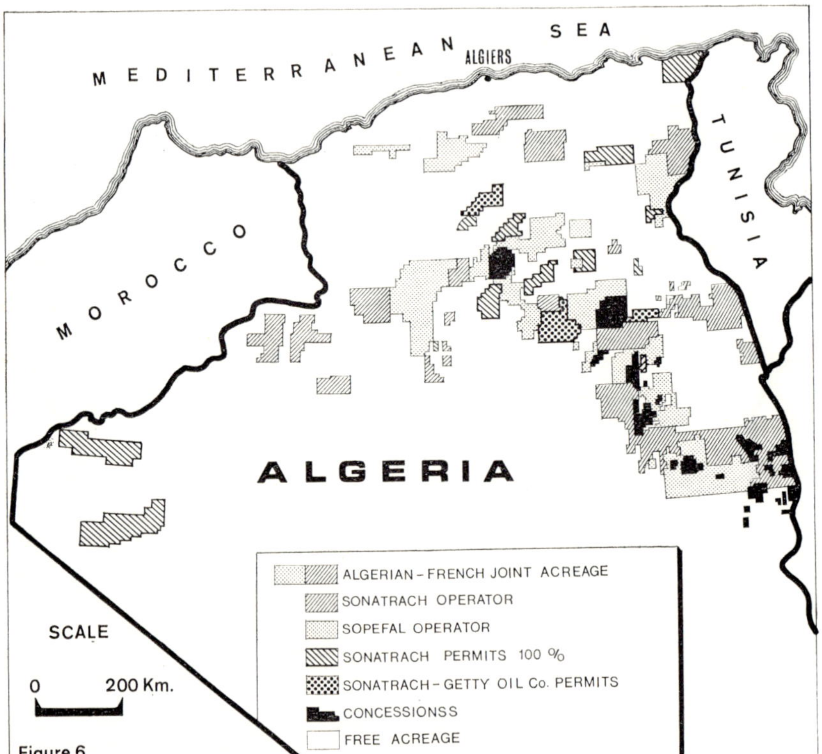

Figure 6

Again to the E and SE are the Triassic and Illizi (ex-Polignac) Basins, from which most of Algeria's present production comes, and where the great majority of the exploration effort has centred since 1956. We refer to this basin complex as the Saharan shelf area.

At the time of writing, two important wildcat discoveries have been made in the Triassic Basin this year.

Fig 3 gives an idealized picture of the Palaeozoic sediments in their sub-crop position under the Hercynian Unconformity, according to an interpretation of the available subsurface control.

As shown, all the Palaeozoics, from Carboniferous to Cambrian (and, in the NW part of the Saharan Platform, even the pre-Cambrian) are unconformably overlain by sediments of Mesozoic age. There were some arguments among our geologists in the preparation for this figure, in the interpretation of Carboniferous, Devonian, and Gothlandian sediments finding themselves so far N on the Saharan Platform. However, since we had no solid proof through subsurface control that they were not there, we decided to present it as such, until future deep well control demands a revision.

Again this figure suggests to the explorationist the immense possibilities that exist in such an area of thick sedimentary series, which are virtually unknown to the drilling bit.

Along the N edge of the Saharan shelf area and extending the entire width of Algeria is the Saharan flexure zone, which separates the Saharan platform from the Cretaceous Jurassic basin of the N part of Algeria.

In Fig 4 we have attempted to show, very schematically (the area involved is some 350,000 km^2) a geological map of the area just N of the Saharan flexure.

Here we find extremely thick deposits of Cretaceous and Jurassic rocks. Their exact thickness is unknown due to the paucity of subsurface control: however, one recent wildcat just N of the flexure drilled almost 4000 m of Lower Cretaceous, and we believe the Jurassic may be as much as 5000 m thick in this area.

To date, there have been no hydrocarbon accumulations of commercially exploitable size in this vast region. However, three small fields exist, and many oil and gas shows have been reported in the few wells that have been drilled (a total of 41 wells below 1600 m).

Sonatrach believes that this area has a large potential for major hydrocarbon accumulations, and our present and future exploration effort in N Algeria reflects this thinking. Five deep wildcats are now being drilled, with other possible locations being refined for the bit by field geology and seismic and surface evaluation.

N Algeria is a complicated geological province, with virtually no deep subsurface control, whereas literally hundreds of reports have been published on its surface geological conditions.

Fig 5 is an effort at portraying or "suggesting" an idealized geological section from the coast to the Saharan Platform.

Fig 6 shows the present-day permit and concession status in Algeria, with the exception of a few minor changes which have been made since this figure was prepared.

The interesting portion on this figure is the large blank area. This denotes open acreage where exploration permits are available.

Fig 7 is a composite geological column or section, indicating hydrocarbon production and/or shows found throughout Algeria, along with the lithological characteristics of the various formations. The great majority of the known production comes from the Cambrian quartzites and the sands of the Devonian and Triassic.

The impact of this figure is the observed presence of oil and gas throughout the geological column.

Fig 8 shows the wells either started or completed in 1968. As mentioned earlier, one can see that by far the most of the activity was concentrated in the Triassic and Illizi Basins. We plan to show an increased exploration effort in the other promising areas of Algeria in the immediate future.

Fig 9 is a combination of several statistical curves and is included simply as a graphic indication that the Algerian oil industry is entering a new "peaking" cycle after the "slump" from 1962 to 1966.

Activity decreased during this period from 54 exploration wells in 1962 to 21 exploration wells in 1966. Our present projection is for approximately 50 wildcats in 1969.

This peaking cycle is the result of the expansion of Sonatrach into exploration and the signing of the Franco—Algerian Accords of July 1965, which created joint interest exploration parcels of the Association Cooperative, referred to as "ASCOOP", with some 180,000 km^2 of prospective areas.

It was further accelerated by the entry of Sonatrach into 100 per cent operation of its own exploration permits, comprising 40,052 km^2, and the Getty–Sonatrach agreement signed in October 1968 for four permits totalling 11,500 km^2.

Today our direction operates approximately one half of the ASCOOP acreage, plus the eleven 100 per cent permits and the four SONAGET permits. In addition, geological and geophysical studies and campaigns are being conducted in all of the possible productive areas, including offshore Algeria.

The present policy of Algeria is to create further joint ventures such as the Sonatrach–Getty, with major and independent companies of all nationalities, whether or not they are now established in Algeria.

Before summarizing this very non-technical paper, it was thought that a few statistics might be of interest to those who have read little about the Algerian oil industry:

1. Number of producing oilfields is 35, which produce about 130,000 tons of oil per day, with several others waiting to be connected to evacuation facilities.

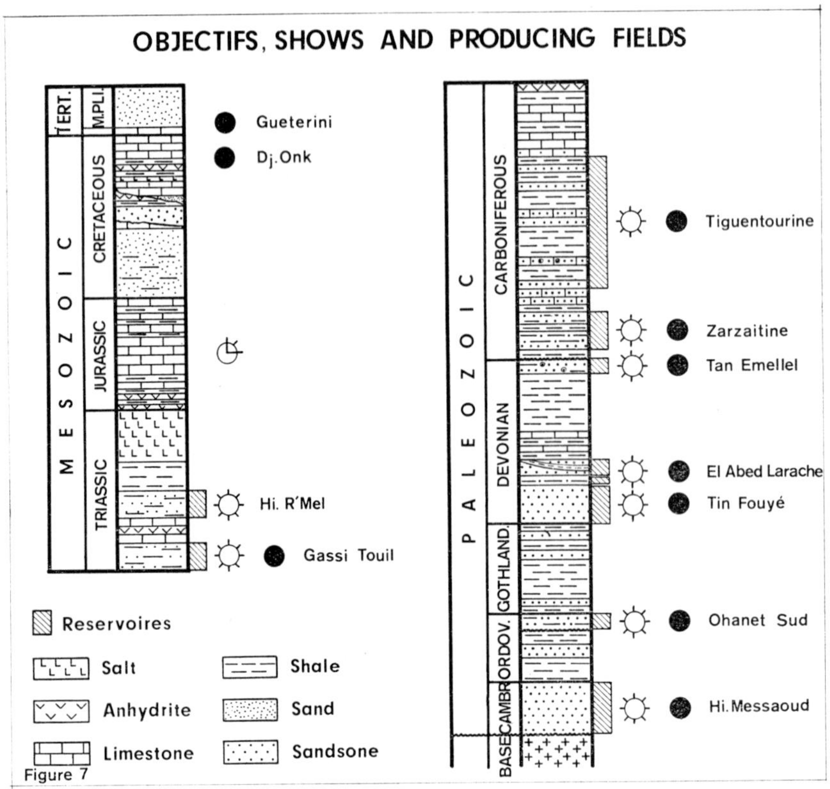

Figure 7

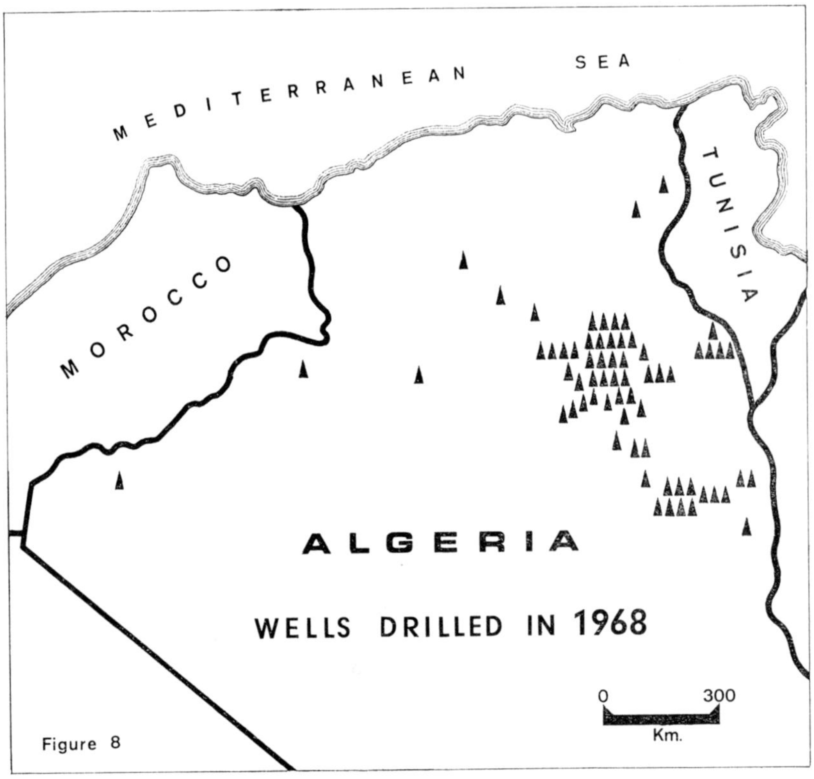

Figure 8

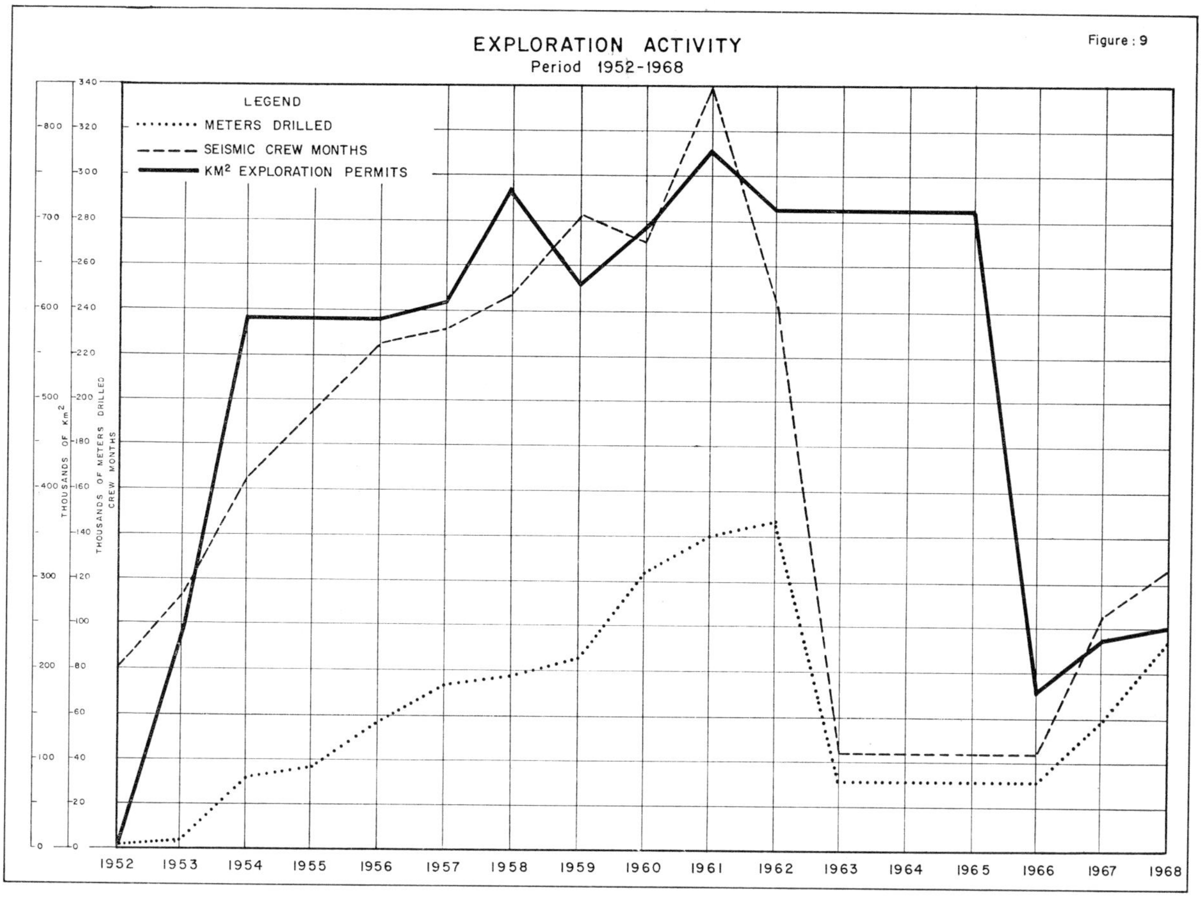

Total estimated remaining recoverable reserves are 690,000,000 tons.

Of these Hassi-Messaoud is by far the largest, and in 1968 produced a total of 17,000,000 tons of oil from the Cambrian reservoir, with a cumulative production of some 100 million tons of oil from 1958 through 1968. Primary reserves are estimated at over 300 million tons.

2. The number of producing gas fields is three, with many reserves waiting on evacuation facilities and market. In 1968 Algeria produced $3\frac{1}{3}$ billion cu m of gas, with $2\frac{1}{3}$ billion coming from the Hassi R'Mel Triassic and reservoirs. Gas reserves in this giant accumulation are some 61 trillion cu ft, plus 170 million tons of condensate.

In summary, let us say that Algeria is a vast region, virtually untouched by the drill, with all the prerequisites of a very bright exploration future.

It is a fairly well established fact in the history of oil exploration that outstanding achievements in this field of endeavour usually come from high risk ventures in geological provinces where little is known, or in old producing areas where much is known.

Presently 15 wildcat wells and five development wells are drilling. There are now 13 seismic crews, two geological field crews, and one aero-magnetic crew working in Algeria.

In order to live up to the future of the hydrocarbon potential in Algeria, it is estimated that over $100,000,000 will be spent on oil and gas exploration in the year 1969 by the companies now involved in this important task.

MS received 14 August 1969

Discussion

Rustam Byramjee (CFP): What are your targets for oil and gas in the open areas?

A. Megateli: Are you meaning by "target" in oil production?

Rustam Byramjee: In reservoirs.

A. Megateli: Actually, the reservoirs in the open area are the ones we have shown in the geological column which was shown in one of the figures. It depends naturally where you are, but if you are in the Tindouf Basin, this is more of a Devonian and Cambro-Ordovician province. If you are in the Triassic Basin, as its name indicates, the main target is Triassic, and also the Cambrian. In the Illizi Basin you have Devonian, Carboniferous, and also Ordovician production.

M. D. Beltrandi: Have any Devonian reefs ever been found while drilling in the Tindouf Basin?

A. Megateli: There were no reefs found in the basin of Tindouf in the eight wells which have been drilled in that area, but these eight wells are quite scattered and very far apart from each other. In the present well that we are drilling, which is going to be a deep well, over 5500 m, we have not penetrated a reef section. Some of our exploration geologists would be delighted to discuss this with you in more detail, but to our knowledge there is no Devonian reef present in any wells drilled in the Tindouf Basin.

Walter Alvarez (American Overseas Petroleum Ltd): You are certainly to be thanked for this look at a country which has not often been described in English. Those of us who work in Libya can only stand in awe of the foresight of whoever drew the E boundary of Algeria. This is the border that precisely follows the E margin of the easternmost oilfields of the Illizi Basin—what used to be called the Polignac Basin—although there is no obvious geological change across the boundary.

Perhaps you could give us some thoughts on why these oilfields stop so abruptly at the Libyan border.

A. Megateli: To be frank with you, when the border was drawn in Algeria I was not there and Algeria was not independent. As far as the Polignac Basin is concerned, we now call it the Illizi Basin.

Dr P. E. Kent (British Petroleum Co. Ltd): The speaker has referred to new Triassic discoveries in the Sahara. Could he elaborate this statement by stating in particular the type of reservoir and the presumed source-rock for the oil?

A. Megateli: The first discovery was made in the area of the Erg Oriental in the E part of the country, not too far from the El Borma field, which was discovered by the Tunisians and proved to extend to the Algerian side of the border. This discovery is in the Triassic sand and the characteristics of the sand are basically like the sands found in El Borma. I do not know if you are aware of them or not.

As far as the other discovery is concerned, it is also in the Triassic sand, not too far from Hassi R'Mel. This is a very interesting discovery of gas, very wet gas. The reservoir is interesting because this is probably the first time in Algeria that we find a very clean, coarse, well-rounded sand reservoir with a very high porosity, about 25 to 30 per cent, and very high permeability going all the way up to 4000 mD. This is going to be a very interesting discovery to analyse. I am not trying to tear down the other discovery that we have in Hassi Keskessa, not too far from El Borma, as both of them are very important discoveries. They just prove that there is a lot of oil to be found yet.

Dr P. E. Kent: And the date of the origin? The source of it?

A. Megateli: It is very really difficult to answer your questions. We are trying to answer that for ourselves, you see.

Monty G. Martin (Texas Gas Transmission Corpn): Your gas reserves, which are appreciable, were not distinguished as to category.

Do you have non-associated gas reserves in any quantity or are most of your reserves made up of solution gas?

A. Megateli: We have a very important amount of reserves in gas. Actually, the gas that I was referring to is mostly coming from gas fields, either dry gas or wet gas. There have been tremendous gas discoveries made very deep in the Ohanet Basin. The gas I was referring to is produced from Hassi R'Mel which is just a gas field, wet gas, and we have naturally a lot of associated gas with some important fields with a big gas cap like the Rhourde Nouss field of El Paso, and the Hassi Touareg permit which is on the El Biod trend S of Hassi Messaoud. These are tremendous gas reserves, and are just gas fields, so really it is not solution gas to which I was referring.

Closing Address

By KENNETH CRANDALL
(*President, American Association of Petroleum Geologists*)

As I listened to the speeches and observed the proceedings here, I was once again impressed with the truly international character of so much of our business. What we have witnessed was the creative interplay of scientific and business forces in and from many nations examining petroleum exploration in Europe and North Africa—working in a globe-circling industry that affects the economic life of every human being.

Ralph Waldo Emerson once wrote that "the greatest meliorator of the world is ... trade". This meeting, I believe, offers contemporary proof that he was right.

Perhaps the biggest lesson we are all learning is the fact that free commerce has no nationality or political beliefs. The process of fulfilling human needs—of matching the product to the demand—will ultimately surmount any national boundary. We are fond of saying "Oil is found in the minds of men". There is a line from the UNESCO charter that says:

> "Since wars begin in the minds of men, it is in the minds of men that the defences of peace must be constructed."

I am sure that no one in this room considers the establishment of world peace his principal mission. But I am equally sure that one of the most valuable by-products of our work in international petroleum geology is greater understanding and unity among nations. For once any nation, large or small, finds, produces, and sells oil, it joins the world marketplace. It is forced to realize that its future lies not in old, geographical hatreds, not in national jealousy or ethnic feuds—but in the economic growth that natural resource development brings.

Petroleum is one of the three juggernauts that are relentlessly levelling national economic borders. The others, I believe, are air transportation and electronic communications.

We are, of course, by now all very familiar with the miracle of Telstar, which can flash the significant events of world history around the earth in an instant. Perhaps even more potent, however, is the televised face of war.

The news films that we view in our living rooms, of course, are whisked to the TV studios on giant, 600-mile-an-hour jets—the same jets that are shrinking our world in a variety of ways. The number of airlines and the countries they originate in that offer intercontinental or round-the-world service is steadily increasing.

Anyone who doubts that petroleum supports most of modern industrial development need only check the figures on jet fuel demand. In 1950 there was virtually none, but last year free world consumption of jet fuel was more than a million barrels a day.

The oil we find through the practice of geology, therefore, is more than just a profit item. It is helping the world to grow up and stop its bickering.

A good example of this is the growing interdependency of oil producing and consuming countries. The economically and scientifically advanced nations need energy, and many of the young, emerging nations have energy for sale. Each needs the other desperately, and this dynamic attraction of supply and demand may prove to be one of the great stabilizing influences of our time. Wars are caused not because of beneficial ties of commerce but possibly in spite of them or because of their absence. Truly mutually beneficial commercial dependence is the stuff that peace is made of.

It is true, of course, that nationalists and demagogues have generated and used oil disputes to accomplish their own internal political aims. But no matter how they have stirred up false emotions, no matter how they fanned the flames of sovereignty, reason has generally prevailed. During the darkest days of the Sukarno regime in Indonesia, for example, the oil never stopped flowing, because it was to the self-interest of that nation to maintain harmonious, reciprocal relationships with its customers and oil is even now fuelling her recovery. The same may be said of the Arab–Israeli crisis in 1967. Despite the guns and bombs, the Middle East Nations produced and shipped oil, because it was to their long-range advantage to do so.

The modern, truly international oil companies—and the international oilmen who operate them—play a big role in maintaining these harmonious relationships. For the producing countries, our companies provide multiple markets; for the consuming nations, we offer multiple sources. By absorbing the jarring effects of political, economic, or catastrophic dislocations, the companies make stability possible. They do it efficiently and economically with prices that yield rates of return not much higher than yields on high grade bonds. Without them the world could not be assured of a continuous supply of low-cost energy to meet the almost insatiable demands of today's economy. It is to their very self-interest that they so operate in a statesmanlike manner. The Bible says that the love of money is the root of all evil—it can also be the root of good when it operates to blunt the emotions, the passions of nationalism.

The geological sciences and the exploration function—which are the essential keys to needed raw materials and which supply the creative thinking and creative action required—are to my mind the most international part of this most international of businesses. Geology ignores national borders because, to do its job, it must look at all the evidence everywhere.

Unlike the more basic sciences, geology cannot assume that what appears to be the same effect in two different areas must have the same cause. Because of the infinite variables in nature, there is always a good chance that each cause will be quite distinct. Just as the most complete knowledge of biochemistry does not help much in predicting human reaction, so knowledge of only some observed data cannot alone accurately solve the vagaries of Mother Nature.

A geologist therefore must take refuge in empirical solutions. Experience is the most valuable tool in his kit. And today the experience of thousands of practitioners in every country has multiplied into a flood of available data. This new wealth of observations, of analyses, of evidence, has made it incumbent on all of us to learn fast, or be counted out. We must continually add to our knowledge and experience, and our need to know is not restricted to the countries in which we live and work.

In fact, geological knowledge is no longer even restricted to the land masses with which we are so familiar. For the first time, we are now obtaining voluminous and critical data from the other three-fourths of the globe—from the continental shelves, the oceans, and the seas.

Furthermore, we are no longer restricted to surface or near-surface data. Modern geophysics is busily furnishing information on the character of the rocks and their structure and stratigraphy at new, great depths.

We are in the midst of an avalanche of new technology, ideas, and hypotheses, and are gaining understanding of basic geologic processes undreamed of only a few years ago. We are entering what I think will be the golden age of geology.

Whether the age fulfils its promise will depend to a large extent on two conditions: first, all geologists everywhere—especially those with international scope and function—must continue their professional educations in a diligent, disciplined fashion (many must reawaken their interest in and love for geology); secondly, the geological profession, in every oil-producing nation, must never accept a background seat in oil industry councils.

The two, of course, are interrelated. If we keep pace with scientific and technological developments, we will continue to earn the respect and confidence of the rest of the industry, and the vital process of exploration will remain in the mainstream of petroleum development.

For an international geologist, sheer distance compounds the problem of continuing education. Geologists assigned to areas far removed from major oil centres are apt to lose their pipeline to this new accumulation of important professional data.

I am sure everyone here remembers when the standard home leave for an international geologist came once every three years. It was really only during that leave that he had a chance to refresh his professional viewpoint.

Today the man coming home three years later would find a totally new science, and a completely different industry. A three-year lapse in professional updating would be crippling indeed.

To insure international field geologists of an opportunity to remain close to their profession, I believe we must go beyond the present, one-country continuing education programmes. A structure must be created to allow the two-way flow of information on the international level as well. In addition to frequent home leaves, one solution might be for the larger geological societies, such as those hosting this meeting, to amplify the projection of their present programmes to locations all over the world. Certainly some such measure will soon be required if international geologists are to avoid becoming technologically sterile.

Meetings such as this one, meanwhile, are an essential step towards that point, and are vital to the international transmission of the latest in geologic science.

To those of us who are fortunate enough to attend them, they provide one other major benefit almost as important. That is an opportunity to renew and cement old friendships, and to make new contacts that will bring personal and professional rewards for years to come.

Certainly, among the many outstanding features of the gathering we are now concluding was the gracious and generous hospitality of our British colleagues. The fine papers, excellent organization, the entertainment, and fine weather, too, were such that I know that these days and this place will have a fond place in our memories throughout the future.

Finally, I know I reflect the thoughts of each of you when I express our heartfelt thanks to all the socieities which laboured so long and hard to make this event possible—and especially to Dan Ion and Mike Halbouty. They had the initial vision and their energies, together with those of Frank Conselman, swept the venture to fulfilment. Then there was the very careful planning work on the papers side by the Papers Committee, led by A. N. Thomas, in which J. McGinty played such an active part in planning the technical programme. Our thanks are also due to N. L. Falcon, Chairman, and the members of the Excursions Sub-Committee, and to Peter Hepple and the very willing band of helpers from the Institute of Petroleum staff, both in the meetings and in the Conference Office. It will be Peter Hepple's responsibility to see that the book of the proceedings is published in approximately three months' time, and if he can manage this publication date, it really will be a tremendous achievement.

Finally, our thanks are most particularly due to the two people, Mrs Joan Cummings and Colin Maynard, who have, between them, organized the whole of the administrative arrangements, starting from fitting 750 people into hotels, arranging the social events, and in general looking after us with such willing helpfulness during our stay here in Brighton.

What was done here so successfully will certainly point the way to many more international exchanges in coming years. Might I point out that the 1970 Annual Meeting of the AAPG will also be away from the U.S.A.—in Calgary, Canada, in June when the Canadian Rockies are at their loveliest. Come, won't you?

Much as I hate to surrender the charm of this lovely resort, and ring down the curtain on this most fruitful meeting, it falls to my lot to do so. I hope this is just the start of many such get-togethers.

WORLD WIDE
CAPABILITIES